Drosophila Neurobiology

A LABORATORY MANUAL

OTHER TITLES FROM CSHL PRESS

RELATED LABORATORY MANUALS

Drosophila: *A Laboratory Handbook*, 2nd edition
Drosophila: *A Laboratory Manual*
Drosophila *Protocols*
Imaging in Neuroscience and Development: A Laboratory Manual
Live Cell Imaging: A Laboratory Manual, 2nd edition
Molecular Cloning: A Laboratory Manual, 3rd edition
Single-Molecule Techniques: A Laboratory Manual

OTHER RELATED TITLES

Experimental Design for Biologists
Fly Pushing: The Theory and Practice of Drosophila *Genetics*, 2nd edition

HANDBOOKS

Lab Math: A Handbook of Measurements, Calculations, and Other Quantitative Skills for Use at the Bench
Lab Ref, Volume 1: A Handbook of Recipes, Reagents, and Other Reference Tools for Use at the Bench
Lab Ref, Volume 2: A Handbook of Recipes, Reagents, and Other Reference Tools for Use at the Bench
Statistics at the Bench: A Step-by-Step Handbook for Biologists

WEBSITE

 *Cold Spring Harbor Protocols*
www.cshprotocols.org

Drosophila
Neurobiology

A LABORATORY MANUAL

EDITED BY

Bing Zhang
University of Oklahoma

Marc R. Freeman
University of Massachusetts Medical School

Scott Waddell
University of Massachusetts Medical School

Also published in *Cold Spring Harbor Protocols*
www.cshprotocols.org/drosophilaneurobiology

COLD SPRING HARBOR LABORATORY PRESS
Cold Spring Harbor, New York • www.cshlpress.com

Drosophila Neurobiology
A LABORATORY MANUAL

Publisher	John Inglis
Acquisition Editor	David Crotty
Director of Development, Marketing, and Sales	Jan Argentine
Developmental Editors	Kaaren Janssen, Catriona Simpson, and Michael Zierler
Project Manager	Mary Cozza
Production Editor	Kathleen Bubbeo
Desktop Editor	Susan Schaefer
Production Manager	Denise Weiss
Book Marketing Manager	Ingrid Benirschke
Sales Account Managers	Jane Carter and Elizabeth Powers
Cover Designer	Ed Atkeson

Front cover artwork: The neurons of the giant fiber system. This simple "escape" circuit in adult *Drosophila* relays sensory information to a large "jump" muscle and the indirect flight muscles in the thorax. Image designed and created by Marcus J. Allen, University of Kent, and Robin Konieczny.

Back cover artwork: Frontal view of a model of the adult *Drosophila* brain. Image produced using Amira software (Visage Imaging, Inc.) by Wolf Hütteroth from primary confocal microscope data collected by Shamik DasGupta.

Library of Congress Cataloging-in-Publication Data

Drosophila neurobiology : a laboratory manual / edited by Bing Zhang, Marc
R. Freeman, Scott Waddell.
      p. cm.
   Includes index.
   ISBN 978-0-87969-904-8 (hardcover : alk. paper) -- ISBN 978-0-87969-905-5
(pbk. : alk. paper)
   1. Neurobiology--Laboratory manuals. 2. Drosophila--Laboratory manuals.
I. Zhang, Bing, 1962- II. Freeman, Marc R., 1970- III. Waddell, Scott, 1970-
IV. Title.

   QP357.D76 2010
   612.8--dc22

2009054102

10 9 8 7 6 5 4 3 2 1

Contents

ACCOMPANYING MOVIES

Movies are freely available online at www.cshprotocols.org/drosophilaneurobiology.

CHAPTER 28
Studying Aggression in *Drosophila*

MOVIE 28.1. One male fly approaches the other and lifts his wings in a threatening pose. This wing threat extends for >10 sec although the movie is clipped before the end. Movie kindly provided by Olga Alekseyenko.

MOVIE 28.2. The male fly on the left lunges three times in quick succession at the male on the right. Lunging is when one male fly rears up on his hind legs and snaps down on the other fly. Movie kindly provided by Olga Alekseyenko.

MOVIE 28.3. This extended encounter starts with the male fly on the left approaching a second male and lunging twice in quick succession. The second male turns and retaliates with lunges that escalate to boxing, tussling, holding, and many lunges. Movie kindly provided by Olga Alekseyenko.

CHAPTER 30
Measurement of Courtship Plasticity in *Drosophila*

MOVIE 30.1. Introduction of flies into a wheel chamber using a mouth aspirator.

Preface

STARTED IN 1984 BY RALPH GREENSPAN, LILY JAN, YUH-NUNG JAN, AND PATRICK O'FARRELL, the Neurobiology of *Drosophila* course at Cold Spring Harbor Laboratory has now run every summer for the last 25 years. Those of us that have been fortunate enough to direct the course have been honored to inherit this rich tradition.

Admission to the Neurobiology of *Drosophila* course is very competitive with only 12 students selected each year. This small intimate group allows for a fantastic interactive and hands-on experience as well as for a more personal intellectual interaction between students and with the individual instructors of the course. However, the editors of this book, who themselves were unavoidably responsible for denying the entry of 100s of potential students, recognized that we could temper the guilt by making the course accessible to all through the production of this book, a course companion. Therefore, beyond being a valuable reference for those students lucky enough to have taken the course, the book allows others to "attend the course by proxy," as the format directly follows that of the course. The content includes a brief introduction to the areas of *Drosophila* neurobiology research that are covered each year in addition to detailed protocols for the techniques that are taught in the laboratory.

The course is broken into three week-long blocks focusing on Development, Physiology, and Behavior, respectively. Most of the contributing authors were instructors of the course during the time that the three editors were Directors of the course. Each instructor teaches for one day with the usual format being lectures in the morning and real experiments in the laboratory in the afternoon/evening/night. It is often difficult to convey the complexity of experimental setup within a day (especially true for experiments that ordinarily take several days) and therefore part of the course unavoidably only provides a snapshot of the research. In this book the instructors, as contributing authors, face the additional difficulty of converting their snapshot into a coherent chapter that conveys the flavor of the research that they discussed—all within a fairly strict page limit. This book is not a comprehensive review of the course lectures or *Drosophila* neurogenetics. Each chapter has a short introduction to the relevant "subfield," but the heart of each chapter provides detailed experimental protocols. We applaud and thank the contributors for their invaluable contributions both to the course and to this book. Neither project would have happened without their efforts and the efforts of numerous students and postdocs that accompany them as course aids.

The course exists for the good of the students and we are extremely proud of the achievements of the many students that have graduated. A remarkable number now have their own research programs in *Drosophila* neurobiology and many are very distinguished names in their respective area.

The course would not function without the generous input of a number of large scientific companies and a large number of people at Cold Spring Harbor Laboratory. Some investigators lug their sophisticated gear to Cold Spring Harbor for a day or two of experimentation, but the rest of the large high-tech equipment arrives on loan from a number of companies. Of particular note, Zeiss, Olympus, Nikon, and Leica provided various microscopes, and Axon Instruments (now called Molecular Devices) lent amplifiers and software for electrophysiology. All the loan agreements, ordering, delivery, and handling of participant travel and housing is the realm of the Cold Spring

Harbor Meetings and Courses people. We particularly wish to thank Barbara Zane, Andrea Stephenson, Andrea Newell, and David Stewart for their dedication and energy. David Stewart is also responsible for obtaining and maintaining external funding for the Cold Spring Harbor courses, and without that effort and the financial support generated, this course would not exist.

Lastly, we have been ably assisted in the production of this book by the enthusiastic and professional help of those at Cold Spring Harbor Laboratory Press. We would never have finished the project without them. We are indebted to our Publisher, John Inglis; Acquisition Editor, David Crotty; Developmental Editors, Kaaren Janssen, Catriona Simpson, and Michael Zierler; Project Manager, Mary Cozza; Director of Development, Marketing, and Sales, Jan Argentine; Production Manager, Denise Weiss; Production Editor, Kathleen Bubbeo; and Desktop Editor, Susan Schaefer. We thank them for their enthusiasm, patience, and professionalism throughout the entire process.

— SCOTT WADDELL, BING ZHANG, AND MARC FREEMAN

Some Feedback from the Course

Taking this course had an absolutely enormous impact on my career. Because the course teaches the Latest and Greatest, and the people who take the course are typically in the top labs and the people who teach the course are among the top in the field, the course has a huge impact on research directions.

— NANCY BONINI (1988), now Professor, Howard Hughes Medical Institute and University of Pennsylvania

As a grad student from a relatively small university participating in this course had an enormously positive impact on my subsequent career in science.

— SHELAGH CAMPBELL (1989), now Associate Professor, University of Alberta, Canada

Eleven years after taking the course, I was a course instructor. I am still in the field and have had the pleasure of taking the course, sending my own students to the course, and teaching the course. I think it is an invaluable resource for our community.

— AARON DIANTONIO (1991), now Associate Professor, Washington University

The course was decisive to continue my research as a postdoc in Cambridge. It has inspired me enormously to hear the history of scientists teaching the course thanks to the time spent with each of them.

— ANDREAS PROKOP (1991), now Senior Lecturer, University of Manchester, United Kingdom

I still benefit from the experience in the course, both in terms of the useful contacts as well as with the breadth of techniques that I was exposed to and are still used in the lab.

— PAUL GARRITY (1992), now Associate Professor, Brandeis University

Studying Neural Development in *Drosophila melanogaster*

An Introduction to Section 1

1

Marc R. Freeman

Department of Neurobiology, University of Massachusetts Medical School, Worcester, Massachusetts 01605

THE NERVOUS SYSTEM IS THE MOST SOPHISTICATED tissue in the human body. It comprises the complex networks of neural circuits (each composed of thousands of morphologically and functionally spectacular cell types) that drive human behavior. The question of how this tissue is assembled has interested biologists for more than a century because an array of fascinating biological questions present themselves when one observes the developing nervous system. How are unique cell fates determined at the right time in the right place? What molecular pathways and cell–cell interactions mediate morphogenesis of specific cell types? How are these interactions integrated together with neural activity to generate functional circuits?

Undertaking the formidable task of trying to understand how an unpatterned population of simple epithelial cells is ultimately transformed into mature neural tissues is no small task. Over the last few decades, *Drosophila melanogaster* has proven itself a workhorse in this challenge. Fly biologists have often led the charge in expanding our cellular and molecular understanding of fundamental processes in the developing nervous system. Notable achievements in the field would certainly include elucidation of the mechanisms of neuroepithelial patterning, neuronal cell-fate specification, asymmetric cell division, specification of neuronal temporal identity, axon guidance, and neuromuscular junction (NMJ) morphogenesis.

The following chapters describe the primary tissue preparations and techniques used to examine these and other topics in neural development using *Drosophila*. They are, of course, only a starting point and are intended to provide the reader with the basic tools needed to get their hands dirty in studies of fly neurogenesis. However, together these chapters cover approaches that can be used to study the majority of nervous system tissues at most developmental stages. The creativity of the reader is the only limit in using these methods to ask novel questions designed to unravel the complexities of the developing *Drosophila* nervous system.

A wide range of *Drosophila* preparations have proven extremely useful in studying neural development, and each has its own advantages depending on the questions one wishes to explore. The *Drosophila* embryo is a superb system in which to analyze central nervous system (CNS) stem cells (termed neuroblasts) and their progeny, the formation of the peripheral nervous system (PNS), and initial events in axon pathfinding. Meticulous work from a number of groups has provided us with a comprehensive map of precisely how many CNS neuroblasts will form, exactly what progeny they will generate, and the axonal projection patterns of a large number of uniquely identifiable motorneurons and interneurons. At embryonic stages, much of neural development seems hardwired. In fact, neuroblast lineages are so invariant the embryonic CNS has been referred to as "the worm in the fly." Similarly, in the embryonic PNS, developmental events are highly stereotyped, and we have a nearly complete understanding of the position and identity of neurons and glia that will form in the body wall and the patterns of neuronal projections into the CNS.

The embryo has been studied so intensely that there is an extensive array of molecular tools available (e.g., antibodies, Gal4 drivers, and enhancer traps) to facilitate the identification of CNS and PNS cell fates, often with single-cell resolution. These tools, together with the hardwired nature of the embryonic nervous system, allow one to ask incisive questions about cell-fate specification, cell morphogenesis, and neural connectivity (Chapters 3 and 4). Furthermore, based on the ease of their collection and the availability of marked balancer chromosomes, embryos are highly amenable to forward genetic screens using very simple staining procedures. Such approaches have been used with great success to define critical signaling pathways driving CNS morphogenesis.

The larval nervous system, especially at later stages, is larger than that of the embryo. It is therefore more easily dissected and visualized, and it exhibits more extensive elaborations of neuronal and glial cell processes (Chapters 2, 5, and 6). When defining the role of specific genes in neural development in larvae there is less concern about maternal contribution of gene products, which can complicate the exploration of gene function in the embryo, and the possibility of using sophisticated genetic mosaic approaches such as MARCM (mosaic analysis with a repressible cell marker; see below) becomes highly feasible. However, the collection of useful molecular markers available for identifying specific cell fates (especially in the CNS) is much smaller in larvae than in embryos.

The study of uniquely identifiable peripheral synapses becomes quite tractable during larval stages, and it can include not only imaging of synaptic structures or specific molecules but also electrophysiological approaches. NMJ synaptic contacts between motorneurons and muscles show tremendous growth (~100x in volume) from L1 to late L3 stages, and this expansion of synapses is modulated by activity. Many researchers have exploited this preparation to investigate the molecular and electrophysiological factors that modulate synapse growth, stabilization, retraction, or homeostasis. How these studies will relate to the function of CNS synapses in the fly remains an open question, and a major future goal for the field is to perform similar high-resolution imaging and electrophysiological studies of CNS synapses.

In addition to mediating animal behavior and physiology during larval stages, the larval brain also serves as a scaffold on which the adult brain is assembled during late larval and pupal stages. A number of populations of neuroblasts have been identified in the larval brain that will give rise to the neurons and glia that ultimately populate the adult brain. Because of the diversity of subtypes of neuroblasts present and the variations in their patterns and timing of proliferation, the larval brain has become an increasingly popular tissue in which to study basic questions in neuroblast biology (e.g., proliferation vs. quiescence, asymmetric division, or the neuroblast stem cell niche).

Although most adult brain neurons are generated before the end of the larval stage, assembly of the adult brain occurs primarily during pupal stages. The histological complexity of the brain increases by an order of magnitude at this time and by developmental criteria the adult fly brain becomes much more akin to that of mammals. The adult brain is physically subdivided into well-defined lobes, each seemingly with distinct functions in information processing, and the CNS becomes increasingly more vascularized with trachea than in embryo or larval brains. Wiring of the adult nervous system includes circuits that use well-characterized continuous (e.g., the visual system) or discrete (e.g., antennal lobe) neural maps. Thus, the subtleties of axon targeting along the anteroposterior and dorsoventral axes can be explored, along with target layer selection in complex multilayered brain lobes such as the lamina and medulla of the visual system. Furthermore, pupae offer the opportunity to study precise steps in circuit rewiring as both dendrites and axons undergo extensive pruning to break down larval connections and generate the new connections needed in the adult nervous system (Chapters 6 and 7).

Drosophilists are famously technophilic and the field has benefited enormously from the creative design of novel strategies to mark and manipulate cells (Chapters 8–10). Common and indispensable tools now available include the *Gal4/UAS* and *LexA/LexAOp* binary expression systems, which allow for precise control of gene expression in cell types of interest, and Flp/FRT mosaic techniques for clonal analysis. One technique of particular importance is MARCM, which allows one to generate small, GFP+ (green fluorescent protein) homozygous mutant clones in an otherwise wild-type nervous system. This approach has revolutionized the way we ask questions about the nervous system because now it is possible to knock out genes even in single cells and ask how that affects cell morphology, migration, wiring, or other aspects of differentiation in an in vivo setting. Such tools are unique to *Drosophila* and make it an extraordinarily powerful system for studying neural development. Not surprisingly, as these new tools have become available and allowed us to study the nervous system with increasing precision, it has become clear that we have only begun to appreciate the complexity of nervous system morphogenesis. In the near future we can expect optogenetic approaches along with the ever-increasing use of live-imaging techniques to help augment our grasp of the cellular events that drive nervous system formation. Here again, *Drosophila* laboratories will undoubtedly help lead the charge.

The accumulated wealth of information regarding *Drosophila* nervous system morphogenesis at all stages of the life cycle, coupled with the wealth of tools and genetic techniques available, make this an ideal time to study neural development in *Drosophila*. Although we fancy ourselves quite savvy to the molecular details driving many major neurodevelopmental events, in truth we are only at the beginning of our understanding of nervous system morphogenesis. A range of major questions remains. How are stem cells made and maintained? What progeny do central brain neuroblasts generate and how? How do CNS synapses form? How plastic is the fly CNS or PNS during development? What role do nonneuronal cell types play in brain and PNS formation? What is the developmental basis of stereotyped behaviors? Much remains to be learned about fundamental aspects of nervous system biology, and the humble fly holds great promise for continuing to advance the field rapidly in important ways.

2

Molecular and Cellular Analyses of Larval Brain Neuroblasts in *Drosophila*

Aric L. Daul,[1] Hideyuki Komori,[1] and Cheng-Yu Lee[1–3]

[1]Center for Stem Cell Biology, Life Sciences Institute, University of Michigan, Ann Arbor, Michigan 48109; [2]Division of Molecular Medicine & Genetics, Department of Internal Medicine, University of Michigan Medical School, Ann Arbor, Michigan 48109; [3]Department of Cell and Developmental Biology, University of Michigan Medical School, Ann Arbor, Michigan 48109

ABSTRACT

Polarized localization of proteins is an evolutionarily conserved mechanism for establishing asymmetry within a cell and producing daughter cells with distinct fates. Neuroblasts (neural stem cells) from *Drosophila melanogaster* are an established paradigm for examining cortical cell polarity and its effects on asymmetric cell divisions. The larval fly brain is ideally suited for studies of asymmetric stem cell division because the larval brain maintains a stable number of about 100 neuroblasts throughout larval development. This chapter describes procedures for the collection and processing of *Drosophila* larval brains for examination by immunolocalization of cell-fate and cell-polarity markers (Protocol 1), 5-ethynyl-2′-deoxyuridine (EdU) labeling of mitotic cells (Protocol 2), and RNA in situ localization (Protocol 3).

OVERVIEW

The *Drosophila* larval brain is a well-established model for investigating the role of stem cells in development. Neuroblasts must be competent to generate many thousands of differentiated neurons through asymmetric divisions during normal development. Given the wide array of genetic and molecular tools available for studying flies, the *Drosophila* larval brain provides a powerful in vivo model system for examining the regulation of neuroblast self-renewal versus differentiation (Wu et al. 2008). Studies in fly neuroblasts have been instrumental in identifying how the establishment and maintenance of cell polarity influence cell fate, and they have produced a wide array of molecular cell-polarity markers. Moreover, neuroblasts and their progeny can be positively identified using a variety of cell-fate markers, which will be discussed in a following section.

In this chapter, we focus on techniques for examining neuroblasts in the larval brain. The larval brain maintains a steady population of approximately 100 neuroblasts, making it possible easily to identify mutants with atypical expansion or premature loss of neuroblast populations, both of which are indicative of disrupted asymmetric cell division (Rolls et al. 2003; Lee et al. 2006a,b,c). It was recently discovered that a small population of larval brain neuroblasts generate transit-amplifying daughter cells capable of limited rounds of asymmetric divisions. This is a particularly intriguing finding given that transit-amplifying cells are commonly seen during development of vertebrate nervous systems (Morrison and Kimble 2006; Nakagawa et al. 2007; Boone and Doe 2008; Bowman et al. 2008). Genes regulating neuroblast polarity and cell fate are evolutionarily conserved between flies and mammals, thus *Drosophila* provides a powerful model system for identifying molecular mechanisms of asymmetric cell division, potentially advancing therapeutic applications in neurology, stem cell biology, and even cancer biology (Rolls et al. 2003; Lee et al. 2006b; Wu et al. 2008).

MOLECULAR MARKERS OF NEUROBLASTS AND THEIR PROGENY

Neuroblasts in the larval brain can usually be identified by their rounded morphology and relatively large size, and can be unambiguously identified by molecular markers including Deadpan (Dpn), Worniu (Wor), and Miranda (Mira) (Lee et al. 2006b). Recent work has revealed that there are at least two classes of larval brain neuroblasts. More than 90% of larval brain neuroblasts are type I (by far the best-characterized class of fly neuroblasts), which divide asymmetrically to self-renew a daughter neuroblast and produce a ganglion mother cell (GMC) that will divide to generate two differentiated neurons. Type I neuroblasts can be unambiguously identified by coexpression of Dpn and Asense (Ase), and their GMCs can be positively identified by nuclear localization of Prospero (Pros) (Brand et al. 1993; Lee et al. 2006b; Bowman et al. 2008). Differentiating neurons can be detected by expression of the neuronal marker Elav (Embryonic lethal, abnormal vision), which is not detected in neuroblasts or GMCs (Fig. 1; Table 1) (Lee et al. 2006b).

The second class of neuroblasts in the larval brain (type II) have tremendous potential to generate many differentiated neurons via transit-amplifying cells (Boone and Doe 2008; Bowman et al. 2008). There are eight type II neuroblasts mostly located in the dorsomedial region of the larval brain lobe that divide asymmetrically to self-renew and generate. Immature intermediate neural progenitors (INPs) commit to the INP fate through maturation, a differentiation process necessary for specification of the INP identity. INPs express similar molecular markers as Type I neuroblasts,

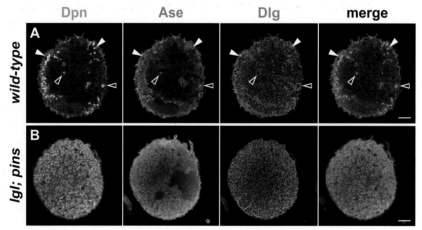

FIGURE 1. Cell-fate markers in larval brain neuroblasts. Third-instar larval brains stained with antibodies against the neuroblast markers Deadpan (Dpn; types I and II; green) and Asense (Ase; type I only; blue), and the cortical marker Discs large (Dlg; red). (*A*) Wild-type brain showing Dpn⁺ Ase⁺ type I neuroblasts (white arrowheads) and Dpn⁺ Ase⁻ type II neuroblasts (black arrowheads). (*B*) An *lgl; pins* double mutant brain showing overproliferation of both type I and type II neuroblasts as determined by coexpression of Dpn and Ase. Anterior is to the *top* in all images. Scale bars, 20 μm. (Images courtesy of J. Haenfler, University of Michigan.)

TABLE 1. Cell-fate markers in neuroblast lineages

Type I	
Neuroblasts	Dpn$^+$ Ase$^+$ Pros$^{+cytoplasmic}$
GMCs	Dpn$^-$ Ase$^+$ Pros^{+nuc}
Neurons	Dpn$^-$ Ase$^-$ Pros^{+nuc}
Type II	
Neuroblasts	Dpn$^+$ Ase$^-$ Pros
Immature INPs	Dpn$^-$ Ase$^-$ Pros$^-$
INPs	Dpn$^+$ Ase$^+$ Pros$^{+cytoplasmic}$
GMCs	Dpn$^-$ Ase$^+$ Pros^{+nuc}
Neurons	Dpn$^-$ Ase$^-$ Pros^{+nuc}

GMCs, ganglion mother cells; INPs, intermediate neural progenitors.

including Dpn, Ase, and cytoplasmic Pros, and divide asymmetrically to regenerate and to produce GMCs. Although expression of Dpn is readily detectable in both type II neuroblasts and INPs, these two cell types can be distinguished by size (>10 μm for neuroblasts vs. <6 μm for INPs). Furthermore, Dpn is expressed in all brain neuroblasts, whereas Ase is specific to type I lineages (Dpn$^+$ Ase$^+$) and thus provides a key reagent to discriminate type I from type II neuroblasts (Fig. 1; Table 1) (Bowman et al. 2008).

MARKERS OF NEUROBLAST CELL POLARITY

In larval brain neuroblasts, cell polarity is established by two protein complexes that localize to the apical cortex: the Par complex (Bazooka–Par6–atypical protein kinase C [aPKC]) and the Partner of Inscuteable (Pins)–G-protein subunit α (Gαi) complex (Rolls et al. 2003; Lee et al. 2006b). Simply put, these protein complexes restrict localization of the neuronal determinants to the basal cortex where they will be inherited by the differentiating GMC. The complexes remain in the apical daughter (the self-renewing neuroblast) during mitosis and provide reliable apical markers of cell polarity (Fig. 2; Table 2). Aurora A kinase (AurA) provides a link between mitosis and asymmetric distribution of fate determinants, initiating cell polarity by phosphorylation of Par6 and activation of aPKC (Lee et al. 2006a; Wirtz-Peitz et al. 2008). The active form of aPKC phosphorylates Lethal giant larvae (Lgl), eliminating it from the apical complex and thereby allowing Bazooka to associate with aPKC and Par6 (Betschinger et al. 2003). The presence of Bazooka in the Par complex facilitates binding to Numb (Nb), promoting phosphorylation of Nb by aPKC and, subsequently, release of Nb from the apical cortex (Smith et al. 2007). Although Nb is required for differentiation of neurons and the end result of this kinase cascade appears to be restriction of Nb to the basal cortex and eventually the GMC, it is not known how Nb might be acting to specify neuronal fate. The most likely mechanism is through modulation of Notch (N) signaling, because Nb is a known repressor of the N receptor (Yoon and Gaiano 2005). Intriguingly, N signaling is active in larval neuroblasts but not GMCs, reinforcing the possibility of this mechanism.

The importance of aPKC in establishing neuroblast cell polarity is apparent when considering aPKC loss-of-function and gain-of-function phenotypes. When aPKC activity is compromised, neuroblasts lose the ability to self-renew and larval brains contain fewer neuroblasts. In contrast, abnormal uniform cortical localization of aPKC in *lgl; pins* double mutants results in the formation of large tumors as both type I and type II neuroblasts execute symmetric divisions to generate ectopic sister neuroblasts (Figs. 1B and 2C).

In addition to regulating distribution of Nb, aPKC is also required for establishing basal localization of additional neuronal determinants Pros and Brain tumor (Brat) to restrict inheritance of these proteins to the GMC during mitosis. The transcription factor Pros is present in neuroblasts, but it remains in the cytoplasm. In GMCs, however, Pros localizes to the nucleus after mitosis where it pre-

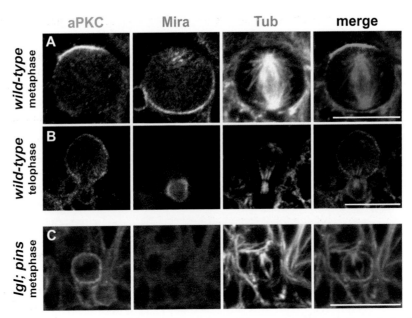

FIGURE 2. Cell-polarity markers in neuroblasts. (A) Metaphase wild-type neuroblasts display apical polar localization of aPKC (green) and basal polar localization of Miranda (Mira; blue). The mitotic spindle is visualized with anti-α-tubulin (red). (B) During telophase, apical proteins such as aPKC (green) are retained in the neuroblast (apical daughter) and basal proteins such as Mira (blue) are segregated into the ganglion mother cell (GMC; basal daughter). The mitotic spindle is visualized with anti-α-tubulin (red). (C) Cell polarity is disrupted in *lgl; pins* double mutants. aPKC (green) is localized uniformly around the cortex, displacing Mira to the cytoplasm (blue). Apical is oriented to the *top*, basal to the *bottom* in all images. Scale bars, 10 μm. (A, Images courtesy of C. Gamble, University of Michigan; C, images courtesy of J. Haenfler, University of Michigan.)

sumably activates transcription of genes required for neuronal differentiation. Brat is a protein that contains NHL, coiled-coil, and B-box protein–protein interaction domains, and it is predicted to act as a posttranscriptional regulator that likely acts together with Pros to specify neuronal wild-type differentiation, although the exact molecular mechanisms are not known. The cargo-binding protein Mira plays a critical role in localizing Pros and Brat to the basal cortex. Molecular interactions between Mira, Pros, and Brat suggest that Mira acts as a scaffold to retain these neuronal determinants at the basal cortex of the mitotic neuroblast. Mira accumulates on the basal cortex in mitotic neuroblasts, and provides a key marker of basal cortical polarity (Fig. 2; Table 2). The basal localization of Mira is unaffected by loss of Pros or Brat, but loss of Mira causes uniform cytoplasmic localization of Pros and Brat.

Although aPKC is responsible for inactivation of Lgl in the apical cortex, Lgl inhibits aPKC activity at the basal cortex. This mutual inhibition ensures aPKC is restricted to the apical cortex where it will be inherited only by the neuroblast at telophase. Lgl, Scribble (Scrib), and Discs large (Dlg) are all potent tumor suppressors with well-conserved homologs in mammals. Lgl, Scrib, and Dlg are each required for the formation of septate junctions in the basal portion of the cell and for proper segregation of cell-polarity proteins (Albertson and Doe 2003; Humbert et al. 2008). Immunolocalization of Lgl, Scrib, or Dlg provides a useful marker for the cortex and will effectively outline cells in the lar-

TABLE 2. Neuroblast cell-polarity markers

Apical	Basal	Cortical
Atypical protein kinase C (aPKC)	Miranda (Mira)	Scribble (Scrib)
Par6	Brain tumor (Brat)	Discs large (Dgl)
Bazooka (Baz)	Prospero (Pros)	Lethal giant larvae (Lgl)
Inscuteable (Insc)	Numb (Nb)	
Partner of Insc (Pins)		
G-protein subunit α (Gαi)		

val brain, allowing some cells to be identified solely on size or morphology in the absence of specific markers such as Wor or Dpn.

Regulation of symmetric versus asymmetric cell division requires more than proper apical–basal localization of fate determinants: it also requires that cytokinesis occurs in the appropriate plane. Not surprisingly, spindle position is closely tied to cell polarity as Inscuteable (Insc) physically links the Pins–Gαi complex to the Par complex and also connects the mitotic spindle to the apical cortex (Kraut et al. 1996; Wu et al. 2008). Symmetric division of a polarized cell requires that each daughter receive equal amounts of both apical and basal determinants, thus it is easy to see the importance of spindle orientation relative to the apical–basal axis in regulating daughter cell fates. The mitotic spindle is easily visualized by immunolocalization of tubulin, which can be costained with cell polarity markers to assay spindle alignment. Mutations in the Gαi complex cause misalignment of the spindle and ectopic symmetric division of neuroblasts, likely because of improper segregation of fate determinants (Izumi et al. 2004; Nipper et al. 2007). As such, it is easy to see how loss of proper spindle alignment can lead to inappropriate symmetrical division of stem cells and possible tumor formation.

When examining mutants affecting cell fate due to altered asymmetric division patterns, it is important to determine whether cells are mitotically active. Phosphohistone H3 provides a useful immunological M-phase marker for assaying mitotic defects (Lee et al. 2006c). Chemical labeling of newly synthesized DNA (S-phase marker) by incorporation of BrdU (5-bromo-2′-deoxyuridine) can be more informative because this thymidine analog can be used to pulse-label dividing cells and chased to identify the progeny of dividing cells (Lee et al. 2006c). Such pulse-chase experiments can provide additional insight by distinguishing actively dividing cells from those that might be arrested at a mitotic checkpoint. EdU provides a more sensitive and practical alternative to BrdU that can be visualized without the need for harsh DNA denaturation and additional rounds of antibody staining; methodology for EdU labeling is described in Protocol 2 (see also Fig. 3) (Kolb et al. 2001; Rostovtsev et al. 2002; Breinbauer and Köhn 2003; Agard et al. 2004).

EXPERIMENTAL DESIGN NOTES

Equipment and Reagents

Analysis of larval brains does not require much specialized equipment beyond what is needed for standard culturing of flies (Ashburner and Roote 2000). Standard fly food media, egg collection bottles with fruit juice agar caps, and a well-calibrated incubator will be needed in addition to the few specific items described in each protocol below. It is strongly recommended that researchers use a thermometer to monitor carefully the interior temperature of incubators containing samples, particularly when working with temperature-sensitive strains, because the air temperature can vary considerably throughout an incubator.

Genetic Considerations

When designing experiments, it is necessary to be able to distinguish the genotypes of interest during larval stages. It is therefore recommended that researchers use balancer chromosomes carrying GFP (green fluorescent protein) or larval morphological markers such as *tubby*. Also, keep in mind that neuroblasts play a critical role during embryogenesis, and therefore when working with transgenes it is important to avoid causing severe defects (e.g., lethality) too early in development.

Synchronization of Larval Populations

Synchronizing populations of larvae for analysis will ensure that larvae are at the appropriate stage of development for the experiment and that strains are examined at comparable time points. Collection periods as short as 1 h will yield the most tightly synchronized populations, but embryos laid over a span of ~8 h generally provide a good balance between synchronization and having a

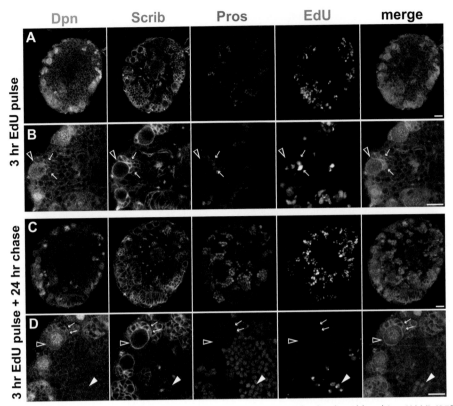

FIGURE 3. EdU pulse-chase in larval brains. (A) Wild-type larval brain 72 h after larval hatching (ALH) (25°C) after a 3 h pulse of EdU by feeding. (B) High-magnification image of wild-type brain after 3 h pulse. EdU is detected in mitotically active neuroblasts (black arrowheads; Dpn+; green) and GMCs (white arrows), but not differentiating neurons (Pros+; violet). Cells are outlined by the cortical marker Scrib (red). (C) Wild-type larval brain 96 h ALH (25°C) after a 3 h pulse of EdU by feeding and a 24 h EdU-free chase. (D) High-magnification image of wild-type brain after 3 h pulse and 24 h chase. EdU is no longer detected in neuroblasts (black arrowheads; Dpn+; green) and GMCs (white arrows), but it is now present in differentiating neurons (white arrowheads; Pros+; violet). Cells are outlined by the cortical marker Scrib (red). Scale bars, 10 μm.

large enough population of embryos to work with. Incubate collection plates for ~24 h to allow larvae to hatch before transferring to standard fly food media (meal caps) stored in 60-mm dishes with a moistened tissue to prevent them from drying out. Larvae will generally be in the third-instar stage ~96 h after larval hatching (ALH) at 25°C or 72 h ALH at 30°C.

Protocol 1

Immunofluorescent Antibody Staining of Larval Tissues

This protocol can be used for dissecting, fixing, and staining brains from larvae at any developmental stage. The number of brains processed using this method is limited only by how many brains can be dissected in 20 min, which is the maximum amount of time dissected tissues should remain in buffer before fixation. This protocol can be used for simultaneous costaining of multiple proteins.

MATERIALS

CAUTION: See Appendix for proper handling of materials marked with <!>.
See the end of the chapter for recipes for reagents marked with <R>.

Reagents

Block solution <R>
Fix solution <R>
Glycerol (70%)
Phosphate-buffered saline (PBS) containing bovine serum albumin (BSA) and Triton X-100
    (PBSBT) <R>
PBS containing Triton X-100 (PBST) <R>
PBS stock solution (10x) <R>
ProLong Gold antifade mounting medium (Invitrogen)
    Schneider's insect medium (Sigma-Aldrich)
Triton X-100 <!> (10%)

Equipment

Coverslips (22 x 22-mm [#1 thickness] and 24 x 40-mm)
Dissection dishes
Fine-tipped forceps (two pairs)
Microfuge tubes (0.5-mL)
Microscope slides
Nutator mixer or rocker
Pipettes and sterile tips

METHOD

Dissection of Larvae

1. Fill the wells of dissection dishes with 200–400 µL of cold Schneider's insect medium.

2. Dissect the larvae by rolling them onto their dorsal side so that the denticle belts are facing upward.

3. Using a pair of forceps, gently grasp the larva just posterior of the midpoint. With a second pair of forceps, grasp the anterior end of the larva at the base of the mouth hooks.

4. Carefully tear the cuticle behind the mouth hooks using side-to-side motion while slowly drawing the mouthparts out away from the body. The brain will remain attached to the head and be clearly visible among the gut and salivary glands. Remove any excess tissue, but leave the brain attached to the mouth hooks.

> Leaving the brains connected to the mouth hooks will help the brains sink to the bottom of the tube during washing steps below. Moreover, the mouth hooks are dark in color, which makes it easier to see the brains during experimental manipulations.

5. After dissection, place the brains in a 0.5-mL tube containing cold Schneider's insect medium.

> Do not let the tissue sit in Schneider's insect medium for >20 min.

Fixation and Staining

6. Remove the Schneider's insect medium from the samples.

7. Add 500 µL of fix solution and incubate with rocking for 23 min at room temperature.

8. Quickly wash the brains twice in ~500 µL of PBST at room temperature. Wash again in PBST twice for 20 min each at room temperature.

> Once fixed, samples can be held in extended washes to synchronize them before proceeding with further processing.

9. Incubate the samples in ~500 µL of block solution for at least 30 min at room temperature.

10. Incubate in primary antibody diluted in PBSBT for 4 h at room temperature or overnight at 4°C.

> Conditions are dependent on the specific antibody being used. For example, Dpn staining is better when incubated for 3–4 h at room temperature rather than overnight at 4°C.

11. Quickly wash the brains twice in PBSBT at room temperature. Wash again in PBSBT twice for 30 min each at room temperature.

12. Incubate the samples in secondary antibody for 1.5 h at room temperature or overnight at 4°C. Protect the samples from light after this point.

> Secondary antibodies are typically diluted in PBSBT.

13. Quickly wash the brains twice in PBST at room temperature. Wash again in PBST twice for 30 min each at room temperature.

14. Equilibrate the brains in ProLong Gold at room temperature. Samples can be stored in the dark at room temperature.

Mounting Samples

15. Adhere two 22 x 22-mm coverslips to a microscope slide using a small amount of 70% glycerol, leaving a ~5-mm space between them.

> These coverslips act as spacers to prevent the brains from being deformed by the 24 x 40-mm coverslip in Step 19.

16. Transfer the brains to the slide using a pipette with the tip cut off.

17. Using forceps, remove all excess tissue including the optic discs from each brain. Be sure to leave the ventral nerve cord intact, as it will aid in proper orientation of the brain on the slide.

> *See Troubleshooting.*

18. Orient the brains ventral side down.

> If the ventral cord is intact, the brain will sit in the appropriate upright position. Without the ventral cord, it is difficult to keep the brain in the proper position and it will tend to end up resting on its anterior or posterior surface.

19. Place a 24 × 40-mm coverslip over the samples and backfill the space between the slide and the coverslip by pipetting a small amount of mounting medium along the edge of the coverslip. Backfilling will reduce the formation of air bubbles trapped in the slide.

See Troubleshooting.

TROUBLESHOOTING

Problem (Step 17): The ventral nerve cord breaks off during dissection.

Solution: Keeping the ventral nerve cord intact requires that you grasp the larva at the right place on its body. Holding the larva at a "sweet spot" near the 4th or 5th abdominal segment will typically allow clean dissection of the brain. Take care to gently break away attached tissues as you tear the head away from the body. The ventral cord is connected to the body by many axons and will likely break off if the head is carelessly pulled from the body.

Problem (Step 19): There is poor signal-to-noise ratio.

Solution: High levels of background staining can result from several steps in this protocol. Ensure that all solutions are at the correct pH because high or low pH levels can negatively affect antibody binding. It is critical to use both primary and secondary antibodies at the appropriate dilution specific for each antibody. The specificity of secondary antibodies should be tested by staining a sample with secondary antibody alone. Thorough washing of samples is also important for reducing background signals, particularly after incubation in primary antibodies. Placing a small, fine pipette tip over a larger 1000-µL tip will help you to remove as much of the wash solutions as possible without losing or damaging the samples. However, note that excessive washing can also lead to weak signal strength. Antibodies can be sensitive to the duration and temperature of incubation. Anti-Dpn, for example, will typically yield cleaner staining when incubated for 3–4 h at room temperature than when incubated overnight at 4°C. Testing different incubation conditions might be necessary to determine the optimal conditions for a particular antibody.

EdU Labeling of Mitotic Neuroblasts

Like BrdU, EdU is a thymidine analog that is incorporated into newly synthesized DNA during S-phase and it provides an efficient method for identifying mitotic cells. Incorporation of EdU is detected through its reaction with an azide dye that is small enough to penetrate tissues efficiently. This method is highly sensitive and does not require the harsh denaturation of DNA that is necessary for staining with antibodies (Kolb et al. 2001; Rostovtsev et al. 2002; Breinbauer and Köhn 2003; Agard et al. 2004). Visualization of EdU is rapid and does not interfere with subsequent antibody staining. EdU can be used to pulse-label mitotic cells and chased to identify their progeny, just like BrdU. This protocol was modified from BrdU-feeding procedures described by Truman and Bate (1988) and Ito and Hotta (1992). Methods for the detection of EdU are described in Invitrogen product manuals.

MATERIALS

CAUTION: See Appendix for proper handling of materials marked with <!>.
See the end of the chapter for recipes for reagents marked with <R>.
Refer also to the Materials list for Protocol 1.

Reagents

Bromophenol blue <!>
Click-iT EdU imaging kit (Invitrogen)
EdU (Invitrogen)
Kankel–White medium <R>

METHOD

Preparation of Medium

1. Prepare EdU and the detection reagents as instructed by the manufacturer (Invitrogen).
2. Prepare Kankel–White medium and heat to dissolve all components. Add a few granules of bromophenol blue to the medium.
3. Allow the medium to cool to 50°C–60°C. Add EdU to give a final concentration of 0.2 mM.
4. Pour the mixture into plates and allow it to solidify.

Feeding EdU to *Drosophila* Larvae

5. Allow the larvae to feed on EdU-containing medium for 3–4 h.

   The presence of bromophenol blue in the medium will make food in the gut visible. If the larvae are eating, they should be taking up EdU.

6. If no EdU-free chase is required, proceed directly to Steps 7–16 below. If an EdU-free chase is required, transfer the larvae to standard fly food and allow them to recover for the desired amount of time under appropriate experimental conditions.

   If using bromophenol blue in the medium, select larvae with blue food visible in their guts.

Dissection and Staining

7. Dissect the larvae in Schneider's insect medium and remove the brains following the procedure described in Protocol 1, Steps 1–6.

8. Add 500 μL of fix solution to the brains and incubate with rocking for 23 min at room temperature.

9. Quickly wash the brains twice in ~500 μL of PBST at room temperature. Wash again in PBST twice for 20 min each at room temperature.

10. Incubate the samples in ~500 μL of block solution for at least 30 min at room temperature.

11. Quickly wash the brains in PBST at room temperature.

12. Prepare the Click-iT reaction mix as instructed by the manufacturer. To prevent photobleaching, protect the samples from light after this point.

13. Add 500 μL of Click-iT reaction mix and incubate with rocking for 30 min at room temperature.

14. Quickly wash the brains twice in PBST at room temperature.

15. Quickly wash the brains in PBSBT at room temperature.

16. Mount a few brains and scan them to check the efficiency of EdU labeling. Process the remaining brains for antibody staining as described in Protocol 1.

    See Troubleshooting.

TROUBLESHOOTING

Problem (Step 16): EdU signal is too bright, thereby saturating detection.

Solution: Detection of EdU is more sensitive than detecting BrdU. The concentration of EdU in the medium should be lower than is necessary in experiments with BrdU. Also, larvae can be fed for shorter periods of time to reduce the pulse of EdU labeling in cells.

Multicolor Fluorescence In Situ Hybridization

RNA in situ hybridization is a useful method for determining the transcriptional expression pattern of a gene when antibodies are not available. Using this technique, it is possible to assay the expression of multiple RNA species using distinct labels on RNA probes, or simultaneously examine RNA and protein localization within larval tissues (Fig. 4). This protocol utilizes a fluorophore-conjugated tyramide that is easily made in the laboratory for a fraction of the cost of the commercially produced product. The method was adapted from B. Pearson (University of Utah; pers. comm..) with modifications by H. Komori and A. Daul (University of Michigan). Additional modifications were derived from D. Kosman (http://superfly.ucsd.edu/~davek/) (Kosman et al. 2004).

MATERIALS

CAUTION: See Appendix for proper handling of materials marked with <!>.
See the end of the chapter for recipes for reagents marked with <R>.
Refer also to the Materials list for Protocol 1.

Reagents

Anti-digoxigenin (DIG)-POD or other antibody for detection of riboprobe
Carbonate buffer (2x) <R>
DIG RNA-labeling kit (Roche)
DNA template for riboprobe
Ethanol <!> (70% and 100%)
Horse serum (heat-inactivated)
Hybridization buffer <R>
Hydrogen peroxide <!> (30%, v/v)
Hydrolysis stop buffer <R>
Imidazole <!>
LiCl <!> (4 M)
MABT <R>
PBST <R>
RNase-free water
SSC stock solution (20x) <R>
Tyramide (fluorescently labeled) <R>

Equipment

Heat block (set at 90°C)
Ice bucket
Incubator (set at 37°C)
Microfuge tubes (0.5-mL; RNase-free)
Nutator mixer or rocker
Pipettes and tips (RNase-free)
Water bath (set at 55°C or 65°C)

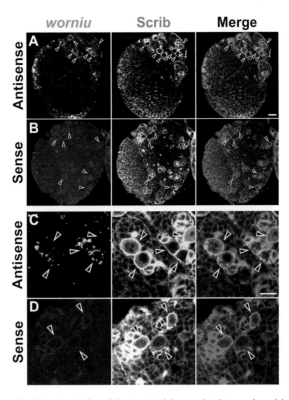

FIGURE 4. Fluorescent in situ localization in larval brains. Wild-type third-instar larval brains hybridized with anti-*worniu* riboprobe and costained with anti-Scrib antibodies. (*A*) *worniu* antisense probe (green) is detected in central brain neuroblasts (arrows). Neuroblasts can be identified by morphology as outlined by the cortical marker Scrib (red). (*B*) No signal is detected when using a *worniu* sense probe. (*C,D*) High-magnification images of *A* and *B*, respectively. (Images courtesy of H. Komori, University of Michigan.)

METHODS

Riboprobe Synthesis

Producing riboprobes requires template DNA flanked by distinct RNA polymerase promoter sites, typically T3, T7, or Sp6. Templates can be easily produced by inserting cDNA or a genomic subclone with minimal intronic sequence into pBluescript or other cloning vector with suitable promoter sites. Having distinct promoters on either end of the template allows the template to be used for transcription of sense and antisense probes. Use a restriction enzyme to cut the plasmid at the opposite end of the template so that transcription of the probe will stop at the cut and not proceed into the vector sequence. Alternatively, template DNA can be made by polymerase chain reaction (PCR) amplification. In this case, the amplified fragment must contain the desired RNA polymerase promoter site at each end in order to transcribe the probe. Regardless of the preparation method, template DNA should cleaned up following the enzymatic reactions and resuspended in RNase-free water. Probes can be labeled with a variety of haptens, including digoxigenin (DIG), dinitrophenol (DNP), biotin (BIO), and fluorescein isothiocyanate (FITC). DIG, BIO, and FITC RNA-labeling kits are available from Roche. DNP-11-UTP is available from PerkinElmer and can be mixed with unlabeled ribonucleotides for use in a similar transcription reaction. Antibodies for the detection of hapten-labeled riboprobes are available from a variety of commercial sources.

1. Combine the following in an RNase-free microfuge tube:

    ~1.5 mg of DNA template
    2 µL of 10x dNTP labeling mix*
    2 µL of 10x transcription buffer*
    1 µL of RNase inhibitor*
    2 µL of RNA polymerase*
    RNase-free water to give a total volume of 20 µL

    * These components are included in the Roche DIG RNA-labeling kit. The amounts required could vary when not using this kit.

2. Incubate for 2 h at 37°C.

    At this point, a small amount of probe can be run on an agarose gel to evaluate the efficiency of the transcription reaction.

3. Add 5 µL of RNase-free water and 25 µL of 2x carbonate buffer. Incubate for 45 min at 65°C.

4. Add 50 µL of hydrolysis stop buffer.

5. Add 10 µL of 4 M LiCl and 330 µL of 100% ethanol. Precipitate at –70°C for at least 30 min. Spin at top speed for 15 min. Quickly wash the pellet in 70% ethanol. Resuspend the pellet in 100 µL of hybridization buffer. Store at –80°C. Avoid repeated freezing and thawing.

RNA In Situ Hybridization with Antibody Costaining

6. Dissect the larvae in Schneider's insect medium and remove the brains following the procedure described in Protocol 1.

7. Add 500 µL of fix solution and incubate with rocking for 23 min at room temperature.

8. Quickly wash the brains twice in ~500 µL of PBST at room temperature. Wash again in PBST twice for 20 min each at room temperature.

9. Incubate in PBST containing 3% hydrogen peroxide for 1 h at room temperature. Seal the tubes tightly with parafilm to prevent the caps from blowing open during this reaction!

    This step eliminates endogenous peroxidase activity.

10. Quickly wash the brains twice in ~500 µL of PBST at room temperature. Wash again in PBST twice for 20 min each at room temperature.

11. Incubate the brains in 400 µL of a 1:1 mixture of PBST and hybridization buffer for 30 min at room temperature.

12. Wash three times for 10 min each in hybridization buffer at room temperature.

13. Incubate the brains in 400 µL of hybridization buffer for 2 h at 55°C. At 1.5 h after the start of this incubation, denature 400 ng (~4 µL) of riboprobe in 400 µL of hybridization buffer for 5 (step 14)min at 90°C. Cool on ice for 5 min, and then place at 55°C.

14. Incubate the brains in 150–200 µL of the probe/hybridization buffer mixture for at least 16 h at 55°C.

15. Remove the probe/hybridization buffer mixture.

    The mixture can be stored at –20°C and reused.

16. Wash the brains twice in a prewarmed 1:1 mixture of hybridization buffer and 2x SSC for 30 min at 55°C.

17. Wash twice in prewarmed 2x SSC for 30 min at 55°C.

18. Wash twice in prewarmed 0.2x SSC for 30 min at 55°C.

19. Quickly wash the brains twice in MABT at room temperature.

20. Block in MABT containing 10% heat-inactivated horse serum for 1 h at room temperature.

    BSA has been reported to reduce the intensity of the tyramide reaction.

21. Incubate the brains in anti-DIG-POD (diluted 1:1000 in MABT containing 10% heat-inactivated horse serum) overnight at 4°C.

    If performing in situ hybridization of multiple RNA species using tyramide reactions, each probe must be detected and developed in sequence to avoid cross-reactivity.

22. Wash the brains six times in MABT for 20 min at room temperature.

23. Incubate the brains in PBST containing 10 mM imidazole for 30 min at room temperature.

24. Incubate the brains in the dark in FITC-tyramide solution diluted 1:1000 in PBST containing 10 mM imidazole (1:500 for Cy3-tyramide) for 30 min at room temperature. Protect the samples from light after this point.

25. Develop the signal by adding hydrogen peroxide to a final concentration of 0.002%–0.01%. Incubate for 45 min at room temperature.

    Signal strength can be enhanced by repeating Steps 24–25 up to twice with fresh fluorescently labeled tyramide solution.

26. Quickly wash the brains twice in ~500 μL of PBST at room temperature. Wash again in PBST twice for 30–60 min each at room temperature.

27. If the samples are to be costained with antibodies for protein localization, proceed as described in Step 5 of Protocol 1.

    If the samples are to be processed for the detection of a second riboprobe, quench the residual peroxidase activity by incubating in PBST containing 3% hydrogen peroxide for 1 h at room temperature. Quickly wash the brains twice in ~500 μL of PBST at room temperature. Wash again in PBST twice for 20 min each at room temperature. Proceed by incubating with the second riboprobe and repeat Steps 11–25 above.

RECIPES

CAUTION: See Appendix for proper handling of materials marked with <!>.
Recipes for reagents marked with <R> are included in this list.

Block Solution

Reagent	Quantity (for 5 mL)	Final concentration
PBSBT <R>	5 mL	1x
Normal goat serum (NGS)*	5 μL	0.1% (v/v)
Glycine (1 M in PBS containing 2% sodium azide <!>)	50 μL	1% (w/v)

Prepare fresh and keep cold. *Omit if using antigoat antibodies.

Carbonate Buffer (2x)

Reagent	Quantity (for 100 mL)	Final concentration
Na_2CO_3 <!>	0.636 g	60 mM
$NaHCO_3$	0.336 g	40 mM

Adjust the pH to 10.2. Store at room temperature.

Fix Solution

Reagent	Quantity (for 10 mL)	Final concentration
Formaldehyde <!> (37%, v/v)	1.1 mL	4% (v/v)
PIPES (1 M, pH 6.9)	1.0 mL	0.1 M
Triton X-100 <!> (10%, v/v)	0.3 mL	0.3% (v/v)
EGTA (0.1 M, pH 8.0)	0.2 mL	20 mM
$MgSO_4$ <!> (1 M)	10 μL	1 mM
dH_2O	7.4 mL	

Prepare fresh every time.

Hybridization Buffer

Reagent	Quantity (for 50 mL)	Final concentration
Deionized formamide <!>	25 mL	50%
SSC (20x) <R>	10 mL	4x
Tween-20 (10%, v/v)	0.5 mL	0.1% (v/v)
Heparin <!> (50 mg/mL)	50 μL	0.05 mg/mL
RNase-free dH_2O	14.5 mL	

Hydrolysis Stop Buffer

Reagent	Quantity (for 10 mL)	Final concentration
Sodium acetate	0.166 g	200 mM
Acetic acid <!>	100 μL	1% (v/v)

Adjust the pH to 6.0 with acetic acid. Store at –20°C

Kankel–White Medium

Reagent	Quantity (for 10 mL)
Agar	80 mg
Sucrose	500 mg
Yeast extract	500 mg
Dried yeast	200 mg
dH_2O	10 mL

Heat to dissolve. Do not boil excessively.

MABT

Reagent	Quantity (for 1 L)	Final concentration
Maleic acid <!>	11.6 g	100 mM
NaCl	8.8 g	150 mM
Tween-20 (10%, v/v)	1 mL	0.1% (v/v)

Dissolve the components and adjust the pH to 7.5 with concentrated NaOH <!>.
Adjust the volume to 1 L with dH_2O and sterilize. Store at room temperature.

PBS Stock Solution (10x)

Reagent	Quantity (for 1 L)	Final concentration
NaCl	80 g	1.37 M
KCl <!>	2 g	27 mM
Na$_2$HPO$_4$	14.4 g	100 mM
KH$_2$PO$_4$	2.4 g	20 mM
dH$_2$O	to 1 L	

Dissolve the components in 400 mL of dH$_2$O and adjust the pH to 7.4 with concentrated HCl <!>. Adjust the volume to 1 L with dH$_2$O and sterilize. Store at room temperature.

PBSBT

Reagent	Quantity (for 25 mL)	Final concentration
PBST <R>	25 mL	1x
Bovine serum albumin (BSA)	0.25 g	1% (w/v)

Make fresh and keep cold. Can be stored short-term at 4°C.

PBST

Reagent	Quantity (for 500 mL)	Final concentration
PBS stock solution (10x) <R>	50 mL	1x
Triton X-100 <!> (10%, v/v)	15 mL	0.3% (v/v)
dH$_2$O	435 mL	

Store at room temperature.

SSC Stock Solution (20x)

Reagent	Quantity (for 1 L)	Final concentration
NaCl	175.3 g	3.0 M
Sodium citrate <!>	88.2 g	0.3 M
H$_2$O	800 mL	

Adjust the pH to 7.0 with a few drops of 14 N HCl <!>. Adjust the volume to 1 L with H$_2$O.

Tyramide (Fluorescently Labeled)

Work should be performed in the hood. For best results, use fresh reagents. *N*-hydroxysuccinimide (NHS) esters are unstable and the coupling reaction should be kept anhydrous.

1. Dissolve 40 mg of fluorescently labeled NHS ester in 4 mL of dimethylformamide (DMF) <!>.
2. Add 10 μL of triethylamine (TEA) <!> to 1 mL of DMF.
3. Dissolve 10 mg of tyramide in 1 mL of TEA–DMF solution.
4. Mix 4 mL of fluorescently labeled NHS ester in DMF with 1.37 mL of tyramide solution. Incubate in the dark for 2 h at room temperature.
5. Add 4.6 mL of ethanol.

Keep protected from light. Store at 4°C or −20°C. The solution is stable for at least 8 mo at 4°C.

REFERENCES

Agard NJ, Prescher JA, Bertozzi CR. 2004. A strain-promoted [3 + 2] azide-alkyne cycloaddition for covalent modification of biomolecules in living systems. *J Am Chem Soc* **126:** 15046–15047.

Albertson R, Doe CQ. 2003. Dlg, Scrib and Lgl regulate neuroblast cell size and mitotic spindle asymmetry. *Nat Cell Biol* **5:** 166–170.

Ashburner M, Roote J. 2000. Laboratory culture of *Drosophila*. In Drosophila *Protocols* (ed. Sullivan W, et al.), Chapter 35, pp. 585–599. Cold Spring Harbor Press, Cold Spring Harbor, NY.

Betschinger J, Knoblich JA. 2004. Dare to be different: Asymmetric cell division in *Drosophila*, *C. elegans* and vertebrates. *Curr Biol* **14:** R674–685.

Betschinger J, Mechtler K, Knoblich JA. 2003. The Par complex directs asymmetric cell division by phosphorylating the cytoskeletal protein Lgl. *Nature* **422:** 326–330.

Boone JQ, Doe CQ. 2008. Identification of *Drosophila* type II neuroblast lineages containing transit amplifying ganglion mother cells. *Dev Neurobiol* **68:** 1185–1195.

Bowman SK, Rolland V, Betschinger J, Kinsey KA, Emery G, Knoblich JA. 2008. The tumor suppressors Brat and Numb regulate transit-amplifying neuroblast lineages in *Drosophila*. *Dev Cell* **14:** 535–546.

Brand M, Jarman AP, Jan LY, Jan YN. 1993. *asense* is a *Drosophila* neural precursor gene and is capable of initiating sense organ formation. *Development (Camb)* **119:** 1–17.

Breinbauer R, Köhn M. 2003. Azide–alkyne coupling: A powerful reaction for bioconjugate chemistry. *Chembiochem* **4:** 1147–1149.

Humbert PO, Grzeschik NA, Brumby AM, Galea R, Elsum I, Richardson HE. 2008. Control of tumourigenesis by the Scribble/Dlg/Lgl polarity module. *Oncogene* **27:** 6888–6907.

Ito K, Hotta Y. 1992. Proliferation pattern of postembryonic neuroblasts in the brain of *Drosophila melanogaster*. *Dev Biol* **149:** 134–148.

Izumi Y, Ohta N, Itoh-Furuya A, Fuse N, Matsuzaki F. 2004. Differential functions of G protein and Baz-aPKC signaling pathways in *Drosophila* neuroblast asymmetric division. *J Cell Biol* **164:** 729–738.

Kolb H, Finn M, Sharpless K. 2001. Click chemistry: Diverse chemical function from a few good reactions. *Angew Chem Int Ed Engl* **40:** 2004–2021.

Kosman D, Mizutani CM, Lemons D, Cox WG, McGinnis W, Bier E. 2004. Multiplex detection of RNA expression in *Drosophila* embryos. *Science* **305:** 846.

Kraut R, Chia W, Jan LY, Jan YN, Knoblich JA. 1996. Role of inscuteable in orienting asymmetric cell divisions in *Drosophila*. *Nature* **383:** 50–55.

Lee CY, Andersen RO, Cabernard C, Manning L, Tran KD, Lanskey MJ, Bashirullah A, Doe CQ. 2006a. *Drosophila* Aurora-A kinase inhibits neuroblast self-renewal by regulating aPKC/Numb cortical polarity and spindle orientation. *Genes Dev* **20:** 3464–3474.

Lee CY, Robinson KJ, Doe CQ. 2006b. Lgl, Pins and aPKC regulate neuroblast self-renewal versus differentiation. *Nature* **439:** 594–598.

Lee CY, Wilkinson BD, Siegrist SE, Wharton RP, Doe CQ. 2006c. Brat is a Miranda cargo protein that promotes neuronal differentiation and inhibits neuroblast self-renewal. *Dev Cell* **10:** 441–449.

Morrison SJ, Kimble J. 2006. Asymmetric and symmetric stem-cell divisions in development and cancer. *Nature* **441:** 1068–1074.

Nakagawa T, Nabeshima Y, Yoshida S. 2007. Functional identification of the actual and potential stem cell compartments in mouse spermatogenesis. *Dev Cell* **12:** 195–206.

Nipper RW, Siller KH, Smith NR, Doe CQ, Prehoda KE. 2007. Gαi generates multiple Pins activation states to link cortical polarity and spindle orientation in *Drosophila* neuroblasts. *Proc Natl Acad Sci* **104:** 14306–14311.

Rolls MM, Albertson R, Shih HP, Lee CY, Doe CQ. 2003. *Drosophila* aPKC regulates cell polarity and cell proliferation in neuroblasts and epithelia. *J Cell Biol* **163:** 1089–1098.

Rostovtsev V, Green L, Fokin V, Sharpless K. 2002. A stepwise Huisgen cycloaddition process: Copper(I)-catalyzed regioselective "ligation" of azides and terminal alkynes. *Angew Chem Int Ed Engl* **41:** 2596–2599.

Smith CA, Lau KM, Rahmani Z, Dho SE, Brothers G, She YM, Berry DM, Bonneil E, Thibault P, Schweisguth F, et al. 2007. aPKC-mediated phosphorylation regulates asymmetric membrane localization of the cell fate determinant Numb. *EMBO J* **26:** 468–480.

Truman JW, Bate M. 1988. Spatial and temporal patterns of neurogenesis in the central nervous system of *Drosophila melanogaster*. *Dev Biol* **125:** 145–157.

Wirtz-Peitz F, Nishimura T, Knoblich JA. 2008. Linking cell cycle to asymmetric division: Aurora-A phosphorylates the Par complex to regulate Numb localization. *Cell* **135:** 161–173.

Wu PS, Egger B, Brand AH. 2008. Asymmetric stem cell division: Lessons from *Drosophila*. *Sem Cell Dev Biol* **19:** 283–293.

Yoon K, Gaiano N. 2005. Notch signaling in the mammalian central nervous system: Insights from mouse mutants. *Nat Neurosci* **8:** 709–715.

3 | Cell-Fate Determination in the Embryonic Central Nervous System

Heather Broihier

Department of Neurosciences, Case Western Reserve University School of Medicine, Cleveland, Ohio 44106

ABSTRACT

The embryonic ventral nerve cord (VNC) in *Drosophila melanogaster* is an established system for the analysis of questions central to developmental neurobiology. In addition to the obvious genetic and experimental advantages of the fly, several features of the VNC make it well suited for the study of neuronal development. Decades of careful descriptive work coupled with the relatively limited complexity of the embryonic VNC allow us to make meaningful connections between transcription factor hierarchies acting within individual lineages and specific aspects of neuronal fate and differentiation. To provide a reference for investigators wishing to characterize the expression patterns of novel factors in the central nervous system (CNS), this chapter includes a table showing the expression patterns of a large number of markers expressed in postmitotic neuronal subsets. There is also a protocol describing techniques for the simultaneous fluorescent detection of RNA and protein.

INTRODUCTION

A fertilized *Drosophila* egg develops into a first-instar larva with a functional nervous system in only 24 h. In this remarkably short period of time, a rapidly executed series of developmental events culminates in the establishment of interconnected neuronal circuits. For a detailed review of embryonic CNS development, see Skeath and Thor (2003). Briefly, the neuroectoderm is first patterned, and proneural clusters form at defined positions. After a neural stem cell/neuroblast is specified within each cluster, it delaminates into the embryo interior and initiates a stereotyped series of stem cell divisions to give rise to near invariant clones of neurons and glia. Immediately after neurons are generated, they extend processes in defined pathways that contact and synapse with appropriate cellular partners. Finally, embryonic neurons become electrically active and the circuits regulating larval behavior are established.

ORGANIZATION AND DEVELOPMENT OF THE EMBRYONIC VENTRAL NERVE CORD

Neuroblasts (NBs) are identified by their relatively superficial position within the embryo and by their large size. They delaminate from the neuroectoderm following the coordinated actions of proneural genes and the Notch pathway leading to the specification of a single NB from each proneural cluster. NBs delaminate in five stages or waves (SI–SV), beginning at embryonic stage 9 and concluding at stage 11. In each half-segment, 30 NBs are specified in an invariant two-dimensional array. Each NB has a unique identity as a result of its particular anteroposterior and dorsoventral position within the array (for review, see Skeath 1999) and expresses distinct combinations of transcriptional regulators based on the position at which it forms. The specific complement of transcription factors expressed by individual NBs has been carefully described (Broadus et al. 1995) and provides an invaluable resource for identifying NBs and thus for characterizing the expression patterns of novel NB-expressed genes.

Whereas spatial cues play a central role in generating NB diversity, temporal cues are essential for regulating cellular diversity within NB clones. NBs divide asymmetrically to generate ganglion mother cells (GMCs), which divide once to yield a pair of neurons or glia. DiI lineage and birthorder data have shown that NBs generate distinct progeny in an invariant order (Bossing et al. 1996; Schmidt et al. 1997; Schmid et al. 1999). In fact, the vast majority of fly embryonic NBs are multipotent and give rise to mixed clones of interneurons (INTs), motor neurons (MNs), and/or glia. For example, all MN-producing lineages also generate INTs (Schmid et al. 1999). The stereotypy of NB clones strongly suggests that genetic regulatory hierarchies drive the ordered production of neurons within lineages. Perhaps surprisingly, the sequential activation of the same transcription factor cassette regulates the temporal transitions in many NB lineages (Kambadur et al. 1998; Isshiki et al. 2001; Pearson and Doe 2004). These regulators, Hunchback (Hb), Krüppel (Kr), Pdm1/Pdm2 (Pdm), and Castor (Cas) are sequentially activated in NBs and are required for proper neuronal fate acquisition of their progeny. Whereas the NB expression of individual regulators is transient, gene expression is inherited and maintained by the neuronal progeny born during each gene expression window. Thus, a GMC born during the Hb expression window inherits Hb as does its neuronal daughters. The Hb→Kr→Pdm→Cas cascade is responsible for the generation of successive fates within a lineage. For example, Hb and Kr are both necessary and sufficient to specify the early temporal identities in several posterior NB lineages (Isshiki et al. 2001). The spatial identity cues that uniquely identify each NB are likely to intersect with the temporal identity cues that control temporal transitions to regulate the identity of specific neuronal populations arising within individual NB lineages.

Interestingly, distinct NBs generate clones of widely divergent sizes. The shortest lineage contains only three GMCs (six neurons), whereas the longest is composed of 20 GMCs (40 neurons) (Schmid et al. 1999). By the end of embryogenesis, NB divisions have fully populated the VNC, giving rise to roughly 300 neurons/glia in abdominal hemisegments and 400 neurons/glia in thoracic hemisegments. Of these, 32–35 are MNs, 30 are glia, roughly 60 are intersegmental INTs, and the remainder are local/axonless INTs. DiI and lineage data have shown that there are not clear relationships between NB identity and neuronal identity, so that functionally related neurons are found in many different NB lineages. Because cells in the VNC do not migrate far from their birthplace, neurons of disparate functional identities are intermingled. For example, one-half of all NB lineages generate MNs, and these lineages arise at all anteroposterior and dorsoventral positions in the VNC. Although there is not an obvious link between spatial signals and MN cell fate, the temporal gene cascade may play a crucial function in regulating MN identity because the vast majority of MNs are early born (Schmid et al. 1999). The trajectories of a few neuronal populations, particularly MNs, have been quite well described, but relatively little is known about the projections of distinct interneuron populations or about the central pattern-generating circuits that regulate MN activity and mediate locomotion. In recent years, a number of genes have been identified that are expressed in distinct neuronal populations (Table 1). However in most cases these markers label neuronal nuclei. Efforts to generate promoter reporter lines (e.g., LacZ and Gal4) to follow the processes of different INT subsets are critical to mapping the circuits that control VNC output.

Drosophila MNs provide a tractable system in which to investigate developmental mechanisms regulating cell-fate determination, axon pathfinding, and neuromuscular junction development (for review, see Landgraf and Thor 2006). Recently, studies of MN development have been advanced by a growing number of MN markers, allowing us to uncover the transcription factor codes dictating aspects of neuronal fate and differentiation. To date, the only two molecular markers that are apparently expressed in all MNs are Zfh-1 and pMad, although neither factor is expressed exclusively in the MN population (Table 1 and references therein). MNs may be subdivided based on their gene-expression profiles and axon-projection patterns. The majority of motor axons leave the VNC in one of two major nerve roots, the intersegmental nerve (ISN) and the segmental nerve (SN). Axons extending in the SN innervate external, largely transverse muscles, whereas axons in the ISN innervate internal, longitudinal muscles. MNs projecting to the ventral and lateral body wall express a cocktail of transcriptional regulators, including Nkx6, Hb9, Islet, and Lim3. In contrast, MNs innervating dorsal muscles express Even-skipped (Eve), Grain, and Lim1 (Table 1 and references therein). These complementary expression patterns are maintained by potent cross-repressive interactions (Fig. 1) (Thor and Thomas 1997; Broihier and Skeath 2002; Broihier et al. 2004).

As might be anticipated, these transcription factors are required for MN differentiation. With the exception of Lim1, all of the single loss-of-function (LOF) mutants display motor axon pathfinding defects. Interestingly, many of these factors continue to be expressed in MNs well after axon pathfinding is complete, suggesting that at least some of these "combinatorial code" transcription factors likely regulate later aspects of MN differentiation. This hypothesis is supported by the findings of Pym et al. (2006), who found that electrical properties of a subset of Eve-positive neurons are reciprocally affected in *eve* LOF and GOF (gain-of-function) flies. To further clarify the mechanisms regulating MN differentiation and function, it is necessary to define the circuits regulating and coor-

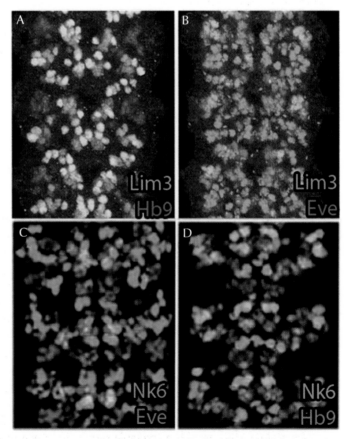

FIGURE 1. Wild-type embryos stained with the indicated antibodies. By double-labeling embryos with antibodies that recognize distinct neuronal subsets, postmitotic neurons are resolved at the level of the individual neuron. (A,B: Reprinted from Broihier and Skeath 2002; C,D: reprinted from Broihier et al. 2004.)

TABLE 1. Molecular markers for subsets of CNS neurons

Marker	CNS expression pattern	Type(s) of antibody	Dilution(s) for embryo staining	Reference(s)
ap^{LacZ}, ap^{Gal4} (apterous)	Three INTs/HS, one of which (dAp) is peptidergic and expresses Nplp1. In T1-3, ap is also expressed in a lateral cluster of four INTs, including the peptidergic Tv and Tvb neurons, which express the FMRFamide and Nplp1 neuropeptides	NA	NA	Lundgren et al. 1995; Allan et al. 2005
Anti-Barh1/h2 (B-H1 and B-H2)	Five SNa-projecting MNs, unpaired ventral midline dopaminergic neuron	Rb polyclonal	1:100	Higashijima et al. 1992; Garces et al. 2006
$B\text{-}H1^{Gal4}$	Five MNs projecting in SNa	NA	NA	Sato et al. 1999; Garces et al. 2006
Cha-Gal4 (choline acetyltransferase)	Cholinergic INTs	NA	NA	Salvaterra and Kitamoto 2001
Anti-Connectin (Conn)	SN MNs and the external muscles they innervate; 4–6 longitudinal glia/HS; INTs	Mouse monoclonal (CI.427)	1:100	Meadows et al. 1994; Nose et al. 1994
Anti-Dachshund2-3 (Dac2-3)	Large population of INTs. In T1-3, also expressed in the Ap+ cluster.	Mouse monoclonal	1:25	Mardon et al. 1994; Miguel-Aliaga et al. 2004
Anti-Dbx	~20 INTs/HS; the majority are GABAergic.	GP polyclonal	1:1500	Lacin et al. 2009
dMP2-Gal4 (Vap-Gal4)	dMP2 visceral MNs in A6–A8.	NA	NA	Allan et al. 2003; Miguel-Aliaga and Thor 2004
Anti-Dimmed (Dimm)	Large peptidergic neurosecretory neurons	GP polyclonal	1:500	Allan et al. 2005; Park et al. 2008
c929-Gal4	dimm-promoter reporter; expressed in 90% of Dimm+ cells	NA	NA	Hewes et al. 2003; Park et al. 2008
Anti-Drifter (Dfr)	ISNb-projecting MNs (including RP1, 3,4,5); two serotonergic EW neurons	Rat polyclonal	1:2000	Anderson et al. 1995; Certel and Thor 2004
$eagle^{P289}$ enhancer trap	NB 7-3 progeny (including ISNd-projecting GW MN and serotonergic EW INT)	NA	NA	Dittrich et al. 1997
Anti-Engrailed (En)	INTs	Mouse monoclonal (4D9)	1:10	Bossing and Brand 2006; Patel et al. 1989
Anti-Even-skipped (Eve)	~20 neurons/HS; dMNs (aCC, RP2, U1-5) and INTs (ELs, pCC, fpCC)	Rb polyclonal	1:3000	Frasch et al. 1987; Landgraf et al. 1999
RN2-Gal4	eve-promoter reporter driving expression in aCC/RP2 MNs	NA	NA	Fujioka et al. 2003
CQ2-Gal4	eve-promoter reporter driving expression in U/CQ MNs	NA	NA	Fujioka et al. 2003
Eyes Absent (Eya)	In T1–T3, expressed in four Ap+ cluster neurons	Mouse monoclonal (10H6)	1:250	Miguel-Aliaga et al. 2004
Anti-Fas2	At stage 16, all MNs as well as a subset of ipsilaterally projecting INTs	Mouse monoclonal (1D4)	1:10	Van Vactor et al. 1993
Anti-Fas3	RP1,3,4,5 MNs, also several lateral populations of neurons	Mouse monoclonal (7G10)	1:20	Chiba et al. 1995; Patel et al. 1989
Anti-Glutamic acid decarboxylase (Gad)	GABAergic INTs	Rabbit polyclonal	1:1000	Jackson et al. 1990
Anti-Grain (Grn)	dMNs (RP2, aCC, and U/CQs); subset of INTs	Rb and Rat polyclonals	1:200	Garces and Thor 2006
grn^{LacZ}, grn^{Gal4}	Identical to anti-Grain	NA	NA	Garces et al. 2006
Anti-Hb9	~30 neurons/HS; many vMNs (including RP1,3,4,5; GW; LC); also some INTs (including serotonergic EWs)	GP, Rb, Rat polyclonals	1:1000 (GP) 1:5000 (Rb) 1:1000 (Rat)	Broihier and Skeath 2002; Odden et al. 2002
$Hb9^{Gal4}$	Identical to Hb9 protein expression pattern	NA	NA	Broihier and Skeath 2002

(Continued on facing page.)

TABLE 1. (*Continued*)

Marker	CNS expression pattern	Type(s) of antibody	Dilution(s) for embryo staining	Reference(s)
Anti-Islet/Tailup (Isl)	~30 neurons/HS; many vMNs (RP1,3,4,5/VUM/LC); also some INTs (including serotonergic EWs)	Rat polyclonal	1:500	Broihier and Skeath 2002
islH-taumyc	ISNb-projecting MNs (including RP1,3,4,5); TN-projecting MNs; ISNd-projecting MNs; also EW INTs	NA	NA	Thor and Thomas 1997
Anti-Late bloomer	MNs	Mouse monoclonal (10C9)	1:4	Kopczynski et al. 1996
Anti-Lim1	dMNs (aCC, RP2, U1-5); also some INTs	GP polyclonal	1:500	Lilly et al. 1999
Anti-Lim3	~40 neurons/HS; many vMNs (RP1,3,4,5, GW, and LC); also some INTs (ELs)	GP polyclonal	1:500	Broihier and Skeath 2002
lim3A-taumyc, *lim3B-Gal4*	ISNb-projecting MNs (including RP1,3,4,5); two TN MNs	NA	NA	Thor et al. 1999; Certel and Thor 2004
Anti-Neuromancer1 (Nmr1)	Subset of Nmr2-positive neurons	GP and Rb polyclonals	1:3000	Leal et al. 2009
Anti-Neuromancer2 (Nmr2)	~20 neurons/HS; majority are INTs; a single U/CQ MN is Nmr2-positive.	Rb polyclonal	1:1000	Leal et al. 2009
Anti-Odd-skipped (Odd)	MP2 precursor; vMP2 INT through st12; sdMP2 visceral MN through st15; also MP1 neurons	Rb polyclonal	1:1000	Spana et al. 1995
Anti-Repo	All glia, aside from midline glia	Mouse monoclonal (8D12)	1:10	Alfonso and Jones 2002
Anti-phospho-Mad	35–40 neurons/HS; all vMNs and dMNs; also Tv, Va, and Vap peptidergic neurons	Rb polyclonal	1:300	Allan et al. 2003; Marques et al. 2002; Tanimoto et al. 2000
Anti-Nkx6	~40 neurons/HS; many vMNs (RP1,2,4,5/ GW/LC); also some INTs	Rat polyclonal	1:1000	Broihier et al. 2004
Nkx6^{Gal4}	All Nkx6$^+$ INTs and MNs; expression is not restricted to Nkx6$^+$ neurons.	NA	NA	Broihier et al. 2004
Anti-Serotonin	Serotonergic INTs	Rabbit polyclonal	1:300	Taghert and Goodman 1984
squeeze/sqz^{LacZ}, *sqz^{Gal4}*	At late st17, expression is restricted to FMRF$^+$ Tv neuron.	NA	NA	Allan et al. 2003
Vesicular glutamate transporter (DVGLUT)	Glutamatergic neurons (MNs and INTs)	Rb polyclonal	1:10,000	Daniels et al. 2004
OK371-Gal4 (*DVGlut-Gal4*)	Glutamatergic neurons (MNs and INTs)	NA	NA	Mahr and Aberle 2006
Anti-Vestigial (Vg)	In T1–T3, 41–43 INTs; in A1–A8, 27–29 INTs; all Vg$^+$ neurons are INTs with the exception of the ventral unpaired median MNs (mVUMs).	Rabbit polyclonal	1:25	Guss et al. 2008; Williams et al. 1991
Anti-Wrapper	High levels in anterior midline glia (AMG); low levels in posterior midline glia (PMG)	Mouse monoclonal (10D3) GP polyclonal	1:10 (mouse)	Noordermeer et al. 1998; Wheeler et al. 2006, 2009
Anti-Vnd (ventral nervous system defective)	Ventral NBs at stage 9; at stage 11, ~20 GMCs and neurons/HS; at stage 14, ~10 INTs/HS	Rabbit polyclonal	1:10	Broihier et al. 2004; McDonald et al. 1998
Anti-Zfh1	All somatic MNs (vMN and dMN); Apterous$^+$ AD INT, exit and surface glia	Rb and GP polyclonals	1:5000 (Rb) 1:300 (GP)	Lai et al. 1991; Tian et al. 2004; Layden et al. 2006

Abbreviations: CNS, central nervous system; INTs, interneurons; HS, hemisegment; MNs, motor neurons; vMNs, ventrally projecting motor neurons; dMNs, dorsally projecting motor neurons; LC, lateral MN cluster; TN, transverse nerve; SN, segmental nerve; ISN, intersegmental nerve; NA, not applicable.

dinating MN activation. Efforts to map the central pattern generating circuit are still in their infancy, although it has been shown that the MNs are activated by cholinergic INTs (Baines and Bate 1998). The underlying functional organization of the motor system has also been clarified by the studies of Landgraf et al. (2003) who found that the dendritic arbors of MNs are partitioned in the neuropil according to the identity of the somatic muscles they innervate. As additional interneuron markers are generated, this work will greatly simplify mapping the excitatory and inhibitory inputs regulating MN activity in the VNC.

The utility of the embryonic VNC for studies of nervous system development and function will grow as additional cell types are identified and their connections mapped. An ever-increasing set of Gal4 lines expressed in discrete neuronal populations allows us to ask precise questions addressing both the roles of individual genes in specific cells and the relationship of different neuronal populations with each other. Table 1 is included to provide a summary of the expression patterns of a number of commonly used markers for MN and INT subtypes in the embryonic VNC.

Typically, when mapping the cell types that express a particular gene of interest, the first step is to evaluate expression of the endogenous RNA. Whole-mount RNA in situ hybridizations with digoxigenin-conjugated probes and alkaline phosphatase biochemistry have been used widely for many years to map expression pattern domains in the embryo. To capitalize on the number of molecular markers in the CNS and to enable expression analysis at the single-cell level, fluorescence in situ hybridization procedures are becoming standard. This chapter contains a protocol for fluorescence in situ hybridization and antibody staining.

Whole-Mount Embryo Fluorescence In Situ Hybridization and Antibody Staining

This protocol describes methods for the simultaneous detection of RNA and protein using fluorescence. It uses the tyramide signal amplication (TSA) system from PerkinElmer to amplify a horseradish peroxidase (HRP) signal. By combining this technology with an HRP-conjugated antidigoxigenin antibody, we can detect standard antidigoxigenin RNA probes fluorescently.

MATERIALS

CAUTION: See Appendix for proper handling of materials marked with <!>.
See the end of the chapter for recipes for reagents marked with <R>.

Reagents

Antidigoxigenin conjugated to HRP (anti-DIG-POD; Roche cat. no. 11 207 733 910)
Bleach (Clorox) <!>
Chloroform <!>
DIG RNA labeling mix (10X) (Roche cat. no. 11 277 073 910)
Ethanol <!> (100%)
Formaldehyde <!> (histological grade; 37% [w/v])
Glycerol (70%)
Heptane <!>
Hybridization buffer (RNase-free) <R>
Methanol <!> (100%)
Phenol/chloroform (1:1) <!>
Phosphate-buffered saline (PBS) containing 0.1% Triton X-100 <!> (PTX) <R>
RNase-free H_2O (DEPC [diethylpyrocarbonate]-treated)
Sodium acetate (3 M)
TNB blocking buffer <R>
TNT wash buffer <R>
Tyramide signal amplification (TSA) plus fluorescein system (PerkinElmer NEL741)

Equipment

Microcentrifuge tubes (1.5-mL)
Mini-platform rocker (e.g., Denville Scientific mini-rocker, model 135)
Nytex mesh
Paintbrush (small)
Pasteur pipettes
Scintillation vials with open-top lids
Squirt bottle

METHOD

Collection and Fixation of Embryos

1. Make the embryo collection vial. Fit a round piece of Nytex mesh into an open-top lid and screw the lid onto a glass scintillation vial that has had the bottom removed. Place this collection vial with the open end at the top inside a 50-mL plastic beaker.

2. Using a small paintbrush or squirt bottle, transfer the embryos from an egg collection plate into the collection vial.

3. Perform the following steps by pouring solutions into the open end of the vial. Catch the washes in the 50-mL beaker.

   i. Wash the embryos several times with H_2O or PTX.

   ii. Dechorionate the embryos by immersing in 50% bleach for 3 min.

   iii. Wash the embryos thoroughly in H_2O until they no longer smell like bleach.

4. Transfer the mesh and embryos into a 10-mL glass tube containing 2 mL of heptane and 2 mL of 37% formaldehyde. Fix for 2 min with shaking on a mini-platform rocker.

   The embryos should settle in the interface between the aqueous and organic phases. This "fast" 2-min fix may be better at preserving fragile cytoskeletal structures than the traditional 20-min fix.

5. Remove the lower formaldehyde phase with a pasteur pipette. Take care to remove as much fixative as possible because this will increase the percentage of devitellinized embryos.

6. Add 4 mL of 100% methanol to the tube. There must be an interface present at this stage to separate the devitellinized embryos. If necessary, add a few more milliliters of fresh heptane to the tube to maintain the aqueous/organic interface. Shake the tube vigorously for 20 sec. Tap the tube firmly against your palm to shake the devitellinized embryos off the sides and to the bottom of the tube.

7. Transfer the devitellinized embryos to a microcentrifuge tube using a pasteur pipette and rinse them several times with 100% methanol.

Preparation of Digoxigenin-Labeled Antisense RNA Probe

8. Clone the gene of interest into a vector containing RNA polymerase binding sites (SP6, T3, or T7).

   In our hands, RNA probes synthesized with T7 polymerase are best as they have a stronger signal and less background.

9. Linearize the plasmid with a restriction enzyme that cuts 5′ to the insert. Calculate amount of total DNA (vector + insert) to cut so that you will have ~1 µg of insert (e.g., for a 1-kb insert in a 3-kb vector, linearize 4 µg of plasmid). Run out 1 µL on an agarose gel to check digestion. Bring the volume up to 100 µL with RNase-free dH_2O.

10. Purify the DNA by extracting once with 1:1 phenol/chloroform and once with chloroform. Vortex each extraction for 1 min and spin in a microcentrifuge for 5 min at maximum speed.

11. Measure the volume and precipitate the DNA with 0.1 volume of 3 M sodium acetate and 2.5 volumes of 100% ethanol. Allow the DNA to dry well and resuspend in 13 µL of RNase-free H_2O.

12. Synthesize the antisense riboprobe in a 20-µL reaction volume using the RNA polymerase with a binding site located 3′ to the insert. Set up the following reaction:

    13 µL of linearized template DNA
    2 µL of 10x transcription buffer (supplied with the Roche DIG RNA labeling mix)
    2 µL of 10x DIG RNA labeling mix
    1 µL of RNase inhibitor
    2 µL of RNA polymerase

    Mix gently and incubate for 2 h at 37°C.

13. Bring the volume up to 50 μL with RNase-free H$_2$O. Add 0.1 volume of 3 M sodium acetate and 2.5 volumes of cold 100% ethanol and precipitate overnight at −30°C.

14. Spin for 15 min at 4°C. Wash the RNA pellet once in cold 70% ethanol and dry completely. Resuspend the pellet in 100 μL of RNase-free hybridization buffer.

> We do not treat RNA probes with carbonate to shorten their length because this usually decreases the signal. As a control, a sense probe may be synthesized by linearizing with a restriction enzyme that cuts 3′ to the insert and using the RNA polymerase with a binding site 5′ to the insert. Alternatively, it is feasible to PCR-amplify your gene of interest using a 3′ primer with a RNA polymerase site added and to use this PCR product as template for the transcription reaction.

In Situ Hybridization

If not otherwise specified, volumes are 1.0 mL. Samples should be mixed on the platform rocker. Aim for at least 50 μL of embryos/tube, but do not exceed 100 μL/tube. The embryos are virtually transparent in hybridization buffer, so there is often significant embryo attrition in the course of the experiment.

15. Wash the RNA sample once in 1:1 methanol/PTX for 5 min.

16. Wash twice in PTX for 5 min each.

17. Wash once in 1:1 PTX/RNase-free hybridization buffer for 5 min (0.5 mL total volume).

18. Wash once in RNase-free hybridization buffer for 5 min (0.2 mL total volume). Do not use the rocker for this step. Mix by inverting the tube several times and place it flat to incubate.

19. Prehybridize the embryos in 0.2 mL of RNase-free hybridization buffer for 1 h at 55°C–60°C, depending on the probe.

> We typically test probes first at 55°C and increase the temperature, if needed, to decrease the background. There is some loss of specific signal at higher temperatures.

20. Use 1 μL of RNA probe per tube of embryos. Denature the probe at 85°C–95°C for 5 min and briefly spin down. Place the probe on ice immediately after the denaturation step and add 100 μL of prewarmed hybridization buffer to each probe. Remove the RNase-free hybridization buffer from the embryos and add 100 μL of probe in hybridization buffer.

21. Hybridize overnight at 55°C–60°C.

> It is important to hybridize for at least 16 h; longer is often better up to ~40 h. At about this point, the embryo morphology begins to suffer.

22. Wash in hybridization buffer (not RNAse-free) at 58°C. All wash volumes are 0.2 mL. Minimum wash time is 4 h; longer is generally better. Use one of the following wash schedules:

    i. 4 x 15 min; 4 x 30 min; 1 x 1 h or

    ii. 3 x 20 min; 3 x 40 min; 1 x 1 h

    iii. 6 x 20 min; 2 x 30 min; 1 x 60 min.

23. Wash once in 1:1 hybridization buffer/TNT wash buffer for 5 min at room temperature (0.5 mL total volume).

24. Equilibrate in TNT wash buffer by washing three times for 5 min each at room temperature (1.0 mL total volume).

25. Block in TNB blocking buffer by washing three times for 10 min each at room temperature (1.0 mL total volume).

26. Incubate the embryos in anti-DIG-POD for 1 h at room temperature. Use 200 μL of diluted anti-DIG-POD per tube. Mix gently and place the tubes flat to incubate.

> We use a dilution between 1:50 and 1:250 of anti-DIG-POD in TNB blocking buffer, depending on the probe.

27. Wash three times in TNB blocking buffer for 10 min each at room temperature (1.0 mL total volume).

28. Dilute the fluorescein tyramide (amplification reagent) 1:50 in 1x plus amplification diluent (which is included in the kit). Prepare the dilution immediately before use. Use 200 μL/tube and incubate for 1–2 h at room temperature.

29. Wash three times in TNT wash buffer for 5 min each at room temperature (1.0 mL total volume). After washing, check the fluorescence under a microscope.

Antibody Staining

This is a variation of a standard antibody staining protocol, with TNT wash buffer substituted for PTX and TNB blocking buffer substituted for PBS containing Triton X-100 and BSA. We use 2-h incubations for both the primary and secondary antibodies to finish the staining in 1 d and to preserve as much RNA fluorescence as possible.

30. Block the embryos in TNB blocking buffer by incubating four times for 15 min each. Use 1 mL of TNB blocking buffer per tube.

31. Incubate the embryos in primary antibody for 2 h at room temperature. Use 200 μL per tube. Mix gently and place the tube flat to incubate.

32. Block the embryos in TNB blocking buffer by washing four times for 15 min each. Use 1 mL of TNB blocking buffer per tube.

33. Incubate the embryos in secondary antibody with appropriate species specificity conjugated to the desired fluorophore (e.g., anti-Rabbit, Alexa 568). Incubate for 2 h at room temperature. Use 200 μL per tube.

34. Wash the embryos four times in TNT wash buffer in a light-shielded tube for 5 min each.

35. Mount the samples in 70% glycerol and image by conventional fluorescence or confocal microscopy.

RECIPES

CAUTION: See Appendix for proper handling of materials marked with <!>.
Recipes for reagents marked with <R> are included in this list.

Hybridization Buffer (RNase-free)

Reagent	Quantity (for 250 mL)	Final concentration
Formamide <!>	125 mL	50% (v/v)
SSC stock solution (20x) <R>	62.5 mL	25% (v/v)
Herring sperm DNA (10 mg/mL) (extract twice with 1:1 phenol/ chloroform for RNase-free hybridization buffer)	2.5 mL	100 mg/L
Heparin (50 mg/mL) <!>	250 μL	50 mg/L
Tween-20 (10%, v/v)	2.5 mL	0.1% (v/v)
H_2O (DEPC-treated)	57.25 mL	

PBS Stock Solution (10x; pH 7.0)

Reagent	Quantity (for 1 L)	Final concentration
NaCl	74 g	1.27 M
Na$_2$HPO$_4$.7H$_2$O	18.78 g	70 mM
NaH$_2$PO$_4$.H$_2$O	4.15 g	30 mM

Dissolve the components in dH$_2$O and adjust the pH to 7.0 with NaOH <!> if necessary. Adjust the volume to 1 L with dH$_2$O

PTX

Reagent	Quantity (for 1 L)	Final concentration
PBS stock solution (10x) <R>	100 mL	1x
Triton X-100 (10%, v/v) <!>	10 mL	0.1% (v/v)

Add dH$_2$O to give a final volume of 1 L.

SSC Stock Solution (20x)

Reagent	Quantity (for 1 L)	Final concentration
NaCl	175.3 g	3.0 M
Sodium citrate <!>	88.2 g	0.3 M
H$_2$O	800 mL	

Adjust the pH to 7.0 with a few drops of 14 N HCl <!>. Adjust the volume to 1 L with H$_2$O.

TNB Blocking Buffer

Reagent	Final concentration
Tris-HCl, pH 7.5	0.1 M
NaCl	0.15 M
Blocking reagent (PerkinElmer cat. No. 1020)	0.5% (w/v)

Add blocking reagent slowly while stirring. Heat gradually to 60°C with continuous stirring to completely dissolve the blocking reagent. (This may take up to several hours.) Aliquot and store at –20°C.

TNT Wash Buffer

Reagent	Final concentration
Tris-HCl, pH 7.5	0.1 M
NaCl	0.15 M
Tween-20	0.05% (v/v)

REFERENCES

Alfonso TB, Jones BW. 2002. *gcm2* promotes glial cell differentiation and is required with *glial cells missing* for macrophage development in *Drosophila*. *Dev Biol* **248:** 369–383.

Allan DW, St Pierre SE, Miguel-Aliaga I, Thor S. 2003. Specification of neuropeptide cell identity by the integration of retrograde BMP signaling and a combinatorial transcription factor code. *Cell* **113:** 73–86.

Allan DW, Park D, St Pierre SE, Taghert PH, Thor S. 2005. Regulators acting in combinatorial codes also act independently in single differentiating neurons. *Neuron* **45:** 689–700.

Anderson MG, Perkins GL, Chittick P, Shrigley RJ, Johnson WA. 1995. *drifter*, a *Drosophila* POU-domain transcription factor, is required for correct differentiation and migration of tracheal cells and midline glia. *Genes Dev* **9:** 123–137.

Baines RA, Bate M. 1998. Electrophysiological development of central neurons in the *Drosophila* embryo. *J Neurosci* **18:** 4673–4683.

Bossing T, Brand AH. 2006. Determination of cell fate along the anteroposterior axis of the *Drosophila* ventral midline. *Development* **133:** 1001–1012.

Bossing T, Udolph G, Doe CQ, Technau GM. 1996. The embryonic central nervous system lineages of *Drosophila melanogaster*. I. Neuroblast lineages derived from the ventral half of the neuroectoderm. *Dev Biol* **179:** 41–64.

Broadus J, Skeath JB, Spana EP, Bossing T, Technau G, Doe CQ. 1995. New neuroblast markers and the origin of the aCC/pCC neurons in the *Drosophila* central nervous system. *Mech Dev* **53:** 393–402.

Broihier HT, Skeath JB. 2002. *Drosophila* homeodomain protein dHb9 directs neuronal fate via crossrepressive and cell-nonautonomous mechanisms. *Neuron* **35:** 39–50.

Broihier HT, Kuzin A, Zhu Y, Odenwald W, Skeath JB. 2004. *Drosophila* homeodomain protein Nkx6 coordinates motoneuron subtype identity and axonogenesis. *Development* **131:** 5233–5242.

Certel SJ, Thor S. 2004. Specification of *Drosophila* motoneuron identity by the combinatorial action of POU and LIM-HD factors. *Development* **131:** 5429–5439.

Chiba A, Snow P, Keshishian H, Hotta Y. 1995. Fasciclin III as a synaptic target recognition molecule in *Drosophila*. *Nature* **374:** 166–168.

Daniels RW, Collins CA, Gelfand MV, Dant J, Brooks ES, Krantz DE, DiAntonio A. 2004. Increased expression of the *Drosophila* vesicular glutamate transporter leads to excess glutamate release and a compensatory decrease in quantal content. *J Neurosci* **24:** 10466–10474.

Dittrich, R., Bossing, T., Gould, A.P., Technau, G.M., and Urban, J. 1997. The differentiation of the serotonergic neurons in the *Drosophila* ventral nerve cord depends on the combined function of the zinc finger proteins Eagle and Huckebein. *Development* **124:** 2515–2525.

Frasch M, Hoey T, Rushlow C, Doyle H, Levine M. 1987. Characterization and localization of the *even-skipped* protein of *Drosophila*. *EMBO J* **6:** 749–759.

Fujioka M, Lear BC, Landgraf M, Yusibova GL, Zhou J, Riley KM, Patel NH, Jaynes JB. 2003. Even-skipped, acting as a repressor, regulates axonal projections in *Drosophila*. *Development* **130:** 5385–5400.

Garces A, Thor S. 2006. Specification of *Drosophila* aCC motoneuron identity by a genetic cascade involving *even-skipped*, *grain* and *zfh1*. *Development* **133:** 1445–1455.

Garces A, Bogdanik L, Thor S, Carroll P. 2006. Expression of *Drosophila* BarH1-H2 homeoproteins in developing dopaminergic cells and segmental nerve a (SNa) motoneurons. *Eur J Neurosci* **24:** 37–44.

Guss KA, Mistry H, Skeath JB. 2008. Vestigial expression in the *Drosophila* embryonic central nervous system. *Dev Dyn* **237:** 2483–2489.

Hewes RS, Park D, Gauthier SA, Schaefer AM, Taghert PH. 2003. The bHLH protein Dimmed controls neuroendocrine cell differentiation in *Drosophila*. *Development* **130:** 1771–1781.

Higashijima S, Michiue T, Emori Y, and Saigo K. 1992. Subtype determination of *Drosophila* embryonic external sensory organs by redundant homeobox genes *BarH1* and *BarH2*. *Genes Dev* **6:** 1005–1018.

Isshiki T, Pearson B, Holbrook S, Doe CQ. 2001. *Drosophila* neuroblasts sequentially express transcription factors which specify the temporal identity of their neuronal progeny. *Cell* **106:** 511–521.

Jackson FR, Newby LM, Kulkarni SJ. 1990. *Drosophila* GABAergic systems: Sequence and expression of glutamic acid decarboxylase. *J Neurochem* **54:** 1068–1078.

Kambadur R, Koizumi K, Stivers C, Nagle J, Poole SJ, Odenwald WF. 1998. Regulation of POU genes by *castor* and *hunchback* establishes layered compartments in the *Drosophila* CNS. *Genes Dev* **12:** 246–260.

Kopczynski CC, Davis GW, Goodman CS. 1996. A neural tetraspanin, encoded by late bloomer, that facilitates synapse formation. *Science* **271:** 1867–1870.

Lacin H, Zhu Y, Wilson BA, Skeath JB. 2009. *dbx* mediates neuronal specification through cross-repressive, lineage-specific interactions with *eve* and *hb9*. *Development* **136:** 3257–3266.

Lai ZC, Fortini ME, Rubin GM. 1991. The embryonic expression patterns of *zfh-1* and *zfh-2*, two *Drosophila* genes encoding novel zinc-finger homeodomain proteins. *Mech Dev* **34**: 123–134.

Landgraf M, Thor S. 2006. Development of *Drosophila* motoneurons: Specification and morphology. *Semin Cell Dev Biol* **17**: 3–11.

Landgraf M, Roy S, Prokop A, VijayRaghavan K, Bate M. 1999. *even-skipped* determines the dorsal growth of motor axons in *Drosophila*. *Neuron* **22**: 43–52.

Landgraf M, Jeffrey V, Fujioka M, Jaynes JB, Bate M. 2003. Embryonic origins of a motor system: Motor dendrites form a myotopic map in *Drosophila*. *PLoS Biol* **1**: E41.

Layden MJ, Odden JP, Schmid A, Garces A, Thor S, Doe CQ. 2006. Zfh1, a somatic motor neuron transcription factor, regulates axon exit from the CNS. *Dev Biol* **291**: 253–263.

Leal SM, Qian L, Lacin H, Bodmer R, Skeath JB. 2009. Neuromancer1 and Neuromancer2 regulate cell fate specification in the developing embryonic CNS of *Drosophila melanogaster*. *Dev Biol* **325**: 138–150.

Lilly B, O'Keefe DD, Thomas JB, Botas J. 1999. The LIM homeodomain protein dLim1 defines a subclass of neurons within the embryonic ventral nerve cord of *Drosophila*. *Mech Dev* **88**: 195–205.

Lundgren SE, Callahan CA, Thor S, Thomas JB. 1995. Control of neuronal pathway selection by the *Drosophila* LIM homeodomain gene *apterous*. *Development* **121**: 1769–1773.

Mahr A, Aberle H. 2006. The expression pattern of the *Drosophila* vesicular glutamate transporter: A marker protein for motoneurons and glutamatergic centers in the brain. *Gene Expr Patterns* **6**: 299–309.

Mardon G, Solomon NM, Rubin GM. 1994. *dachshund* encodes a nuclear protein required for normal eye and leg development in *Drosophila*. *Development* **120**: 3473–3486.

Marques G, Bao H, Haerry TE, Shimell MJ, Duchek P, Zhang B, O'Connor MB. 2002. The *Drosophila* BMP type II receptor Wishful Thinking regulates neuromuscular synapse morphology and function. *Neuron* **33**: 529–543.

McDonald JA, Holbrook S, Isshiki T, Weiss J, Doe CQ, Mellerick DM. 1998. Dorsoventral patterning in the *Drosophila* central nervous system: The *vnd* homeobox gene specifies ventral column identity. *Genes Dev* **12**: 3603–3612.

Meadows LA, Gell D, Broadie K, Gould AP, White RA. 1994. The cell adhesion molecule, connectin, and the development of the *Drosophila* neuromuscular system. *J Cell Sci* **107** (Pt 1) : 321–328.

Miguel-Aliaga I, Thor S. 2004. Segment-specific prevention of pioneer neuron apoptosis by cell-autonomous, postmitotic Hox gene activity. *Development* **131**: 6093–6105.

Miguel-Aliaga I, Allan DW, Thor S. 2004. Independent roles of the *dachshund* and *eyes absent* genes in BMP signaling, axon pathfinding and neuronal specification. *Development* **131**: 5837–5848.

Noordermeer JN, Kopczynski CC, Fetter RD, Bland KS, Chen WY, Goodman CS. 1998. Wrapper, a novel member of the Ig superfamily, is expressed by midline glia and is required for them to ensheath commissural axons in *Drosophila*. *Neuron* **21**: 991–1001.

Nose A, Takeichi M, Goodman CS. 1994. Ectopic expression of connectin reveals a repulsive function during growth cone guidance and synapse formation. *Neuron* **13**: 525–539.

Odden JP, Holbrook S, Doe CQ. 2002. *Drosophila* HB9 is expressed in a subset of motoneurons and interneurons, where it regulates gene expression and axon pathfinding. *J Neurosci* **22**: 9143–9149.

Park D, Veenstra JA, Park JH, Taghert PH. 2008. Mapping peptidergic cells in *Drosophila*: Where DIMM fits in. *PLoS ONE* **3**: e1896.

Patel NH, Martin-Blanco E, Coleman KG, Poole SJ, Ellis MC, Kornberg TB, Goodman CS. 1989. Expression of engrailed proteins in arthropods, annelids, and chordates. *Cell* **58**: 955–968.

Pearson BJ, Doe CQ. 2004. Specification of temporal identity in the developing nervous system. *Annu Rev Cell Dev Biol* **20**: 619–647.

Pym EC, Southall TD, Mee CJ, Brand AH, Baines RA. 2006. The homeobox transcription factor Even-skipped regulates acquisition of electrical properties in *Drosophila* neurons. *Neural Dev* **1**: 3.

Salvaterra PM, Kitamoto T. 2001. *Drosophila* cholinergic neurons and processes visualized with Gal4/UAS-GFP. *Brain Res Gene Expr Patterns* **1**: 73–82.

Sato M, Kojima T, Michiue T, Saigo K. 1999. *Bar* homeobox genes are latitudinal prepattern genes in the developing *Drosophila* notum whose expression is regulated by the concerted functions of *decapentaplegic* and *wingless*. *Development* **126**: 1457–1466.

Schmid A, Chiba A, Doe CQ. 1999. Clonal analysis of *Drosophila* embryonic neuroblasts: Neural cell types, axon projections and muscle targets. *Development* **126**: 4653–4689.

Schmidt H, Rickert C, Bossing T, Vef O, Urban J, Technau GM. 1997. The embryonic central nervous system lineages of *Drosophila melanogaster*. II. Neuroblast lineages derived from the dorsal part of the neuroectoderm. *Dev Biol* **189**: 186–204.

Skeath JB. 1999. At the nexus between pattern formation and cell-type specification: The generation of individual neuroblast fates in the *Drosophila* embryonic central nervous system. *Bioessays* **21**: 922–931.

Skeath JB, Thor S. 2003. Genetic control of *Drosophila* nerve cord development. *Curr Opin Neurobiol* **13**: 8–15.

Spana EP, Kopczynski C, Goodman CS, Doe CQ. 1995. Asymmetric localization of numb autonomously determines sibling neuron identity in the *Drosophila* CNS. *Development* **121**: 3489–3494.

Taghert PH, Goodman CS. 1984. Cell determination and differentiation of identified serotonin-immunoreactive neurons in the grasshopper embryo. *J Neurosci* **4:** 989–1000.

Tanimoto H, Itoh S, ten Dijke P, Tabata T. 2000. Hedgehog creates a gradient of DPP activity in *Drosophila* wing imaginal discs. *Mol Cell* **5:** 59–71.

Thor S, Thomas JB. 1997. The *Drosophila islet* gene governs axon pathfinding and neurotransmitter identity. *Neuron* **18:** 397–409.

Thor S, Andersson SG, Tomlinson A, Thomas, J.B. 1999. A LIM-homeodomain combinatorial code for motor-neuron pathway selection. *Nature* **397:** 76–80.

Tian X, Hansen D, Schedl T, Skeath JB. 2004. Epsin potentiates Notch pathway activity in *Drosophila* and *C. elegans*. *Development* **131:** 5807–5815.

Van Vactor D, Sink H, Fambrough D, Tsoo R, Goodman CS. 1993. Genes that control neuromuscular specificity in *Drosophila*. *Cell* **73:** 1137–1153.

Wheeler SR, Kearney JB, Guardiola AR, Crews ST. 2006. Single-cell mapping of neural and glial gene expression in the developing *Drosophila* CNS midline cells. *Dev Biol* **294:** 509-524.

Wheeler SR, Banerjee S, Blauth K, Rogers SL, Bhat MA, Crews ST. 2009. Neurexin IV and Wrapper interactions mediate *Drosophila* midline glial migration and axonal ensheathment. *Development* **136:** 1147–1157.

Williams JA, Bell JB, Carroll SB. 1991. Control of *Drosophila* wing and haltere development by the nuclear vestigial gene product. *Genes Dev* **5:** 2481–2495.

Analysis of Axon Guidance in the Embryonic Central Nervous System of *Drosophila*

Greg J. Bashaw

Department of Neuroscience, University of Pennsylvania School of Medicine, Philadelphia, Pennsylvania 9104

ABSTRACT

How axons in the developing nervous system navigate to their correct targets is a fundamental question in neuroscience. Studies of axon guidance in the embryonic central nervous system (CNS) of *Drosophila melanogaster* have proven instrumental to the identification of molecules and mechanisms that regulate wiring. The relative simplicity of the embryonic CNS, the advantages of genetic approaches, and the ability to analyze the pathfinding decisions of well-defined axon groups or even individual axons have all contributed to our understanding of the fundamental mechanisms of axon guidance and target selection. To assist researchers who wish to use the fly embryo to test candidate molecules for roles in regulating connectivity, this chapter describes the major guidance paradigms that are studied in the embryo, as well as the molecular pathways that are implicated in the regulation of these events. Simple protocols for visualizing axons in the embryonic CNS using immunohistochemistry and immunofluorescence are included. In addition, a method is presented for specifically detecting the cell-surface expression of receptors and secreted factors.

INTRODUCTION

To ensure correct and efficient wiring of the nervous system, an intricately coordinated sequence of events must occur. First, neurons and their surrounding target tissues must be specified to express the correct complement of receptors and guidance cues. Second, receptors must be assembled into the appropriate complexes and localized to the axonal or dendritic growth cones, and guidance cues must be correctly trafficked to and localized within the extracellular environment. Third, signaling mechanisms must be in place to integrate signals from the surface receptors and transmit them into changes in the growth-cone actin cytoskeleton, which result in stereotyped steering decisions. Each of these steps provides many potential levels for the regulation of axon guidance decisions, and although recent work has enriched our understanding of the complexities of guidance regulation, many questions remain. For comprehensive discussions of the mechanisms that mediate axon guidance in *Drosophila* and other model systems, see reviews by Yu and Bargmann (2001), Dickson (2002), and Garbe and Bashaw (2004).

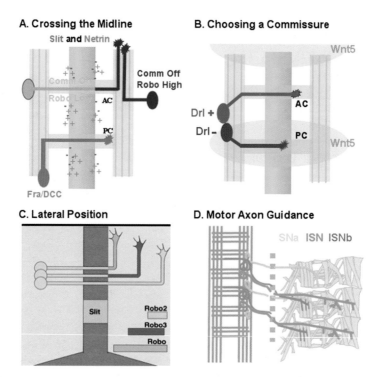

FIGURE 1. Guidance decisions in the fly embryo. (*A*) Midline guidance. Glial cells in the ventral midline secrete Netrin, which attracts axons expressing the Frazzled receptor. Midline cells also secrete Slit, which prevents abnormal midline crossing by repelling axons expressing Robo receptors. Robo repulsion is inhibited in precrossing axons by Commissureless (Comm). (*B*) Commissure choice is regulated by secreted Wnt5, which is expressed in a region near the posterior commissure. Axons expressing Derailed (Drl) avoid the Wnt5 region and cross the midline in the anterior commissure (AC), whereas axons that do not express Drl cross in the posterior commissure (PC). (*C*) A schematic diagram of the role of Slit and its Robo receptors in specifying lateral position. Lateral position is specified by the complement of Robo receptors that a given axon expresses. Axons that occupy more lateral positions express more Robo family receptors. (*D*) Motor axon guidance. Three of the major motor branches as well as their muscle targets are depicted: Segmental nerve a (SNa) (yellow), intersegmental nerve (ISN) (green), and ISNb (blue). Ventral views are presented in all panels. Anterior is up in all panels.

The period of axon growth and guidance in the *Drosophila* embryo commences at around embryonic stage 12, after the majority of neuroblasts have delaminated from the overlying neuroectoderm. Neurons extend axons that make a series of stereotyped decisions determined by the molecular identity of the neuron, interactions with neighboring axons, and the influence of secreted or transmembrane molecules that are distributed in strategic locations within the developing embryo. A number of distinct guidance events have been extensively characterized in the fly embryo including (1) the decision of whether or not to cross the midline; (2) the choice of which axon commissure to follow across the midline; (3) the specification of lateral position within the longitudinal connectives of the CNS; and (4) the guidance of motor axons to their appropriate muscle targets (Fig. 1). Each of these guidance paradigms is discussed briefly below, with an emphasis on the major ligand and receptor systems that have been implicated in regulating these events. The goal is to provide investigators with a simple framework for considering how novel factors may contribute to these guidance events.

CROSSING THE MIDLINE

Establishing precise midline circuitry is essential to control rhythmic and locomotor behaviors. In diverse organisms, the midline is an important intermediate target for many classes of navigating axons, which must decide whether or not to cross it (Kaprielian et al. 2001). Most axons in both invertebrate and vertebrate nervous systems cross the midline once and then project on the contralateral side of the CNS, never to cross again. A small percentage of axons remain on their own side of the midline. Frazzled (Fra) is the deleted in colorectal cancer (DCC)-like Netrin receptor in *Drosophila* that mediates attraction, and Roundabout (Robo) is a Slit receptor that mediates repulsion (Harris et al. 1996; Kolodziej et al. 1996; Mitchell et al. 1996; Kidd et al. 1998a, 1999; Battye et al. 1999). There is striking conservation in both the structure and midline guidance functions of these proteins from nematodes to vertebrate species (Chan et al. 1996; Keino-Masu et al. 1996; Serafini et al. 1996; Zallen et al. 1998; Brose et al. 1999; Hao et al. 2001). Netrin-Fra mediates the initial attraction of many commissural axons to the midline. However, significant numbers of axons still cross in *Netrin* and *fra* mutants, suggesting that additional unidentified factors contribute to midline attraction (Brankatschk and Dickson 2006; Garbe and Bashaw 2007). Once across the midline, commissural axons are prevented from recrossing by the Slit-Robo system. In *slit* or *robo* mutants, too many axons cross and recross the midline.

These two ligand receptor systems must act in a coordinated manner to ensure proper axon responses at the midline. The *commissureless* (*comm*) gene plays an important role in coordinating guidance at the midline (Tear et al. 1996; Kidd et al. 1998b). Comm is expressed in commissural neurons during a narrow window of time as their axons approach the midline, where it acts to downregulate Robo to prevent crossing axons from prematurely responding to Slit (Keleman et al. 2002, 2005) (Fig. 1A). Thus, midline crossing is dependent on the precise spatial and temporal regulation of Comm. Recent work aimed at understanding the regulation of Comm expression suggests that there is an important cross talk between attractive and repulsive pathways at the fly midline. Specifically, the Fra receptor has two genetically separable functions in regulating midline guidance. First, Fra mediates canonical chemoattraction in response to Netrin, and, second, it functions independently of Netrin to activate *comm* transcription, allowing attraction to be coupled to the downregulation of repulsion in precrossing commissural axons (Yang et al. 2009).

Select axonal markers that are useful for evaluating midline crossing include (1) MAb BP102, which labels most CNS axons; (2) Eagle-Gal4 driving UASTauMycGFP (Callahan and Thomas 1994; Callahan et al. 1998), which labels two bundles of commissural axons, the EW neurons that project their axons across the midline in the posterior commissure and the EG neurons that project their axons in the anterior commissure; (3) anti-FasII, which labels subsets of ipsilateral (noncrossing) interneuron axons and all motor axons; and (4) Apterous-Gal4 driving UASTauMycGFP, which labels three ipsilateral neurons in each abdominal hemisegment (Fig. 2). MAb BP102 and especially Eagle-Gal4 (which can be easily quantified) are particularly suited for assessing defects in midline attraction, whereas anti-FasII and Apterous-Gal4 are well suited to evaluate deficits in midline repulsion. In addition, antibodies to Slit, Robo, and Frazzled can be used to determine if new mutations exert their effect on midline guidance by affecting the expression of these proteins. It is important to remember that both Slit and Netrin are secreted by midline glial cells. Thus, in evaluating new mutations that affect guidance at the midline, it is imperative to determine that midline glia are not disrupted by the mutation; anti-Wrapper is an excellent marker for midline glia (Noordermeer et al. 1998). Additional markers for midline glia and other subsets of neurons are discussed in Chapter 3.

CHOICE OF COMMISSURE

In addition to deciding whether or not to cross the midline, commissural axons are faced with the choice of projecting their axons across the midline in either the anterior or posterior commissure of the segment (Fig. 1B). The soluble Wnt5 ligand and its Ryk-family receptor encoded by the *derailed* (*drl*) gene have been shown to dictate commissure selection (Callahan et al. 1995; Bonkowsky et al. 1999; Yoshikawa et al. 2003). Specifically, Wnt5 is expressed in a broad domain spanning the region of the posterior commissure, and the Drl receptor is expressed in a complementary pattern in neurons of the anterior commissure. Wnt5 acts to repel axons that express the Drl receptor, effectively shunting them into the anterior commissure (Fig. 1B) (Yoshikawa et al. 2003; Wouda et al. 2008). In *wnt5* or *drl*

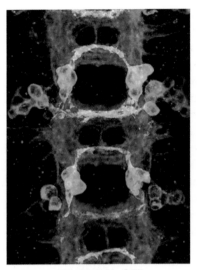 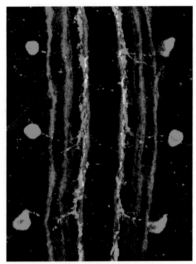

BP102/EGL-GFP   FasII/ AP-GFP

FIGURE 2. Visualization of subsets of neurons in fly embryos. Stage 16 *Drosophila* embryos expressing UASTauMycGFP under the control of either Eagle-Gal4 (*left*) or Apterous-Gal4 (*right*) are stained with anti-GFP (green) and MAb BP102 (magenta; *left*) or anti-FasII (magenta; *right*). EglGal4 specifically labels two clusters of commissural neurons. One cluster, the EG neurons, consists of approximately 10–12 cells that project their axons across the midline in the anterior commissure. The second cluster, the EW neurons, consists of three to four cells that project their axons in the posterior commissure. The BP102 and Egl markers can both be used to assess midline crossing, and Egl is also a good marker for examining commissure choice. ApGal4 labels three ipsilateral interneurons per hemisegment and FasII labels several bundles of longitudinal interneurons as well as motor neurons. Both Ap and FasII can be used to evaluate lateral positioning and test for abnormal midline crossing; FasII is also an excellent marker to look at motor axon guidance. Anterior is up in both panels.

mutants, there is a selective loss of midline crossing in the anterior commissure and a corresponding increase in the number of axons that cross in the posterior commissure. In contrast, misexpression of Drl in axons of the posterior commissure can reroute them through the anterior commissure.

There are a number of markers available that allow investigators to test for a specific role for genes of interest in the process of commissure selection. Antibodies to Drl and Wnt can be used to assess upstream effects on this pathway, and markers such as BP102 and Eagle-Gal4 can be used to assess commissure formation. Eagle-Gal4 is particularly useful in this context because discrete bundles of commissural axons in each commissure can be observed, making misrouting phenotypes straightforward to detect (Fig. 2). The reader is referred to Table 1 in Chapter 3, in which a veritable encyclopedia of known markers for specific cell populations is presented.

LATERAL POSITION

After choosing the appropriate side of the midline, most growth cones make an anterior or posterior turn and select a specific longitudinal pathway, fasciculating with other axons in a highly specific fashion. Loss or gain of Robo2 or Robo3 can alter the pathway choice of longitudinal axons, redirecting them to incorrect positions closer to or farther from the midline (Rajagopalan et al. 2000; Simpson et al. 2000). Notably, *Drosophila* Robo2 and Robo3 are not equivalent to vertebrate Robo2 and Robo3/Rig-1, because these receptors have been independently derived in flies and vertebrates. However, lateral positioning defects have been reported for axons in the ventral and lateral funiculi in *robo1* and *robo2* mutant mice (Long et al. 2004), suggesting that control of lateral position is a common function for Robo receptors in addition to their more canonical roles in midline repulsion. Whether this additional role represents an ancestral activity or one that has been independently acquired in insects and vertebrates remains to be determined.

The related but distinct roles of the three *Drosophila* Robo receptors are reflected in their expression patterns and mutant phenotypes in the embryonic CNS. Robo is expressed on nearly all CNS axons and can be detected across the entire width of the longitudinal connectives; in contrast, Robo3 and Robo2 are restricted to axons residing within the outermost two-thirds and one-third of the axon scaffold, respectively (Rajagopalan et al. 2000; Simpson et al. 2000). These overlapping domains divide the connectives into three broad zones of qualitatively distinct Robo receptor expression: Axons closest to the midline express only Robo; axons occupying the intermediate zone express Robo and Robo3; and axons in the most lateral zone express Robo, Robo3, and Robo2. Loss- and gain-of-function experiments revealed that changing the Robo receptors that were expressed in neurons, either broadly or in restricted subsets, could cause axons to shift between these zones. Loss of Robo2 resulted in lateral axons shifting to intermediate positions, whereas loss of Robo3 shifted intermediate axons to medial positions. Similarly, ectopic expression of Robo2 or Robo3 in medial axons could force them to more lateral positions, whereas increased levels of Robo could not. These observations led to the "Robo code" model, which posits that a qualitative combinatorial code of Robo receptor expression determines the lateral position of CNS axons (Rajagopalan et al. 2000; Simpson et al. 2000) (Fig. 1C).

Surprisingly, since the original reports in 2000, no other factors involved in the specification of lateral position have been identified. Anti-FasII, anti-connectin, ApGal4, and other markers can all be used to assess the fidelity of longitudinal pathway selection, and these tools, along with studies to evaluate genetic interactions with *robo2* and *robo3*, should shed further light on this process.

GUIDANCE OF MOTOR AXONS

There are a number of features that make the *Drosophila* embryonic nervous system a very attractive model in which to define the mechanisms of motor axon guidance. First, the pattern of motor connectivity is stereotyped and extremely well defined. Individual motor neurons and their postsynaptic muscle targets have been extensively characterized (Sink and Whitington 1991b; Van Vactor et al. 1993) (Fig. 1D). Second, there are excellent markers and tools available that allow labeling and genetic manipulation of subsets of motor neurons or, in some cases, even individual neurons. Accordingly, studies of motor axon guidance in the *Drosophila* embryo have revealed a large number of transcriptional regulators, guidance cues, receptors, and cell adhesion molecules that are important for finding the correct path, including Netrins and their Unc-5 and Frazzled receptors, Semaphorins and their Plexin receptors, Sidestep, members of the Beat family, Connectins, Fasciclins, and receptor tyrosine phosphatases (RPTPs) (Desai et al. 1996; Fambrough and Goodman 1996; Krueger et al. 1996; Nose et al. 1997; Winberg et al. 1998; Yu et al. 1998, 2000; Pipes et al. 2001; Sink et al. 2001). A full discussion of the roles of these molecules is well beyond the scope of this chapter, which instead focuses on the intersection between the transcriptional regulation of motor neuron subtype identity and the expression of specific guidance molecules that are required to execute differential guidance decisions.

Considerable evidence indicates that combinatorial codes of transcription factors function to specify motor neuron subtype identity and axon-projection patterns in the vertebrate spinal cord and in the *Drosophila* and *Caenorhabditis elegans* ventral nerve cords (Hobert et al. 1998; Jessell 2000; Shirasaki and Pfaff 2002; Thor and Thomas 2002). Little is known about how these transcriptional codes are read out at the level of specific axon guidance receptors to control motor axon pathfinding. However, recent studies have begun to forge direct functional links between motor neuron identity and the transcriptional regulation of specific guidance receptors and cell adhesion molecules. For example, in the vertebrate spinal cord, Lim1 has been shown to regulate the expression of EphA4 receptors to control the guidance of subsets of LMC (lateral motor column) neurons to appropriate domains in the limb (Kania and Jessell 2003). A second example comes from studies of motor axon guidance in *Drosophila*, in which Nkx6 has been shown to regulate the expression of the cell adhesion molecule Fasciclin III (FasIII) in ventrally projecting motor neurons (Broihier et al.

2004). It is unclear whether the effects of Lim1 on EphA4 expression and the effects of Nkx6 on FasIII expression are exerted directly at the level of the respective promoters or instead whether the transcriptional regulation is indirect. Future studies in *Drosophila* promise to provide important insight into how specific guidance programs are implemented in different classes of motor neurons.

The 1D4 monoclonal antibody FasII is the classic motor neuron marker for studies of motor axon guidance in the *Drosophila* embryo (Van Vactor et al. 1993). A cartoon of some of the major motor neuron projections labeled by this antibody is depicted in Figure 1D. Additional markers include Gal4 enhancer traps and promoter fusions that allow precise visualization of particular sub-populations of motor axons. For a catalog of these markers, see Chapter 3.

Antibody Staining of Embryos

This protocol is similar to the one described in Chapter 3 with some small modifications. Methods for fluorescent antibody staining as well as horseradish peroxidase (HRP) immunohistochemistry are described. HRP forms a brown, stable precipitate that allows visualization of axon populations under white light. This can be quite advantageous when teaching basic embryonic anatomy and dissection techniques, because unlike fluorescent antibody staining, HRP can be visualized in a lighted room. HRP detection does not allow precise colocalization of signals, nor can multiple colors be visualized simultaneously. These objectives are easily achieved with fluorescent antibody staining. Examples of two-color fluorescent staining with markers used to assess midline crossing, commissure choice, lateral position, and motor axon guidance are shown in Figure 2.

MATERIALS

CAUTION: See Appendix for proper handling of materials marked with <!>.
See the end of the chapter for recipes for reagents marked with <R>.

Reagents

Antibodies available for the detection of subsets of neurons and glia in the embryonic CNS are extensively cataloged in Chapter 3. The antibodies listed below are those used to stain the embryos visualized in Figure 2.

Bleach (Clorox) <!>

Blocking solution (PTX containing 5% normal goat serum)

Diaminobenzidine (DAB) <!> (Sigma-Aldrich cat. no. D5905; supplied as 10-mg tablets)

Formaldehyde <!> (histological grade; 37%, w/v)

Glycerol (70%)

Heptane <!>

Hydrogen peroxide <!> (3%, v/v)

Methanol <!>

Normal goat serum (GIBCO cat. no. 16210-064)

Heat the goat serum for 30 min at 65°C. Divide into 5-mL aliquots and store at –20°C.

PBS containing 0.1% Triton X-100 (PTX) <R>

Phosphate-buffered saline (PBS) <R>

Primary antibodies (diluted to final concentrations in blocking solution):

Anti-BP102 (1:50 dilution; mouse monoclonal; labels most CNS axons) (DSHB)

Anti-FasII (1:50 dilution; mouse monoclonal; labels subsets of interneurons and all motor neurons) (DSHB)

Anti-GFP (1:500 dilution; rabbit polyclonal) (Molecular Probes)

Secondary antibodies (diluted to final concentrations in blocking solution):

Cy3-conjugated goat anti-mouse (1:1000 dilution) (Jackson ImmunoResearch)

Alexa488-conjugated goat anti-rabbit (1:500 dilution) (Molecular Probes)

Tween-20

Equipment

Juice plates for *Drosophila* embryos (apple) <R>
Microcentrifuge tubes (1.5-mL)
Nytex mesh
Pasteur pipettes
Rotating platform (e.g., Nutator mixer)
Scintillation vials (glass)
Squirt bottle

METHOD

Collection and Fixation of Embryos

1. Collect the embryos (for 20–24 h at 25°C) on small grape or apple juice plates.

2. Dechorionate the embryos by immersing in 50% bleach for 3–5 min. Squirt the bleach directly onto the plate.

3. Strain the embryos through mesh and rinse with lots of water.

4. Transfer the embryos to glass scintillation vials and fix with 10 mL of 3.7% formaldehyde (diluted in either water or PBS) and 10 mL of heptane by rocking on a rotating platform for 15 min.

    The embryos should float at the interface.

5. Remove the lower formaldehyde phase with a pasteur pipette. Take care to remove as much fixative as possible.

6. Add 10 mL of methanol. Shake and vortex for 1 min.

    Many embryos should sink.

    See Troubleshooting.

7. Remove the methanol/heptane and, using a P1000 pipette, transfer the embryos to a microcentrifuge tube.

8. Rinse the embryos two to three times with methanol. Store at –20°C or proceed to staining.

    See Troubleshooting.

Basic Antibody Staining

In all steps (except rinses), mix the reagents by rocking on a rotating platform.

9. Rinse the embryos twice in PTX.

    See Troubleshooting.

    Some antibodies (e.g., anti-Robo) do not perform well when incubated in Triton X-100. Try Tween-20 instead.

10. Wash once in PTX for 5 min.

11. Add 250 µL of blocking solution. Rock for 5–10 min.

12. Add the primary antibody. Use dilutions of between 1:25 and 1:100 for monoclonals and between 1:500 and 1:2000 for polyclonals. Incubate overnight at 4°C.

13. Rinse three times in PTX.

14. Wash twice in PTX for 5 min each.

15. Add the secondary antibody. Use dilutions of 1:250 for HRP and of between 1:500 and 1:2000 for fluorescent conjugates.

16. Incubate for ≥30 min at room temperature.
17. Wash as in Steps 13 and 14.

HRP Immunohistochemistry

For fluorescent staining, omit Steps 18–21 and proceed with clearing.

18. Incubate in 300 µL of DAB (0.3 mg/mL in PBS containing 0.1% Tween-20).
19. Add 5 µL of 3% hydrogen peroxide.
20. Develop for 2–30 min.
    See Troubleshooting.
21. Stop by rinsing several times in PTX.
22. Rinse once in PBS.
23. Clear in 70% glycerol (in PBS) for at least 2 h. Embryos can be stored forever in glycerol at 4°C. *Do not freeze!*

TROUBLESHOOTING

Problem (Steps 6 and 8): When the methanol is added, the solution turns milky.
Solution: In Step 5, it is important to remove as much of the lower fixative phase as possible. If too much fixative is left behind, the solution will turn milky when the methanol is added, obscuring the view of the embryos. This can be remedied by adding more methanol.

Problem (Step 9): The embryos clump together when PTX solution is added.
Solution: Unfortunately, if this happens the embryos must be discarded and a fresh collection obtained. This happens when residual heptane is left on the embryos. It is important to rinse the embryos adequately with several changes of methanol after fixation.

Problem (Step 20): A weak signal is observed despite a long developing time.
Solution: The most common reason for a weak signal in HRP-staining experiments is inferior hydrogen peroxide, which has been stored for too long or exposed to light. Use a new bottle.

Protocol 2

Live Dissection and Surface Labeling of Proteins in *Drosophila* Embryos

This protocol allows visualization of surface pools of receptor proteins and also allows the investigator to determine the surface distribution of secreted cues. In addition, live dissection of embryos can be used to improve the ability to detect signals from suboptimal antibodies, where conventional staining gives poor results.

MATERIALS

CAUTION: See Appendix for proper handling of materials marked with <!>.
See the end of the chapter for recipes for reagents marked with <R>.

Reagents

Antibodies (see Protocol 1)
Bleach (Clorox) <!>
Glycerol (70%)
Normal goat serum (GIBCO cat. no. 16210-064)
　　Heat the goat serum for 30 min at 65°C. Divide into 5-mL aliquots and store at –20°C.
Paraformaldehyde <!>
PBS <R>

Equipment

Double-sided sticky tape
Juice plates for *Drosophila* embryos (apple or grape) <R>
Microscope slides (positively charged) (e.g., Fisher Scientific)
　　Embryos stick better to charged slides than they do to standard microscope slides.
Nytex mesh
PAP pen (for creating a hydrophobic boundary on the surface of a microscope slide)
Pasteur pipettes
Plastic probe (for transferring embryos from juice plate to microscope slide)
　　Use a 3-mL syringe and a plastic pipette tip. Melt the end of the pipette tip with a Bunsen burner to create a rounded, sealed plastic probe.
Squirt bottle
Tungsten dissecting needle

METHOD

Preparation of the Dissection Arena

1. Use a PAP pen to draw a boundary that conforms to the edges of a microscope slide.
   This hydrophobic boundary retains the staining solution and wash buffer.

2. Place a small piece of double-sided sticky tape (0.5-cm-wide and 3-cm-long) inside the area marked by the PAP pen. Ensure the tape is completely inside the boundary.

Collection of Embryos

3. Collect the embryos (for 20–24 h at 25°C) on small grape or apple juice plates.

4. Dechorionate the embryos by immersing in 50% bleach for 3–5 min. Squirt the bleach directly on to the plate.

5. Strain the embryos through mesh and rinse with lots of water.

6. Transfer the embryos to a fresh juice plate by gently blotting the mesh that contains the embryos onto the plate. Avoid transferring the embryos in large clumps, as this makes it more difficult to select the desired animals.

7. Under a dissecting microscope, select the desired stage of embryos and transfer them with a plastic probe to the sticky tape in the dissection arena (see Step 2). Transfer the embryos so that their anterior ends are pointing toward the top of the slide. Not all of the embryos are likely to make it through the entire procedure, so be sure to transfer more than you think you will need.

    To be sure to have 10–12 stained embryos at the end of the procedure, you will need to transfer about 20 embryos.

Dissection

8. Using a plastic probe or the blunt side of a tungsten dissecting needle, position the embryos ventral side down on the sticky tape. Gently roll the embryo, if necessary.

9. Using a P1000 pipette, cover the embryos with ~0.5 mL of 1x PBS.

10. Under the dissecting microscope, gently insert the tungsten needle into the posterior third of the embryo and pierce the vitelline membrane.

11. With a smooth motion, push the needle toward the anterior end of the embryo.

    This should release the embryo from the vitelline membrane and the embryo should now be attached to your needle.

12. Place the embryo on the glass surface of the dissecting slide and push down gently with the needle to allow the embryo to adhere to the glass.

    Late-stage embryos (stage 16 or greater) do not stick well to the slide surface, so it is important to use embryos that are stage 15 or younger.

13. Repeat these steps for each of the embryos on the sticky tape in the dissection arena.

14. Fillet each embryo by slicing down the dorsal midline with the needle and roll out the body wall until it makes contact with the slide.

15. Remove the guts and any other debris that obscures your view of the ventral nerve cord.

    The nerve cord is relatively thick and should be easily visible under the light microscope in embryos from stage 13 to stage 16.

Antibody Staining

It is important that the dissected embryos remain covered by buffer solution through all subsequent steps. Depending on the size of the dissection arena, use a volume between 0.5 mL and 1 mL. Exact volumes for wash steps are not important as long as the embryos remain covered by buffer solution. To change solutions, you can either use two pipettes to add and remove solutions simultaneously or you can use a P1000 pipette to add solution and a vacuum aspirator (set on low) mounted with either a pasteur pipette or a P200 tip to remove the old solution as you add the new. Antibody concentrations are the same as for standard embryo staining (see Protocol 1 and Table 1 in Chapter 3). Add the solutions gently or the embryos can be washed away!

16. Replace the PBS used during dissection with fresh PBS.

17. Replace the PBS with PBS containing 5% goat serum.

18. Incubate the slide for 15 min at 4°C.

19. Prepare 2 mL of the desired primary antibody diluted in PBS containing 5% goat serum chilled to 4°C. Remove as much of the blocking solution as possible without uncovering the embryos and then slowly add the antibody using a P1000 pipette until you have used the entire 2 mL. At the same time as you add antibody, you must also slowly remove antibody solution with the vacuum aspirator or second pipette. Leave ~0.5 mL of primary antibody solution on the embryos.

20. Incubate in primary antibody for 30 min at 4°C.

21. Wash with 5 mL of cold PBS to remove the primary antibody.

    This should take 2–3 min with constant addition and removal of the PBS until the 5 mL is depleted.

22. Fix the embryos with 4% paraformaldehyde in 1x PBS for 15 min at 4°C.

23. Wash with 5 mL of cold PBS.

    The following steps can be performed at room temperature.

24. Incubate in secondary antibody diluted in PBS containing 5% goat serum.

25. Wash with 5 mL of PBS. Leave for 5 min.

26. Repeat the wash.

27. Remove as much of the wash solution as possible and place a small drop of 70% glycerol in PBS over the embryos.

28. Cover with an appropriate sized glass coverslip.

29. The embryos are now ready for viewing.

    When performing surface staining experiments, always include a control to ensure that the embryos have not been made permeable and that primary antibodies have not been internalized. Antibodies against known cytoplasmic or nuclear proteins can be used for this purpose (see Chapter 3).

RECIPES

CAUTION: See Appendix for proper handling of materials marked with <!>.
Recipes for reagents marked with <R> are included in this list.

Juice Plates

Reagent	Amount
Agar	105 g
ddH$_2$O	3 L
Sucrose	100 g
Methyl paraben <!> (Tegosept)	6 g
Apple juice	1 L

1. In a 6-L flask, stir 105 g agar into 3 L of dd H$_2$O.

2. In a 2-L flask, stir 100 g sucrose and 6 g methyl paraben into 1 L apple juice and shake well.

3. Autoclave the flasks separately for 45 min and sterilize five 1-L flasks by autoclaving (these flasks will be used for pouring the plates).

4. Slowly pour the sucrose/apple juice solution into the agar solution, add a stir bar, and stir for about 15 min (use settings of stir speed 9, heat 6).

5. Pour ~800 ml of the combined solution into each of the autoclaved 1-L flasks. Stopper each flask so a skin does not form on the solution and take care not to agitate them to avoid formation of bubbles. Once cooled slightly, pour the agar solution into the plates.

PBS Stock Solution (10x; pH 7.0)

Reagent	Quantity (for 1 L)	Final concentration
NaCl	74 g	1.27 M
$Na_2HPO_4 \cdot 7H_2O$	18.78 g	70 mM
$NaH_2PO_4 \cdot H_2O$	4.15 g	30 mM

Dissolve the components in dH_2O and adjust the pH to 7.0 with NaOH if necessary. Adjust the volume to 1 L with dH_2O

PTX

Reagent	Quantity (for 1 L)	Final concentration
PBS stock solution (10x) <R>	100 mL	1x
Triton X-100 <!> (10%, v/v)	10 mL	0.1% (v/v)

Add dH_2O to give a final volume of 1 L.

REFERENCES

Battye R, Stevens A, Jacobs JR. 1999. Axon repulsion from the midline of the *Drosophila* CNS requires slit function. *Development* **126:** 2475–2481.

Bonkowsky JL, Yoshikawa S, O'Keefe DD, Scully AL, Thomas JB. 1999. Axon routing across the midline controlled by the *Drosophila* Derailed receptor. *Nature* **402:** 540–544.

Brankatschk M, Dickson BJ. 2006. Netrins guide *Drosophila* commissural axons at short range. *Nat Neurosci* **9:** 188–194.

Broihier HT, Kuzin A, Zhu Y, Odenwald W, Skeath JB. 2004. *Drosophila* homeodomain protein Nkx6 coordinates motoneuron subtype identity and axonogenesis. *Development* **131:** 5233–5242.

Brose K, Bland KS, Wang KH, Arnott D, Henzel W, Goodman CS, Tessier-Lavigne M, Kidd T. 1999. Slit proteins bind Robo receptors and have an evolutionarily conserved role in repulsive axon guidance. *Cell* **96:** 795–806.

Callahan CA, Thomas JB. 1994. Tau-β-galactosidase, an axon-targeted fusion protein. *Proc Natl Acad Sci* **91:** 5972–5976.

Callahan CA, Muralidhar MG, Lundgren SE, Scully AL, Thomas JB. 1995. Control of neuronal pathway selection by a *Drosophila* receptor protein-tyrosine kinase family member. *Nature* **376:** 171–174.

Callahan CA, Yoshikawa S, Thomas JB. 1998. Tracing axons. *Curr Opin Neurobiol* **8:** 582–586.

Chan SS, Zheng H, Su MW, Wilk R, Killeen MT, Hedgecock EM, Culotti JG. 1996. UNC-40, a *C. elegans* homolog of DCC (Deleted in Colorectal Cancer), is required in motile cells responding to UNC-6 netrin cues. *Cell* **87:** 187–195.

Desai CJ, Gindhart JG Jr, Goldstein LS, Zinn K. 1996. Receptor tyrosine phosphatases are required for motor axon guidance in the *Drosophila* embryo. *Cell* **84:** 599–609.

Dickson BJ. 2002. Molecular mechanisms of axon guidance. *Science* **298:** 1959–1964.

Fambrough D, Goodman CS. 1996. The *Drosophila beaten path* gene encodes a novel secreted protein that regulates defasciculation at motor axon choice points. *Cell* **87:** 1049–1058.

Garbe D, Bashaw G. 2004. Axon guidance at the midline: From mutants to mechanisms. *Crit Rev Biochem Mol Biol* **39:** 319–341.

Garbe DS, Bashaw GJ. 2007. Independent functions of Slit-Robo repulsion and Netrin-Frazzled attraction regulate axon crossing at the midline in *Drosophila*. *J Neurosci* **27:** 3584–3592.

Hao JC, Yu TW, Fujisawa K, Culotti JG, Gengyo-Ando K, Mitani S, Moulder G, Barstead R, Tessier-Lavigne M, Bargmann CI. 2001. *C. elegans* Slit acts in midline, dorsal–ventral, and anterior–posterior guidance via the SAX-3/Robo receptor. *Neuron* **32**: 25–38.

Harris R, Sabatelli LM, Seeger MA. 1996. Guidance cues at the *Drosophila* CNS midline: Identification and characterization of two *Drosophila* Netrin/UNC-6 homologs. *Neuron* **17**: 217–228.

Hobert O, D'Alberti T, Liu Y, Ruvkun G. 1998. Control of neural development and function in a thermoregulatory network by the LIM homeobox gene *lin-11*. *J Neurosci* **18**: 2084–2096.

Jessell TM. 2000. Neuronal specification in the spinal cord: Inductive signals and transcriptional codes. *Nat Rev Genet* **1**: 20–29.

Kania A, Jessell TM. 2003. Topographic motor projections in the limb imposed by LIM homeodomain protein regulation of ephrin-A:EphA interactions. *Neuron* **38**: 581–596.

Kaprielian Z, Runko E, Imondi R. 2001. Axon guidance at the midline choice point. *Dev Dyn* **221**: 154–181.

Keino-Masu K, Masu M, Hinck L, Leonardo ED, Chan SS, Culotti JG, Tessier-Lavigne M. 1996. Deleted in Colorectal Cancer (DCC) encodes a netrin receptor. *Cell* **87**: 175–185.

Keleman K, Rajagopalan S, Cleppien D, Teis D, Paiha K, Huber LA, Technau GM, Dickson BJ. 2002. Comm sorts Robo to control axon guidance at the *Drosophila* midline. *Cell* **110**: 415–427.

Keleman K, Ribeiro C, Dickson BJ. 2005. Comm function in commissural axon guidance: Cell-autonomous sorting of Robo in vivo. *Nat Neurosci* **8**: 156–163.

Kidd T, Brose K, Mitchell KJ, Fetter RD, Tessier-Lavigne M, Goodman CS, Tear G. 1998a. Roundabout controls axon crossing of the CNS midline and defines a novel subfamily of evolutionarily conserved guidance receptors. *Cell* **92**: 205–215.

Kidd T, Russell C, Goodman CS, Tear G. 1998b. Dosage-sensitive and complementary functions of Roundabout and Commissureless control axon crossing of the CNS midline. *Neuron* **20**: 25–33.

Kidd T, Bland KS, Goodman CS. 1999. Slit is the midline repellent for the Robo receptor in *Drosophila*. *Cell* **96**: 785–794.

Kolodziej PA, Timpe LC, Mitchell KJ, Fried SR, Goodman CS, Jan LY, Jan YN. 1996. *frazzled* encodes a *Drosophila* member of the DCC immunoglobulin subfamily and is required for CNS and motor axon guidance. *Cell* **87**: 197–204.

Krueger NX, Van Vactor D, Wan HI, Gelbart WM, Goodman CS, Saito H. 1996. The transmembrane tyrosine phosphatase DLAR controls motor axon guidance in *Drosophila*. *Cell* **84**: 611–622.

Long H, Sabatier C, Ma L, Plump A, Yuan W, Ornitz DM, Tamada A, Murakami F, Goodman CS, Tessier-Lavigne M. 2004. Conserved roles for Slit and Robo proteins in midline commissural axon guidance. *Neuron* **42**: 213–223.

Mitchell KJ, Doyle JL, Serafini T, Kennedy TE, Tessier-Lavigne M, Goodman CS, Dickson BJ. 1996. Genetic analysis of Netrin genes in *Drosophila*: Netrins guide CNS commissural axons and peripheral motor axons. *Neuron* **17**: 203–215.

Noordermeer JN, Kopczynski CC, Fetter RD, Bland KS, Chen WY, Goodman CS. 1998. Wrapper, a novel member of the Ig superfamily, is expressed by midline glia and is required for them to ensheath commissural axons in *Drosophila*. *Neuron* **21**: 991–1001.

Nose A, Umeda T, Takeichi, M. 1997. Neuromuscular target recognition by a homophilic interaction of connectin cell adhesion molecules in *Drosophila*. *Development* **124**: 1433–1441.

Pipes GC, Lin Q, Riley SE, Goodman CS. 2001. The Beat generation: A multigene family encoding IgSF proteins related to the Beat axon guidance molecule in *Drosophila*. *Development* **128**: 4545–4552.

Rajagopalan S, Vivancos V, Nicolas E, Dickson BJ. 2000. Selecting a longitudinal pathway: Robo receptors specify the lateral position of axons in the *Drosophila* CNS. *Cell* **103**: 1033–1045.

Serafini T, Colamarino SA, Leonardo ED, Wang H, Beddington R, Skarnes WC, Tessier-Lavigne M. 1996. Netrin-1 is required for commissural axon guidance in the developing vertebrate nervous system. *Cell* **87**: 1001–1014.

Shirasaki R, Pfaff SL. 2002. Transcriptional codes and the control of neuronal identity. *Annu Rev Neurosci* **25**: 251–281.

Simpson JH, Bland KS, Fetter RD, Goodman CS. 2000. Short-range and long-range guidance by Slit and its Robo receptors: A combinatorial code of Robo receptors controls lateral position. *Cell* **103**: 1019–1032.

Sink H, Whitington PM. 1991. Pathfinding in the central nervous system and periphery by identified embryonic *Drosophila* motor axons. *Development* **112**: 307–316.

Sink H, Rehm EJ, Richstone L, Bulls YM, Goodman CS. 2001. *sidestep* encodes a target-derived attractant essential for motor axon guidance in *Drosophila*. *Cell* **105**: 57–67.

Tear G, Harris R, Sutaria S, Kilomanski K, Goodman CS, Seeger MA. 1996. *commissureless* controls growth cone guidance across the CNS midline in *Drosophila* and encodes a novel membrane protein. *Neuron* **16**: 501–514.

Thor S, Thomas J. 2002. Motor neuron specification in worms, flies and mice: Conserved and "lost" mechanisms. *Curr Opin Genet Dev* **12**: 558–564.

Van Vactor D, Sink H, Fambrough D, Tsoo R, Goodman CS. 1993. Genes that control neuromuscular specificity in

Drosophila. Cell **73:** 1137–1153.

Winberg ML, Noordermeer JN, Tamagnone L, Comoglio PM, Spriggs MK, Tessier-Lavigne M, Goodman CS. 1998. Plexin A is a neuronal semaphorin receptor that controls axon guidance. *Cell* **95:** 903–916.

Wouda R, Bansraj M, De Jong A, Noordermeer J, Fradkin L. 2008. Src family kinases are required for WNT5 signaling through the Derailed/RYK receptor in the *Drosophila* embryonic central nervous system. *Development* **135:** 2277–2287.

Yang L, Garbe DS, Bashaw GJ. 2009. A frazzled/DCC-dependent transcriptional switch regulates midline axon guidance. *Science* **324:** 944–947.

Yoshikawa S, McKinnon RD, Kokel M, Thomas JB. 2003. Wnt-mediated axon guidance via the *Drosophila* Derailed receptor. *Nature* **42:** 583–588.

Yu HH, Araj HH, Ralls SA, Kolodkin AL. 1998. The transmembrane Semaphorin Sema I is required in *Drosophila* for embryonic motor and CNS axon guidance. *Neuron* **20:** 207–220.

Yu HH, Huang AS, Kolodkin AL. 2000. Semaphorin-1a acts in concert with the cell adhesion molecules fasciclin II and connectin to regulate axon fasciculation in *Drosophila. Genetics* **156:** 723–731.

Yu TW, Bargmann CI. 2001. Dynamic regulation of axon guidance. *Nat Neurosci* (suppl) **4:** 1169–1176.

Zallen JA, Yi BA, Bargmann CI. 1998. The conserved immunoglobulin superfamily member SAX-3/Robo directs multiple aspects of axon guidance in *C. elegans. Cell* **92:** 217–227.

5 Analysis of Glial Cell Development and Function in *Drosophila*

Tobias Stork, Rebecca Bernardos, and Marc R. Freeman

Department of Neurobiology, University of Massachusetts Medical School, Worcester, Massachusetts 01605

ABSTRACT

Glial cells are the most abundant cell type in our brains, yet we understand very little about their development and function. An accumulating body of work over the last decade has revealed that glia are critical regulators of nervous system development, function, and health. Based on morphological and molecular criteria, glia in *Drosophila melanogaster* are very similar to their mammalian counterparts, suggesting that a detailed investigation of fly glia has the potential to add greatly to our understanding of fundamental aspects of glial cell biology. In this chapter, we provide an overview of the subtypes of glial cells found in *Drosophila* and discuss our current understanding of their functions, the development of a subset of well-defined glial lineages, and the molecular-genetic tools available for manipulating glial subtypes in vivo.

INTRODUCTION

Neurons are not the only cell type in the nervous system; ~90% of the cells in our brain are glia. Although long thought of as simple support cells, work over the last two decades has revealed a number of critical roles for glia in the development, function, and maintenance of the nervous system. Glia perform diverse roles in the developing nervous system such as modulating neural stem cell proliferation (Ebens et al. 1993), regulating the differentiation of neural precursors guiding axon pathfinding (Hidalgo and Booth 2000; Sepp et al. 2001; Gilmour et al. 2002), ensheathing nerves and individual axons (Barres 2008; Nave and Trapp 2008), delineating and isolating distinct lobes of the brain and their subcompartments (Oland and Tolbert 2003; Awasaki et al. 2008), supplying trophic support for neurons (Xiong and Montell 1995; Booth et al. 2000), engulfing neurons and debris that are eliminated during development (Sonnenfeld and Jacobs 1995; Freeman et al. 2003; Awasaki and Ito 2004; Watts et al. 2004), and promoting synapse formation and maturation (Barres 2008). In the mature nervous system, glia maintain a proper ionic balance in the central nervous system (CNS), take up neurotransmitters after synaptic signaling (Danbolt 2001), isolate and protect neurons by forming or regulating the blood–brain barrier (BBB; Abbott 2005), associate closely with synapses

53

and likely modulate their activity (Barres 2008), and act as the major immune cell type in the brain. These lists that attempt to catalog the breadth of glial functions undoubtedly only partially represent the true range of glial roles in the developing and mature nervous system. Whenever glial biology is carefully explored in any new developmental or functional context, it seems glia reveal additional unexpected depths to their governance of nervous system morphogenesis and function.

Despite the widespread importance of glia in the nervous system, we know surprisingly little about the underlying molecular pathways that mediate glial biology in any organism. For example, excellent studies have described the morphology of glia as they ensheath individual synapses, but we understand almost nothing about how glia and synapses acquire this intimate relationship or the functional significance of such associations. Why are the majority of synapses covered by glial membranes? What drives glia to ensheath synapses? What is the in vivo function for glia at a glutamatergic, cholinergic, or GABAergic synapse? Do they aid in information processing through direct modulation of synaptic signaling?

There are a number of reasons to explain our ignorance of the molecular pathways involved in most aspects of glial biology. First, glia have not been studied as intensively as their excitable neighbors, neurons. The focus on neurons is due partly to their ability to fire action potentials, in contrast to most glia, which remain electrophysiologically silent. Second, mammalian glia are difficult to analyze in vivo. This is attributable mainly to a dearth of markers and reagents for manipulating glia. Instead, much of our understanding of neuron–glia interactions has been gleaned from studies of primary cultures. This approach can certainly provide an excellent first step toward identifying genes that mediate neuron–glia signaling. For example, recent work in primary cultures has identified glial-secreted molecules that are required for synapse formation and maturation, some of which have been verified by functional analysis in vivo (Christopherson et al. 2005; Stevens et al. 2007). Nevertheless, the vast majority of in vitro observations have not been repeated in experiments with the living organism. Confirming observations from in vitro studies in vivo is important because glia and neurons are in close association with each other soon after differentiation, both cell types become interdependent for survival, and they mature morphologically, and likely molecularly, in response to cues from each other. It is abundantly clear that when glia are dissociated and plated in culture, they rapidly acquire properties that are different from those of glia in the intact brain. For example, astrocytes in vivo are tightly coupled to one another through gap junctions and essentially form a syncytium within the brain that acts as a sink for K^+ ions (Kofuji and Newman 2009). Dissociation of astrocytes destroys this coupling and results in dramatic changes in the electrophysiological properties of these cells (Kimelberg 2009). Likewise, astrocytes display a tufted morphology and are profusely branched in three-dimensional space in vivo, but when dissociated and plated for culture in vitro, astrocytes lose this complex morphology and take on a fibroblast-like, flattened appearance. As such, it is reasonable to assume that the physiology of astrocytes in vitro may also be dramatically different from astrocytes in vivo. Until in vitro observations are also assayed for functional significance in the intact animal, the relevance of in vitro neuron–glia signaling events will remain in question.

Although the *Drosophila* nervous system is relatively simple in structure, it shares a number of quite sophisticated glial characteristics with its mammalian counterparts including (1) mechanisms mediating reciprocal trophic support between neurons and glia; (2) glial pathways for recycling of synaptic neurotransmitters (e.g., glutamate and GABA [γ-butyric acid]); (3) subtypes of glia that are capable of parsing axons into distinct fascicles and individually ensheathing axons (similar to Remak bundles in mammals); (4) a subtype of glia that bears striking morphological and molecular similarity to mammalian astrocytes; and (5) glia that act as the primary immune cell in the brain and can respond immunologically to neural trauma. As such, *Drosophila* glia seem well positioned to provide exciting insights into glial biology that will be relevant to glial functions in mammals. This chapter is meant to provide an introduction to the study of glial cell biology in *Drosophila*. Flies offer the opportunity to study a diverse array of neuron–glia interactions in the intact organism while also exploiting the powerful molecular-genetic tools available in *Drosophila* to address fundamental questions in glial cell biology. We have divided this chapter into two broad sections: (1) glia in the embryonic nervous system and (2) glia in the larva and adult. We provide a list of methods and tools we find particularly useful for studying these cell types. Much of the detailed discussion of lineages, morphogenesis, and function focuses on embryonic glia because they have been studied much more extensively than glia in the larva, pupa, or adult. However, postembryonic stages offer a number of advantages with respect to molecular-genetic approaches to assay glial gene function or glial cell biology, and these are discussed.

GLIAL CELL DEVELOPMENT AND FUNCTION IN THE EMBRYONIC CNS

The embryonic ventral nerve cord (VNC) is an excellent model for studying early glial developmental events such as glial cell-fate specification and migration. The locations of and lineages generated by each of the 30 neuroblasts that form in each hemisegment of the VNC have been extremely well characterized at the cellular level (Doe 1992; Bossing et al. 1996; Schmid et al. 1997, 1999). We refer the reader to Chapters 2 and 3 of this volume for reviews of embryonic VNC development and neuroblast biology. Of the 30 neuroectoderm-derived neuroblasts per hemisegment, eight (the so-called neuroglioblasts [NGBs]) produce mixed lineages of neurons and glia (NGB1-1A, NGB1-3, NGB2-2T, NGB2-5, NGB3-5, NGB5-6, NGB6-4A, and NGB7-4 [A refers to abdominal and T to thoracic; 6-4A is purely glial]) and one precursor produces only glia (Bossing et al. 1996; Schmid et al. 1997, 1999). The *Drosophila* VNC contains on average 25–30 glia per hemisegment, and embryonic glia can be assigned to three broad categories based on their position and morphology: (1) surface glia, which form a layer around the CNS or peripheral nerves; (2) cortex glia, which ensheath neuronal cell bodies; and (3) neuropil glia, which associate directly with the neuropil (Ito et al. 1995). As discussed below, these broad subtypes of glia appear to be present throughout development and into the adult stage of *Drosophila*.

Midline glia are also present in the *Drosophila* embryonic CNS. These cells represent a unique class of mesectoderm-derived glia that have a critical role in axon pathfinding at the CNS midline (see Chapter 4) and commissure formation in the VNC. Midline glia are a small subset of glia in the embryo (only about three per hemisegment at embryonic stage 17) and will not be discussed further in this chapter. Instead, we refer the reader to the excellent review by Jacobs (2000).

The earliest known marker for cells that will become glia in the embryonic nervous system is the novel transcription factor encoded by *glial cells missing* (*gcm*). Shortly after Gcm is activated, *reversed polarity* (*repo*) is expressed; this is a direct transcriptional target of Gcm that encodes a homeodomain transcription factor. Both of these (along with their enhancer traps and *Gal4* driver lines) are extremely useful tools for labeling glia (except midline glia, which do not express Gcm or Repo) very early in their development (Table 1). The first Gcm-expressing glial cells appear in the developing embryonic VNC at late stage 10 or early stage 11, and Repo-positive glial cells become detectable shortly thereafter, at approximately stage 11 (Campbell et al. 1994; Halter et al. 1995). The majority of Gcm/Repo-positive embryonic VNC glia are likely present by early stage 14.

Gcm is most useful for labeling or manipulating early embryonic glial lineages, but is not an appropriate glial-specific marker at later developmental stages. Gcm is expressed very early in glial development, but its expression fades by embryonic stage 14. Consequently, Gcm and *gcm-Gal4* will not label late-stage embryonic glia. In addition, Gcm is expressed in cells outside the nervous system that are not glia (i.e., macrophages and apodemal cells). Finally, Gcm has recently been shown to be expressed in neural precursors in the larval/pupal brain and is required for the formation of many neurons that populate the adult brain (Chotard et al. 2005). These findings should be considered when attempting to label glia at larval or pupal stages.

Nevertheless, *gcm-Gal4* is the earliest and strongest glial-specific driver in the embryonic CNS and it remains the best tool for experiments in which one wishes to ectopically express genes under *UAS* control at the earliest possible time point in embryonic glial development. Repo, in contrast to Gcm, appears to be expressed specifically in glia, and this holds true throughout the *Drosophila* life cycle (to our knowledge no Repo[+] cells have been described that are not glia). Therefore, Repo anti-

TABLE 1. Useful tools in *Drosophila* glial research

Markers/lines	CNS expression pattern	Type(s) of antibodies	Dilution(s) for embryo staining	Reference
Anti-Repo	All glia except midline glia	Mouse	1:5, 1:10	Campbell et al. 1994
Anti-Gcm	All glia at early stages (except midline glia)	Rat	1:400	Alfonso and Jones 2002
Anti-Moody-β	Subperineurial glia	Rabbit	1:15	Bainton et al. 2005
Anti-Moody-α		Rat	1:10	
Anti-Gs2	Subset of longitudinal glia	Mouse	1:10	Thomas and van Meyel 2007
Anti-NrxIV	Subperineurial glia, neurons, epithelia	Rabbit	1:500	Baumgartner et al. 1996
nrxIV-GFP exon trap	Subperineurial glia, neurons, epithelia	NA	NA	Edenfeld et al. 2006

Driver line	CNS expression pattern	Comment	Reference(s)
repo-Gal4	All glia except midline glia	Enhancer trap	Sepp et al. 2001
repo-Gal4-4.3	All glia except midline glia (note that some variability in expression in different subsets of glia has been observed)	Promoter fusion	Lee and Jones 2005
repo:LexA::GAD	All glia except midline glia	Promoter fusion	Lai and Lee 2006
gcm-Gal4	All glia (except midline glia), apodemal cells, and macrophages	Enhancer trap	Paladi and Tepass 2004
htl-Gal4	Longitudinal glia and other glia	Promoter fusion	Shishido et al. 1997
rl82-Gal4 (gliotactin-Gal4)	Pronounced in subperineurial glia, weaker in perineurial glia	Enhancer trap	Sepp and Auld 1999
Spg-Gal4	Subperineurial glia	Promoter fusion	Stork et al. 2008; Mayer et al. 2009
alrm-Gal4	Astrocyte-like glia	Promoter fusion	Doherty et al. 2009
deaat1-Gal4	Astrocyte-like glia, some cortex, weak in neurons	Promoter fusion	Rival et al. 2004, 2006
Mz709-Gal4	Ensheathing glia, neurons	Enhancer trap	Ito et al. 1995
nrv2-Gal4	Cortex, subperineurial, ensheathing? astrocytes?	Promoter fusion	Sun et al. 1999
moody-Gal4	Subperineurial glia	Promoter fusion	Schwabe et al. 2005
NP6293-Gal4	Perineurial glia, subset neurons	Enhancer trap	Hayashi et al. 2002; Awasaki et al. 2008
NP2276-Gal4	Subperineurial glia	Enhancer trap	Hayashi et al. 2002; Awasaki et al. 2008
NP577-Gal4	Cortex glia	Enhancer trap	Hayashi et al. 2002; Awasaki et al. 2008
NP2222-Gal4	Cortex glia	Enhancer trap	Hayashi et al. 2002; Awasaki et al. 2008
NP3233-Gal4	Astrocyte-like glia	Enhancer trap	Hayashi et al. 2002; Awasaki et al. 2008
NP1243-Gal4	Astrocyte-like glia and weaker in ensheathing glia and cortex glia	Enhancer trap	Hayashi et al. 2002; Awasaki et al. 2008
NP6520-Gal4	Ensheathing glia and weaker in cortex glia	Enhancer trap	Awasaki et al. 2008

This table gives a brief list of markers and tools that will be helpful for labeling and manipulating glial cells in *Drosophila*. Note that most markers, especially Gal4 driver lines, are often not exclusive to glia but show expression in other tissues as well. NA, not applicable.

bodies remain the best marker to uniquely label glia at all developmental stages, and *repo-Gal4* remains the best tool to specifically label and/or manipulate glia.

There is tremendous molecular heterogeneity in *Drosophila* embryonic glia, with probably 10–15 molecularly distinct subtypes of glia being present per hemisegment. One can assay this diversity in glial cell-fate specification in the embryo in exquisite detail by using a combination of cell position in the VNC along with an extensive list of available markers that label specific glial subtypes (e.g., Ito et al. 1995; Beckervordersandforth et al. 2008; von Hilchen et al. 2008). Here we will not delve deeply into this list of markers, rather we will briefly discuss the above-described broad categories of glia (surface, cortex, and neuropil) and show how they can be identified using simple combinations of antibody markers or *Gal4* driver lines (e.g., Repo) in conjunction with position and cell morphology.

Embryonic glia can be successfully manipulated by a number of approaches including the *Gal4/UAS* system, "flip-out" clonal strategies, or through the functional analysis of mutations in specific glial genes. The embryo is also well suited to techniques such as immunohistochemistry, in situ hybridization, and live imaging because of preparation simplicity and tissue transparency. These techniques allow for incisive in vivo studies of the function of specific glia and glial-expressed genes in CNS development and physiology. Gene misexpression, or the expression of dominant negatives, via *Gal4/UAS* appears to be the best approach for assaying mosaic animals (e.g., altering the fate of a subset of glia with *Gal4/UAS*). Loss-of-function genetic mosaic methods, such as mosaic analysis with a repressible cell marker (MARCM; Lee and Luo 1999), have not proven highly successful in the embryo. It is thought that MARCM likely fails to reveal clones because of perdurance of the maternally contributed Gal80 repressor through late stages of embryogenesis. Likewise, although there are a few studies in the literature that report successful use of RNA interference (RNAi) in the embryonic CNS (both injected and transgenic RNAi), the vast majority of researchers report poor results with such methods. RNAi is generally slow to knock down target genes, and such a delayed knockdown, coupled with the speed in which the entire embryonic CNS develops (~18 h), may explain the lack of robust RNAi expression in the embryonic CNS. Attempts at transgenic RNAi (e.g., *UAS*-regulated RNAi constructs) likely exacerbates this issue because *Gal4* itself will take additional time to activate high levels of the transgene encoding the double-stranded RNAi (dsRNAi).

Assaying Glial Cell-Fate Specification in the Embryonic Nervous System

Is the correct number of glia generated in the mutant background that I am studying? A simple experiment to determine whether the appropriate number of glia have been generated in an embryonic preparation (see the Protocol) would be to stain an embryo with an anti-Repo antibody (available from the Developmental Studies Hybridoma Bank at the University of Iowa) and simply count the number of glia per hemisegment. Segment boundaries can be delineated by costaining with anti-HRP (horseradish peroxidase), anti–Fasciclin II, or any other marker that has a segmentally reiterated pattern. Repo, being a nuclear transcription factor, will label all glial nuclei, thereby allowing for the identification of total glial number. Although the total number and subtypes of glia are nearly identical from embryo to embryo, the precise position of glial nuclei can be somewhat variable. As such, it is often difficult to identify exactly the same glial cell from animal to animal unless additional markers for specific glial subtypes are used (Beckervordersandforth et al. 2008), positions are scored very carefully in the CNS, or glial cell fate is assayed at very specific developmental stages. Simple examples are given below that use only Repo antibody and cell position at specific developmental stages.

NGB6-4T Glia

Glia produced by NGB6-4 can be uniquely identified by simply staining embryos with an anti-Repo antibody and scoring glial cell fates at embryonic stage 13. In the thorax, NGB6-4 (NGB6-4T) first produces three glia, and then only generates neurons. NGB6-4T-derived glia are generated by late stage 12, and include three glia per hemisegment, which appear to become either cortex or surface glia (Schmid et al. 1997, 1999). By stage 13, two of these glia have migrated from the lateral position of NGB6-4T toward the midline and take on a stereotyped morphology when viewed from the ventral surface—they immediately flank the midline, positioned along the anteroposterior axis, and form a "box-like" structure with contralateral NGB6-4T-derived glia (Fig. 1A). To identify all NGB6-4T-derived glia in embryonic thoracic segments one can also stain with Eagle antibody. NGB6-4 is one of only four NBs that express Eagle, and Eagle expression is also present in NGB6-4T progeny (Higashijima et al. 1996; Freeman and Doe 2001). The only cells in the embryo that costain with Repo and Eagle antibodies are the three NGB6-4T-derived glia.

NGB6-4A Glia

NGB6-4 in abdominal segments (NGB6-4A) generates only two glia and no neurons. One of these glia migrates to a position adjacent to the midline, similar to that of the NGB6-4T-derived glia. Instead of appearing as a box of four cells per segment, NGB6-4A glia appear in pairs (Fig. 1A). Conveniently, both glia from NGB6-4A can be identified uniquely by using an *eagle-Kinesin-LacZ* reporter and scoring only abdominal segments. Although Eagle antibody does not label NGB6-4A-derived glia, the *eagle-Kinesin-LacZ* reporter does, perhaps because of the loss of a NGB6-4A-specific repressor for *eagle* when the construct was generated (Higashijima et al. 1996).

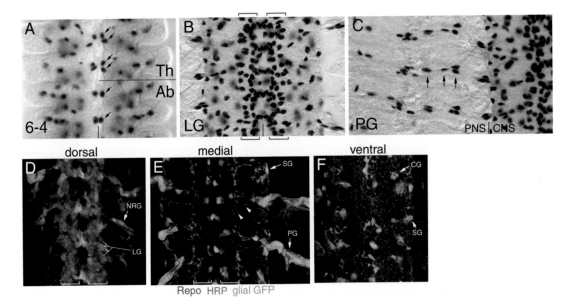

FIGURE 1. Glial cell subtypes in the *Drosophila* embryonic central nervous system (CNS). Whole-mounted embryos processed by immunohistochemistry with anti-Repo (glial nuclei) and visualized with DAB histochemistry (*A–C*) or with anti-Repo, anti-HRP (neuropil), and anti-GFP (glia) and fluorescent secondary antibodies (blue, red, and green, respectively) (*D–F*). (*A*) Stage 13 embryo depicting NGB6-4 glial cell nuclei (arrows) in the thoracic (Th) and abdominal (Ab) hemisegments of the CNS. The lines mark the boundary between Th and Ab segments (horizontal) and location of the midline (vertical). (*B*) Longitudinal glia (LG) nuclei in an early stage 16 embryo. LG flank the midline (vertical line) and are present in columns along the length of the CNS (between brackets). (*C*) Peripheral glia (PG) nuclei labeled in an early stage 16 embryo. The boundary between the CNS and peripheral nervous system (PNS) is marked with a vertical line. PG migrate away from the CNS along motor neuron tracts (arrows). (*D–F*) Stage 16 transgenic embryos expressing UAS-cytoplasmic/membrane-bound-GFP driven by *repo-Gal4*. Each image represents a single slice from the same z-stack. Neuron processes are labeled in red to show their spatial relationship with glia. (*D*) Dorsal slice of the CNS showing examples of nerve root glia (NRG; arrow) and longitudinal glia (LG; lines). (*E*) Medial slice of the CNS with examples of surface glia (SG; arrow), peripheral glia (PG; arrow), and glial membrane processes adjacent to the neuropil and within the cortex (arrowheads). (*F*) Ventral slice of the CNS showing examples of cortex glia (CG; arrow) and surface glia (SG; arrowhead).

Longitudinal Glia

At late embryonic stages, the longitudinal glia (LG) reside at the dorsal surface of the CNS (Fig. 1B) and are thought to ensheath the longitudinal axon tracts. LG are derived exclusively from the glial precursor (GP), a stem cell that gives rise only to glia (in all segments), beginning at late stage 12. This precursor divides once, begins migrating medially, and continues to divide until it ultimately gives rise to seven to nine glia per hemisegment (Schmidt et al. 1997). These glia can be identified based on their dorsal position in the CNS, but there are a few additional glia that reside very close to the LG, and additional markers are essential to identify dorsal glial cells definitively as LG. Two extremely specific markers for LG are the enhancer trap line F236 (Jacobs et al. 1989) and the recently published *Gal4* driver line *alrm-Gal4* (Doherty et al. 2009; O Tasdemir, T Stork, and MR Freeman, unpubl.). LG can also be subdivided further at late embryonic stages by either of two available molecular markers: (1) a commercially available antibody to mammalian glutamine synthetase that cross-reacts with *Drosophila* glutamine synthetase 2 (GS2) and uniquely labels six LG per hemisegment, or (2) anti-Prospero antibody, which labels the same six anterior LG as GS2 (Thomas and van Meyel 2007). Thus, LG can be subdivided into Repo$^+$/Pros$^+$/GS2$^+$ or Repo$^+$/Pros$^-$/GS2$^-$ subtypes.

Peripheral Glia

Embryonic peripheral glial are primarily derived from the CNS precursors NGB1-3 and NGB2-5, but quickly migrate out of the CNS along motor neuron tracts to the periphery in which they associate with motor neurons that project out of the CNS and sensory neurons that project into the CNS.

Peripheral glia can be identified definitively based on their expression of Repo and position along the peripheral nerve (Fig. 1C), but they include a number of glial subtypes: The outermost layer is composed of perineurial glia, beneath these are subperineurial glia that form the BBB of the peripheral nerve, and the deepest cells are the wrapping glia that wrap axons (Stork et al. 2008). Leiserson et al. (2000) have also referred to wrapping glia as "ensheathing" glia. To avoid confusion we will refer to these as wrapping glia, which in the peripheral nerve refers specifically to their surrounding individual axons with membrane. In the adult, a morphologically distinct glial subtype that compartmentalizes brain regions has also been termed ensheathing glia (Awasaki et al. 2008; Doherty et al. 2009), although whether these glia wrap individual axons remains to be determined. In total, 12 glial cells per abdominal hemisegment are classified as peripheral glial cells. Of these, seven to nine are derived from CNS neuroblasts, and the remaining are derived from peripheral sensory organ precursor (SOP) lineages (Schmidt et al. 1997). A number of markers for further identification of specific peripheral glia subtypes have been described in detail elsewhere (von Hilchen et al. 2008).

Assaying Glial Cell Migration in the Embryonic Nervous System

Most glia are not generated at the site in which they will ultimately function within the nervous system. Instead, they have to migrate, sometimes significant distances, to their final positions within the developing nervous system. The vast majority of this migration occurs in the CNS/peripheral nervous system during embryonic stages 12–15. The lineages described above provide excellent examples of lineages that can be assayed for defects in glial migration in different mutant or misexpression backgrounds. For example, glia derived from NGB6-4T and NGB6-4A migrate medially along the ventral surface of the embryonic CNS; GP-derived LG migrate medially to the dorsalmost region of the CNS; and all peripheral glia migrate laterally out of the CNS to very specific positions along the peripheral nerve. Based on our knowledge of the positions in which these glia are generated, their migration pattern, and when they should arrive at their final destination, a specific mispositioning of selected subtypes of glia can be used to argue strongly for a defect in glial cell migration. For example, extensive screens for mutants with defects in peripheral glial migration along the nerve have led to the identification of key factors required to specify peripheral glial fates (Edenfeld et al. 2006, 2007). In addition, through the use of simple glial-specific drivers, one can force the expression of green fluorescent protein (GFP) and assay the dynamic movements of glia in live preparations by confocal microscopy (von Hilchen et al. 2008).

The analysis of stereotyped lineages such as those described above in combination with simple anti-Repo stains provides a good initial indicator of glial development in any mutant of interest. For example, if in a given mutant or misexpression background, NB6-4T glia are properly positioned near the midline at stage 13 and found in the appropriate numbers, then one can reasonable conclude that NB6-4T was formed, it generated glia, and those glia migrated medially and proliferated properly during development.

Visualizing Glial Cell Morphogenesis and Functions in the Embryonic Nervous System

After cell-fate specification and migration to the appropriate position within the nervous system are complete, glial cells elaborate fine membrane processes, which are presumably essential for functional maturation. During this developmental transition, glia undergo dramatic changes in shape that ultimately result in a striking array of specific glial morphological subtypes (Fig. 1D–F; Box 1). In this section we outline a number of interesting changes in glial morphology or function that occur during embryonic neurogenesis, and describe simple assays for their visualization or functional dissection. However, owing to developmental and technical limitations, not all glial morphogenetic events are easily assayed in the embryo. For example, in some cases the elaboration of glial membranes does not occur until embryonic stage 17, which for technical reasons is difficult to study, or until larval stages, and some elaborations can be difficult to visualize because of the extremely small size of the embryonic CNS. As such, it is appropriate to study a number of neuron–glia associations specifically in the larva or adult,

BOX 1. MAIN MORPHOLOGICAL CLASSES OF GLIAL CELLS IN *DROSOPHILA*

From embryonic to adult stages, three main classes of glial cells have been distinguished in *Drosophila*: surface-associated glia, cortex-associated glia, and neuropil-associated glia. These main classes can be further subdivided into morphologically distinct groups. The most common subtypes are exemplified in the third-instar larva (shown in Fig. 2) but can be found with very similar characteristics in the late embryo or in the adult.

Surface-Associated Glia

Perineurial glia: Perineurial glia form the outermost cellular sheath surrounding the nervous system. These Repo-positive cells are believed to be of mesodermal origin (Edwards et al. 1993) and their

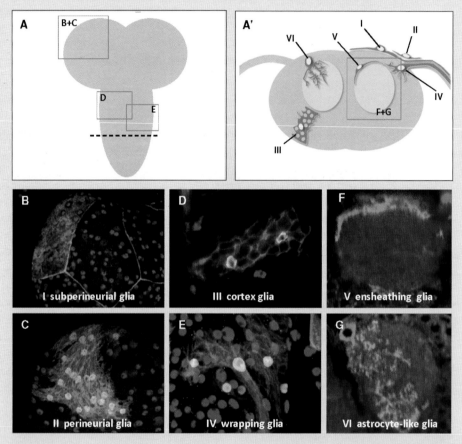

FIGURE 2. Main morphological subclasses of glial cells in *Drosophila*. (A) Schematic overview of the central nervous system (CNS) of a third-instar larva with the two brain lobes and the unpaired ventral nerve cord shown in gray. The dashed line indicates the plane of a schematic cross section shown in panel A′. (A′) Dark gray depicts the cortex, the region of cell bodies, and light gray depicts the synaptic neuropil. The different subtypes of glial cells described in the text are highlighted as single cells in red and numbered I–VI. The boxes refer to the approximate regions in the CNS in which the images shown in panels B–G were taken. (B–G) Confocal images of glial subtype morphology highlighted with different "flip-out" strategies. Glial nuclei are visualized with α-Repo staining (B–F; blue), and glial membranes are shown with the use of Gal4-driven mCD8-Cherry (B–E) or mCD8-GFP (F,G). Additional markers include a NrxIV-GFP exon trap, which labels the cell borders of subperineurial glia (B; green); nuclear β-galactosidase, which labels the cell nuclei of the glial cell clones (C–E; green); and α-HRP to label neuronal membranes (F,G; red). Note that in B a single subperineurial glial cell is labeled and that nearly all Repo-positive nuclei correspond to other, deeper located glial cells. The following genotypes were used: (B) *act5c>CD2>Gal4, repo-flp*, and *UAS mCD8-Cherry/nrxIV-GFP*; (C–E) *act5c>CD2>Gal4, repo-flp*, and *UAS mCD8-Cherry/UAS lacZ::NLS*; (F) *UAS>CD2>mCD8-GFP repo-flp* and *Mz709-Gal4*; (G) *UAS>CD2>mCD8-GFP repo-flp* and *alrm-Gal4*.

BOX 1. *(CONTINUED)*

function remains unknown. Although only very few perineurial cells cover the nervous system at the end of embryogenesis, these cells proliferate extensively to form a continuous monolayer in later larval stages and also in the adult stage (Awasaki et al. 2008; Stork et al. 2008).

Subperineurial glia: Underneath the perineurial sheath lies the thin monolayer of subperineurial glia. These extremely big, flat cells show epithelial characteristics. They ensheath the nervous system and seal their extensive cell–cell contacts with pleated septate junctions to form a paracellular diffusion barrier. This glial BBB is extremely important for neuronal physiology because it helps to exclude high potassium concentrations in the hemolymph from the nervous system and allows for proper action potential propagation (Auld et al. 1995; Baumgartner et al. 1996). Several structural components of pleated septate junctions and regulatory factors such as *moody* have been identified as being essential for BBB function, making subperineurial glia one of the best characterized glial subtypes in *Drosophila*.

Cortex-Associated Glia

This glial subtype surrounds and tightly ensheathes neuronal cell bodies in the CNS, forming a trophospongium (Dumstrei et al. 2003; Pereanu et al. 2005). A single cortex glial cell can ensheath dozens of neuronal cell bodies and typically contacts the subperineurial sheath, the neuropil associated glia, or both. Cortex glia are thought to supply trophic and metabolic support for neurons.

Neuropil-Associated Glia

Wrapping glia or nerve-associated glia: This subtype of *Drosophila* glia is closely associated with nerves and enwraps axonal profiles. In a mature state, a single wrapping glial cell is able to individually enwrap most axons in the diameter of a nerve simultaneously (Stork et al. 2008), resembling nonmyelinating Schwann cells in a Remak bundle in vertebrates (Nave and Salzer 2006; Nave and Trapp 2008). However, in the cortical regions of the CNS, wrapping glia also send out processes to ensheath neighboring neuronal cell bodies, as is seen in cortex glia.

Ensheathing glia: In contrast to wrapping glia, ensheathing glia are not associated with nerves but rather with the synaptic neuropil. Glial processes cover the synaptic neuropil (e.g., in the ventral nerve cord) and can ensheath and subdivide morphologically distinct substructures like the glomeruli of the adult antennal lobe (Awasaki et al. 2008; Doherty et al. 2009). In the adult this subtype of glial cell (probably together with wrapping glia) has been shown to respond to axonal injury by clearing neuronal debris by phagocytosis (Doherty et al. 2009).

Astrocyte-like glia: These glia have their cell bodies closely associated with the synaptic neuropil. In contrast to ensheathing glia, the astocyte-like glial cell processes invade the neuropil and infiltrate the volume densely. Their tufted morphology and the close contact of glial membranes to synapses are reminiscent of vertebrate protoplasmic astrocyte morphology. Additionally, similar to the situation in vertebrates, *Drosophila* astrocyte-like glia cells express neurotransmitter transporters that are able to clear neurotransmitters, such as glutamate or GABA, from the synapse to ensure proper brain function (Rival et al. 2004; T Stork, unpubl.).

in which one can exploit the full array of genetic approaches available in *Drosophila* and glia can be considered to be functionally mature. These are discussed in the final section.

Formation of the BBB by Subperineurial Glia

Surface glia form the BBB that isolates and protects the embryonic CNS. A subset of glia, termed subperineurial glia (SPGs), position themselves along the surface of the CNS and by embryonic stage 15 these cells have flattened to cover the entire surface of the VNC and brain. Subsequently, they form pleated septate junctions (pSJs) with one another to seal off the CNS from the surrounding hemolymph (Auld et al. 1995; Baumgartner et al. 1996; Schwabe et al. 2005). SPGs are unique in their expression of *moody*, a seven-transmembrane-domain receptor required for the regulation of pSJs in SPGs (Bainton et al. 2005; Schwabe et al. 2005). *moody-Gal4* or Spg-*Gal4* can be used to drive markers for pSJs such as GFP::DlgS97 (Bachmann et al. 2004; Stork et al. 2008) and pSJ assembly can be visualized in live preparations (Bainton et al. 2005; Schwabe et al. 2005). Alternatively, GFP gene trap lines like *Nrx-GFP* (Edenfeld et al. 2006) can be used to visualize pSJs in vivo.

A number of groups have used a simple protocol initially designed to test the integrity of pSJs in the salivary gland epithelium (Lamb et al. 1998) to assay BBB integrity in vivo. Briefly, fluorescent dye-dextran conjugates (generally 10-kDa dextrans) are injected into the hemolymph of late-stage embryos, and after ~10–20 min their CNS is visualized by confocal microscopy. If the BBB is intact, dye-dextran conjugates are excluded from the CNS. However, if the BBB is compromised, dye-dextran conjugates accumulate in the CNS (around cell bodies and in the neuropil), indicating that the BBB is permeabilized. This simple assay has been used effectively to assay BBB integrity in a number of mutants affecting pSJs or their formation (Bainton et al. 2005; Schwabe et al. 2005; Banerjee et al. 2006; Stork et al. 2008).

Glial Engulfment of Neuronal Cell Corpses

Approximately 30% of CNS neurons generated during embryonic neurogenesis appear to undergo programmed cell death (Rogulja-Ortmann et al. 2007) and their cell corpses are engulfed and destroyed by glia (Sonnenfeld and Jacobs 1995; Freeman et al. 2003; Kurant et al. 2008). A simple approach to assay the efficiency of glial clearance of neuronal cell corpses is to stain the embryonic nervous system at a defined developmental stage (e.g., stage 15) with 7-aminoactinomycin-D (7-AAD; Franc et al. 1999). 7-AAD is a fluorescent nucleic acid intercalating agent that is used to visualize cellular nucleic acids. The vast majority of neuronal cell bodies in the CNS stain weakly with 7-AAD in fixed preparations, and neuronal cell corpses stain as bright puncta that are ~25% the size of a healthy neuronal cell (Freeman et al. 2003). The number of cell corpses per hemisegment can be quantified by costaining with a marker that delineates segment boundaries (e.g., Fasciclin II). It is important to note that this approach gives an overall assessment of the number of neuronal cell corpses present at a given developmental stage. Care must be taken in interpreting these results and distinguishing between glia function and neuronal survival. For example, a mutant may be suspected to affect engulfment by glia, but its actual role is to give trophic support to neurons. The increase in neuronal cell corpses may be a result of loss of trophic support rather than a decrease in engulfment by glia.

A more incisive method for exploring the dynamics of glial engulfment of neuronal cell corpses is to observe glial morphological changes directly during engulfment activity in live preparations. One can easily express GFP specifically in glia (e.g., *repo-Gal4, UAS-mCD8-GFP*) to visualize glial morphology and neuronal cell corpses can be visualized simultaneously using commercially available markers for apoptotic cells such as annexin V (Kurant et al. 2008). Levels of glial lysosomal activity can be assayed using LysoTracker probes (Watts et al. 2004; Kurant et al. 2008).

Axon Pathfinding and Trophic Support of Neurons by Glia

Based largely on cell ablation studies, embryonic CNS glial cells have been reported to play critical roles in axon pathfinding (Hidalgo and Booth 2000; Sepp et al. 2001; Sepp and Auld 2003) and trophic support of neurons (Booth et al. 2000), topics that are not discussed in detail in this chapter. It is thus expected that a number of glial-expressed genes will modulate axonal outgrowth and neuronal survival. With respect to axon pathfinding, we refer the reader to Chapter 4 for a discussion of methods used to score defects in neuronal wiring of the embryonic CNS and to Sepp and Auld (2003) and Sepp et al. (2001) for discussions of the role of peripheral glia in entry of sensory neuronal processes into the embryonic CNS. A lack of glial trophic support of neurons would be expected to result in the apoptotic death of neurons that would normally survive, thereby generating an increase in overall neuronal cell death in the developing embryonic CNS. A simple initial method helpful for detecting increased neuronal cell death would be to combine the use of antibodies directed toward activated caspase-3 (Kurant et al. 2008) with markers for identifiable subsets of embryonic CNS neurons (i.e., do cells that normally survive inappropriately activate caspase-3?). Alternatively, one could use a battery of markers for uniquely identifiable neurons, count total numbers of neurons, and determine whether neuronal populations are reduced in genetic backgrounds of interest (i.e., are specific cells missing?). We refer the reader to Chapter 3 for a discussion of markers for specific subsets of neurons.

GLIAL CELL DEVELOPMENT AND FUNCTION IN THE LARVAL AND ADULT NERVOUS SYSTEM

Over the last 20 years, the embryo has proven to be an excellent system in which to explore basic aspects of glial cell development. However, embryonic preparations also have their limitations. First, with standard staining procedures, only embryos that are at developmental stage 16 or earlier can be analyzed because the developing cuticle becomes increasingly impermeable to fixatives and antibodies by embryonic stage 17. At 25°C neurogenesis begins about 3–4 h after egg lay (AEL) and continues until larval hatching. Stage 17 represents the final ~6 h of embryonic neurogenesis, or the last 33% of neural tissue development (Campos-Ortega and Hartenstein 1997; Prokop 1999). Therefore, the final stages of embryonic nervous system maturation are largely missed, and this may in fact be the time when some of the most interesting elaborations of glial processes take place to give rise to fully functional glia. Second, a number of additional interesting glial morphogenetic changes do not occur until larval stages. For example, at late embryonic stages peripheral glia appear to wrap entire bundles of axons, but as animals progress through larval stages these glia eventually separate and individually ensheath individual axons (Stork et al. 2008). Third, although it is generally difficult to manipulate embryos with powerful molecular-genetic tools like cell type–specific RNAi or MARCM, these approaches seem to work well in larval, pupal, and adult nervous system glia (Lee and Luo 1999; Aigouy et al. 2004; Dietzl et al. 2007; Doherty et al. 2009). The postembryonic CNS is an attractive system in which the full complement of sophisticated genetic tools available in *Drosophila* can be used to visualize and manipulate glia, even at the single-cell level. As discussed in Box 1, the major subtypes of glia established in the embryo are also found in the larva (Fig. 2A–G) and adult (not shown). Here we describe a number of genetic approaches for marking and manipulating specific subsets, or a very small number of glial cells, and discuss their advantages and limitations.

Molecular Markers for Postembryonic Stages

A number of molecular markers for stereotyped subpopulations of glial cells have been identified in *Drosophila* (e.g., Ito et al. 1995; Beckervordersandforth et al. 2008; von Hilchen et al. 2008). Most of these markers have been used primarily to study glial cell populations in the embryo, but remain poorly characterized at later developmental stages. An array of potential tools exist for studying glial cells at larval stages and beyond, but at the moment we know surprisingly little about molecular identities and fine morphology of glial cells at these later time points. Nevertheless, a reasonable number of genetic tools for the visualization and manipulation of glial cells have been characterized in more detail in postembryonic stages (e.g., Awasaki et al. 2008; Stork et al. 2008; Doherty et al. 2009), and we have listed some of the tools we find particularly useful for the analysis of glia at different developmental stages (Table 1).

Analysis of Glial Gene Function by Cell Type–Specific dsRNAi

Cell type–specific dsRNAi can be accomplished through the use of Gal4-driven *UAS* constructs (Brand and Perrimon 1993) that carry inverted repeats of a gene of interest (Lee and Carthew 2003). On expression, these inverted repeats can fold back to form double-stranded RNA, which is processed by Dicer (a ribonuclease), to give rise to the siRNAs (small interfering RNAs) (Fire et al. 1998). These siRNAs then lead to a sequence-specific knockdown of the corresponding gene of interest. By choosing glial-specific *Gal4* driver lines, expression of genes can be manipulated in all glia, or even in small subsets of glia cells, without perturbing gene function in other tissues (Table 1). Although the *Gal4* lines listed in Table 1 are predominantly expressed in glial cells, additional expression in other tissues is common, and expression patterns may dynamically change at different developmental stages.

Therefore, care must be taken to show rigorously that any phenotype observed using *UAS-dsRNAi* with glial driver lines is in fact the result of glial gene expression manipulation rather than manipulation of gene expression in other tissues. Strong *Gal4* driver lines and prolonged expression of the *UAS-dsRNAi* constructs are normally necessary for efficient knockdown of the targeted gene's expression. Both of these issues are quite manageable in larval, pupal, and adult stages. Recent reports suggest that coexpression of *UAS-dicer2* can enhance the effects of dsRNAi (Dietzl et al. 2007). The efficacy of dsRNAi-mediated knockdown may vary for different genes, *UAS-dsRNAi* constructs, and even different insertions of the same transgene, and therefore efficacy of dsRNAi-mediated knockdown should be empirically verified in each experiment (Dietzl et al. 2007; Ni et al. 2008). Furthermore, one needs to be aware that dsRNAi constructs may show off-target effects in which the resulting siRNAs are not entirely specific to the gene of interest (Kulkarni et al. 2006; Ma et al. 2006; Moffat et al. 2007). Ideally, data obtained by a *UAS-dsRNAi* approach should be complemented by additional genetic data, such as MARCM analysis, whenever possible (see below). Because *UAS-dsRNAi* constructs exist for ~90% of *Drosophila* genes and are readily available from public stock centers (Vienna *Drosophila* RNAi Center, Austria; Dietzl et al. 2007; National Institute of Genetics Fly Stock Center, Japan; Transgenic RNAi Project, USA; Ni et al. 2008, 2009), this approach has the potential to be extremely powerful in the analysis of glial cell biology from larval stages through adulthood.

Genetic Mosaic Approaches to Characterizing Glial Morphology and Function

"Flip-out" Techniques

UAS Flip-Out Constructs: One way to restrict the expression patterns of *Gal4* lines is by using a so-called "flip-out" approach (Basler and Struhl 1994). In this method a promoter element is followed by a cassette, which is flanked by direct repeats of flippase recognition target (FRT) sites. These sites are recognized by the yeast site–specific recombinase, termed flippase (FLP), which leads to a circular excision of the FRT flanked sequences (Fig. 3A). One example of such a construct is *UAS-FRT-CD2-y⁺-FRT-mCD8-GFP* (or shorthand *UAS>CD2>mCD8-GFP*; Wong et al. 2002). Without FLP activity, only the membrane marker CD2 is expressed by a *Gal4* driver line, whereas the mCD8-GFP reporter remains silent. On activation of FLP, the CD2 flip-out cassette is excised and mCD8-GFP comes under direct Gal4 control (Fig. 3A). The excision of the FRT-flanked sequences is stochastic, depending on how strong the FLP activity is in a given cell, and can occur in either a mitotic or postmitotic cell. The latter point is important for studying glia during larval or adult stages because the vast majority of larval glia do not appear to proliferate. Several different FLP sources are available and their specific abilities are summarized in Box 2. Typically, FLP activity leads to clonal expression of the UAS-flip-out GFP reporter so that subtle changes in cell morphology can be visualized, rather than obscured by the full expression pattern normally generated by the Gal4 driver (Fig. 2A–G). Such approaches can lead to very clear images of the morphology of single glial cells. In addition, this labeling technique can be used in a mutant background, or coupled with a *UAS-RNAi* construct to assay changes in glial morphology in response to loss of a particular gene product.

Flip-Out Gal4 Drivers: Without generating additional constructs, the approach described above is, at the moment, mainly restricted to mCD8::GFP as the clonal reporter. However, with other flip-out strategies it is possible to visualize and manipulate single glial cells with any available UAS line. This can be performed, for example, with an *actin5C* promoter-driven construct that conditionally activates Gal4 (*act5c>CD2>Gal4*; Ito et al. 1997; Pignoni and Zipursky 1997). In this case, FLP activity leads to the expression of Gal4 under the control of the ubiquitously expressed *actin5C* promoter, which then can be used to drive any UAS-regulated construct. Because this promoter is not specific to glia cells, one can increase its utility for studying glia by spatially restricting the FLP expression by using either a *repo* promoter–driven version of FLP (*repo-flp*; Silies et al. 2007; Fig. 2) or through a *LexA/lexA* operon–controlled expression of FLP (Lai and Lee 2006; Shang et al. 2008). In general, it is also possible to exchange the ubiquitous promoters in the flip-out *Gal4* constructs or the promoters controlling FLP expression for glial subset-specific promoters, as long these sequences are identified.

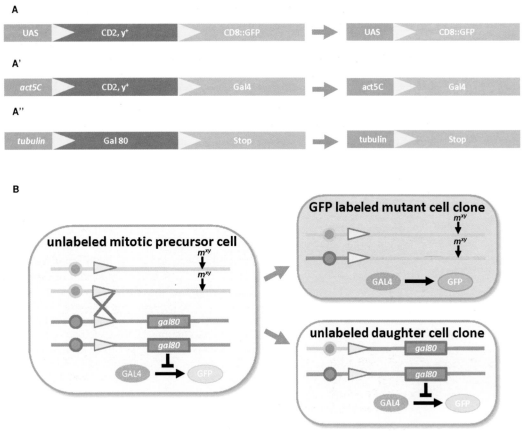

FIGURE 3. "Flip-out" and MARCM techniques. (*A,A',A''*) Schematic representations of flip-out events for three different constructs and (*B*) the genetic basis of the MARCM technique. (*A,A',A''*) The flip-out cassettes (red boxes) are flanked by directly repeated FRT sites (yellow triangles). Expression of genes contained in the flip-out cassette can be under the control of different promoter elements like Gal4 controllable UAS sequences (*A*) or specific promoter elements like the ubiquitously active *actin5c* or *tubulin* promoters (*A',A''*). Expression of Flp (not shown) mediates site-specific recombination between the FRT sites leading to the removal of the flip-out cassette and its respective marker expression. Instead, a second reporter element (green boxes) is now under the control of the promoter, here either CD8-GFP (*A*) or Gal4 (*A'*). For details see text. (*B*) In a MARCM approach, the two homologous chromosome arms are carrying FRT sites (yellow triangles) near to their centrosomes (blue and red circles representing maternal and paternal copy of the chromosome). One of the chromosome arms is equipped with a *tub-Gal80* insertion distal to the FRT site and on the other arm a mutation of interest (m^{xy}) can be placed. Gal80 suppresses Gal4-mediated GFP expression (orange and green oval, respectively; transgenic insertions for these proteins are not depicted and can be placed elsewhere in the genome). Flp-mediated site-specific recombination during the G_2 phase of mitosis between the FRT sites of homologous chromosome arms (red cross; FLP not shown) leads to the separation of *Gal80* and the mutation of interest after the mitotic division. This results in a homozygous mutant daughter cell that lost *Gal80* and therefore starts to express the GFP reporter while the corresponding daughter cell and all other nonclonal cells keep expressing Gal80. GFP-labeled mutant clones can then be analyzed for morphological defects. For details see text.

Flip-Out Gal80 Constructs: Another more flexible approach that recently became available uses the clonal loss of the *Gal4* repressor Gal80 as a way to activate *Gal4* in subsets of cells. In this approach, a *tubulin* promoter drives the expression of a *Gal80* gene that is flanked by FRT sites (*tub>Gal80>*Stop) and ubiquitously represses *Gal4* activity. On FLP-mediated clonal removal of the *Gal80* repressor cassette, Gal4 is activated and any given UAS construct can be driven in the clonal cells (Shang et al. 2008; Gordon and Scott 2009). FLP activity can be induced by heatshock (*hs-flp*), or specifically targeted to glial cells by *repo-flp* or *lexAop-flp* under control of a glial *LexA::VP16* driver (not repressible by Gal80). The advantage of this system is the high degree of experimental freedom because the expression can be controlled ad libitum by combining different Gal4 driver lines with any UAS effector line. Minor drawbacks of this approach might be the perdurance of the Gal80 protein after the cassette has been excised, which might lead to repressed Gal4 activity for *hs-flp* is the workhorse for any flip-out or MARCM strategy. The expression of yeast recombinase FLP under the control of a heat shock–inducible promoter (Golic and Lindquist 1989; Struhl and Basler 1993) allows for high temporal control of clone induction. Additionally, strength of FLP activity can be adjusted by temperature, duration, and repetition of heat shocks. As a starting point a typical heat shock would be conducted for 1 h at 37°C in a water bath at the desired developmental stage. Prolonged exposures or slightly higher temperatures are possible, but will likely lead to elevated levels of lethality. Instead, additional heat shocks can be given after a period of recovery on consecutive days. Furthermore, several *hs-flp* constructs and insertions are available that can differ in their abilities, like maximal FLP activity or basal level of expression, without heat shock induction (Theodosiou and Xu 1998).

repo-flp is a promoter fusion of the *repo* promoter (Lee and Jones 2005) with a FLP open reading frame (Silies et al. 2007). The *repo* promoter is active specifically in glia beginning at early developmental stages and remains on through adulthood, making it applicable for glial-specific flip-out as well as MARCM experiments. Because this construct is constitutively active in glia, FLP activity can only be modulated by copy number or choosing different *repo-flp* insertions, which can differ in FLP expression level quite substantially. Although *repo-flp* tends to work well for FLP-out approaches, activity can be low in early glial precursors, which may result in low clone frequency in some glial lineages when using MARCM. The clone frequency also depends on the efficiency of the FRT sites used (Rooke et al. 2000) and must be tested empirically for each glial MARCM setup. The main advantages of a *repo-flp* source are (1) mutant clones are generated specifically in glial cells and (2) clonal events can be induced without additional heat-shock protocols, which can be especially advantageous when large numbers of different genotypes have to be tested routinely.

UAS-flp/lexAop-flp: In combination with flip-out labeling approaches (e.g., *UAS>CD2>CD8GFP*), *UAS-flp* and *lexAop-flp* constructs can be especially useful for characterizing expression patterns of Gal4 or LexA driver lines with high resolution because typically low numbers of cells will be labeled. For crossing schemes, including *Gal80* constructs like MARCM or a *tub>stop>Gal80* approach, these constructs are only useful if the corresponding *Gal4* or *LexA* line is not repressible by *Gal80* activity. This can be accomplished by use of the VP16 transactivation domain rather than that of Gal4 with Gal4 or LexA transgenes. Using such an approach, a glial *LexA::VP16* driver can be used to drive *lexAop-flp* expression and generate Gal4-labeled clones in a MARCM background. At the moment, suitable glial VP16 fusions are not yet available but should be generated in the near future.

another 24–48 h after excision (Lee and Luo 1999; Lee et al. 2000), or incomplete repression of Gal4 activity before the flip-out event of the *Gal80* cassette, especially with strong Gal4 driver lines.

MARCM Analysis in Glial Cells

Another powerful method used to label and manipulate subsets of glial cells is MARCM. This approach allows positive genetic labeling of wild-type or mutant cell clones (Lee and Luo 1999; see also Chapters 7 and 8) (Fig. 3B). Briefly, mitotic recombination is induced by FLP activity between homologous chromosome arms that carry FRT sites near to their centromeres. *tub-Gal80* is present on one of the chromosome arms to repress activation of Gal4-regulated reporters (e.g., repo-*Gal4*, *UAS-mCD8-GFP*) in all heterozygous progenitor cells. The other chromosome arm either can be wild-type (control) or can carry a mutation of interest. Induction of mitotic recombination can result in production of two different daughter cell clones: one that is homozygous for *tub-Gal80*, and therefore Gal4/UAS remains repressed, and another that loses *Gal80*, which results in the derepression of Gal4/UAS and activation of UAS-driven reporter expression. By placing a mutation on the chromosome arm in *trans* to the *tub-Gal80* (Fig. 3B) one can generate positively marked mutant glia cells that can be analyzed for cell-autonomous morphological defects. MARCM can also be used with any UAS-dsRNAi lines of interest to assay the autonomous effects of a gene in glia because the dsRNAi construct will only be driven in GFP-marked clones.

In contrast to the flip-out techniques described earlier, induction of MARCM clones requires mitotic recombination and therefore the presence of dividing glial precursor cells. Consequently, MARCM leads to lower clone frequencies, and the timing of clone production is restricted to specific developmental time points (Awasaki et al. 2008). For example, to make MARCM clones in larval astrocyte-like glia or subperineurial glia, one would have to induce clones in the embryo, when these populations of glia are still mitotically active. This presents a challenge for glial lineages with low proliferation rates because these will inherently generate fewer clones. Clone production can be enhanced by crossing in additional transgenes of the FLP source (Box 2), or in the case of *hs-flp*, through prolonged and repeated heat shock administrations during development. Finally, it is important to note that depending on the FLP source used, MARCM can generate a substantial fraction of homozygous mutant cells in the nervous system (e.g., neurons) that are not labeled by the Gal4 reporter but could potentially contribute to any phenotypes observed. As with all analyses that involve the stochastic production of clones, these issues can be overcome by assaying for consistent phenotypes in many independently generated clonal animals. An additional possible pitfall of (glial) MARCM can be incomplete or inconsistent suppression of Gal4 activity by *tub-Gal80*. This can lead to reporter expression in small subsets of glial cells even in the absence of any FLP source, which could be mistaken as mutant clones. Awasaki et al. (2008) used an additional *repo-Gal80* construct in their MARCM setup to circumvent this problem.

CONCLUSIONS

The embryonic nervous system has been and will continue to be a superb setting for assaying some of the earliest events in glial cell development including glial cell-fate specification, generation of molecularly and morphologically diverse glial subtypes, glial cell migration, and morphogenesis of the BBB. In addition, the embryo has been quite helpful for understanding glial control of neuronal development including neuronal trophic support, axon pathfinding, nerve fasciculation, and clearance of dead cells from the CNS. Because we know so little about glial cell morphogenesis, there is clearly a wealth of interesting glial developmental biology to be uncovered through studies in the embryo, and our hope is that the methods outlined in this chapter will serve as a starting point for those types of analyses. At the same time, some of the most amazing features of glial cells, such as their massive expansion in size (e.g., along segmental nerves during the L1 to L3 growth), intimate association with highly organized brain structures (e.g., the lamina of the visual system), or fine elaboration of membranes (e.g., the tufted morphology of astrocyte vs. the flattened morphology of subperineurial glia), seem better addressed in the larval, pupal, and adult nervous system. In the second part of this chapter we have attempted to outline a number of useful tools, genetic techniques, and glial markers that should facilitate the exploration of these later steps in glial cell morphogenesis. Finally, a major goal in the field of glial biology, and neuroscience in general, is to understand how glial cells control neural function to ultimately affect animal behavior. There are probably few synapses, neural circuits, or behaviors that are not regulated in some way by glia. The challenge before us now is to understand the cellular and molecular basis of these interactions. We believe the future is bright in this regard for *Drosophila* researchers, and hope that the information provided in this chapter will aid in this endeavor.

Whole-Mount Embryo Fluorescent and Histochemical Antibody Staining

The following protocol describes how to visualize protein expression patterns in fixed, whole-mounted *Drosophila* embryos. First, we explain how to detect primary antibody markers using fluorophore-tagged secondary antibodies. Then, we show how primary antibodies are detected using the diaminobenzidine (DAB) histochemical method.

METHODS

CAUTION: See Appendix for proper handling of materials marked with <!>.
See the end of the chapter for recipes for reagents marked with <R>.

Reagents

Antifade mounting medium <R>
DAB substrate kit <!> (Thermo Scientific cat. no. 34002)
Fluorescent secondary antibodies (Jackson ImmunoResearch Laboratories)
   Dilute 50% in glycerol and store at –20°C.
Glycerol (80%)
Heptane <!>
Horseradish peroxidase (HRP)–conjugated secondary antibodies (Jackson ImmunoResearch Laboratories)
   Dilute 50% in glycerol and store at –20°C.
Methanol (100%) <!>
PBT <R>
PEMFA (fixative solution) <R>
PTx <R>

Equipment

Coverslips (#1.5; 22 x 50-mm)
Dissection dish (glass; three-well)
Glass slides
Microcentrifuge tubes (1.5-mL)
Pasteur pipettes (glass)
Pipette tips (P200)
Rotating platform (e.g., NUTATOR mixer)
Stereomicroscope

METHOD

Fluorescent Antibody Staining

1. Collect and fix the embryos according to Protocol 1 in Chapter 3.

   We typically fix embryos in a 1:1 solution of heptane (500 µL) and fixative solution (PEMFA, 500 µL) for 20 min.

2. After fixation, transfer the embryos to a 1.5-mL microcentrifuge tube and rinse with 100% methanol.

3. Rinse the embryos three times quickly with PBT. Block in PBT and rock on a rotating platform for 30 min.

4. Dilute an appropriate primary antibody in PBT. Add 300–400 µL of antibody solution to the embryos and rock on a rotating platform overnight at 4°C.

5. Wash the embryos six times in PBT for 10 min each. Rock on a rotating platform.

6. Dilute the fluorescent secondary antibody 1:100 in PBT and add to the embryos. Cover the microcentrifuge tubes with foil to protect the samples from light. Rock on a rotating platform for 1.5–2 h at room temperature.

7. Wash the embryos six times in PTx for 10 min each. Cover the tubes with foil to protect the samples from light.

8. Remove the PTx from each tube. Add 80–100 µL of antifade mounting medium.

   Embryos can mounted on slides immediately or they can be protected from light and stored overnight at 4°C.

9. Cut the tip off a 200-µL pipette tip. Transfer ~90 µL of embryos in antifade mounting medium to a glass slide and cover with a coverslip.

Histochemical Antibody Staining Using the DAB Substrate Kit

10. Follow Steps 1–5 described above.

11. Dilute the HRP-conjugated secondary antibody 1:100 in PBT and add to the embryos. Rock on a rotating platform for 2 h at room temperature.

12. Wash the embryos six times in PTx for 10 min each.

13. Prepare a 10% DAB solution in 1x stable peroxide substrate buffer (included in the DAB substrate kit).

14. Transfer the embryos to a three-well glass dissection dish. Rinse once with 1x stable peroxide substrate buffer.

15. Remove the 1x stable peroxide substrate buffer and add ~200 µL of 10% DAB solution per well of embryos. Monitor the staining reaction using a stereomicroscope.

16. When the desired level of staining is attained, rinse the embryos three times quickly in PTx to stop the DAB reaction. Transfer the embryos to a new 1.5-mL microcentrifuge tube.

17. Remove the PTx and add 80% glycerol. Let the embryos settle overnight at 4°C.

    Embryos are easier to "roll" between the coverslip and slide if stored overnight in glycerol.

18. Cut the tip off of a 200-µL pipette tip. Transfer ~90 µL of embryos in glycerol to a glass slide and cover with a coverslip.

RECIPES

CAUTION: See Appendix for proper handling of materials marked with <!>.
Recipes for reagents marked with <R> are included in this list.

Antifade Mounting Medium

Reagent	Quantity (for 50 mL)
Tris-HCl (0.2 M; pH 8.5)	5 mL
n-propyl gallate (NPG) <!>	2.5 g
Glycerol	45 mL

Combine the reagents in a 50-mL conical tube. To dissolve the NPG, rock by hand under a stream of hot water from the faucet. Aliquot in 1.5-mL microcentrifuge tubes. Store at −20°C.

PBS Stock Solution (10x; pH 7.4)

Reagent	Quantity (for 1 L)	Final concentration
NaCl	80 g	1.37 M
KCl <!>	2 g	27 mM
Na_2HPO_4	14.4 g	101.44 mM
KH_2PO_4	2.4 g	17.64 mM

Dissolve the components in dH_2O while stirring continuously, and adjust the pH to 7.4.

PBT (pH 7.4)

Reagent	Quantity (for 500 mL)	Final concentration
PBS stock solution (10x; pH 7.4) <R>	50 mL	1x
Bovine serum albumin (BSA)	5 g	1% (w/v)
Triton X-100 <!>	500 µL	0.1% (v/v)

Add the PBS stock solution to 400 mL of dH_2O. Adjust the pH to 7.4. Add the other reagents and dH_2O to give a final volume of 500 mL. Stir until the BSA dissolves. Store at 4°C.

PEMFA

Reagent	Final concentration
PIPES (pH 6.9)	100 mM
EGTA	2 mM
$MgSO_4$ <!>	1 mM
Formaldehyde <!>	4% (w/v)

The PEMFA solution must be made at the time of use. However, the fixative buffer (PEM) without formaldehyde (FA) can be made ahead of time and stored. Prepare 100 mL of

PEM as a stock solution and store at 4°C. When you are ready to fix embryos, add FA to PEM, at a final concentration of 4%. The addition of FA to PEM will not significantly change the final concentration of PEM. You will need 500 μL of PEMFA per 1.5 mL microfuge tube of embryos.

PTx (pH 7.4)

Reagent	Quantity (for 1 L)	Final concentration
PBS stock solution (10x; pH 7.4) <R>	100 mL	1x
Triton X-100 <!>	1 mL	0.1% (w/v)

Add the PBS stock solution to 950 mL of dH₂O. Adjust the pH to 7.4. Add the Triton X-100 and dH₂O to give a final volume of 1 L.

REFERENCES

Abbott NJ. 2005. Dynamics of CNS barriers: Evolution, differentiation, and modulation. *Cell Mol Neurobiol* **25:** 5–23.

Aigouy B, Van de Bor V, Boeglin M, Giangrande A. 2004. Time-lapse and cell ablation reveal the role of cell interactions in fly glia migration and proliferation. *Development* **131:** 5127–5138.

Alfonso TB, Jones BW. 2002. *gcm2* promotes glial cell differentiation and is required with *glial cells missing* for macrophage development in *Drosophila*. *Dev Biol* **248:** 369–383.

Auld VJ, Fetter RD, Broadie K, Goodman CS. 1995. Gliotactin, a novel transmembrane protein on peripheral glia, is required to form the blood-nerve barrier in *Drosophila*. *Cell* **81:** 757–767.

Awasaki T, Ito K. 2004. Engulfing action of glial cells is required for programmed axon pruning during *Drosophila* metamorphosis. *Curr Biol* **14:** 668–677.

Awasaki T, Lai SL, Ito K, Lee T. 2008. Organization and postembryonic development of glial cells in the adult central brain of *Drosophila*. *J Neurosci* **28:** 13742–13753.

Bachmann A, Timmer M, Sierralta J, Pietrini G, Gundelfinger ED, Knust E, Thomas U. 2004. Cell type-specific recruitment of *Drosophila* Lin-7 to distinct MAGUK-based protein complexes defines novel roles for Sdt and Dlg-S97. *J Cell Sci* **117:** 1899–1909.

Bainton RJ, Tsai LT, Schwabe T, DeSalvo M, Gaul U, Heberlein U. 2005. *moody* encodes two GPCRs that regulate cocaine behaviors and blood-brain barrier permeability in *Drosophila*. *Cell* **123:** 145–156.

Banerjee S, Pillai AM, Paik R, Li J, Bhat MA. 2006. Axonal ensheathment and septate junction formation in the peripheral nervous system of *Drosophila*. *J Neurosci* **26:** 3319–3329.

Barres BA. 2008. The mystery and magic of glia: A perspective on their roles in health and disease. *Neuron* **60:** 430–440.

Basler K, Struhl G. 1994. Compartment boundaries and the control of *Drosophila* limb pattern by hedgehog protein. *Nature* **368:** 208–214.

Baumgartner S, Littleton JT, Broadie K, Bhat MA, Harbecke R, Lengyel JA, Chiquet-Ehrismann R, Prokop A, Bellen HJ. 1996. A *Drosophila* neurexin is required for septate junction and blood-nerve barrier formation and function. *Cell* **87:** 1059–1068.

Beckervordersandforth RM, Rickert C, Altenhein B, Technau GM. 2008. Subtypes of glial cells in the *Drosophila* embryonic ventral nerve cord as related to lineage and gene expression. *Mech Dev* **125:** 542–557.

Booth GE, Kinrade EF, Hidalgo A. 2000. Glia maintain follower neuron survival during *Drosophila* CNS development. *Development* **127:** 237–244.

Bossing T, Udolph G, Doe CQ, Technau GM. 1996. The embryonic central nervous system lineages of *Drosophila melanogaster*. I. Neuroblast lineages derived from the ventral half of the neuroectoderm. *Dev Biol* **179:** 41–64.

Brand AH, Perrimon N. 1993. Targeted gene expression as a means of altering cell fates and generating dominant phenotypes. *Development* **118:** 401–415.

Campbell G, Goring H, Lin T, Spana E, Andersson S, Doe CQ, Tomlinson A. 1994. RK2, a glial-specific homeodomain protein required for embryonic nerve cord condensation and viability in *Drosophila*. *Development* **120:** 2957–2966.

Campos-Ortega JA, Hartenstein V. 1997. *The embryonic development of* Drosophila melanogaster. Springer-Verlag, Berlin.

Chotard C, Leung W, Salecker I. 2005. *glial cells missing* and *gcm2* cell autonomously regulate both glial and neuronal development in the visual system of *Drosophila. Neuron* **48:** 237–251.

Christopherson KS, Ullian EM, Stokes CC, Mullowney CE, Hell JW, Agah A, Lawler J, Mosher DF, Bornstein P, Barres BA. 2005. Thrombospondins are astrocyte-secreted proteins that promote CNS synaptogenesis. *Cell* **120:** 421–433.

Danbolt NC. 2001. Glutamate uptake. *Prog Neurobiol* **65:** 1–105.

Dietzl G, Chen D, Schnorrer F, Su KC, Barinova Y, Fellner M, Gasser B, Kinsey K, Oppel S, Scheiblauer S, et al. 2007. A genome-wide transgenic RNAi library for conditional gene inactivation in *Drosophila. Nature* **448:** 151–156.

Doe CQ. 1992. Molecular markers for identified neuroblasts and ganglion mother cells in the *Drosophila* central nervous system. *Development* **116:** 855–863.

Doherty J, Logan MA, Tasdemir OE, Freeman MR. 2009. Ensheathing glia function as phagocytes in the adult *Drosophila* brain. *J Neurosci* **29:** 4768–4781.

Dumstrei K, Wang F, Hartenstein V. 2003. Role of DE-cadherin in neuroblast proliferation, neural morphogenesis, and axon tract formation in *Drosophila* larval brain development. *J Neurosci* **23:** 3325–3335.

Ebens AJ, Garren H, Cheyette BN, Zipursky SL. 1993. The *Drosophila* anachronism locus: A glycoprotein secreted by glia inhibits neuroblast proliferation. *Cell* **74:** 15–27.

Edenfeld G, Altenhein B, Zierau A, Cleppien D, Krukkert K, Technau G, Klambt C. 2007. Notch and Numb are required for normal migration of peripheral glia in *Drosophila. Dev Biol* **301:** 27–37.

Edenfeld G, Volohonsky G, Krukkert K, Naffin E, Lammel U, Grimm A, Engelen D, Reuveny A, Volk T, Klambt C. 2006. The splicing factor crooked neck associates with the RNA-binding protein HOW to control glial cell maturation in *Drosophila. Neuron* **52:** 969–980.

Edwards JS, Swales LS, Bate M. 1993. The differentiation between neuroglia and connective tissue sheath in insect ganglia revisited: The neural lamella and perineurial sheath cells are absent in a mesodermless mutant of *Drosophila. J Comp Neurol* **333:** 301–308.

Fire A, Xu S, Montgomery MK, Kostas SA, Driver SE, Mello CC. 1998. Potent and specific genetic interference by double-stranded RNA in *Caenorhabditis elegans. Nature* **391:** 806–811.

Franc NC, Heitzler P, Ezekowitz RA, White K. 1999. Requirement for croquemort in phagocytosis of apoptotic cells in *Drosophila. Science* **284:** 1991–1994.

Freeman MR, Doe CQ. 2001. Asymmetric Prospero localization is required to generate mixed neuronal/glial lineages in the *Drosophila* CNS. *Development* **128:** 4103–4112.

Freeman MR, Delrow J, Kim J, Johnson E, Doe CQ. 2003. Unwrapping glial biology: Gcm target genes regulating glial development, diversification, and function. *Neuron* **38:** 567–580.

Gilmour DT, Maischein HM, Nusslein-Volhard C. 2002. Migration and function of a glial subtype in the vertebrate peripheral nervous system. *Neuron* **34:** 577–588.

Golic KG, Lindquist S. 1989. The FLP recombinase of yeast catalyzes site-specific recombination in the *Drosophila* genome. *Cell* **59:** 499–509.

Gordon MD, Scott K. 2009. Motor control in a *Drosophila* taste circuit. *Neuron* **61:** 373–384.

Halter DA, Urban J, Rickert C, Ner SS, Ito K, Travers AA, Technau GM. 1995. The homeobox gene *repo* is required for the differentiation and maintenance of glia function in the embryonic nervous system of *Drosophila melanogaster. Development* **121:** 317–332.

Hayashi S, Ito K, Sado Y, Taniguchi M, Akimoto A, Takeuchi H, Aigaki T, Matsuzaki F, Nakagoshi H, Tanimura T, et al. 2002. GETDB, a database compiling expression patterns and molecular locations of a collection of Gal4 enhancer traps. *Genesis* **34:** 58–61.

Hidalgo A, Booth GE. 2000. Glia dictate pioneer axon trajectories in the *Drosophila* embryonic CNS. *Development* **127:** 393–402.

Higashijima S, Shishido E, Matsuzaki M, Saigo K. 1996. *eagle*, a member of the steroid receptor gene superfamily, is expressed in a subset of neuroblasts and regulates the fate of their putative progeny in the *Drosophila* CNS. *Development* **122:** 527–536.

Ito K, Awano W, Suzuki K, Hiromi Y, Yamamoto D. 1997. The *Drosophila* mushroom body is a quadruple structure of clonal units each of which contains a virtually identical set of neurones and glial cells. *Development* **124:** 761–771.

Ito K, Urban J, Technau GM. 1995. Distribution, classification and development of *Drosophila* glial cells during late embryogenesis. *Roux's Arch Dev Biol* **204:** 284–307.

Jacobs JR. 2000. The midline glia of *Drosophila*: A molecular genetic model for the developmental functions of glia. *Prog Neurobiol* **62:** 475–508.

Jacobs JR, Hiromi Y, Patel NH, Goodman CS. 1989. Lineage, migration, and morphogenesis of longitudinal glia in the *Drosophila* CNS as revealed by a molecular lineage marker. *Neuron* **2:** 1625–1631.

Kimelberg HK. 2009. Astrocyte heterogeneity or homogeneity? In *Astrocytes in (patho)physiology of the nervous system* (ed. V Parpura, PG Haydon), pp. 1–26. Springer Science+Business Media, LLC, New York.

Kofuji P, Newman EA. 2009. Regulation of potassium by glial cells in the central nervous system. In *Astrocytes in (patho)physiology of the nervous system* (ed. V Parpura, PG Haydon), pp. 151–176. Springer Science and Business Media, LLC, New York.

Kulkarni MM, Booker M, Silver SJ, Friedman A, Hong P, Perrimon N, Mathey-Prevot B. 2006. Evidence of off-target effects associated with long dsRNAs in *Drosophila melanogaster* cell-based assays. *Nat Methods* **3:** 833–838.

Kurant E, Axelrod S, Leaman D, Gaul U. 2008. Six-microns-under acts upstream of Draper in the glial phagocytosis of apoptotic neurons. *Cell* **133:** 498–509.

Lai SL, Lee T. 2006. Genetic mosaic with dual binary transcriptional systems in *Drosophila*. *Nat Neurosci* **9:** 703–709.

Lamb RS, Ward RE, Schweizer L, Fehon RG. 1998. *Drosophila coracle*, a member of the protein 4.1 superfamily, has essential structural functions in the septate junctions and developmental functions in embryonic and adult epithelial cells. *Mol Biol Cell* **9:** 3505–3519.

Lee BP, Jones BW. 2005. Transcriptional regulation of the *Drosophila* glial gene repo. *Mech Dev* **122:** 849–862.

Lee T, Luo L. 1999. Mosaic analysis with a repressible cell marker for studies of gene function in neuronal morphogenesis. *Neuron* **22:** 451–461.

Lee T, Winter C, Marticke SS, Lee A, Luo L. 2000. Essential roles of *Drosophila* RhoA in the regulation of neuroblast proliferation and dendritic but not axonal morphogenesis. *Neuron* **25:** 307–316.

Lee YS, Carthew RW. 2003. Making a better RNAi vector for *Drosophila*: Use of intron spacers. *Methods* **30:** 322–329.

Leiserson WM, Harkins EW, Keshishian H. 2000. Fray, a *Drosophila* serine/threonine kinase homologous to mammalian PASK, is required for axonal ensheathment. *Neuron* **28:** 793–806.

Ma Y, Creanga A, Lum L, Beachy PA. 2006. Prevalence of off-target effects in *Drosophila* RNA interference screens. *Nature* **443:** 359–363.

Mayer F, Mayer N, Chinn L, Pinsonneault RL, Kroetz D, Bainton RJ. 2009. Evolutionary conservation of vertebrate blood-brain barrier chemoprotective mechanisms in *Drosophila*. *J Neurosci* **29:** 3538–3550.

Moffat J, Reiling JH, Sabatini DM. 2007. Off-target effects associated with long dsRNAs in *Drosophila* RNAi screens. *Trends Pharmacol Sci* **28:** 149–151.

Nave KA, Salzer, JL. 2006. Axonal regulation of myelination by neuregulin 1. *Curr Opin Neurobiol* **16:** 492–500.

Nave KA, Trapp BD. 2008. Axon-glial signaling and the glial support of axon function. *Annu Rev Neurosci* **31:** 535–561.

Ni JQ, Markstein M, Binari R, Pfeiffer B, Liu LP, Villalta C, Booker M, Perkins L, Perrimon N. 2008. Vector and parameters for targeted transgenic RNA interference in *Drosophila melanogaster*. *Nat Methods* **5:** 49–51.

Oland LA, Tolbert LP. 2003. Key interactions between neurons and glial cells during neural development in insects. *Annu Rev Entomol* **48:** 89–110.

Paladi M, Tepass U. 2004. Function of Rho GTPases in embryonic blood cell migration in *Drosophila*. *J Cell Sci* **117:** 6313–6326.

Pereanu W, Shy D, Hartenstein V. 2005. Morphogenesis and proliferation of the larval brain glia in *Drosophila*. *Dev Biol* **283:** 191–203.

Pignoni F, Zipursky SL. 1997. Induction of *Drosophila* eye development by Decapentaplegic. *Development* **124:** 271–278.

Prokop A. 1999. Integrating bits and pieces: Synapse structure and formation in *Drosophila* embryos. *Cell Tissue Res* **297:** 169–186.

Rival T, Soustelle L, Cattaert D, Strambi C, Iche M, Birman S. 2006. Physiological requirement for the glutamate transporter dEAAT1 at the adult *Drosophila* neuromuscular junction. *J Neurobiol* **66:** 1061–1074.

Rival T, Soustelle L, Strambi C, Besson MT, Iche M, Birman S. 2004. Decreasing glutamate buffering capacity triggers oxidative stress and neuropil degeneration in the *Drosophila* brain. *Curr Biol* **14:** 599–605.

Rogulja-Ortmann A, Luer K, Seibert J, Rickert C, Technau GM. 2007. Programmed cell death in the embryonic central nervous system of *Drosophila melanogaster*. *Development* **134:** 105–116.

Rooke JE, Theodosiou NA, Xu T. 2000. Clonal analysis in the examination of gene function in *Drosophila*. *Methods Mol Biol* **137:** 15–22.

Schmid A, Chiba A, Doe CQ. 1999. Clonal analysis of *Drosophila* embryonic neuroblasts: Neural cell types, axon projections and muscle targets. *Development* **126:** 4653–4689.

Schmidt H, Rickert C, Bossing T, Vef O, Urban J, Technau GM. 1997. The embryonic central nervous system lineages of *Drosophila melanogaster*. II. Neuroblast lineages derived from the dorsal part of the neuroectoderm. *Dev Biol* **189:** 186–204.

Schwabe T, Bainton RJ, Fetter RD, Heberlein U, Gaul U. 2005. GPCR signaling is required for blood-brain barrier formation in *Drosophila*. *Cell* **123:** 133–144.

Sepp KJ, Auld VJ. 1999. Conversion of lacZ enhancer trap lines to *GAL4* lines using targeted transposition in *Drosophila melanogaster*. *Genetics* **151:** 1093–1101.

Sepp KJ, Auld VJ. 2003. Reciprocal interactions between neurons and glia are required for *Drosophila* peripheral nervous system development. *J Neurosci* **23:** 8221–8230.

Sepp KJ, Schulte J, Auld VJ. 2001. Peripheral glia direct axon guidance across the CNS/PNS transition zone. *Dev Biol* **238:** 47–63.

Shang Y, Griffith LC, Rosbash M. 2008. Light-arousal and circadian photoreception circuits intersect at the large PDF cells of the *Drosophila* brain. *Proc Natl Acad Sci* **105:** 19587–19594.

Shishido E, Ono N, Kojima T, Saigo K. 1997. Requirements of DFR1/Heartless, a mesoderm-specific *Drosophila* FGF-receptor, for the formation of heart, visceral and somatic muscles, and ensheathing of longitudinal axon tracts in CNS. *Development* **124:** 2119–2128.

Silies M, Yuva Y, Engelen D, Aho A, Stork T, Klambt C. 2007. Glial cell migration in the eye disc. *J Neurosci* **27:** 13130–13139.

Sonnenfeld MJ, Jacobs JR. 1995. Macrophages and glia participate in the removal of apoptotic neurons from the *Drosophila* embryonic nervous system. *J Comp Neurol* **359:** 644–652.

Stevens B, Allen NJ, Vazquez LE, Howell GR, Christopherson KS, Nouri N, Micheva KD, Mehalow AK, Huberman AD, Stafford B, et al. 2007. The classical complement cascade mediates CNS synapse elimination. *Cell* **131:** 1164–1178.

Stork T, Engelen D, Krudewig A, Silies M, Bainton RJ, Klambt C. 2008. Organization and function of the blood-brain barrier in *Drosophila*. *J Neurosci* **28:** 587–597.

Struhl G, Basler K. 1993. Organizing activity of wingless protein in *Drosophila*. *Cell* **72:** 527–540.

Sun B, Xu P, Salvaterra PM. 1999. Dynamic visualization of nervous system in live *Drosophila*. *Proc Natl Acad Sci* **96:** 10438–10443.

Theodosiou N A, Xu T. 1998. Use of FLP/FRT system to study *Drosophila* development. *Methods* **14:** 355–365.

Thomas GB, van Meyel DJ. 2007. The glycosyltransferase Fringe promotes Delta-Notch signaling between neurons and glia, and is required for subtype-specific glial gene expression. *Development* **134:** 591–600.

von Hilchen CM, Beckervordersandforth RM, Rickert C, Technau GM, Altenhein B. 2008. Identity, origin, and migration of peripheral glial cells in the *Drosophila* embryo. *Mech Dev* **125:** 337–352.

Watts RJ, Schuldiner O, Perrino J, Larsen C, Luo L. 2004. Glia engulf degenerating axons during developmental axon pruning. *Curr Biol* **14:** 678–684.

Wong AM, Wang JW, Axel R. 2002. Spatial representation of the glomerular map in the *Drosophila* protocerebrum. *Cell* **109:** 229–241.

Xiong WC, Montell C. 1995. Defective glia induce neuronal apoptosis in the *repo* visual system of *Drosophila*. *Neuron* **14:** 581–590.

6

Methods for Exploring the Genetic Control of Sensory Neuron Dendrite Morphogenesis

Brikha R. Shrestha[1] and Wesley B. Grueber[1,2]

[1]Department of Neuroscience, Columbia University Medical Center, New York, New York 10032; [2]Department of Physiology and Cellular Biophysics, Columbia University Medical Center, New York, New York 10032

ABSTRACT

Dendrite morphological diversity helps to define the properties of neural circuits by influencing circuit organization and information processing. Dendrite development has been shown to be driven by a combination of cell-intrinsic and -extrinsic factors. However, much remains to be discovered about the cellular and molecular mechanisms that give rise to unique yet highly stereotyped dendrite arbors of diverse neuronal types. Work in the last decade has established the *Drosophila* system as an excellent model for studies of dendrite morphogenesis. Work has been performed primarily in three systems: the dendritic arborization sensory neurons, the adult olfactory system, and the embryonic and larval motor system. Each of these systems offers a manageable number of neurons that are easily accessible, both optically and physically, and can be genetically manipulated to address a broad range of questions regarding dendrite development and patterning. Here, we provide a brief overview of the organization of these systems together with aspects of dendrite development that have been elucidated in recent years and methods for studying dendrite morphogenesis in embryonic and larval sensory neurons.

INTRODUCTION

Neurons are highly polarized cells that consist of dendrites, which are specialized to receive and process inputs, and axons, which are specialized to transfer information to other neurons or nonneuronal cells. In drawings of Golgi-stained neurons first published more than a century ago, Santiago Ramón y Cajal illustrated the critical differences in morphology between dendrites and axons as well as the tremendous diversity in dendritic arborization patterns (Ramón y Cajal 1995). Addressing how diversity in dendritic morphology arises during development was prohibitively difficult in *Drosophila* until the last decade or so. Two major technical achievements occurred that opened the door to detailed genetic analysis of dendrite patterning, and these continue to be critical methodologies common to nearly all ensuing studies.

First, live imaging strategies were established for carrying out large-scale forward genetic screens for genes regulating dendrite development in embryonic stages (Gao et al. 1999). This screening strategy and subsequent screens using analogous methodology have led to the identification of hundreds of new candidate loci affecting dendrite morphology (Gao et al. 1999; Parrish et al. 2006; Grueber et al. 2007; Zheng et al. 2008). Second, the mosaic analysis with a repressible cell marker (MARCM) approach was developed, which permits resolution of dendrites at a single-cell level and genetic manipulation of individual neurons to assess gene function during neuronal morphogenesis (Lee and Luo 1999; Wu and Luo 2006). This system has led to rapid progress in understanding the cell-autonomous basis of dendritic morphological development. In recent years, scientists have begun to identify the factors that determine the complexities and shapes of particular dendritic arbors, thus piecing together the molecular codes that generate different arbor types. Below, we provide brief introductions to the different neuronal systems being studied in *Drosophila*, as we discuss key transcription factors, cell adhesion molecules, and signaling proteins that are known to be important for dendrite morphogenesis in those neurons.

MAJOR SYSTEMS IN WHICH DENDRITE MORPHOGENESIS IS STUDIED IN *DROSOPHILA*

Multidendritic Neurons of the Larval Peripheral Nervous System

Forward genetic screens for dendrite morphology mutants were first performed using the multidendritic (md) group of neurons that lie just beneath the larval body wall. The key experimental advantage of this group of neurons is their accessibility for immunostaining, imaging, and genetic manipulation. The md neurons comprise about half of the 45 sensory neurons in each abdominal hemisegment of *Drosophila* embryos and larvae, the other major types being the external sensory (es) neurons and chordotonal (ch) neurons. Each es and ch neuron has a single unbranched dendrite, whereas md neurons have more highly branched arborizations. The md neurons have been categorized into three subtypes based on their morphology and substrate: The most numerous are the dendritic arborization (da) neurons, but, in addition, there are bipolar dendrite (bd) neurons and tracheal dendrite (td) neurons (Bodmer et al. 1987). The 15 da neurons in each hemisegment are further divided into classes I–IV based on the increasing complexities of their dendritic arbors (Grueber et al. 2002). The diversity in their dendrite arborization patterns and their accessible position along the body wall make the da neurons of the larval peripheral nervous system (PNS) an excellent system for studying the control of dendrite branching and territory formation (Gao et al. 1999; Sweeney et al. 2002; Grueber et al. 2003b; Orgogozo and Grueber 2005; Corty et al. 2009).

Olfactory Projection Neurons

The mechanisms of dendrite development have been studied extensively in the context of olfactory circuit formation. Olfactory information received by olfactory receptor neurons (ORNs) in antennae and maxillary palps is relayed to projection neurons (PNs) at the antennal lobe (AL). At the AL, ORN axons synapse with PN dendrites at approximately 50 discrete glomeruli. There are three major lineages arranged around the AL that produce adult PNs: the anterodorsal (adPNs), lateral (lPNs), and ventral (vPNs) lineages. For information to be transferred properly at the AL, PNs from these lineages must project dendrites to distinct and stereotyped glomeruli. Developmental studies indicate that dendrites project to protoglomeruli before ORN axon arrival, suggesting an important role for dendrite targeting in AL wiring (Jefferis et al. 2004). Dendrite targeting has been shown to depend on three major mechanisms: (1) graded cues that determine general dendrite positioning in the AL (Komiyama et al. 2007); (2) intrinsic combinatorial codes of transcription factors that impact precise glomerular targeting (Komiyama et al. 2003; Komiyama and Luo 2007); and (3) dendrite–dendrite interactions that determine glomerular coverage (Zhu and Luo 2004; Zhu et al. 2006).

Motor Neurons

The *Drosophila* embryonic motor system is comprised of a segmentally repeated array of 36 motor neurons per half-segment of the central nervous system (CNS) (Landgraf and Thor 2006). Motor neuron dendrites are situated in the dorsal region of the neuropil segregated from the sensory afferents that project to more ventral regions. Pairwise retrograde labeling of motor neurons revealed that the dendrites of motor neurons are arranged in a "myotopic map," in which dendritic territories represent centrally the location of cognate axon terminals in the peripheral muscle field (Landgraf et al. 2003). Different types of motor neurons show distinct responses to midline cues. Some dendritic arbors remain ipsilateral, some project to the contralateral neuropil, and others have both ipsilateral and contralateral arborizations. The motor system has thus provided important insights into how dendritic targeting is regulated by midline-derived guidance signals.

MOLECULAR CONTROL OF DENDRITE MORPHOGENESIS

Dendrite Branching Diversity

Genetic screens and studies of candidate genes have begun to reveal how dendrites develop characteristic branching morphologies. Sensory neurons generate either single-dendrite or multiple-dendrite morphologies in the PNS. The alternate dendrite morphologies appear to be specified by the gene *hamlet* (*ham*), which encodes a zinc-finger protein that is required for a single-dendrite fate (Moore et al. 2002). Loss-of-function analysis using the MARCM system and gain-of-function approaches have identified a group of transcription factors, including Cut, Abrupt, Spineless, and Knot, that in turn help to specify class-specific branching complexities in da neurons (Grueber et al. 2003a; Li et al. 2004; Sugimura et al. 2004; Kim et al. 2006; Hattori et al. 2007; Jinushi-Nakao et al. 2007; Crozatier and Vincent 2008). The factors that are regulated by these various transcriptional regulators are largely unknown.

The microtubule cytoskeleton plays a critical role in regulating dendritic branch patterning. Dendritic microtubules are either mixed plus-end distal and minus-end distal (in vertebrates) or primarily minus-end distal (several types of neurons in *Drosophila*) (Sharp et al. 1997; Stone et al. 2008). Forward genetic screens have identified components of the minus-end-directed microtubule motor dynein as having a critical role in the patterning of branches along da neuron dendrite arbors (Satoh et al. 2008; Zheng et al. 2008). Dynein appears to deliver cargoes to distal dendrites that are necessary for dendrite branching, including fractions of a satellite secretory system, the Golgi outposts (Ye et al. 2007), and endosomes (Satoh et al. 2008).

Specifying Dendritic Targeting

The territories of dendrites define the region from which they can receive input; thus an important focus of current research in the field is to identify how dendritic territories are specified. This problem can be separated into two major questions: First, how are dendrites targeted to particular regions of the nervous system and, second, once dendritic targeting is specified, how is dendrite growth limited so that branches grow into appropriate territories?

Significant progress in this area has been made in *Drosophila* motor and olfactory systems. In the olfactory system, combinatorial expression of different transcription factors appears to provide an instructive code for diverse dendritic targeting by mediating both general and glomerular-specific targeting (Komiyama et al. 2003, 2007). Both PN dendrite and motor dendrite targeting also require guidance cues that have been shown to be important for axonal targeting, including semaphorins, Slit/Robo and Netrin/Frazzled (Komiyama et al. 2007a). Although Sema proteins typically act as ligands for plexin receptors in *Drosophila*, the use of MARCM to verify cell-autonomous requirements suggests that Sema-1a acts in this case as a receptor (the identity of the presumptive ligand is as yet unknown) (Komiyama et al. 2007a). The midline guidance cues Slit and Netrin act as repulsive and attractive cues, respectively, for axons in the CNS (for review, see Garbe and Bashaw 2004), and play analogous roles in the guidance of motor dendrites (Furrer et al. 2003, 2007).

Self-Avoidance and Tiling

Two principles of dendritic territory patterning that have been studied in flies are self-avoidance and tiling. Self-avoidance refers to the capacity of sister dendrites to repel each other and ensure nonredundant territory coverage. Tiling is an analogous phenomenon that occurs between different cells within the same functional class to achieve complete, but nonredundant, coverage of a receptive area. Clones lacking functional Dscam (Down syndrome cell adhesion molecule) transmembrane receptors show disrupted self-avoidance (Hughes et al. 2007; Matthews et al. 2007; Soba et al. 2007). Self-avoidance is also controlled in at least some da neurons by the Tricornered (Trc) kinase and Furry (Fry). trc and fry mutant neurons show self-crossing phenotypes, likely because of a defective like-repels-like turning response. Trc signaling similarly impacts tiling of dendritic arbors (Emoto et al. 2004).

Dendrite Remodeling

At the end of the larval stage, the nervous system of *Drosophila* undergoes tremendous changes to coincide with the different life history and remodeled body of the adult fly. Some larval-specific neurons perish, whereas others persist to become functional adult neurons after substantial remodeling of dendritic arbors. The widespread anatomical changes that occur during insect metamorphosis are orchestrated by two hormones, juvenile hormone (JH) and ecdysone. Metamorphic remodeling is observed among the larval da neurons, and their proximity to the body surface has allowed high-resolution analysis of the cellular and molecular events associated with dendritic pruning and growth (Williams and Truman 2004; Kuo et al. 2005; Williams and Truman 2005a,b; Kuo et al. 2006; Williams et al. 2006). Interestingly, this process involves selective loss of dendrites but little or no change in the axonal arbors. Williams and Truman followed single MARCM clones from larval to adult stages using two-photon microscopy to study morphological alterations in identified remodeling arbors (Williams and Truman 2004). The transition to adult forms involves removal of larval branches by local degeneration and branch retraction. Regrowth of the adult arbor involves laying down of the adult scaffold, which is accomplished by dynamic branch additions and retractions, and then filling in of the stabilized scaffold with higher-order branches. Ecdysone and JH are required, respectively, for pruning of dendritic branches by local degeneration and for controlling the balance between extension and retraction programs at different phases of arbor rebuilding (Williams and Truman 2004, 2005a).

Metamorphic remodeling of dendrites has been studied in several different neuronal types, including mushroom body neurons and olfactory projection neurons (Zheng et al. 2003; Marin et al. 2005). In the mushroom body, a genetic screen has identified the ecdysone receptor B1 isoform (EcR-B1) (Zheng et al. 2003) as critical for remodeling. The transforming growth factor β (TGF-β) type-I receptor Baboon is required for initiation of EcR-B1 expression and dendrite remodeling (Zheng et al. 2003; Marin et al. 2005).

METHODS FOR STUDYING DENDRITIC MORPHOGENESIS

The identification of molecules important for dendrite development has been aided greatly by recent advances in imaging and manipulation of single neurons in *Drosophila*. There are numerous techniques that are applicable to the study of dendrite biology. Many genetic studies described for the da neuron system analyze phenotypes in mutant embryos using transgenic markers of subsets of sensory neurons and also use mosaic analysis of loss-of-function phenotypes. Here we present our implementation of embryonic (Protocol 1) and larval (Protocol 2) techniques for studying wild-type and mutant da neuron dendritic arbors. In studies of da neuron dendritic development that use these methodologies, dendrites are imaged by confocal microscopy in embryonic and larval stages. The advantage of studies of embryonic stages is the relative ease of screening candidate genes and the ability to examine large numbers of neurons for phenotypes. Mosaic systems for examining mutant phenotypes are advantageous for studies of larval stages because of the ability to discern cell-autonomous phenotypes at very high resolution. However, there is a considerable time investment required for examining each individual gene of interest. In this chapter, we provide protocols that should allow interested researchers to use the da neurons as models for studying genetic and molecular determinants of dendrite patterning. We discuss these protocols and provide information that should help in determining the best experimental approach for a particular line of research.

QUANTIFICATION OF DENDRITIC ARBOR MORPHOLOGY

Quantification of dendritic arbor morphology is a critical step for understanding the role of a particular gene in dendrite morphogenesis. Some of the more commonly used morphometric parameters include total branch length, branch number, branch order, branch interval, and field size, which together provide a good starting point for any analysis of da neuron branching (Table 1). Sholl analysis is a simple, but powerful technique for determining the spatial geometry of a dendritic arbor. Parameters such as number and order of branches intersecting or enclosed between evenly spaced concentric circles/spheres centered at the soma are measured to quantify differences in spatial distribution of dendrite branching (e.g., see Libersat and Duch 2002; Uylings and van Pelt 2002; Zheng et al. 2008). Standard Sholl analysis can be adapted to reveal information about dendrite orientation by grouping morphometric parameters at each sphere or circle into cones or quadrants, respectively.

Several software packages have tools that allow measurements of these and numerous other parameters. Different software packages suit different budgets and needs. Programs such as ImageJ or Fiji offer powerful and platform-independent image analysis. Both ImageJ and Fiji are free for download and offer a large library of plug-ins for expansion, including NeuronJ for tracing and morphometric analysis, and the Simple Neurite Tracer plug-in for semiautomated tracing in Fiji. Neurolucida is a package commercially available from MBF Bioscience, Inc., which offers a large number of tools for reconstruction, mapping, and morphometric analysis of dendrites. Tracing and full reconstruction of dendritic arbors in Neurolucida is followed by quantitative analysis with Neurolucida Explorer, which is included with the Neurolucida package. Additionally, Amira software has an interactive automated three-dimensional tracing function that can be used effectively for reconstruction and analysis of some da neurons. The largest time investment in quantitative analysis is the phase of manual arbor reconstruction (Fig. 1). However, tools for automated and semiautomated reconstruction are likely to continue to improve in the coming years. Three-dimensional analysis of arbor geometry can also provide insights into the volume occupied by dendritic arbors,

TABLE 1. Publications that illustrate and exemplify the methods for quantification of different aspects of dendrite morphology

Dendrite property	Quantitative measure	Publication(s)
Branching	Branch order, number, length	Grueber et al. 2003a
Branch organization	Sholl analysis	Hartwig et al. 2008; Zheng et al. 2008a
Dynamics	Extension, retraction	Parrish et al. 2007
Maintenance	Branch points, branch length, extension, retraction	Emoto et al. 2006; Parrish et al. 2007
Remodeling	Extension, retraction, organization	Williams and Truman 2004; Kuo et al. 2006
Self-avoidance	Branch (tip) overlap	Hughes et al. 2007; Matthews et al. 2007; Soba et al. 2007
Targeting	Dendritic distribution	Komiyama and Luo 2007
Tiling/territory	Territory size, overlap	Grueber et al. 2002; Emoto et al. 2006

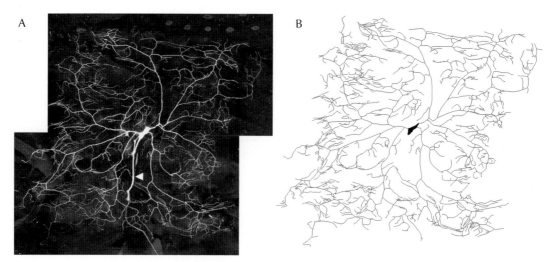

FIGURE 1. (*A*) A MARCM clone of ddaC, a class IV da neuron, was imaged on a confocal microscope. Four high-resolution images of different parts of the dendritic arbor were captured and merged using Adobe Photoshop to produce this image of the entire dendritic field. The arrowhead indicates the axon. (*B*) The neuron in *A* was traced in Neurolucida by hand using a drawing tablet.

the spatial relationships between arbors, and relationships between arbors and other neurons or tissue. For example, dendritic growth of different genetically manipulated motor neurons has been assessed using volumetric measurements (Hartwig et al. 2008). There are several software packages available to support three-dimensional analysis of confocal images, including Amira, Volocity, MetaMorph, and Fiji. We recommend trials of these and other software packages to determine which one(s) best fit your specific or general needs.

Image analysis software does not usually include tools for basic (or sophisticated) statistical analyses. Quantitative results can be exported to software with such a capability, including R, which is an open source environment for data analysis that allows simple and advanced statistical operations, or analyzed with commercially available statistical programs such as Stata (StataCorp LP), SPSS (SPSS, Inc.), and JMP (SAS Institute, Inc.).

Analysis of Dendrite Development in *Drosophila* Embryos

This method is a basic and indispensable component of the set of techniques used for screening genes involved in dendrite morphogenesis (Gao et al. 1999, 2000; Parrish et al. 2006; Grueber et al. 2007; Ye et al. 2007).

MATERIALS

CAUTION: See Appendix for proper handling of materials marked with <!>.
See the end of the chapter for recipes for reagents marked with <R>.

Reagents

Bleach (100%) <!>
ddH$_2$O
Embryo wash <R>
Glycerol (90% in PBS [phosphate-buffered saline] <R>)
Grape juice agar <R>

Equipment

Collection mesh inserts in a 12-well plate (e.g., Netwell Set from Electron Microscopy Sciences; cat. no. 64740-00)
Coverslips (22 x 50-mm)
Fly culture bottles or vials
Paintbrush or cotton swab
Pipettes (glass)
Rubber bulb
Slides (glass)

METHOD

The initial steps of this method are similar to those described in Chapters 3 and 4.

Collection of Embryos

1. Collect embryos in a plastic fly culture bottle with several small holes pierced on each side or in a vial containing grape juice agar.

2. Place approximately 50 flies in each collection bottle or vial. If using a collection bottle, seal it with a grape juice agar plate.

   The length of collection and aging of collected embryos will vary depending on the stage of dendrite development needed. Primary dendrite growth of multidendritic neurons initiates at 13 h after egg laying (AEL), followed by dorsal dendrite extension (until 15–16 h AEL), secondary branch extension (16–20 h AEL), and tertiary branch extension (20–24 h AEL) (Gao et al. 1999). Thus, enrichment of later embryonic stages is desirable for imaging dendrites. Long collections at

25°C (i.e., 15–20 h AEL) can be followed by an aging period (6–8 h) if late embryonic/early larval stages are required. Embryos within a narrower developmental window may be desired for
more precise timing of dendrite growth, in which case precollections can be performed and the
collection and aging period can be adjusted accordingly.

Processing of Embryos for Live Imaging

Allow 10–30 min for this procedure, depending on the number of lines.

3. Add 1x embryo wash to the collection bottle or agar vial and gently dislodge the embryos with
   a paintbrush or cotton swab. Pour the wash and the embryos into a collection mesh.

4. Rinse the embryos well with ddH$_2$O. Pat the underside of the mesh screen dry with a paper
   towel.

5. Place the mesh dishes into a 12-well plate and immerse each mesh in 100% bleach for 2 min to
   dechorionate the embryos.

6. Rinse the embryos with 1x embryo wash to remove all traces of bleach. Place the dish back into
   a clean well and fill with embryo wash.

7. Using a clean standard glass pipette, transfer all embryos to a 1.5-mL centrifuge tube and allow
   them to settle to the bottom of the tube.

8. Remove excess embryo wash from the tube, but keep a small amount covering the embryos.

9. Draw up 70–80 µL of 90% glycerol into a large orifice pipette tip.

   If only a narrow orifice tip is available, cut off ~0.5 cm of the pointed end of the tip before using.

10. Drop a small volume of glycerol onto the embryos (~20–30 µL) and take the embryos up into
    the tip of the pipette.

11. Expel just the embryos and a small amount of glycerol (not the entire contents of the pipette)
    onto the middle of a labeled glass slide.

    This will concentrate the embryos and make screening easier.

12. Repeat Step 10 to retrieve the remaining embryos from the centrifuge tube.

    It is useful to add these remaining embryos in a ring around the high concentration of embryos made
    in Step 11.

13. Cover the embryos with a 22 x 50-mm coverslip. The embryos are now ready for imaging (Fig. 2).

    See Troubleshooting.

TROUBLESHOOTING

There should be few complications with carrying out this protocol, but some that might be encountered are listed below.

Problem (Step 13): The embryos on the slide are damaged or shrunken.
Solution: This problem might arise if too little glycerol is used in Step 10 or if the embryos have been
    under the coverslip for too long. In general, do not prepare more slides than can be imaged within
    ~30 min. This number will vary depending on familiarity with the preparation. Try one or two
    slides initially; with experience, we can screen six genotypes in 60 min (including preparation time).

Problem (Step 13): Dendrites are not visible under the confocal microscope.
Solution: Because dendrites are often thin processes, they can be faint when viewed under the con-
    focal microscope, particularly relative to the cell body. To ensure that the arbors are imaged to
    the highest possible resolution, try imaging samples with increased pinhole, gain, or laser power.
    These might not be optimal imaging conditions for cell bodies.

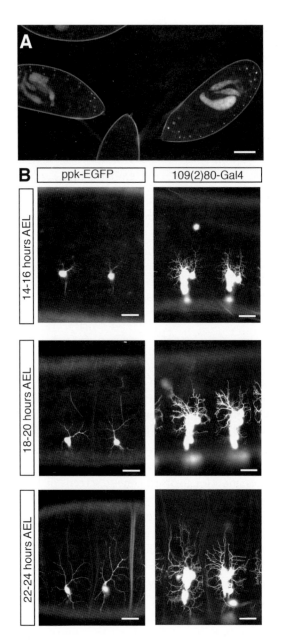

FIGURE 2. Live imaging of *Drosophila* sensory neuron dendrites in embryos. (*A*) A typical view through the confocal microscope after mounting embryos expressing the transgene *ppk-EGFP* 14–16 h after egg laying (AEL). Scale bar, 100 μm. (*B*) da neurons expressing the transgenes *ppk-EGFP* or *109(2)80Gal4> UAS-mcd8GFP* at various stages of development. Embryos were collected for 2 h and aged at 25°C for 14 h (*top*), 18 h (*middle*), or 22 h (*bottom*) before processing and mounting as described in Protocol 1. Scale bar, 20 μm. Dorsal is up and anterior is to the *left* in images. All images were acquired using either a 10x (*A*) or 40x (*B*) objective on a laser scanning confocal microscope.

Problem (Step 13): The embryos are in the wrong orientation.

Solution: Roll the embryos by pushing on one end of the coverslip until sensory neurons are clearly visible. If the embryos do not roll there may be too much glycerol on the slide or too many animals in the larval stage, which can prevent the coverslip from resting close to the embryos.

DISCUSSION

This assay is a good first approach for examining roles for genes in dendrite morphogenesis. The power of the approach is its simplicity and ease of implementation. Researchers should quickly become familiar with the stages of morphogenesis, the look of different parts of the embryo (more information on this can be found in Ashburner et al. 2004), and the normal morphology of embryonic dendrites. They should then be able to assess deviations from this pattern. One drawback of the approach is that cell-autonomous versus nonautonomous effects on dendrites cannot be inferred at the outset. However, embryonic neurons are often imaged by the Gal4-UAS system; thus cell auton-

TABLE 2. Some Gal4 drivers commonly used for studying dendrite morphogenesis in the *Drosophila* peripheral nervous system

Gal4 line	Expression in da neuron				Advantages	Disadvantages
	I	II	III	IV		
109(2)80	●	●	●	●	Drives strong expression in all da neurons; comes on in da neurons before dendritogenesis commences (Gao et al. 1999)	Difficult to resolve arbors of different classes
221	●	○	○	◐	Comes on early	Widespread expression in the CNS
477	◐	○	○	●	Strong expression in class IV	Widespread expression in the CNS
C161	●	●	●	○	Drives strong expression in da neurons (Williams and Truman 2005a)	
ppk	○	○	◐	●	Strong in class IV, and no expression in the ventral nerve cord in larvae	

Expression levels: Strong ● Weak ◐ None ○

Further information about these and other Gal4 lines that drive expression in da neurons can be obtained from FlyPNS (http://www.normalesup.org/~vorgogoz/FlyPNS/page1.html) (Orgogozo and Grueber 2005). Abbreviations: da, dendritic arborization; CNS, central nervous system.

omy can be interpreted if phenotypes are rescued by introducing upstream activating sequence (UAS)-rescuing transgenes only in neurons (e.g., Gao et al. 2000). Another potential drawback is that several genetic crosses are required to introduce the mutation into the appropriate marked background, so there is more of a delay at the outset of the assay compared with the MARCM approach (see Protocol 2). However, the hands-on time required for MARCM analysis is significantly greater, so for screening genes of as-yet-unknown function, analysis of embryonic stages can be a good first step. The ease of this approach and the ability to examine genes that are required for viability also make embryonic stages good for forward genetic screens or RNA interference–based approaches in which many genotypes are examined for changes in dendrite morphologies (Gao et al. 1999, 2000; Parrish et al. 2006; Grueber et al. 2007; Ye et al. 2007). Descriptions of how to perform such screens are given in Gao et al. (1999) and Parrish et al. (2006).

One experimental consideration is the selection of the marker to be used. There are lines available that drive reporter expression in all md neurons, such as *Gal4-109(2)80* and *IG1-2*, as well as lines that drive expression in specific subsets of neurons, such as *Gal4-221*, *NP2225*, and *ppk-Gal4* (Table 2) (Gao et al. 1999; Grueber et al. 2003a,b; Sugimura et al. 2003). These drivers are expressed early enough to allow imaging of embryonic neurons. If mutant phenotypes are neuron- or class-specific, pan-md markers might not allow detection of the phenotype. In contrast, if a large number of md neurons are affected by a mutation, this might be easier to detect when all neurons are labeled. Thus, for different markers one might expect a trade-off between resolution and sensitivity. Finally, one drawback of imaging dendrites at embryonic stages is the difficulty in obtaining reliable quantitative data for certain morphological parameters, particularly when using pan-md labeling methods.

Protocol 2

Generation and Immunostaining of MARCM Clones

This protocol can be used to generate and label MARCM clones for the analysis of dendritic patterning and branching control.

MATERIALS

CAUTION: See Appendix for proper handling of materials marked with <!>.
See the end of the chapter for recipes for reagents marked with <R>.

Reagents

DPX mounting medium (Electron Microscopy Sciences; cat. no. 13510) <!>
Ethanol (200 proof; i.e., anhydrous) <!>
Grape juice agar <R>
Paraformaldehyde (PFA; 4% v/v) <!> <R>
PBS-TX <R>
Phosphate-buffered saline (PBS) <R>
Sylgard (184 Silicone Elastomer Kit; Dow Corning Corporation)
Xylenes <!>

Equipment

Circulating water bath
Coverslips (22 x 22-mm; poly-L-lysine-coated)
Dissection dishes (glass)
Fine dissecting scissors (e.g., Fine Science Tools, cat. no. 15000-00)
Fine-tipped forceps (e.g., Dumont #5, Fine Science Tools, cat. no. 11252-20)
Fluorescence dissecting microscope
Fly culture bottles
Insect pins (0.1-mm-diameter) (e.g., Fine Science Tools, cat. no. 26002-10)
Parafilm
Petri dishes (35 x 10-mm)
Slides (glass)
Snap-cap plastic tubes (e.g., Fisher Scientific, 14-956-1D)
Staining dishes
Tabletop platform shaker
Tube shaker/rotator

METHODS

Collection of Embryos and Heat Shock

Allow 10 h for this procedure.

1. Prepare an embryo collection chamber as described in Protocol 1. Place about 50 adult flies in the chamber and use at least 30 females to ensure that plenty of eggs are laid within a short collection period. Seal the chamber with a grape juice agar plate.

2. Collect the embryos for 3 h. Age the embryos for 4–5 h at 25°C before application of heat shock.

3. Place the grape juice agar plate on an inverted empty lid of the same size and wrap in two layers of Parafilm to make the join between the plates watertight. Place this "sandwich" in a ziplock plastic bag together with a weight.

4. Place the bag in a circulating water bath set at 38°C.

   The addition of the weight in Step 3 ensures that the plate remains submerged in the water bath.

5. Subject the embryos to heat shock either for 1 h at 38°C or for 30 min at 38°C, 30 min at 25°C, and 45 min back at 38°C.

   We tend to obtain more clones with the second paradigm, but it is not clear whether this is because of a more efficient heat shock or the broader time window within which cells can be caught in mitosis. These are good parameters to start your experiments.

   Animals that are heat shocked on a Wednesday evening are typically ready to select and dissect the following Monday.

Selection of Clones

Allow up to 2 h for this procedure, depending on experience and the number of larvae screened.

6. Pick crawling third-instar larvae with a pair of forceps and transfer them to a large glass dissection dish filled with 1x PBS.

   All larvae from one grape juice agar plate can usually be placed in a single dissection dish, but do not overcrowd. There should not be multiple layers of larvae in the dish.

7. Rinse the larvae several times with 1x PBS to clear food debris from the dish.

   Food or yeast can make the dish cloudy and hinder the identification of clones.

8. Select clones under a fluorescence dissecting microscope.

   This process can take a very long time, especially when just starting. It is essential to select clones knowing what genotypes will arise from the cross and what the expected fluorescence pattern in larvae will be.

   Using this heat-shock protocol, clones are usually observed in relative isolation, often one to three per segment in segments that have clones. We typically keep larvae with clones on the ventral side of the body separate from larvae with clones on the dorsal side of the body so that the dissection can be performed in the correct orientation.

   See Troubleshooting.

Dissection

Allow 10–12 min for this procedure.

9. Prepare the 4% PFA solution.

10. Place the larvae that were selected in Step 8 in a dissection dish containing Sylgard. Use at least six insect pins per animal.

    We aim to limit the time between dissection and fixation to 5–10 min and dissect no more than six animals per dissection dish.

11. Orient the larvae with side-of-interest down (i.e., dorsal side down if you want to observe or image dorsal neurons) and place one pin near the anterior end and one near the posterior end of each larva (Fig. 3A).

12. Cut a small slit at both anterior and posterior ends with fine dissecting scissors (see arrows in Fig. 3B).

13. Starting at the posterior slit, make an incision along the length of the animal, taking care not to damage the body wall.

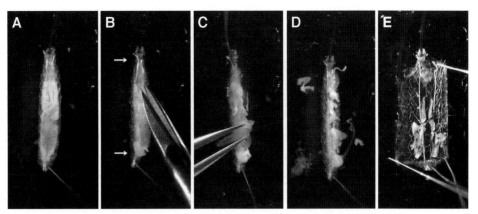

FIGURE 3. Snapshots showing the dissection of a third-instar larva. (*A*) The larva is pinned at the anterior and posterior ends. (*B*) A pair of fine scissors is used to make two horizontal incisions (see arrows) and one vertical (relative to the larva's length) incision. Ensure that the scissors do not penetrate too deeply into the animal while making the vertical incision. (*C*) To allow more accurate pinning, larval gut can be removed by pulling with fine forceps. (*D*) Fat bodies can be left behind. (*E*) The four corners are then pinned down to flatten the larval body wall.

It can be useful at this point to remove the excess gut tissue. However, be sure to avoid contact between the forceps and the body wall (Fig. 3C,D).

14. Pin the corners of the larvae. Flatten the preparation by pulling the corners away from the center of the animal as you go (Fig. 3E).

Fixation and Staining

Allow 24–48 h for this procedure.

15. After all dissections are complete, remove the PBS from the dissection dish and replace with enough 4% PFA to cover all the animals (~2–3 mL is usually enough). Place the dish on a tabletop platform shaker and fix the larvae for 15 min at room temperature.

16. Remove all pins.

17. Using forceps or a wide-mouth plastic transfer pipette, transfer the larvae to a 5-mL snap-cap tube filled with PBS-TX. Add no more than six larvae to each tube. Wash three times for 5 min each on a rotator.

18. For each tube, prepare 250 μL of 5% blocking solution (i.e., add 12.5 μL of normal serum that matches the species of the secondary antibody to 237.5 μL of PBS-TX). Remove the PBS-TX from the tube and replace with blocking solution. Rock for 1 h at 4°C.

19. Remove the blocking solution and add 250 μL of primary antibody (or antibodies) at an appropriate dilution. Rock overnight at 4°C.

    Volumes can be decreased for antibodies in limited supply.

20. Remove the primary antibody and fill the tubes with PBS-TX. Place the tubes on a rotator and wash in PBS-TX three to five times for at least 60 min in total.

21. After the final rinse, remove all PBS-TX and replace with 250 μL of secondary antibody in PBS-TX. Tubes can be wrapped in aluminum foil to avoid prolonged exposure to light. Rock overnight at 4°C. Alternatively, rock for 2 h at room temperature.

22. Rinse off the secondary antibody by washing in PBS-TX three to five times for at least 60 min in total.

Mounting

Allow about 1 h for this procedure.

23. Transfer the larval preparations from each tube to a dish filled with 1x PBS. Place a poly-L-lysine-coated coverslip under a dissecting microscope. (It may be useful to have the cap of the 5-mL tube under the coverslip so that it can be transferred easily at a later step.) Using a transfer pipette, place a few drops of PBS on the coverslip. Using forceps or a pipette, pick the preparations individually from the dish and place them on the coverslip, muscle side down and cuticle side up. The larval preparations should stick to the coverslip fairly well. Press down gently with the side of the forceps to flatten them.

24. Remove the PBS by wicking it away with a paper towel or tissue and immerse the larval preparation in 30% ethanol for 5 min. After this initial dehydration, the larvae should be securely affixed to the coverslip.

25. Place the coverslips vertically in a series of five staining dishes containing 50%, 70%, 95%, 100%, and 100% ethanol. Keep the coverslips immersed in each dish for 5 min.

26. Immerse the dehydrated larval preparations successively in two staining dishes containing xylenes for 10 min each.

    This makes the preparations optically clear for subsequent imaging.

27. Remove the larvae from the xylenes and, holding the coverslip with forceps, rapidly drizzle the preparations with DPX until they are covered. Place the coverslip on a glass slide, animal-side down.

28. After ~1 h, check to see that the DPX still completely covers the coverslip. Keep overnight on a slide holder with a weight (~30 g) placed on the coverslip.

    See Troubleshooting.

TROUBLESHOOTING

Problem (Step 8): Very few clones are obtained.
Solution: This problem could occur for several reasons. The heat shock may not be timed properly or the fluorescent dissection microscopes may not be optimal. Double-check your cross to verify genotypes. Check that you can visualize single neurons in *C155-Gal4, UAS-mCD8-GFP* heterozygotes to check the suitability of your microscope. Optimize the heat shock strategy for your equipment by trying heat shocks at different periods after egg laying is complete. For further troubleshooting advice on creating MARCM clones, see Wu and Luo (2006).

Problem (Step 28): Dendrites have excessive "blebbing" or otherwise appear unhealthy.
Solution: This problem may arise for several reasons, including damage during dissection or excessively long dissections.

Problem (Step 28): Preparations are cloudy or have high levels of autofluorescence.
Solution: This problem most likely arises during Steps 17–20, so make sure the suggested timing is followed closely. Before clearing with xylenes (Step 26), the preparations must be completely dehydrated. Cloudy preparations could occur as a result of incomplete dehydration, possibly because the 100% ethanol is no longer anhydrous (Step 25). We replace the 100% ethanol regularly to ensure that it is anhydrous.

DISCUSSION

MARCM analysis of neuron morphology at late larval stages is generally complemented by analysis of dendrite morphology in homozygous mutant embryos (Gao et al. 1999; Grueber et al. 2003a; Li and Gao 2003; Ye et al. 2007). Discrimination of cell-autonomous effects is one clear advantage of using the MARCM system. In addition, by adding a UAS-rescuing construct, cell-autonomous rescue of clone phenotypes can be showed. Another approach for demonstrating cell autonomy is to express double-stranded RNA using the *Gal4-UAS* system. Here, however, inferences of cell autonomy are limited by the specificity of transgene expression from the Gal4 driver line throughout development. One caveat of MARCM analysis that has been addressed extensively in the literature is the possibility of wild-type protein perdurance in mutant clones and the resultant rescue of mutant phenotypes (Lee and Luo 1999). Likewise, Gal80 perdurance in mutant clones prevents analysis in earlier stages of development. Using the dissection and fixation protocol above, however, it is not always feasible to examine these early stages owing to the small size of the animals and the difficulty of dissection. Another consideration of MARCM is the efficiency of clone production. Systematic variation of the time between egg laying and heat shock so as to enrich for certain cells or cell types could be one approach for optimizing this protocol in the future.

Neuron morphology can be examined in live or fixed preparations. The protocol described here is to be used for fixed preparations. When a single stage of development is examined, there are some advantages to using dissected and fixed preparations rather than temporary whole mounts of flattened larvae. These advantages include good preservation of the preparations for several months or even years, the ability to consistently image entire arbors at high resolution, unambiguous identification of cell identity (this can be difficult in live preparations particularly when dendrite morphological identity is under study), and the ability to detect multiple protein epitopes by immunocytochemistry. The drawbacks of such preparations include the relatively large time investment involved in dissection, fixation, staining and mounting, as well as the possibility of dissection-induced damage to dendritic arbors. With proper care and dissection techniques, the latter concerns can be abrogated.

RECIPES

CAUTION: See Appendix for proper handling of materials marked with <!>.
Recipes for reagents marked with <R> are included in this list.

Embryo Wash (10x Stock Solution)

Reagent	Quantity (for 1 L)	Final concentration
NaCl	70 g	7% (w/v)
Triton X-100 <!>	5 mL	0.5% (v/v)

Dissolve the reagents in 500 mL of ddH$_2$O. Adjust the final volume to 1 L with ddH$_2$O.

Grape Juice Agar

Reagent	Quantity
ddH$_2$O	800 mL
Bacto Agar	24.0 g
Sucrose	26.4 g
Grape juice (100%)	200 mL
Methylparaben <!>	0.5 g
Ethanol (95%) <!>	20 mL

Mix Bacto Agar and water, then microwave until the agar dissolves completely (take care to prevent overflowing). Remove solution from the microwave, stir in sucrose and grape juice, and allow it to cool to 50°C. Dissolve methylparaben in 95% ethanol, and add to the agar solution. Stir well, and pour to plastic vials or plates (we typically use lids of Petri dishes) as desired. Let the agar cool and solidify before storing or using.

Paraformaldehyde (PFA; 4% v/v)

Mix 10 mL of 16% paraformaldehyde <!> (EM-grade obtained in 10-mL ampoules from Electron Microscopy Sciences, cat. no. 15710) with 30 mL of 1x PBS <R>. Store at 4°C when not in use.

PBS Stock Solution (10x)

Reagent	Quantity (for 1 L)	Final concentration
NaCl	80 g	1.37 M
KCl <!>	2 g	27 mM
Na$_2$HPO$_4$	14.4 g	100 mM
KH$_2$PO$_4$	2.4 g	20 mM
dH$_2$O	to 1 L	

Dissolve the components in 400 mL of dH$_2$O and adjust the pH to 7.4 with concentrated HCl. Adjust the volume to 1 L with dH$_2$O and sterilize. Store at room temperature.

PBS-TX

Reagent	Quantity (for 500 mL)	Final concentration
PBS stock solution (10x) <R>	50 mL	1x
Triton X-100 (10%, v/v) <!>	15 mL	0.3% (v/v)
dH$_2$O	435 mL	

Store at room temperature.

ACKNOWLEDGMENTS

We thank members of our laboratory and other colleagues who have helped in improving these protocols. Work in our laboratory is funded by the National Institutes of Health, Searle Scholars Program, Gatsby Initiative in Brain Circuitry, McKnight Endowment Fund, The Esther A. and Joseph Klingenstein Fund, and The Irma T. Hirschl/Monique Weill-Caulier Trust.

REFERENCES

Ashburner M, Golic KG, Hawley SR. 2004. Drosophila: *A laboratory handbook.* Cold Spring Harbor Laboratory Press, Cold Spring Harbor, NY.

Bodmer R, Barbel S, Sheperd S, Jack JW, Jan LY, Jan YN. 1987. Transformation of sensory organs by mutations of the cut locus of *D. melanogaster. Cell* **51:** 293–307.

Corty MM, Matthews BJ, Grueber WB. 2009. Molecules and mechanisms of dendrite development in *Drosophila. Development* **136:** 1049–1061.

Crozatier M, Vincent A. 2008. Control of multidendritic neuron differentiation in *Drosophila:* The role of Collier. *Dev Biol* **315:** 232–242.

Emoto K, He Y, Ye B, Grueber WB, Adler,PN, Jan LY, Jan YN. 2004. Control of dendritic branching and tiling by the Tricornered-kinase/Furry signaling pathway in *Drosophila* sensory neurons. *Cell* **119:** 245–256.

Emoto K, Parrish JZ, Jan LY, Jan YN. 2006. The tumour suppressor Hippo acts with the NDR kinases in dendritic tiling and maintenance. *Nature* **443:** 210–213.

Furrer MP, Kim S, Wolf B, Chiba A. 2003. Robo and Frazzled/DCC mediate dendritic guidance at the CNS midline. *Nat Neurosci* **6:** 223–230.

Furrer MP, Vasenkova I, Kamiyama D, Rosado Y, Chiba A. 2007. Slit and Robo control the development of dendrites in *Drosophila* CNS. *Development* **134:** 3795–3804.

Gao FB, Brenman JE, Jan LY, Jan YN. 1999. Genes regulating dendritic outgrowth, branching, and routing in *Drosophila*. *Genes Dev* **13:** 2549–2561.

Gao FB, Kohwi M, Brenman JE, Jan LY, Jan YN. 2000. Control of dendritic field formation in *Drosophila:* The roles of Flamingo and Competition between homologous neurons. *Neuron* **28:** 91–101.

Garbe DS, Bashaw GJ. 2004. Axon guidance at the midline: From mutants to mechanisms. *Crit Rev Biochem Mol Biol* **39:** 319–341.

Grueber WB, Jan LY, Jan YN. 2002. Tiling of the *Drosophila* epidermis by multidendritic sensory neurons. *Development* **129:** 2867–2878.

Grueber WB, Jan LY, Jan YN. 2003a. Different levels of the homeodomain protein cut regulate distinct dendrite branching patterns of *Drosophila* multidendritic neurons. *Cell* **112:** 805–818.

Grueber WB, Ye B, Moore AW, Jan LY, Jan YN. 2003b. Dendrites of distinct classes of *Drosophila* sensory neurons show different capacities for homotypic repulsion. *Curr Biol* **13:** 618–626.

Grueber WB, Ye B, Yang CH, Younger S, Borden K, Jan LY, Jan YN. 2007. Projections of *Drosophila* multidendritic neurons in the central nervous system: Links with peripheral dendrite morphology. *Development* **134:** 55–64.

Hartwig CL, Worrell J, Levine RB, Ramaswami M, Sanyal S. 2008. Normal dendrite growth in *Drosophila* motor neurons requires the AP-1 transcription factor. *Dev Neurobiol* **68:** 1225–1242.

Hattori Y, Sugimura K, Uemura T. 2007. Selective expression of Knot/Collier, a transcriptional regulator of the EBF/Olf-1 family, endows the *Drosophila* sensory system with neuronal class-specific elaborated dendritic patterns. *Genes Cells* **12:** 1011–1022.

Hughes ME, Bortnick R, Tsubouchi A, Bäumer P, Kondo M, Uemura T, Schmucker D. 2007. Homophilic Dscam interactions control complex dendrite morphogenesis. *Neuron* **54:** 417–427.

Jefferis GS, Vyas RM, Berdnik D, Ramaekers,A, Stocker RF, Tanaka NK, Ito K, Luo L. 2004. Developmental origin of wiring specificity in the olfactory system of *Drosophila*. *Development* **131:** 117–130.

Jinushi-Nakao S, Arvind R, Amikura R, Kinameri E, Liu AW, Moore AW. 2007. Knot/Collier and cut control different aspects of dendrite cytoskeleton and synergize to define final arbor shape. *Neuron* **56:** 963–978.

Kim MD, Jan LY, Jan YN. 2006. The bHLH-PAS protein Spineless is necessary for the diversification of dendrite morphology of *Drosophila* dendritic arborization neurons. *Genes Dev* **20:** 2806–2819.

Komiyama T, Luo L. 2007. Intrinsic control of precise dendritic targeting by an ensemble of transcription factors. *Curr Biol* **17:** 278–285.

Komiyama T, Johnson WA, Luo L, Jefferis GS. 2003. From lineage to wiring specificity. POU domain transcription factors control precise connections of *Drosophila* olfactory projection neurons. *Cell* **112:** 157–167.

Komiyama T, Sweeney LB, Schuldiner O, Garcia K, Luo L. 2007. Graded expression of semaphorin-1a cell-autonomously directs dendritic targeting of olfactory projection neurons. *Cell* **128:** 399–410.

Kuo CT, Jan LY, Jan YN. 2005. Dendrite-specific remodeling of *Drosophila* sensory neurons requires matrix metalloproteases, ubiquitin-proteasome, and ecdysone signaling. *Proc Natl Acad Sci* **102:** 15230–15235.

Kuo CT, Zhu S, Younger S, Jan LY, Jan YN. 2006. Identification of E2/E3 ubiquitinating enzymes and caspase activity regulating *Drosophila* sensory neuron dendrite pruning. *Neuron* **51:** 283–290.

Landgraf M, Thor S. 2006. Development and structure of motoneurons. *Int Rev Neurobiol* **75:** 33–53.

Landgraf M, Jeffrey V, Fujioka M, Jaynes JB, Bate M. 2003. Embryonic origins of a motor system: Motor dendrites form a myotopic map in *Drosophila*. *PLoS Biol* **1:** e41. doi: 10.1371/journal.pbio.0000041.

Lee T, Luo L. 1999. Mosaic analysis with a repressible cell marker for studies of gene function in neuronal morphogenesis. *Neuron* **22:** 451–461.

Li W, Gao FB. 2003. Actin filament-stabilizing protein tropomyosin regulates the size of dendritic fields. *J Neurosci* **23:** 6171–6175.

Li W, Wang F, Menut L, Gao FB. 2004. BTB/POZ-zinc finger protein abrupt suppresses dendritic branching in a neuronal subtype-specific and dosage-dependent manner. *Neuron* **43:** 823–834.

Libersat F, Duch C. 2002. Morphometric analysis of dendritic remodeling in an identified motoneuron during postembryonic development. *J Comp Neurol* **450:** 153–166.

Marin EC, Watts RJ, Tanaka NK, Ito K, Luo L. 2005. Developmentally programmed remodeling of the *Drosophila* olfactory circuit. *Development* **132:** 725–737.

Matthews BJ, Kim ME, Flanagan JJ, Hattori D, Clemens JC, Zipursky SL, Grueber WB. 2007. Dendrite self-avoidance is controlled by Dscam. *Cell* **129:** 593–604.

Moore AW, Jan LY, Jan YN. 2002. Hamlet, a binary genetic switch between single- and multiple-dendrite neuron morphology. *Science* **297:** 1355–1358.

Orgogozo V, Grueber WB. 2005. FlyPNS, a database of the *Drosophila* embryonic and larval peripheral nervous system. *BMC Dev Biol* **5:** 4. doi: 10.1186/1471-213X-5-4.

Parrish JZ, Kim MD, Jan LY, Jan YN. 2006. Genome-wide analyses identify transcription factors required for proper morphogenesis of *Drosophila* sensory neuron dendrites. *Genes Dev* **20:** 820–835.

Parrish JZ, Emoto K, Jan LY, Jan YN. 2007. Polycomb genes interact with the tumor suppressor genes *hippo* and *warts* in the maintenance of *Drosophila* sensory neuron dendrites. *Genes Dev* **21**: 956–972.

Ramón y Cajal S. 1995. *Histology of the nervous system of man and vertebrates*. Oxford University Press, Oxford.

Satoh D, Sato D, Tsuyama T, Saito M, Ohkura H, Rolls MM, Ishikawa F, Uemura T. 2008. Spatial control of branching within dendritic arbors by dynein-dependent transport of Rab5-endosomes. *Nat Cell Biol* **10**: 1164–1171.

Sharp DJ, Yu W, Ferhat L, Kuriyama R, Rueger DC, Baas, PW. 1997. Identification of a microtubule-associated motor protein essential for dendritic differentiation. *J Cell Biol* **138**: 833–843.

Soba P, Zhu S, Emoto K, Younger S, Yang SJ, Yu HH, Lee T, Jan LY, Jan YN. 2007. *Drosophila* sensory neurons require Dscam for dendritic self-avoidance and proper dendritic field organization. *Neuron* **54**: 403–416.

Stone MC, Roegiers F, Rolls MM. 2008. Microtubules have opposite orientation in axons and dendrites of *Drosophila* neurons. *Mol Biol Cell* **19**: 4122–4129.

Sugimura K, Yamamoto M, Niwa R, Satoh D, Goto S, Taniguchi M, Hayashi S, Uemura T. 2003. Distinct developmental modes and lesion-induced reactions of dendrites of two classes of *Drosophila* sensory neurons. *J Neurosci* **23**: 3752–3760.

Sugimura K, Satoh D, Estes P, Crews S, Uemura T. 2004. Development of morphological diversity of dendrites in *Drosophila* by the BTB-zinc finger protein abrupt. *Neuron* **43**: 809–822.

Sweeney NT, Li W, Gao FB. 2002. Genetic manipulation of single neurons in vivo reveals specific roles of Flamingo in neuronal morphogenesis. *Dev Biol* **247**: 76–88.

Uylings HB, van Pelt J. 2002. Measures for quantifying dendritic arborizations. *Network (Bristol, England)* **13**: 397–414.

Williams D, Truman JW. 2004. Mechanisms of dendritic elaboration of sensory neurons in *Drosophila*: Insights from in vivo time lapse. *J Neurosci* **24**: 1541–1550.

Williams D, Truman JW. 2005a. Cellular mechanisms of dendrite pruning in *Drosophila*: Insights from in vivo time-lapse of remodeling dendritic arborizing sensory neurons. *Development* **132**: 3631–3642.

Williams D, Truman JW. 2005b. Remodeling dendrites during insect metamorphosis. *J Neurobiol* **64**: 24–33.

Williams DW, Kondo S, Krzyzanowska A, Hiromi Y, Truman JW. 2006. Local caspase activity directs engulfment of dendrites during pruning. *Nat Neurosci* **9**: 1234–1236.

Wu JS, Luo L. 2006. A protocol for mosaic analysis with a repressible cell marker (MARCM) in *Drosophila*. *Nat Protoc* **1**: 2583–2589.

Ye B, Zhang Y, Song W, Younger SH, Jan LY, Jan YN. 2007. Growing dendrites and axons differ in their reliance on the secretory pathway. *Cell* **130**: 717–729.

Zheng X, Wang J, Haerry TE, Wu AY, Martin J, O'Connor MB, Lee CH, Lee T. 2003. TGF-β signaling activates steroid hormone receptor expression during neuronal remodeling in the *Drosophila* brain. *Cell* **112**: 303–315.

Zheng Y, Wildonger J, Ye B, Zhang Y, Kita A, Younger,SH, Zimmerman S, Jan LY, Jan YN. 2008. Dynein is required for polarized dendritic transport and uniform microtubule orientation in axons. *Nat Cell Biol* **10**: 1172–1180.

Zhu H, Luo L. 2004. Diverse functions of N-cadherin in dendritic and axonal terminal arborization of olfactory projection neurons. *Neuron* **42**: 63–75.

Zhu H, Hummel T, Clemens JC, Berdnik D, Zipursky SL, Luo L. 2006. Dendritic patterning by Dscam and synaptic partner matching in the *Drosophila* antennal lobe. *Nat Neurosci* **9**: 349–355.

7

Embryonic and Larval Neuromuscular Junction
An Overview with Selected Methods and Protocols

Preethi Ramachandran and Vivian Budnik

Department of Neurobiology, University of Massachusetts Medical School, Worcester, Massachusetts 01605

ABSTRACT

Over the last two decades, the *Drosophila* larval neuromuscular junction (NMJ) has gained immense popularity as a model system for the study of synaptic development, function, and plasticity. With this model, it is easy to visualize synapses and manipulate the system genetically with a high degree of temporal and spatial control, which makes it ideal for resolving problems in synaptic physiology and development. In this chapter, we summarize some of the features of this preparation and provide a variety of protocols for the cell-biological investigation of the NMJ.

INTRODUCTION

The *Drosophila* larval NMJ has emerged as one of the most popular genetic model systems to study a wide variety of questions fundamental to the field of neuroscience (Griffith and Budnik 2006; Collins and DiAntonio 2007; Speese and Budnik 2007; Levitan 2008). These studies include investigations of ion channel function, mechanisms of neurotransmitter release, the components of different signal transduction pathways, synapse physiology, and functional and structural plasticity. Many of the molecules involved in these processes are strongly conserved across phyla, making the study of this system highly relevant for all organisms.

The features that make the larval NMJ ideal for studies of synapse development are its relative simplicity, easy accessibility, and the presence of large muscles with well-defined synapses. An important hallmark of the system is the fact that synapses at the body-wall muscles are highly plastic, having the ability to respond to a variety of perturbations. Modifications in muscle size (Gorczyca et al. 1993), electrical activity (Budnik et al. 1990; Ataman et al. 2008), and experience (Schuster 2006a,b) lead to functional and structural short-term and long-term changes that can be

93

observed and quantified using a variety of different techniques (Johansen et al. 1989; Budnik et al. 1990; Broadie and Bate 1993; Sigrist et al. 2003; Yoshihara et al. 2005; Ataman et al. 2008). Nevertheless, the overall development of the NMJ is a tightly orchestrated process that involves the precise regulation of numerous structural, functional, and signaling components. Loss of any of these interdependent components can severely disrupt NMJ growth and function.

The inherent ability to manipulate the system combined with the availability of many cell-biological and genetic tools has made the study of the larval NMJ a rapidly advancing field of research. With the advent of techniques such as the bipartite UAS/Gal4 system (Brand and Perrimon 1993), the related LexA/LexAop system (Lai and Lee 2006), and RNA interference (RNAi) (Fjose et al. 2001; Armknecht et al. 2005), it is now possible to express proteins or knock down protein levels, either partially or completely, in a regulated manner in motor neurons or muscles. An additional level of precise spatiotemporal regulation can be achieved by combining the above techniques with the Gal80 (McGuire et al. 2004) and GeneSwitch systems (Osterwalder et al. 2001; McGuire et al. 2004). At the cellular level, the visualization of proteins has been made much easier by the availability of a continuously growing battery of antibodies that label different compartments of the NMJ. These include a number of pre- and postsynaptic proteins such as vesicle proteins (Schwarz 2006; Levitan 2008), cytoskeletal proteins (Ruiz-Canada and Budnik 2006), neurotransmitter receptors (Griffith 2004; DiAntonio 2006), and proteins belonging to a number of different signal transduction pathways (Marques 2005; Griffith and Budnik 2006). Another major development has been the use of low-toxicity, genetically encoded fluorescent markers, such as green fluorescent protein (GFP) and its variously colored relatives, which have enabled preparations to be imaged live for considerable periods of time (Rasse et al. 2005; Ataman et al. 2008) and have made possible the visualization in real time of processes such as receptor trafficking (Rasse et al. 2005; Fuger et al. 2007; Schmid et al. 2008) and bouton formation (Zito et al. 1999; Ataman et al. 2008). Furthermore, the ability to alter motor neuron activity acutely by expressing light-activated channels, like channelrhodopsin-2 (ChR2) (Nagel et al. 2003), temperature-activated channels such as TRPM8 (Lima and Miesenbock 2005; Peabody et al. 2009), and caged ligands (Lima and Miesenbock 2005), or through externally applied activity paradigms, combined with the ability to live-image a preparation over time, has allowed us to distinguish broad developmental changes from acute activity-dependent changes (Ataman et al. 2008).

GENERAL ANATOMY OF THE BODY-WALL MUSCLES AND THEIR INNERVATION

A great advantage of using the larval NMJ as a model system is the ability to observe the same identified muscle and innervating nerve endings from animal to animal, thus decreasing the degree of noise derived from looking at populations of cells. Larval body-wall muscles are composed of 30 uniquely identifiable skeletal supercontractile muscle fibers per hemisegment, repeated in each of the seven abdominal segments with some variation at the three thoracic segments and single terminal segment (Crossley 1978). These multinucleated muscles are large and readily identifiable because of their shape, size, and specific site of insertion at the larval cuticle.

Muscle fibers are innervated by approximately 36 motor neurons per central nervous system (CNS) hemisegment (Landgraf and Thor 2006). These motor neurons are located in the ventral ganglion from which they send axons that extend through the segmental and transverse nerves (Halpern et al. 1991; Sink and Whitington 1991). Once they exit the CNS, the axons navigate and branch over the body wall in a stereotypic fashion (Keshishian and Chiba 1993). Each motor neuron ending innervates specific muscle cells with high fidelity and a characteristic branching pattern. These NMJs are composed of varicosities or synaptic boutons, which have been classified into three types depending on their neurotransmitter content characteristics (Anderson et al. 1988; Johansen et al. 1989; Gorczyca et al. 1993; Monastirioti et al. 1995).

The great majority of body-wall muscle cells are poly-innervated by at least two motor neurons that release glutamate as the primary excitatory transmitter, the so-called type I big (type Ib) and type I small (type Is) NMJs (Jan and Jan 1976b; Johansen et al. 1989). In addition to type I boutons, two other bouton types have been identified. Type II endings contain small synaptic boutons, but their branches are much longer than those of type I, usually running through the length of the muscle fibers (Johansen et al. 1989). These boutons contain both glutamate and octopamine (Johansen et al. 1989; Monastirioti et al. 1995), but their exact function is still poorly understood. Unlike type I motor neurons, which tend to individually innervate one muscle, just three octopamine-containing neurons that give rise to type II endings innervate all but eight muscles in the body wall (Monastirioti et al. 1995). Thus, in contrast to type I boutons that are likely to mediate rapid excitation–contraction, octopamine-containing type II terminals seem to be more modulatory in nature. Type III boutons are less defined subsets of motor endings, which contain peptides and innervate specific muscles in a segment-specific fashion (Cantera and Nassel 1992; Gorczyca et al. 1993). The type III endings in muscle 12, for example, contain an insulin-like peptide and innervate muscle 12 at abdominal segments 2–5. These endings have elliptical boutons, with branches of intermediate length between type I and type II branches (Gorczyca et al. 1993; Jia et al. 1993).

ULTRASTRUCTURE OF NMJs

Type I NMJs

During late embryogenesis and the onset of the first-instar larval stage, type I endings are normally found sitting in a depression formed on the muscle surface, but as synaptic boutons and muscles enlarge, the muscle completely envelops the endings and forms a complex system of folded membranes, the subsynaptic reticulum (SSR) (Atwood et al. 1993; Jia et al. 1993; Guan et al. 1996). The SSR around type Ib terminals is more voluminous and densely packed than that at type Is terminals (Atwood et al. 1993; Jia et al. 1993). The exact significance of the SSR is still poorly understood. However, the SSR contains a number of signaling molecules (Griffith and Budnik 2006; Marques and Zhang 2006), scaffolding (Ataman et al. 2006b) or cytoskeleton-associated proteins (Ruiz-Canada and Budnik 2006), neurotransmitter receptors (Griffith 2004; DiAntonio 2006), and cell adhesion molecules (Griffith and Budnik 2006). Furthermore, the identification of polysomal profiles within the SSR and the discovery of proteins that play a role in translational control at the SSR suggest that it may also be a seat of subsynaptic protein synthesis (Koh et al. 2000; Schuster 2006a,b).

Type I boutons typically have a rounded shape, and in addition to a few mitochondria and endosomes, these structures contain a large number of clear, rounded neurotransmitter vesicles ~30–40 nm in diameter that occupy most of the bouton volume (Atwood et al. 1993; Jia et al. 1993). These vesicles increase dramatically in number during larval development, forming the readily releasable and reserve pools (Kuromi and Kidokoro 2003). They are either derived from axonal transport of vesicle precursors (Hannah et al. 1999; Murthy et al. 2003) or through vesicle recovery from the plasma membrane via endocytosis (Estes et al. 1996; Koenig and Ikeda 1996), which can be visualized by FM-143 dye imaging (Betz and Bewick 1993; Roche et al. 2002).

The presynaptic membrane is separated from the postsynaptic membrane by a synaptic cleft that is 15–20-nm wide (Jia et al. 1993). At the presynaptic membrane, active zones can be distinguished from the adjacent perisynaptic membrane by their denser appearance, highly organized dense material within the synaptic cleft, and the presence of presynaptic electron-dense structures referred to as T-bars (Atwood et al. 1993; Jia et al. 1993). T-bars are dense bodies that are anchored to the presynaptic membrane by a pedestal surmounted by a platform (Atwood et al. 1993; Jia et al. 1993) and are surrounded by a higher density of synaptic vesicles that usually fuse to the presynaptic plasma membrane beneath the T-bars (Jia et al. 1993; Prokop 1999; Verstreken and Bellen 2002). T-bars are thought to play a role in tethering vesicles for release as multiunit packages, making them active sites of neurotransmitter release within synapses (Prokop and Meinertzhagen 2006). Furthermore, their association with clusters of calcium channels suggests that they may play a role in inducing local high Ca^{2+} microdomains, which are required for synaptic release (Kawasaki et al. 2004; Prokop and

Meinertzhagen 2006). Unlike the vertebrate active-zone components, not much is known about the proteins that make up the T-bar. A recent study demonstrated that Nc82/Bruchpilot, a homolog of CAST/ERK, is contained in these structures and is required for their structural integrity (Wagh et al. 2006).

Type II and III NMJs

Type II and III terminals are smaller than type I boutons, lack an SSR, and remain on grooves in the muscle surface covered only by the basement membrane (Jia et al. 1993). Type II boutons contain a "mixed population" of clear vesicles (similar to type I boutons) and dense-core vesicles and have no clear synaptic specialization sites or active zones, whereas type III boutons are filled mostly with discrete populations of spherical dense-core vesicles with varying size and degree of electron density. Furthermore, in addition to dense-core vesicles, type III boutons also have small clear vesicles that are exclusively concentrated at restricted sites of the presynaptic membrane surrounding putative synapses. In contrast to the well-defined synapses of type I boutons, the synaptic release sites for clear vesicles in type III boutons appear as darker thickenings of the presynaptic membrane surrounded by translucent vesicles (Jia et al. 1993).

EXAMINATION OF SYNAPTIC GROWTH IN LIVE AND FIXED SAMPLES

The ability to perform live-imaging techniques on larvae for extended periods of time has led to major advances in understanding the sequence of events that occur during synapse proliferation. By imaging the NMJs live, using noninvasive techniques, it has been possible to observe bouton budding and outgrowth over extended periods of time. For example, by expressing mCD8-Shaker GFP in the muscles and imaging the same larva every 24 h throughout development, from the first-instar to the third-instar larval stage, it has been demonstrated that new boutons are added either between already existing boutons, by bouton division, or at the end of a growing string of boutons (Zito et al. 1999). On a more acute timescale, it has also been demonstrated that spaced stimulation by K+-induced depolarization, motor nerve stimulation, or light activation of neuronally expressed ChR2 induces the formation of dynamic filopodia-like extensions (synaptopods) and ghost boutons (presynaptic terminals without any postsynaptic specializations) 2 h after the beginning of the stimulation cycles (Ataman et al. 2008). These ghost boutons can be observed in live undissected larvae as well as in fixed samples. These boutons also eventually acquire glutamate receptors (GluRs) and active zones, indicating that they represent intermediate synaptic boutons (Ataman et al. 2008). By using techniques such as fluorescence recovery after photobleaching (FRAP) along with live-imaging techniques, it has been possible to demonstrate the formation of new postsynaptic densities by following the incorporation of GluRs into newly forming synapses over almost 20% of larval development, a period of 36 h (Rasse et al. 2005; Schmid et al. 2008). In addition, these studies have also shown activity-dependent changes in GluR subtype incorporation at these newly formed synapses.

Overall, the *Drosophila* NMJ has become a fertile ground for the identification and elucidation of many essential structural and signaling components and their roles in inducing synaptic plasticity. Given the high degree of conservation in many of these mechanisms across phyla, it is likely that the *Drosophila* NMJ will continue to figure prominently in the exploration of these issues.

Protocol 1

Dissection of the Larval Body-Wall Muscles

A number of different techniques have been used to dissect third-instar larval preparations to expose the body-wall muscles. Here we describe a procedure that uses magnetic chambers and pins to allow for fine control in spreading the larval body wall (Fig. 1). Magnetic chambers consist of a thin rectangular magnetic strip with a center hole (the dissecting well) glued to a glass slide, which forms the bottom of the well in which the larva is dissected (Fig. 1A). The larva is held in place and spread

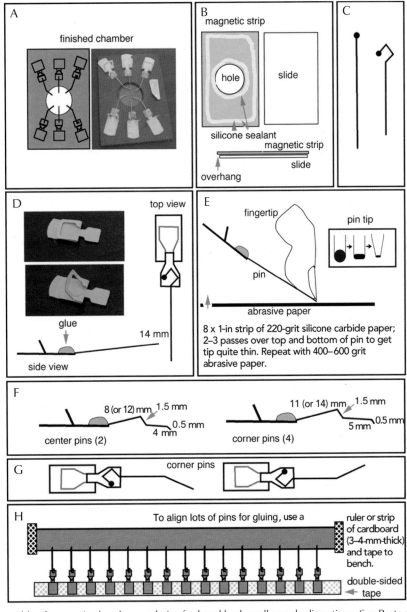

FIGURE 1. Assembly of magnetic chambers and pins for larval body-wall muscle dissections. See Protocol 1 for details. (Reprinted, with permission of Elsevier, from Budnik et al. 2006.)

97

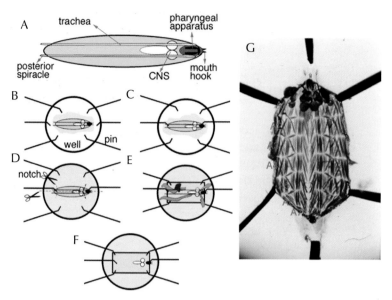

FIGURE 2. Dissection of larval body-wall muscles using magnetic chambers. See Protocol 1 for details. (Reprinted, with permission of Elsevier, from Budnik et al. 2006.)

open by the use of stainless steel insect pins that are optimally bent and shaped to apply pressure to the cuticle and stretch out the edges of the larval body wall. The pins are glued in place on steel tabs, which rest on the magnetic base. For a view of a dissected body-wall muscle preparation using this technique, see Figure 2G. This protocol is based on a method described in Budnik et al. (2006).

MATERIALS

CAUTION: See Appendix for proper handling of materials marked with <!>.

See the end of the chapter for recipes for reagents marked with <R>.

Reagents

Ethanol (95%) <!>
Fixative (either 4% paraformaldehyde <!> <R> or nonalcoholic Bouin's <R>)
    The fixative should be freshly made (<1-wk-old) and kept at 4°C.
HL-3 saline (low-Ca^{2+}; containing 0–0.3 mM Ca^{2+}) <R>

Equipment

Dissecting microscope
Epoxy glue (with working time 5–15 min)
Fine brush (size 00) or needle-tipped probe
Forceps (e.g., Dumont #5; made of Inox alloy)
    These can be used new without any sharpening, especially on second- and third-instar larvae. A little sharpening for first-instar larvae is recommended.
Glass slides (51 x 76-mm; 2 x 3-in)
Hole punch (Small Parts, Inc.), scissors, scalpel, or exacto knife
Iridectomy scissors (e.g., Vannas angled-on-edge scissors with 3-mm cutting edge, Roboz RS-5618)
    File down the tips with abrasive paper.
Magnetic strip (flexible, with one side more magnetic than the other; 76-mm [3-in]-wide and 1.5-mm [0.06-in]-thick) (http://www.ma.com/Flexible/Strip.htm; cat. no. 0648)
Metal tabs (also termed file flags or projecting signals [see Fig. 1D]) (e.g., Nu-Vise metal signals, Advantus Corp.)

Advantus Corp. sell metal tabs only in very large quantities under special order (1000 boxes). A limited number of tabs can be obtained from the Budnik laboratory on request.

Paper cutter

Pasteur pipettes

Pliers (fine-tipped, flat-nose pliers, high-quality linesman pliers, and "micro" pliers fashioned from damaged fine forceps) (http://www.monsterslayer.com/Pages/PliersSherSnip/PlierShrNip.aspx)

Ruler or strip of cardboard (~3–4-mm-thick)

Silicone carbide abrasive paper (grit nos. 200, 400, and 600) (Small Parts Inc.)

Silicone sealant

Spot plates (glass; with three concave depressions each ~22-mm-OD x 7-mm-deep) (Corning, Inc.)

Stainless steel pins (Carolina Biological Supply Company; size 00 for second- and third-instar larvae, cat. no. 65-4331; size 000 for first-instar larvae, cat. no. 65-4330)

METHOD

Assembly of Magnetic Chamber

1. Mark the magnetic side of the strip.

    This will be the top of the strip that will hold the larva and the dissecting pins.

2. Cut the magnetic strip to size using a large paper cutter (52 x 77-mm).

    When the strip is glued to the glass slide, there should be a 1-mm overhang on each side of the slide to allow easy handling of the chamber (Fig. 1B; bottom diagram).

3. Make a 16–18-mm (13/16–7/8-in) hole in the center of the magnetic strip with a hole punch, scissors, scalpel, or exacto knife using a coin (e.g., a U.S. nickel) as a template (Fig. 1B).

4. On the unmarked (back) or the less magnetic side of the strip, place a bead of silicone sealant around the perimeter of the strip and then again right at the rim of the cut hole (Fig. 1B). Ensure the silicone sealant is not interrupted so as to avoid leakage and retention of fluids in between the magnetic strip and the glass slide.

5. Place the 51 x 76-mm glass slide over the back side of the magnetic strip and press it firmly down into the adhesive (Fig. 1B; bottom diagram). Clean up any excess sealant that spilled onto the glass dissecting area. Place a heavy, flat object over the slides to keep the two surfaces securely pressed down together until the sealant dries. Wash thoroughly before use, because the silicone adhesive releases acetic acid during the curing process.

Assembly of Pins

Six pins are required per dissection chamber: two central pins and four corner pins.

6. Bend the beaded end of the pins as shown in Figure 1C (right) using a fine-tipped pair of pliers. Start from the beaded end of the pin and bend as follows: right angle, right angle, then 45° angle. Raise the bead slightly from the horizontal. Rinse the bent end with 95% ethanol.

7. Using a small screwdriver or sandpaper, scratch off some of the paint on the metal tab where the pin will be glued and clean the tab with ethanol to allow the epoxy glue to adhere better. Raise the top part of the tab to ~60° using a screwdriver or fine-tipped pliers (Fig. 1D; this will allow easy handling of the tabs during dissections).

8. Place the bent end of the pin on the metal tab so that the pin tip is raised ~3–4 mm from the horizontal.

    This is done to prevent the saline/fixative from creeping out of the chamber's well via capillary action and spreading all over the top of the chamber.

Check by setting the tip on a strip of cardboard and observing the angle of the pin tip. To glue the pin to the tab, drop one drop of quick-setting epoxy glue from a toothpick or wooden applicator stick over the tab (Fig. 1D).

9. Allow the glue to harden. Using high-quality linesman pliers, cut the two center pins to ~14 mm and the four corner pins to ~18 mm in length (for second and third instars) measured from the glued end of the tab. Sand the pin tips with silicon carbide abrasive paper (grit # 220) until they become flat and thin (Fig. 1E). Finish with grit # 400–600.

10. Bend the pins as in Figure 1F (dimensions are in mm). The rear center pin, over which the initial incision of the larva during dissection is made, should have a shallower bend to its tip to allow for scissor tip access to the posterior end of the larva. Corner pins should also be bent from the side (a little more than halfway down their length) as in Figure 1G. For accuracy purposes, the last bend (0.5 mm) is best done with the "micro" pliers under a low-power dissecting microscope.

11. Readjust the angles of the pins by positioning them on the surface of the chamber or attempting a larval dissection to make sure that they are optimized to hold the larva in the stretched position.

> Ideally the pins should be able to apply moderate tension on the surface of the glass or when stretching the larva. Too much or too little tension will not allow for the fine control of larval stretching and will rip the cuticle. All dimensions are approximate and depend on your chamber and purpose.

Dissection of Larval Body-Wall Muscles (First to Third Instar)

12. Pick a larva (Fig. 2A) from a vial or bottle with a fine brush or needle-tipped probe and place it in a small drop of low-Ca^{2+} HL-3 saline (containing 0–0.3 mM Ca^{2+}) in the middle of the well in the magnetic chamber (Fig. 2B).

> Muscle contractions are completely eliminated by adding 1 mM EGTA. However, some antibodies and proper preservation of membranes require the presence of at least some Ca^{2+}.

13. Pin the larva at the anterior and posterior ends with the center pins (Fig. 2C) right at the tip of the larvae. Straighten the larva out as much as possible such that the dorsal side (with the two tracheal tubes) is visible at the top. Wash with HL-3 saline three times and add enough saline so that it reaches the walls of the well and completely immerses the larva.

14. Using the iridectomy scissors, slightly lift the dorsal cuticle and make a small horizontal incision about one-third of the distance from the rear end. Insert the scissors into the incision and cut the larva all the way to the anterior end along the midline (Fig. 2D). Make a small horizontal incision at the anterior end. Make sure the midline cuts are superficial enough to just pass through the cuticle to avoid cutting through the brain and the muscles. At each end, cut two notches as shown in Figure 2D. Bring the two center pins closer to let the body cavity open. This will help in clearance of the internal organs. Place the corner pins at each side of the dorsal incision and spread the body wall apart (Fig. 2E).

> Throughout the cutting process, keep washing the larva periodically with saline. It is important that the larva not be exposed to air. Therefore, perform the saline exchange by using two pasteur pipettes simultaneously, one to aspirate the old solution and the other to add new saline, ensuring that the preparation always remains submerged.

15. Clean out the internal organs using forceps and saline. Leave the CNS intact. Gently stretch the larva with corner pins until it reaches the shape shown in Figure 2, F and G, making sure you do not tear the muscles during the process. Wash with saline two or three times.

16. Replace the saline with fixative. Place the chamber with the pinned larva in fixative on a horizontal shaker and incubate for 5–10 min with gentle agitation.

17. Remove the pins carefully and transfer the preparation (which now is fairly rigid) into a three-depression glass spot plate for incubation with other solutions.

Protocol 2

Immunocytochemical Staining of Larval Body-Wall Muscles

This protocol is based on a method described in Budnik et al. (2006) and can be used to label various proteins with antibodies in dissected larval fillets and visualize their localization and distribution in the brain, NMJ, and muscle.

MATERIALS

CAUTION: See Appendix for proper handling of materials marked with <!>.
See the end of the chapter for recipes for reagents marked with <R>.

Reagents

Fixative (either 4% paraformaldehyde <!> <R> or nonalcoholic Bouin's <R>)
　　The fixative should be freshly made (<1-wk-old) and kept at 4°C.
HL-3 saline (low-Ca^{2+}; containing 0–0.3 mM Ca^{2+}) <R>
Mounting medium (e.g., Vectashield)
Nail polish (for sealing coverslips)
PBT (0.1 M phosphate buffer containing 0.2% Triton X-100 <!>) <R>
Primary and secondary antibodies

Equipment

Box with lid (e.g., plastic slide box)
Coverslips (18 x 18-mm)
Dissecting microscope
Glass microscope slides
Pasteur pipettes
Spot plates (glass; with three concave depressions each ~22-mm-OD x 7-mm-deep)
　　(Corning, Inc.)

METHOD

1. Dissect the larval body-wall muscles in cold, low-Ca^{2+} HL-3 saline as described in Protocol 1. Fix in 4% paraformaldehyde or nonalcoholic Bouin's fixative for 5–10 min.

2. Remove the preparations from the magnetic chambers and place into the wells of three-depression glass spot plates for washing. Put the spot plates in a small slide box containing a wet paper towel for the rest of the incubations to prevent evaporation. Wash the samples three times with PBT for 15 min each.

3. Incubate the larval samples in primary antibody mixture diluted in PBT (50–100 μL per approximately five preparations) overnight at 4°C with gentle agitation. Cover the three-spot depression wells either with a large coverslip (22 x 22-mm or 25 x 25-mm) or a glass microscope slide to prevent excessive evaporation.

4. Wash the larval samples three times with PBT for 15 min each. Incubate with secondary antibody mixture diluted in PBT for 2 h at room temperature.

   At this time, anti-horseradish peroxidase directly conjugated to a fluorophore can be added to the secondary antibody mixture to label presynaptic arbors.

5. Wash the samples three times with PBT for 15 min each.

6. Mount the samples under a dissecting microscope. Pick the larva up with forceps and gently touch the forceps to the corner of a tissue paper to remove excess PBT. Put a small drop (~20 μL) of mounting medium on a microscope slide and place the sample in it. Orient the larva on the slide such that the muscles (not the cuticle) are facing up. Make sure no air bubbles are generated. Carefully place a coverslip over the preparation and seal the sides of the coverslip with clear nail polish.

   Samples can be kept in slide boxes for a very long time if frozen at –80°C.

Protocol 3

Preparation of Early Embryonic Fillets (before Cuticle Deposition)

This protocol is adapted from a procedure developed by the Keshishian laboratory (Johansen et al. 1989) and can be used to label various proteins with antibodies in dissected early embryonic fillets. To dissect embryos, a modified chamber or well is designed using glass slides and sealant. Embryos are prepared by removing the chorionic (outer) membrane (performed outside the chamber) and then removing the vitelline (inner) membrane (performed inside the chamber and under saline).

MATERIALS

CAUTION: See Appendix for proper handling of materials marked with <!>.
See the end of the chapter for recipes for reagents marked with <R>.

Reagents

Agar plates (regular or enriched with fruit juice) <R>
Ethanol (95%) <!>
HL-3 saline (low-Ca^{2+}; containing 0–0.3 mM Ca^{2+}) <R>
Live yeast paste <R>
Mounting medium (e.g., Vectashield)
Nail polish (for sealing coverslips)
Primary and secondary antibodies (see Protocol 2)

Equipment

Beaker (100-mL; plastic; perforated with tiny holes)
Dissecting microscope
Double-sided tape
Fine brush
Forceps (blunt)
Funnels (plastic; one slightly smaller than the other)

> One small funnel is used to "pour" the flies from bottles into the plastic beaker. The other must have an opening that fits snugly in the top of the plastic beaker with the spout facing outward. This is used to "pour" the flies back from the plastic beaker into the bottles.

Mesh basket (1-in-ID) (*optional*)
Microscope slides (glass)
Nitex nylon mesh (120-mm pore size) (*optional*)
Petri dishes (5 cm in diameter)

> These are used for the agar plates and must fit exactly under the plastic beaker.

Silicone sealant
Syringe with 16–18-gauge needle
Toothpicks (or pipettes)

> These are used to spread yeast paste into the plates.

Tungsten needles (blunt and sharp)

METHOD

Egg Laying

1. Raise 50–200 flies in bottles and then place them in an egg-laying chamber (a small agar plate with a dab of live yeast paste covered by a plastic beaker with tiny holes) (see Fig. 3A).

2. Allow the flies to lay eggs on the agar plate. Incubate for about 30 min to 1 h at 18°C or 25°C depending on the desired degree of embryo staging (Fig. 3C).

    Keep a couple of egg-laying plates going simultaneously and also be sure to change the plastic beaker occasionally.

Construction of Chambers to Process Embryonic Fillets

3. On a glass slide, using silicone sealant in a syringe with a 16–18-gauge needle, create a rectangular well ~1.5 mm high with inside dimensions of ~1 × 2-cm (Fig. 4).

    Dissections and immunocytochemistry are performed within this chamber, so the walls should be thick and must not leak.

4. Clean the inside of the chamber thoroughly with 95% ethanol to optimize fillet adherence. Using a pair of blunt forceps to keep the chamber free of oil from fingertips, place a piece of double-sided tape cut lengthwise (~3-mm-wide) inside the chamber.

5. At the time of dissection, place another piece of double-sided tape (2-cm-long) on the right side of the slide and outside the chamber. This is to be used for dechorionation of embryos (Fig. 4).

Preparation of Embryos

6. Collect stage-16 embryos (13 h after egg laying at 25°C) from agar plates using a fine brush or blunt forceps and place about 10–15 eggs on the double-sided tape outside the chamber for

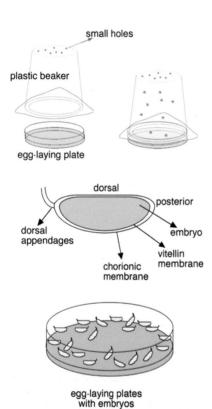

FIGURE 3. Egg-laying plates and collection of embryos for whole-mount or fillet embryo staining. See Protocol 3 for details. (Reprinted, with permission of Elsevier, from Budnik et al. 2006.)

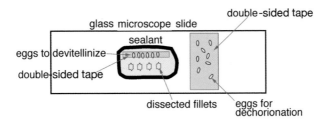

FIGURE 4. Setup for dissection of early embryonic fillets (or flat preparations). See Protocol 3 for details. (Reprinted, with permission of Elsevier, from Budnik et al. 2006.)

dechorionation (Fig. 4). Make sure the agar is wet to prevent the eggs from adhering to one another.

7. Use a pair of dull forceps with the tips closed to dechorionate the eggs. Looking under a dissecting microscope, roll the eggs gently to one side; the chorion should break open because one side will stick to the tape. Gently push with the forceps and the egg should pop out.

8. Transfer these eggs to the tape inside the well using closed forceps (with slight pressure the dechorionated eggs will stick to the tape) and align them near the edge of the tape in the correct orientation inside the chamber (see Fig. 4).

> The anterior pole is more tapered and carries the small outgrowth of the micropyle (the site of sperm entry). The ventral surface of the egg is convex (Fig. 3B).

Make sure the eggs are stuck securely to the tape by pushing them down gently.

9. Gently fill the well with HL-3 saline because the eggs need to be immersed for subsequent steps.

10. For devitellization, press on the anterior end of the embryo and create a "clear space" between the embryo and the vitelline membrane using the blunt end of a tungsten needle. Puncture this space with the needle or a pair of very sharp forceps.

11. Using the side of the needle or forceps, push the embryo out by pressing on the posterior end, opposite to where the puncture was.

> This is somewhat analogous to pushing toothpaste out of a tube. The embryo will often appear misshapen, but this is usually not a problem for the ultimate steps of the dissection.

12. Push the embryo very gently off the tape (as the embryo is extremely fragile without the vitelline membrane) and float it to another point in the well where it will be stuck to the glass. Directing it with the edge of the needle, make sure it contacts the glass on its ventral side.

> At this point, the nervous system should be visible as you move the embryo.

13. Make a dorsal midline incision using a sharp tungsten needle. Start between the brain lobes and cut toward the anterior, and then as far posterior as possible.

14. With the edge of the needle, brush the edges of the body wall onto the slide so they will adhere.

> Pushing the guts into the wall so that the needle does not come into direct contact with the musculature assists this process.

15. After the body walls securely adhere to the glass, use the needle to clean out the guts and the trachea.

> Gently administering drops of saline can help clean out the embryo, but if they are added too forcefully, they will dislodge the embryos from the glass.

16. Dissect several embryos and then remove the tape inside the well. Fix the preparations and incubate with antibodies inside the well.

> It is important to note that if you remove all the fluid in the well as you change solutions, the embryos will lift off the glass and self-destruct. Because the chamber is small, liquid surface tension is substantial, and fluid exchange needs to be performed very carefully to avoid embryo disruption. Fortunately, embryos do not require as much rinsing (or fixation) as third instars.

17. For mounting, use a tungsten needle to dislodge the processed embryos from the glass slide by sliding the tip of the needle all the way under the embryo.

18. Suck up the embryo with a pasteur pipette (keeping it right at the tip of the pipette) and drop it onto a clean slide. Position the embryo right side up by finding the CNS. Remove excess fluid with a tissue. Place a very small drop of mounting medium on the embryo and cover with a coverslip. Do not move the coverslip after this point. Use nail polish to seal the coverslip.

Protocol 4

Preparation of Late Embryonic Fillets

Embryos beyond the age of ~15 h (25°C) become impenetrable to aqueous fixatives and staining methods as a result of cuticle deposition. Flat (fillet) preparations can overcome this problem (Broadie 2000). For flat dissections, use extracellular recording solution (Broadie 2000) designed to keep exposed cells in good physiological condition. This protocol is based on a method described in Budnik et al. (2006).

MATERIALS

CAUTION: See Appendix for proper handling of materials marked with <!>.
See the end of the chapter for recipes for reagents marked with <R>.

Reagents

Histoacryl glue (B. Braun Melsungen AG, Germany)
HL-3 saline (low Ca^{2+}; containing 0.1–0.3 mM Ca^{2+}) <R>
Mounting medium (e.g., Vectashield)
Oil (e.g., mineral oil)
Paraformaldehyde <!> (4% in PB) <R>
PBT (0.1 M phosphate buffer containing 0.2% Triton X-100 <!>) <R>
Primary and secondary antibodies
Sylgard 186 silicone elastomer (Dow Corning Corp.)

Equipment

Blocks to lift coverslips (e.g., rubber block, washer ring, or ring of plasticine)
Coverslips (22 x 22-mm; glass)
Electrode puller
Forceps
Glass electrode tubing without capillary filaments
Microfuge tubes (0.5-mL; lids are used)
Microscope slides (glass)
Oven (set at 60°C)
Pasteur pipettes
Pipette tips (1-mL)
Razor blade splinters
Tubing
Tungsten needle (sharp) (*optional*)

METHOD

Preparation of Sylgard-Coated Coverslips

1. Place a drop of Sylgard in the middle of a glass coverslip and spread it out using the tip of a pasteur pipette (Fig. 5).

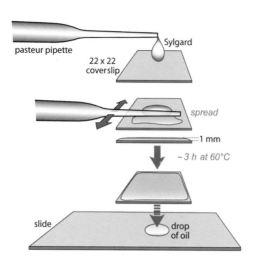

FIGURE 5. Preparation of Sylgard-coated coverslips for the dissection of late embryonic fillets. See Protocol 4 for details. (Reprinted, with permission of Elsevier, from Budnik et al. 2006.)

2. Bake the coverslips for 3–4 h at 60°C (Fig. 5) and let cool.

3. Put a drop of oil on a glass microscope slide. Place the coverslip gently on the oil, and push it down to make sure it stays in place.

Dissection of Late Embryonic Fillets

It is preferable to perform the dissections on a dark surface illuminated with light at a very shallow angle to improve visibility.

4. Put a drop of ~300 µL of saline on a Sylgard-coated coverslip placed on a glass microscope slide.

5. Using forceps, take cleaned late embryos/early larvae and attach them with their posterior ends to the surface of the Sylgard (larvae will stick to the Sylgard) arranging them in rows.

6. Place a small drop of Histoacryl glue in a lid cut from a 0.5-mL microfuge tube.

7. Connect a patch pipette to tubing using an adaptor (e.g., a 1-mL blue pipette tip) and dip it into the microfuge tube lid slightly breaking its tip (to enlarge the tip diameter) on the lid's bottom.

8. Aspirate some glue into the electrode.

> Histoacryl polymerizes on contact with an aqueous solution. Therefore adjust the pressure in the capillary so that glue neither spills out nor withdraws from the pipette tip throughout the following procedure.

9. Glue the embryos/larvae through their anal plates by placing little drops of Histoacryl glue at their posterior tip (Fig. 6, Step 1). Do not glue them through their spiracles.

10. Touch the electrode to the Sylgard surface at the point where you want to attach the anterior end of the larva (Fig. 6, Step 2) and release some glue so that it attaches to the surface.

11. Move the electrode tip immediately to the larva and catch its anterior end (Fig. 6, Step 3) and bring it back to the Sylgard surface where the glue was attached (Fig. 6, Step 4).

> The glue will stay elastic for a few seconds allowing this dynamic action. If the electrode becomes blocked, try to break its tip further by stabbing it through the Sylgard onto the glass surface.

12. Stab a hole into the posterior dorsal midline of the embryo (Fig. 6, Step 5) using either patch electrodes or pointed tungsten needles. Starting from this point, rip open the dorsal midline in small steps (Fig. 6, Step 6). Rubbing a second tungsten needle against the one inserted into the embryo helps to break the cuticle (Fig. 6, Step 7). Make sure that needle stays superficially on the dorsal surface to prevent damage to the muscles on the ventral side.

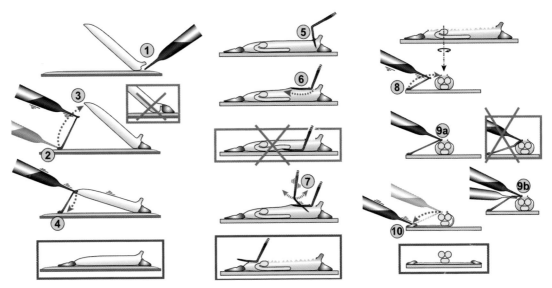

FIGURE 6. Methodology for dissecting late embryonic fillets (flat preparations). See Protocol 4 for details. (Reprinted, with permission of Elsevier, from Budnik et al. 2006.)

13. To spread the larva, release a tiny drop of glue beside it (Fig. 6, Step 8). Pick up the edge of the opened epidermis (Fig. 6, Step 9a) and bring it down to the surface (Fig. 6, Step 10). To ensure that the glue does not adhere all along the lateral epidermis, but only at its edge, hold the string of glue away from the epidermis using a second needle (Fig. 6, Step 9b). Fixing the epidermis at four points (anterior and posterior on each side) is sufficient to spread the larva appropriately.

14. Remove the saline and add fixative. Incubate in 4% paraformaldehyde in PB for 1 h.

15. Wash carefully in PBT for 1 h.

> Before transferring to PBT, remove the coverslip from the slide and place it on an elevated surface (e.g., rubber block, washer ring, ring of plasticine) so that the PBT does not flow off the edges.

16. Stain with primary and secondary antibody solutions.

17. Place a drop of mounting medium on the embryo. Using a glass electrode or a sharp tungsten needle, rip the epidermis from its fixation points on the coverslip, or cut the embryo away using a fine razor blade splinter held by forceps.

18. Transfer with the electrode to a tiny drop of embedding medium on a fresh slide and cover with a coverslip.

Protocol 5

FM1-43 Labeling of Larval NMJs

Styryl dyes such as FM1-43 were initially used by the Betz group to study vesicle recycling (Betz and Bewick 1992, 1993). Since its original use, several FM1-43 variants have been developed (Brumback et al. 2004). Here we describe a simple method for looking at vesicle recycling at the NMJ, which is adapted from previously published procedures (Ramaswami et al. 1994; Kuromi and Kidokoro 1999; Roche et al. 2002).

MATERIALS

See the end of the chapter for recipes for reagents marked with <R>.

Reagents

Ca^{2+}-free saline <R>
FM1-43 dye solution <R>
High-potassium HL-3 saline (containing 90 mM K^+) <R>
Low-Ca^{2+} HL-3 saline (containing 0–0.3 mM Ca^{2+}) <R>
Normal HL-3 saline (containing 2 mM Ca^{2+}) <R>

Equipment

Confocal microscope or epifluorescent microscope with a charge-coupled device (CCD) camera
Water-immersion objectives (40 x –63x)

METHOD

1. Dissect larval body-wall muscle preparations in low-Ca^{2+} HL-3 saline (this diminishes excitation damage during dissection) as described in Protocol 1. Replace the low-Ca^{2+} saline with normal HL-3 saline (containing 2 mM Ca^{2+}).

2. To induce loading with FM1-43 dye, replace the normal saline with high-potassium HL-3 saline containing 4 µM FM1-43. Let the preparation sit in the dye for 5 min.

3. Wash the preparation twice rapidly and then four times for 5 min each with Ca^{2+}-free saline to remove surface-bound dye. Image under the confocal or epifluorescence microscope.

4. To unload the dye, replace the normal saline with high-potassium HL-3 saline (or stimulate the segmental nerve in normal saline) and wash with Ca^{2+}-free saline.

Protocol 6

Internalization and Trafficking Assay

This assay can be used to observe the cycling of transmembrane proteins at the plasma membrane and their trafficking within cells when there are antibodies available that bind cell-surface epitopes in vivo. We have performed such internalization and trafficking assays with three transmembrane proteins, Fasciclin II, DFrizzled-2, and Evi (Mathew et al. 2003. 2005; Ataman et al. 2006a; Korkut et al. 2009).

MATERIALS

CAUTION: See Appendix for proper handling of materials marked with <!>.

See the end of the chapter for recipes for reagents marked with <R>.

Reagents

Antibody that binds to the extracellular domain of a protein in vivo
Low-Ca^{2+} HL-3 saline (containing 0.1 mM Ca^{2+}) <R>
Normal HL-3 saline (containing 2 mM Ca^{2+}) <R>
Paraformaldehyde (4% in PB) <!> <R>
PB (0.1 M phosphate buffer, pH 7.2) <R>
PBT (0.1 M phosphate buffer containing 0.2% Triton X-100 <!>) <R>
Secondary antibodies conjugated to Alexa Fluor 647 and fluorescein isothiocyanate (FITC)
Texas Red–conjugated anti-horseradish peroxidase (HRP)

Equipment

Coverslips
Dissection chambers and tools
Microscope slides
Pasteur pipettes
Spot plates (glass; with three concave depressions each ~22-mm-OD x 7-mm-deep) (Corning Inc.)

METHOD

1. Dissect larvae in low-Ca^{2+} HL-3 saline containing 0.1 mM Ca^{2+}. Keep the larvae in the dissection chamber and incubate with primary antibody diluted in normal HL-3 saline for 30–60 min at 4°C.

2. Wash the samples with normal HL-3 saline at 4°C. Bring the samples to room temperature (to allow protein cycling at the membrane) for selected periods between 5 min and 1.5 h.

3. Fix the samples in 4% paraformaldehyde for 10 min.

4. Transfer the samples to a three-depression glass spot plate and wash them three times with PB for 15 min each.

5. Label plasma membrane surface protein–antibody complexes by incubating the samples for 1 h with Alexa Fluor 647–conjugated secondary antibody at saturating concentrations to ensure

that all antibody-binding sites are completely occupied (1:200 dilution in PB not PBT). Wash the samples again with PB three times for 15 min each.

> Note that until this point all washes are performed in the absence of detergent.

6. Postfix the larvae in 4% paraformaldehyde for 10 min.

7. Permeabilize the samples by washing three times with PBT for 15 min each.

8. Visualize internalized protein–antibody complexes by incubating the samples for 1 h with FITC-conjugated secondary antibody (1:200 dilution in PBT) together with Texas Red–conjugated anti-HRP (1:100) to label the presynaptic arbor.

9. To eliminate the possibility of the FITC-conjugated antibody binding to unoccupied sites of surface (noninternalized) protein–antibody complexes, include a set of controls where the samples are incubated with the FITC-conjugated secondary after incubation with the Alexa Fluor 647–conjugated antibody but before permeabilization.

> No signal should be observed in the green channel in the controls, thus confirming that surface protein is exclusively labeled by the Alexa Fluor 647–conjugated secondary antibody and the FITC-conjugated secondary antibody labels only the protein that is internalized.

10. Wash the samples three times with PBT for 15 min each and mount the samples.

Protocol 7

Activity Paradigm to Induce NMJ Growth

This protocol is adapted from Ataman et al. (2008) and the spaced-stimulation method can be used to stimulate motor neurons for short periods of time (~2 h), fix (or image live), and examine the immediate effects of altering activity on synaptic protein localization as well as NMJ structure and function. In our study, spaced stimulation of the motor neurons using this paradigm induced the formation of dynamic filopodia-like extensions (synaptopods) and ghost boutons (synaptic varicosities devoid of postsynaptic specialization), which were observable in both live and fixed preparations (for more details, see Ataman et al. 2008).

MATERIALS

CAUTION: See Appendix for proper handling of materials marked with <!>.
See the end of the chapter for recipes for reagents marked with <R>.

Reagents

High-potassium HL-3 saline (containing 90 mM K^+ and high Ca^{2+}) <R>
Low-Ca^{2+} HL-3 saline (containing 0.1 mM Ca^{2+} and low K^+) <R>
Paraformaldehyde <!> (4% in PB) <R>
PBT (0.1 M phosphate buffer containing 0.2% Triton X-100 <!>) <R>
Mounting medium (e.g., Vectashield)
Primary antibodies and secondary antibodies conjugated to FITC or Texas Red

Equipment

Coverslips
Dissection chambers and tools
Microscope slides
Pasteur pipettes
Spot plates (glass; with three concave depressions each ~22-mm-OD x 7-mm-deep) (Corning, Inc.)

METHOD

1. Dissect the larva in low-Ca^{2+} HL-3 saline containing 0.1 mM Ca^{2+}, leaving the CNS and the peripheral nerves innervating the body-wall muscles intact.

2. Remove the corner pins leaving the two center pins intact. While keeping the larva in a pinned position, bring the two center pins as close to each other as possible without damaging the larva so that the larva is in a completely unstretched state.

3. Replace the low-Ca^{2+} HL-3 saline in the dissection chamber with high-potassium HL-3 saline. Repeat this step one or two times to ensure complete exchange of solutions.

4. Incubate in high-potassium HL-3 saline for 2 min. Replace the high-potassium HL-3 saline with low-Ca^{2+} HL-3 saline. Repeat this step two or three times to ensure complete exchange of solutions.

5. Incubate in low-Ca^{2+} HL-3 saline for 15 min.

6. Repeat the above steps four times.

> We used 2-, 2-, 4-, and 6-min pulses in the high-potassium HL-3 saline.

7. After the last pulse, incubate in low-Ca^{2+} HL-3 saline for 15 min.

8. Pull the center pins apart gradually to carefully stretch the larva vertically. Then, using the corner pins, carefully stretch the larva in all directions. Process the larva for live imaging or fix with 4% paraformaldehyde for 10–15 min. After fixing, process the larva for immunocytochemistry as described in Protocol 2 or image live.

Protocol 8

Electron Microscopy of Larval NMJs

This protocol is adapted from Jia et al. (1993) and Packard et al. (2002).

MATERIALS

CAUTION: See Appendix for proper handling of materials marked with <!>.
See the end of the chapter for recipes for reagents marked with <R>.

Reagents

Acetone <!>
CBo (osmotically balanced cacodylate buffer) (0.1 M and 0.2 M) <R>
Ethanol <!>
Jan's saline containing 0.1 mM $CaCl_2$ <R>
Lead citrate solution <R>
Modified Trump's universal fixative <R>
NaOH pellets <!>
Osmium tetroxide (2%) <!>
Propylene oxide <!>
Spurr's resin (low-viscosity embedding medium) (Electron Microscopy Sciences, cat. no. 14300)
> Prepare according to manufacturer's instructions using the "hard" modification.

Toluidine blue stain <R>
Uranyl acetate (2% aqueous solution) <!>

Equipment

Copper grids (with thin bar hexagonal mesh; Electron Microscopy Sciences, cat. no. T400H-Cu)
Dissecting scissors and forceps (see Protocol 1)
Embedding molds
Filter paper discs
Glass and/or diamond knife
Magnetic dissection chambers (see Protocol 1)
Parafilm
Pasteur pipettes
Petri dishes (10 cm in diameter)
Plastic pipettes with bulb
> Use instead of pasteur pipettes to avoid contact with osmium tetroxide and embedding resins.

Razor blades (surgical carbon steel; stainless steel Teflon-coated)
Spot plates (glass; with three concave depressions each ~22-mm-OD x 7-mm-deep) (Corning Inc.)
Stick with a wire loop 2 mm in diameter
Transmission electron microscope
Ultramicrotome

METHOD

1. Dissect the body-wall muscles in Jan's saline containing 0.1 mM Ca^{2+}.

    To prevent the sample from curling during the dehydration and embedding process, make sure that muscles are not as stretched for electron microscopy as they would be for immunocyto-chemistry.

2. Fix in modified Trump's universal fixative in the dissection chamber for ~10 min. Transfer the larvae to a three-depression glass spot plate and continue to fix under gentle agitation with Trump's fixative for 30 min at room temperature and then for at least 2 h or overnight at 4°C.

3. Wash three times with 0.1 M CBo for 10 min each. Postfix the samples in a solution consisting of equal parts of 2% osmium tetroxide solution and 0.2 M CBo (~50 μL per group of samples) for 30 min with gentle agitation.

4. Wash three times with 0.1 M CBo for 10 min each. Wash three times with distilled water for 10 min each.

5. Stain the samples en bloc for 20–30 min in aqueous 2% uranyl acetate.

6. Dehydrate by incubating sequentially in a series of ethanol concentrations (30%, 50%, 70%, 85%, 95%, anhydrous 100%) for 3 min each. Repeat the wash in 100% ethanol twice for 10 min each.

    The exchange must be performed gradually between each ethanol step. For example, we exchange just one-third of the ethanol from the previous step during the first minute followed by one-half during the second and the entire solution during the third.

7. Infiltrate in Spurr's resin gradually and sequentially with gentle agitation as follows:

    i. In propylene oxide for 30 min.

    ii. In propylene oxide/Spurr's resin 3:1 for 2 h (or overnight).

    iii. In propylene oxide/Spurr's resin 1:1 for 2 h (or overnight).

    iv. In propylene oxide/Spurr's resin 1:3 overnight. Cover the vial with Parafilm and make holes to allow evaporation of the propylene oxide.

    v. In Spurr's resin twice for at least 3 h each.

8. Mount in embedding molds. To prevent the preparation from sinking to the bottom of the mold, which causes difficulty in trimming, construct the resin blocks beforehand by placing a thin layer of Spurr's resin in the bottom of the molds and incubating at 70°C for 3 h. Carefully place the samples and Spurr's resin in the blocks.

    If the samples curl up, straighten them by pinning them down into the molds using sharpened insect pins.

9. Bake at 70°C overnight. Using a surgical carbon steel razor blade cleaned with acetone, trim the block around the preparation to ~2.5 × 1-mm, until the desired segment to be sectioned is reached.

10. Cut thick sections (0.5–1-μm) and place them on a microscope slide using a stick with a wire loop at the end (~2 mm in diameter). Add a drop of water and leave on a slide warmer at 60°–80°C until the water completely evaporates to eliminate wrinkling. Add a few drops of toluidine blue stain and allow to stand at 60°–80°C for 1–2 min. Make sure that the stain does not dry out by placing a dish containing water close to the slide.

11. Remove the excess stain by squirting the sections with distilled water, and observe under a compound microscope.

    At this point the muscles can be identified according to their profiles and position in cross section. It is also possible to identify terminals, which appear as small irregularities, slightly lighter than their surroundings close to the visceral surface of the muscle.

12. Trim the block further (to ~0.5 x 0.5-mm) using an acetone-cleaned stainless steel Teflon-coated razor blade until the desired region to be sectioned is identified. At this point, cut sections of ~50–80-nm thick (silver gray) and float them onto a clean copper grid.

13. To stain the grids, line the bottom of two Petri dishes (#1 and #2) with Parafilm. Place two or three pellets of NaOH in the sides of Petri dish #2 and keep covered. Place one drop of 2% uranyl acetate for each grid in Petri dish #1, and one drop of lead citrate for each grid in Petri dish #2.

14. Float each grid (section-side-down) in a drop of uranyl acetate for 10–30 min. Pick up the grids with forceps and blot away excess solution with paper tissue. Holding the grids with the forceps, wash them by quickly dipping down and up about 20 times in a 30-mL beaker filled with distilled water.

15. Blot away excess water. Float each grid in a drop of lead citrate for 5–8 min. Wash the grids in a diluted NaOH solution (about three drops of 0.02 M NaOH in 30 mL of distilled water), blot to remove excess solution, and store until further viewing. Keep in a desiccator for long-term storage.

RECIPES

CAUTION: See Appendix for proper handling of materials marked with <!>.
Recipes for reagents marked with <R> are included in this list.

Agar plates (Enriched with Fruit Juice)

Reagent	Quantity (for 1 L)
Agar	32 g
Sucrose	16.6 g
Fruit juice (apple or grape)	300 mL

Add distilled H2O to 1 L. Boil the ingredients while stirring on a hot plate at high setting (the solution will foam and rise and it may be necessary to take it off the heat and put it back a couple of times before fully dissolving the agar). Cool to touch. Add 5 mL of propionic acid <!>/phosphoric acid <!>, 9:1, and stir. Pour into 5-cm Petri dishes. Keep the lids off the plates for 10 min. Cover and store the plates at 4°C.

Ca^{2+}-Free Saline

Dissolve the ingredients of Jan's saline <R> in ~950 mL of distilled H_2O without adding $CaCl_2$ <!>. Add 0.8 g of EGTA (to give a final concentration of 1 mM). Bring the solution to pH 7.2 with NaOH <!>. Note that the EGTA will take a few minutes to dissolve and that it changes the pH as it dissolves. Adjust the final volume to 1 L with distilled H_2O.

CBo (Osmotically Adjusted Cacodylate Buffer) (0.1 M)

Reagent	Quantity (for 100 mL)	Final concentration
Cacodylic acid <!>	1.38 g	0.1 M
Sucrose	4.52 g	132 mM

Bring to 100 mL with distilled H_2O after adjusting the pH to 7.2 with NaOH <!>. This buffer is adjusted to match the osmolarity of modified Trump's universal fixative.

CBo (Osmotically Adjusted Cacodylate Buffer) (0.2 M)

Reagent	Quantity (for 100 mL)	Final concentration
Cacodylic acid <!>	2.76 g	0.2 M
Sucrose	9.04 g	264 mM

Bring to 100 mL with distilled H_2O after adjusting the pH to 7.2 with NaOH <!>. This buffer is adjusted to match the osmolarity of modified Trump's universal fixative.

FM1-43 (N-[3-triethylammoniumpropyl]-4-[p-di-butylaminostyryl]pyridinium dibromide) Dye Solution

Dissolve 1 mg of FM1-43 powder in 625 μL of water to prepare a 1 mM stock solution. Add 40 μL of the stock solution to 10 mL of high-potassium HL-3 saline to prepare 4 μM FM1-43 saline.

High-Potassium HL-3 Saline

Reagent	Quantity (for 1 L)	Final concentration
NaCl	2.34 g	40 mM
KCl <!>	6.71 g	90 mM
$MgCl_2 \cdot 6H_2O$ <!>	4.07 g	20 mM
$NaHCO_3$	0.84 g	10 mM
Sucrose	1.71 g	5 mM
Trehalose	1.89 g	5 mM
HEPES	1.19 g	5 mM

Dissolve the reagents in distilled H_2O. Bring the pH to 7.2 and adjust the final volume to 1 L with distilled H_2O. Add $CaCl_2$ <!> from a 1 M stock solution to give the desired concentration of Ca^{2+} (1.5 mM Ca^{2+} for activity paradigm).

HL-3 Saline (Hemolymph-like Saline-3)

Reagent	Quantity (for 1 L)	Final concentration
NaCl	4.09 g	70 mM
KCl <!>	0.37 g	5 mM
$MgCl_2 \cdot 6H_2O$ <!>	4.07 g	20 mM
$NaHCO_3$	0.84 g	10 mM
Sucrose	39.4 g	115 mM
Trehalose	1.89 g	5 mM
HEPES	1.19 g	5 mM

Dissolve the reagents in distilled H_2O. Bring the pH to 7.2 and adjust the final volume to 1 L with distilled H_2O. Add $CaCl_2$ <!> from a 1 M stock solution to give the desired concentration of Ca^{2+} (normal saline has 2 mM Ca^{2+}) (see Stewart et al. 1994; Broadie 2000).

Jans' Saline (Jan and Jan 1976a)

Reagent	Quantity (for 1 L)	Final concentration
NaCl	7.48 g	128 mM
KCl <!>	0.149 g	2 mM
MgCl$_2$ (1 M) <!>	4 mL	4 mM
Sucrose	12.16 g	35.5 mM
HEPES	1.19 g	5 mM

Dissolve the reagents in distilled H$_2$O. Bring the pH to 7.2 with NaOH <!> and adjust the final volume to 1 L with distilled H$_2$O. Add 1.8 mL of 1M CaCl$_2$ <!> for normal saline or as indicated in each protocol.

Lead Citrate Solution

Dissolve 5 mg of lead citrate <!> in 1 mL of 4.5 mM NaOH <!>. Spin for 5 min at high speed before use.

Live Yeast Paste

Fill a vial (e.g., a scintillation vial) to about one-quarter with live baker's yeast. Add H$_2$O while stirring with a wooden stick until the consistency of peanut butter is attained. Keep refrigerated (otherwise the yeast will rise and spill out of the vial).

Modified Trump's Universal Fixative

Add 0.56 g of cacodylic acid <!> to ~20 mL of water. Adjust the pH to 7.2 with NaOH <!>. Bring the volume to 25 mL with distilled H$_2$O. Add 10 mL of 16% paraformaldehyde <!> (Electron Microscopy Sciences; sealed ampoules; final concentration 4%). Add 5 mL of 8% glutaraldehyde <!> (Electron Microscopy Sciences; sealed ampoules; final concentration 1%). Add 80 μL of 1 M MgCl$_2$ <!> (final concentration 2 mM). Make this fixative fresh immediately before using.

Nonalcoholic Bouin's Fixative

Reagent	Quantity
Saturated picric acid solution (Sigma) <!>	75 mL
Formalin (37% formalin in water with no methanol; Fisher Scientific) <!> (methanol will produce bouton shrinkage and give poor results with certain antibodies)	25 mL
Glacial acetic acid <!>	5 mL

Paraformaldehyde (4%) Fixative

Place 8 g of paraformaldehyde <!> powder and one pellet of NaOH <!> (~0.2 g) in a 250-mL beaker. Add 100 mL of 0.1 M sodium phosphate dibasic (use 20 mL of 0.5 M stock solution <R> and bring to 100 mL with distilled H$_2$O). Stir while heating on a hot plate under the fume hood until all the paraformaldehyde powder is dissolved. Do not boil. Turn off the heat. Slowly add 0.1 M sodium phosphate monobasic until the pH is 7.2 (check with pH paper). (Prepare 100 mL of monobasic solution using 20 mL of 0.5 M stock solution <R> and bring to 100 mL with distilled H_2O.) Measure the volume of 0.1 M sodium phosphate monobasic left unused. Add this volume of PB to the fixative to give a total fixative volume of 200 mL. Store at 4°C and discard after 5 d.

PB (0.1 m phosphate buffer pH 7.2)

Prepare two stock solutions.

1. Sodium phosphate dibasic (0.5 M). Dissolve 35.5 g of sodium phosphate dibasic in a final volume of 500 mL of distilled H_2O. Some crystallization will occur when the solution is stored at 4°C. Warm on a hot plate and stir until the crystals dissolve.

2. Sodium phosphate monobasic (0.5 M). Dissolve 30 g of anhydrous sodium phosphate monobasic in a final volume of 500 mL of distilled H_2O.

Prepare in separate beakers.

1. 80 mL of 0.5 M sodium phosphate dibasic stock solution and distilled H_2O to give a final volume of 400 mL.

2. 30 mL of 0.5 M sodium phosphate monobasic stock solution and distilled H_2O to give a final volume of 150 mL.

Bring the above dibasic solution to pH 7.2 using the monobasic solution.

PBS (Phosphate-Buffered Saline)

Reagent	Quantity (for 1 L)	Final concentration
NaCl	7.6 g	150 mM
PB (0.1 M, pH 7.2) <R>	100 mL	10 mM

Adjust the final volume to 1 L with distilled H_2O.

PBT (0.1 M Phosphate Buffer pH 7.2 Containing 0.2% Triton X-100)

Reagent	Quantity (for 100 mL)	Final concentration
PB (0.1 M, pH 7.2) <R>	98 mL	1x
Triton X-100 (10%, v/v) <!>	2 mL	0.2% (v/v)

Toluidine Blue Stain

1% (w/v) toluidine blue and 1% (w/v) borax in water.

REFERENCES

Anderson MS, Halpern ME, Keshishian H. 1988. Identification of the neuropeptide transmitter proctolin in *Drosophila* larvae: Characterization of muscle fiber–specific neuromuscular endings. *J Neurosci* **8:** 242–255.
Armknecht S, Boutros M, Kiger A, Nybakken K, Mathey-Prevot B, Perrimon N. 2005. High-throughput RNA inter-

ference screens in *Drosophila* tissue culture cells. *Methods Enzymol* **392:** 55–73.

Ataman B, Ashley J, Gorczyca D, Gorczyca M, Mathew D, Wichmann C, Sigrist SJ, Budnik V. 2006a. Nuclear trafficking of DFrizzled-2 during synapse development requires the PDZ protein dGRIP. *Proc Natl Acad Sci* **103:** 7841–7846.

Ataman B, Budnik V, Thomas U. 2006b. Scaffolding proteins at the *Drosophila* neuromuscular junction. *Int Rev Neurobiol* **75:** 181–216.

Ataman B, Ashley J, Gorczyca M, Ramachandran P, Fouquet W, Sigrist SJ, Budnik V. 2008. Rapid activity-dependent modifications in synaptic structure and function require bidirectional wnt signaling. *Neuron* **57:** 705–718.

Atwood HL, Govind CK, Wu CF. 1993. Differential ultrastructure of synaptic terminals on ventral longitudinal abdominal muscles in *Drosophila* larvae. *J Neurobiol* **24:** 1008–1024.

Betz WJ, Bewick GS. 1992. Optical analysis of synaptic vesicle recycling at the frog neuromuscular junction. *Science* **255:** 200–203.

Betz WJ, Bewick GS. 1993. Optical monitoring of transmitter release and synaptic vesicle recycling at the frog neuromuscular junction. *J Physiol (Lond)* **460:** 287–309.

Brand AH, Perrimon N. 1993. Targeted gene expression as a means of altering cell fates and generating dominant phenotypes. *Development* **118:** 401–415.

Broadie KS. 2000. Electrophysiological approaches to the neuromuscular junction. In Drosophila *protocols* (ed. W Sullivan et al.), pp. 273–295. Cold Spring Harbor Laboratory Press, Cold Spring Harbor, NY.

Broadie KS, Bate M. 1993. Development of the embryonic neuromuscular synapse of *Drosophila melanogaster*. *J Neurosci* **13:** 144–166.

Brumback AC, Lieber JL, Angleson JK, Betz WJ. 2004. Using FM1-43 to study neuropeptide granule dynamics and exocytosis. *Methods* **33:** 287–294.

Budnik V, Zhong Y, Wu CF. 1990. Morphological plasticity of motor axons in *Drosophila* mutants with altered excitability. *J Neurosci* **10:** 3754–3768.

Budnik V, Gorczyca M, Prokop A. 2006. Selected methods for the anatomical study of *Drosophila* embryonic and larval neuromuscular junctions. *Int Rev Neurobiol* **75:** 323–365.

Cantera R, Nassel DR. 1992. Segmental peptidergic innervation of abdominal targets in larval and adult dipteran insects revealed with an antiserum against leucokinin I. *Cell Tissue Res* **269:** 459–471.

Collins CA, DiAntonio A. 2007. Synaptic development: insights from *Drosophila*. *Curr Opin Neurobiol* **17:** 35–42.

Crossley CA. 1978. The morphology and development of the *Drosophila* muscular system. In *The genetics and biology of* Drosophila (ed. M Ashburner, TRF Wright), pp. 499–560. Academic Press, New York.

DiAntonio A. 2006. Glutamate receptors at the *Drosophila* neuromuscular junction. *Int Rev Neurobiol* **75:** 165–179.

Estes PS, Roos J, van der Bliek A, Kelly RB, Krishnan KS, Ramaswami M. 1996. Traffic of dynamin within individual *Drosophila* synaptic boutons relative to compartment-specific markers. *J Neurosci* **16:** 5443–5456.

Fjose A, Ellingsen S, Wargelius A, Seo HC. 2001. RNA interference: Mechanisms and applications. *Biotechnol Annu Rev* **7:** 31–57.

Fuger P, Behrends LB, Mertel S, Sigrist SJ, Rasse TM. 2007. Live imaging of synapse development and measuring protein dynamics using two-color fluorescence recovery after photo-bleaching at *Drosophila* synapses. *Nat Protoc* **2:** 3285–3298.

Gorczyca M, Augart C, Budnik V. 1993. Insulin-like receptor and insulin-like peptide are localized at neuromuscular junctions in *Drosophila*. *J Neurosci* **13:** 3692–3704.

Griffith LC. 2004. Receptor clustering: Nothing succeeds like success. *Curr Biol* **14:** R413–415.

Griffith LC, Budnik V. 2006. Plasticity and second messengers during synapse development. *Int Rev Neurobiol* **75:** 237–265.

Guan B, Hartmann B, Kho YH, Gorczyca M, Budnik V. 1996. The *Drosophila* tumor suppressor gene, *dlg*, is involved in structural plasticity at a glutamatergic synapse. *Curr Biol* **6:** 695–706.

Halpern ME, Chiba A, Johansen J, Keshishian H. 1991. Growth cone behavior underlying the development of stereotypic synaptic connections in *Drosophila* embryos. *J Neurosci* **11:** 3227–3238.

Hannah MJ, Schmidt AA, Huttner WB. 1999. Synaptic vesicle biogenesis. *Annu Rev Cell Dev Biol* **15:** 733–798.

Jan LY, Jan YN. 1976a. L-glutamate as an excitatory transmitter at the *Drosophila* larval neuromuscular junction. *J Physiol (Lond)* **262:** 215–236.

Jan LY, Jan YN. 1976b. Properties of the larval neuromuscular junction in *Drosophila melanogaster*. *J Physiol (Lond)* **262:** 189–214.

Jia XX, Gorczyca M, Budnik V. 1993. Ultrastructure of neuromuscular junctions in *Drosophila*: Comparison of wild type and mutants with increased excitability. *J Neurobiol* **24:** 1025–1044 [Erratum: 1994. **25:** 893–895].

Johansen J, Halpern ME, Johansen KM, Keshishian H. 1989. Stereotypic morphology of glutamatergic synapses on identified muscle cells of *Drosophila* larvae. *J Neurosci* **9:** 710–725.

Kawasaki F, Zou B, Xu X, Ordway RW. 2004. Active zone localization of presynaptic calcium channels encoded by the cacophony locus of *Drosophila*. *J Neurosci* **24:** 282–285.

Keshishian H, Chiba A. 1993. Neuromuscular development in *Drosophila:* Insights from single neurons and single genes. *Trends Neurosci* **16:** 278–283.

Koenig JH, Ikeda K. 1996. Synaptic vesicles have two distinct recycling pathways. *J Cell Biol* **135:** 797–808.

Koh YH, Gramates LS, Budnik V. 2000. *Drosophila* larval neuromuscular junction: Molecular components and mechanisms underlying synaptic plasticity. *Microsc Res Tech* **49:** 14–25.

Korkut C, Ataman B, Ramachandran P, Ashley J, Barria R, Gherbesi N, Budnik V. 2009. *Trans*-synaptic transmission of vesicular Wnt signals through Evi/Wntless. *Cell* **139:** 393–404.

Kuromi H, Kidokoro Y. 1999. The optically determined size of exo/endo cycling vesicle pool correlates with the quantal content at the neuromuscular junction of *Drosophila* larvae. *J Neurosci* **19:** 1557–1565.

Kuromi H, Kidokoro Y. 2003. Two synaptic vesicle pools, vesicle recruitment and replenishment of pools at the *Drosophila* neuromuscular junction. *J Neurocytol* **32:** 551–565.

Lai SL, Lee T. 2006. Genetic mosaic with dual binary transcriptional systems in *Drosophila. Nat Neurosci* **9:** 703–709.

Landgraf M, Thor S. 2006. Development and structure of motoneurons. *Int Rev Neurobiol* **75:** 33–53.

Levitan ES. 2008. Signaling for vesicle mobilization and synaptic plasticity. *Mol Neurobiol* **37:** 39–43.

Lima SQ, Miesenbock G. 2005. Remote control of behavior through genetically targeted photostimulation of neurons. *Cell* **121:** 141–152.

Marques G. 2005. Morphogens and synaptogenesis in *Drosophila. J Neurobiol* **64:** 417–434.

Marques G, Zhang B. 2006. Retrograde signaling that regulates synaptic development and function at the *Drosophila* neuromuscular junction. *Int Rev Neurobiol* **75:** 267–285.

Mathew D, Popescu A, Budnik V. 2003. *Drosophila* amphiphysin functions during synaptic Fasciclin II membrane cycling. *J Neurosci* **23:** 10710–10716.

Mathew D, Ataman B, Chen J, Zhang Y, Cumberledge S, Budnik V. 2005. Wingless signaling at synapses is through cleavage and nuclear import of receptor DFrizzled2. *Science* **310:** 1344–1347.

McGuire SE, Mao Z, Davis RL. 2004. Spatiotemporal gene expression targeting with the TARGET and gene-switch systems in *Drosophila. Sci STKE* **2004:** pl6.

Monastirioti M, Gorczyca M, Rapus J, Eckert M, White K, Budnik V. 1995. Octopamine immunoreactivity in the fruit fly *Drosophila melanogaster. J Comp Neurol* **356:** 275–287.

Murthy M, Garza D, Scheller RH, Schwarz TL. 2003. Mutations in the exocyst component Sec5 disrupt neuronal membrane traffic, but neurotransmitter release persists. *Neuron* **37:** 433–447.

Nagel IG, Szellas T, Huhn W, Kateriya S, Adeishvili N, Berthold P, Ollig D, Hegemann P, Bamberg E. 2003. Channelrhodopsin-2: A directly light-gated cation-selective membrane channel. *Proc Natl Acad Sci* **100:** 13940–13945.

Osterwalder T, Yoon KS, White BH, Keshishian H. 2001. A conditional tissue-specific transgene expression system using inducible GAL4. *Proc Natl Acad Sci* **98:** 12596–12601.

Packard M, Koo ES, Gorczyca M, Sharpe J, Cumberledge S, Budnik V. 2002. The *Drosophila* wnt, wingless, provides an essential signal for pre- and postsynaptic differentiation. *Cell* **111:** 319–330.

Peabody NC, Pohl JB, Fengqiu D, Vreede AP, Sandstrom DJ, Wang H, Zelensky PK, White BH. 2009. Characterization of the decision network for wing expansion in *Drosophila* using targeted expression of TRPM8 channel. *J Neuro* **29:** 3343–3353.

Prokop A. 1999. Integrating bits and pieces: Synapse structure and formation in *Drosophila* embryos. *Cell Tissue Res* **297:** 169–186.

Prokop A, Meinertzhagen IA. 2006. Development and structure of synaptic contacts in *Drosophila. Semin Cell Dev Biol* **17:** 20–30.

Ramaswami M, Krishnan KS, Kelly RB. 1994. Intermediates in synaptic vesicle recycling revealed by optical imaging of *Drosophila* neuromuscular junctions. *Neuron* **13:** 363–375.

Rasse TM, Fouquet W, Schmid A, Kittel RJ, Mertel S, Sigrist CB, Schmidt M, Guzman A, Merino C, Qin G, et al. 2005. Glutamate receptor dynamics organizing synapse formation in vivo. *Nat Neurosci* **8:** 898–905.

Roche JP, Packard MC, Moeckel-Cole S, Budnik V. 2002. Regulation of synaptic plasticity and synaptic vesicle dynamics by the PDZ protein Scribble. *J Neurosci* **22:** 6471–6479.

Ruiz-Canada C, Budnik V. 2006. Synaptic cytoskeleton at the neuromuscular junction. *Int Rev Neurobiol* **75:** 217–236.

Schmid A, Hallermann S, Kittel RJ, Khorramshahi O, Frolich AM, Quentin C, Rasse TM, Mertel S, Heckmann M, Sigrist SJ. 2008. Activity-dependent site-specific changes of glutamate receptor composition in vivo. *Nat Neurosci* **11:** 659–666.

Schuster CM. 2006a. Experience-dependent potentiation of larval neuromuscular synapses. *Int Rev Neurobiol* **75:** 307–322.

Schuster CM. 2006b. Glutamatergic synapses of *Drosophila* neuromuscular junctions: A high-resolution model for the analysis of experience-dependent potentiation. *Cell Tissue Res* **326:** 287–299.

Schwarz TL. 2006. Transmitter release at the neuromuscular junction. *Int Rev Neurobiol* **75:** 105–144.

Sigrist SJ, Reiff DF, Thiel PR, Steinert JR, Schuster CM. 2003. Experience-dependent strengthening of *Drosophila* neuromuscular junctions. *J Neurosci* **23:** 6546–6556.

Sink H, Whitington PM. 1991. Pathfinding in the central nervous system and periphery by identified embryonic *Drosophila* motor axons. *Development* **112:** 307–316.

Speese SD, Budnik V. 2007. Wnts: Up-and-coming at the synapse. *Trends Neurosci* **30:** 268–275.

Stewart BA, Atwood HL, Renger JJ, Wang J, Wu C-F. 1994. *Drosophila* neuromuscular preparations in haemolymph-like physiological salines. *J Comp Physiol (A)* **175:** 179–191.

Verstreken P, Bellen HJ. 2002. Meaningless minis? Mechanisms of neurotransmitter-receptor clustering. *Trends Neurosci* **25:** 383–385.

Wagh DA, Rasse TM, Asan E, Hofbauer A, Schwenkert I, Durrbeck H, Buchner S, Dabauvalle MC, Schmidt M, Qin G, et al. 2006. Bruchpilot, a protein with homology to ELKS/CAST, is required for structural integrity and function of synaptic active zones in *Drosophila*. *Neuron* **49:** 833–844.

Yoshihara M, Adolfsen B, Galle KT, Littleton JT. 2005. Retrograde signaling by Syt 4 induces presynaptic release and synapse-specific growth. *Science* **310:** 858–863.

Zito K, Parnas D, Fetter RD, Isacoff EY, Goodman CS. 1999. Watching a synapse grow: Noninvasive confocal imaging of synaptic growth in *Drosophila*. *Neuron* **22:** 719–729.

8 Genetic Mosaic Analysis of *Drosophila* Brain by MARCM

Chih-Fei Kao[1] and Tzumin Lee[1,2]

[1]*Department of Neurobiology, University of Massachusetts Medical School, Worcester, Massachusetts 01605;* [2]*Janelia Farm Research Campus, HHMI, Ashburn, Virginia 20147*

ABSTRACT

The generation and analysis of clones of cells with a genotype different from the rest of an organism has been a powerful tool for studying the molecular mechanisms that underlie development in multicellular organisms. Genetic mosaics in *Drosophila* typically involve derivation of homozygous daughter cells from heterozygous precursors through mitotic recombination. Informative mosaic analyses require unambiguous detection of the homozygous cells in otherwise heterozygous organisms. The genetic mosaic system MARCM (mosaic analysis with a repressible cell marker) couples loss of heterozygosity with derepression of a marker gene. This technique permits unique labeling of specific homozygous daughter cells, and thus makes mosaic analysis possible in the complex nervous system. One can further restrict mosaic analysis to certain brain cells by controlling patterns of mitotic recombination and/or marker gene expression. With MARCM, it is possible to label neurons derived from a common progenitor in a multicellular clone or to target specific neurons in single-cell clones for detailed circuitry mapping. MARCM has facilitated thorough lineage analysis, permitting identification of single neurons based on their developmental origins. Furthermore, one can manipulate gene function specifically in MARCM-labeled cells to determine the molecular mechanisms underlying the development of the targeted cells. Diverse creative applications of MARCM, involving wise selection and proper assembly of multiple transgenic elements plus induction of mitotic recombination in various controlled strategies, will continue to revolutionize the field of *Drosophila* neurobiology.

INTRODUCTION

The brain is a highly complex structure, consisting of an immense number of neuron types as well as diverse glial cells. Each neuron in a concise circuit may be unique, with individually specific gene expression, electrophysiological properties, and connections to other neurons. To comprehend the complexity of the brain and understand its mechanisms of action, it is necessary to analyze individual neurons systematically and determine their positions and functions in the circuitry of the brain. To study the development of the brain, it is important to determine the cellular origins of specific neurons or glia and, if possible, follow individual brain cells through the wiring of the circuit. Finally, to elucidate the molecular mechanisms of brain development and function, it is essential to be able to manipulate gene activity in specific brain cells within an otherwise unperturbed organism. MARCM is a genetic mosaic system that allows one to uniquely label subsets of cells in otherwise unstained brains (positive labeling) for single-cell analysis of development, morphology, and even cell-autonomous gene functions (Lee and Luo 1999, 2001; Wu and Luo 2006). Using MARCM, it is possible to label clones of brain cells or isolated single neurons, reveal the patterns of proliferation of their corresponding precursors, and manipulate gene activity selectively in the marked cells (Lee and Luo 2001; Yu and Lee 2007). MARCM has been applied to determine patterns of neurogenesis and gliogenesis, identify single neurons based on the neural progenitors from which they arise and the birth order in which they are derived, and investigate the mechanisms of neuronal diversification and morphogenesis (Lee et al. 1999, 2000b; Jefferis et al. 2001; Zhu et al. 2006; Awasaki et al. 2008). Diverse creative applications of MARCM will help to resolve the complexity of the *Drosophila* brain and allow us to determine how the complex brain develops.

MECHANISM OF MARCM ANALYSIS

MARCM is a positive-labeling genetic mosaic system. Using MARCM analysis, it is possible to generate clones of cells homozygous for a specific chromosome arm or its homologous counterpart in otherwise heterozygous organisms (Fig. 1A). This process involves flippase (FLP)-mediated mitotic recombination at the FRT (flippase recognition target) sites of specific homologous chromosomes. Following X segregation, one daughter cell inherits the maternal recombinant chromosome (carrying a transposed paternal chromosome arm) as well as the paternal nonrecombinant one, and becomes homozygous for all the paternal alleles of the genes situated distal to the site of recombination. In contrast, the other daughter cell inherits the paternal recombinant chromosome (carrying a transposed maternal chromosome arm) and the maternal nonrecombinant one, and it carries the maternal alleles of the genes on the transposed chromosome arm in duplicate while losing the paternal alleles of the same genes. Using MARCM, it is also possible to label uniquely one of the two homozygous sister cells and its subsequently derived progeny (Fig. 1B). This involves derepression of UAS-marker gene following loss of the GAL4 repressor, GAL80. A ubiquitously expressed GAL80 transgene, *tubP-GAL80*, is placed distal to the FRT site on one of the homologous chromosome arms. Mitotic recombination between the solo *tubP-GAL80*-containing chromosome arm and its homologous counterpart would lead to loss of the GAL80 transgene in the derived daughter cell that has not inherited the *tubP-GAL80*-containing chromosome arm. Clones of GAL80-minus cells can thus be generated among GAL80-containing cells. GAL4 can then drive expression of UAS-marker gene(s) to specifically label the GAL80-minus cells that simultaneously become homozygous for all the genes located distal to the site of mitotic recombination. One can further knock out an essential gene specifically in the GAL80-minus cells by introducing a recessive lethal mutation into the transposed chromosome arm that does not carry *tubP-GAL80* (Fig. 1A). One can also incorporate additional UAS-transgenes to manipulate gene functions in the uniquely labeled GAL80-minus cells of mosaic organisms.

APPLICATIONS OF MARCM ANALYSIS

MARCM greatly facilitates cell lineage analysis in the *Drosophila* brain. Neurons in the *Drosophila* central nervous system (CNS) arise in clones from neural progenitors called neuroblasts (NBs). One NB gives rise to many neurons through multiple rounds of asymmetric cell division that regenerate

A

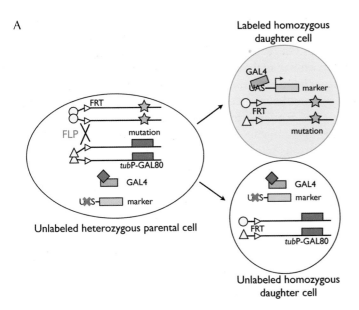

Labeled homozygous
daughter cell

Unlabeled heterozygous parental cell

Unlabeled homozygous
daughter cell

B

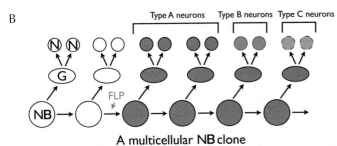

A multicellular NB clone

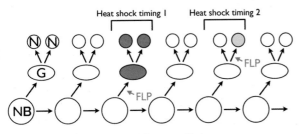

Single-cell/two-cell clones

FIGURE 1. Genetics of the mosaic analysis with a repressible cell marker (MARCM) system. (*A*) The following genetic elements are required for MARCM: FRTs (open triangles), FLP, GAL4 (orange box), tubP-GAL80 (red box), and UAS-reporter (green box). After site-specific mitotic recombination and cell division, the resulting daughter cell, which is devoid of GAL80 activity, is revealed by GAL4-activated UAS-reporter. In addition, if a given mutation (blue star) is introduced *in trans* to FRT, *tubP-GAL80*, the resulting MARCM clones labeled with markers will be homozygous for the mutation in an otherwise heterozygous background. (*B*) Schematic cartoon shows how FRT/FLP-mediated recombination in a developing neuronal lineage can lead to the formation of two mutually exclusive types of labeled clones. When mitotic recombination takes place in the regenerated neuroblast (NB), all postmitotic neurons (N) subsequently generated will be labeled (*upper panel*, green circles), forming an NB clone. The cell-type composition of a lineage will be preserved in an NB clone. If the ganglion mother cell (GMC; shown as G in the figure) loses the repressor gene, only two neurons derived from this GMC will be labeled in the lineage (*bottom panel*; heat shock timing 2), resulting in a two-cell clone. A single-cell clone is generated only when mitotic recombination occurs in a dividing GMC (*bottom panel*; heat shock timing 1). Following transient induction of FLP at a particular time of development, the labeled single-cell/two-cell clones would represent the types of cell being generated at that developmental stage of interest.

the NB while yielding two neurons at one time (in most neuronal lineages) through an intermediate precursor, called the ganglion mother cell (GMC) (Yu and Lee 2007; Knoblich 2008). This principle pattern of neurogenesis is reenforced by the observation that three distinct sizes of neuronal clones can be labeled by MARCM in the *Drosophila* CNS. Mitotic recombination in a dividing NB may give rise to a multicellular NB clone or a two-cell clone, depending on which daughter cell (the regenerated NB or the GMC) has lost the GAL80 transgene (Fig. 1B). In contrast, mitotic recombination in GMCs can produce lone GAL80-minus neurons labeled as single-cell clones. MARCM permits clonal analysis because neurons derived from a common progenitor can be colabeled in a multicellular NB clone. In addition, it allows one to identify neurons with stereotyped projections as revealed in two-cell/single-cell clones. One can selectively label two-cell/single-cell clones of the same lineage origin using lineage-specific GAL4 drivers. One can further determine the temporal pattern of neurogenesis by inducing mitotic recombination with hs-FLP (heat-shock-inducible FLP) at specific times of development. For a given lineage, one can determine the birth order in which distinct neurons are derived through analysis of the two-cell/single-cell clones that were generated at different times of development. Systematic clonal analysis by MARCM will ultimately resolve how the complex *Drosophila* brain develops from a limited number of neural progenitors.

The uniquely labeled GAL80-minus cells can be made homozygous for any mutation located on the GAL80-lacking homologous chromosome arm and distal to the site of mitotic recombination (Fig. 1A). One can also express multiple UAS-transgenes specifically in the GAL4-positive, GAL80-minus cells so as to differentially highlight various subcellular structures and simultaneously modulate gene function or neural activity using a "gain-of-function" UAS transgene. This enables manipulation of gene function in a small subset of cells in an otherwise unperturbed organism, which is essential for determining the cell-autonomous functions of an essential gene in later developmental processes. One can further target specific neurons for such genetic/transgenic studies in mosaic organisms by controlling patterns of mitotic recombination or using cell-type-specific GAL4 drivers. For example, to selectively generate clones of GAL80-minus cells in the *Drosophila* olfactory learning and memory center (the mushroom bodies [MBs]; Fig. 2A) one can transiently induce mitotic recombination in the newly hatched larvae when only five NBs per brain hemisphere (including the

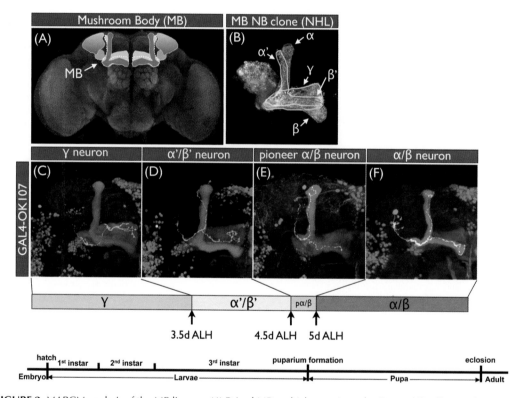

FIGURE 2. MARCM analysis of the MB lineage. (*A*) Paired MBs, which constitute the *Drosophila* olfactory learning and memory center, are located in the upper medial part of an adult fly brain. (*B*) A composite confocal image of an adult MB NB clone generated at the newly hatched larval (NHL) stage, showing five distinct axon bundles, three projecting medially (β, β', and γ) and two projecting dorsally (α and α'). The genotype for the analyzed MARCM fly is *hs-FLP/Y; FRTG13, tubP-GAL80/FRTG13, UAS-mCD8::GFP; GAL4-OK107/+*. (*C–F*) Different types of MB neurons are sequentially produced. Each type of MB neuron, which possess distinctive axon and dendrite projection patterns, is shown in the single-cell MB clones generated at different developmental windows: (*C*) γ neuron, (*D*) α'/β' neuron, (*E*) pioneer α/β neuron, and (*F*) α/β neuron. Temporal windows for the production of each type of MB neuron are summarized at the bottom. (Green) mCD8::GFP signals for the visualization of neurite projection patterns in the clones. (Magenta) Anti-FasII and anti-Dac (*dachshund*) staining to label MB axonal lobes and cell bodies, respectively. Abbreviations: pα/β, pioneer α/β; ALH, after larval hatching; d, days.

four MB NBs) are actively dividing (Fig. 2B). Using an MB-specific GAL4 driver plus induction of mitotic recombination at different developmental stages, one can further target different subtypes of MB neurons, which arise from common progenitors in an invariant sequence, for mosaic analysis (Fig. 2C–F; Lee et al. 1999; Zhu et al. 2006). The ability to reproducibly hit the same subset of neurons makes it possible to conduct genetic mosaic screens in specific neurons for systematically uncovering cell-autonomous genes required for various specific aspects of neural development.

In this chapter, we describe procedures for induction and analysis of wild-type MARCM clones in the *Drosophila* CNS (Protocol 1), generation of homozygous mutant clones in heterozygous organisms (Protocol 2), and use of MB lineages as a model system for forward genetic mosaic screens (Protocol 3). The same principles can be applied to conduct more complex MARCM experiments (Lai and Lee 2006; Lai et al. 2008).

Generation of Standard Wild-Type MARCM Clones

The principle of MARCM involves the generation of GAL80-minus homozygous daughter cells in otherwise heterozygous tissues, therefore allowing GAL4-dependent activation of UAS-reporter specifically in the homozygous cells of interest (Fig. 1A). To make MARCM clones, it is necessary to generate organisms carrying at least five genetic elements (FRTs, FLP, *tubP-GAL80*, GAL4, and UAS-marker) in specific configurations (Lee and Luo 1999, 2001; Wu and Luo 2006). Induction of FLP in neural precursors can be temporally controlled using a heat-shock promoter or spatially regulated using a tissue-specific promoter. Mitotic recombination in an NB may yield a multicellular NB clone (progeny of the renewed NB) or a two-cell clone (progeny of the derived GMC). Single-cell clones can be obtained following mitotic recombination in GMCs (Fig. 1B). Phenotypic analysis of mosaic brains can be greatly simplified using a GAL4 driver that is only expressed in neurons of interest. One can also mark different subcellular structures of the clones using distinct UAS-reporters. Here we describe common genetic strategies used in the study of *Drosophila* neurobiology and then outline the steps involved in standard MARCM analysis.

MATERIALS

Reagents

Antibody against specific UAS-marker
Fly lines (Bloomington *Drosophila* Stock Center, Indiana University)

Equipment

Computer equipped with digital image processing software
Fluorescence or confocal microscope for digital imaging
Fly culture media and vials
Fly incubator (set at 25°C)
Stereoscope for tissue dissection
Water bath for heat-shock-induced FLP expression (set at 37°C)

METHOD

Genetic Requirements for MARCM-Applicable Flies

1. Generate MARCM-applicable parental flies using the following genetic elements.

   i. *FRT and GAL80:* One parental fly carries an FRT site and a *tubP-GAL80* distal to FRT on the same chromosome arm. The other fly carries the same FRT site to allow mitotic recombination (Table 1).

   ii. *FLP:* FLP activity that mediates mitotic recombination can be expressed under a variety of controllable conditions such as a heat-shock promoter to elicit temporal induction or a tissue-specific promoter (enhancer-trap lines) to induce FLP expression in desired neuronal precursors.

TABLE 1. MARCM flies available from the Bloomington *Drosophila* Stock Center

Genotype	Bloomington Stock Center number	Features in MARCM genetics
P{neoFRT}19A, P{tubP-GAL80}LL1, P{hsFLP}1, w[*]	5132–5134	FRT and *tubP-GAL80* recombined on X chromosome
y1 w*; P{tubP-GAL80}LL10 P{neoFRT}40A/CyO	5192	FRT and *tubP-GAL80* recombined on 2L chromosome
w*; P{FRT(whs)}G13 P{tubP-GAL80}LL2	5140	FRT and *tubP-GAL80* recombined on 2R chromosome
y1 w*; P{tubP-GAL80}LL9 P{FRT(whs)}2A/TM3, Sb1	5190	FRT and *tubP-GAL80*0 recombined on 3L chromosome
y1 w*; P{neoFRT}82B P{tubP-GAL80}LL3	5135	FRT and *tubP-GAL80* recombined on 3R chromosome
P{hsFLP}22, w*	8862	Heat-shock promoter-driven FLP on X chromosome
P{hsFLP}1, w[1118]; Adv[1]/CyO	6	Heat-shock promoter-driven FLP on X chromosome
P{ey4x-FLP.Exel}1, y1 w1118	8205	FLP expression is regulated by four copies of *eyeless* enhancers
yd2 w1118 P{ey-FLP.N}2	5580	FLP is expressed in the pattern of the *eyeless* gene
y1 w*; P{tubP-GAL4}LL7/TM3, Sb1	5138	*tublin*-promoter-fused GAL4; a ubiquitous GAL4 driver
y1 w*; P{Act5C-GAL4}17bFO1/TM6B, Tb1	3954	*actin*-promoter-fused GAL4; a ubiquitous GAL4 driver
P{UAS-mCD8::GFP.L}LL4, y1 w*; PinYt/CyO	5136	Membrane-associated GFP marker on X chromosome
y1 w*; P{UAS-mCD8::GFP.L}LL5	5137	Membrane-associated GFP marker on second chromosome
y1 w*; PinYt/CyO; P{UAS-mCD8::GFP.L}LL6	5130	Membrane-associated GFP marker on third chromosome
w1118; P{UAS-myr-mRFP}1	7118	Membrane-targeted monomeric RFP marker on second chromosome
w1118; P{UAS-myr-mRFP}2/TM6B, Tb1	7119	Membrane-targeted monomeric RFP marker on third chromosome

iii. *GAL4 driver:* A GAL4 line of interest is used to drive expression of UAS-transgenes in target cells. A given GAL4 driver may express in tissue-specific (e.g., *GAL4-OK107* for MB labeling) or ubiquitous manners (*actinP-GAL4* or *tubP-GAL4*).

iv. *UAS-marker:* A UAS-marker is used for visualization of MARCM clones. For example, a membrane-tagged form of GFP (e.g., *UAS-mCD8::GFP*) is frequently used to outline neurons and their neuronal processes. There are various molecular markers available in the field (see Table 1). It is possible to use two or more UAS-markers at the same time if desired.

Depending on the experimental design and genetic feasibility, these different genetic elements can be creatively chosen and arranged together. Note that GAL80 needs to be distal to one FRT and that GAL4 and UAS-reporter cannot be on the same chromosome arm as GAL80. A sample set of MARCM-applicable flies is *hs-FLP; FRT G13, tubP-GAL80/Cyo; +; GAL4-OK107* and *FRT G13, UAS-mCD8::GFP. GAL4-OK107* is a pan-MB GAL4 driver and permits selective labeling of MB clones in mosaic brains (see Fig. 2).

See Troubleshooting.

Cross MARCM-Applicable Parental Flies

2. Collect 20–30 virgin female and 10–15 male MARCM-applicable flies to make the cross.

3. Culture the parental flies in a vial containing fresh cornmeal medium with dry yeast extract for 1–2 d to allow fertilization of females and laying of eggs. Keep transferring the parental flies to new vials to expand the progeny.

See Troubleshooting.

Induction of MARCM Clones

FLP activity can be induced either in a tissue-specific manner (Step 4) or under the control of a heat-shock promoter (Steps 5–8). One major advantage of using a heat-shock promoter to control FLP activity is the ability to induce MARCM clones at specific developmental times. To achieve stringent temporal induction, synchronized animals are required. Transient heat shock should be performed before (or when) cells of interest are being produced. The duration of heat shock needed can vary from 10 to 60 min at 37°C, depending on the strength of the hs-FLP line used.

4. The expression of FLP in neuronal precursors is intrinsically controlled, so to induce FLP activity in a tissue-specific manner, just continue to transfer the parental flies to new vials. When enough offspring are obtained, proceed to Step 9.

5. To induce FLP activity under the control of a heat-shock promoter, let the parental flies mate for 1–2 d at 25°C.

6. Transfer the parental flies to a fresh vial containing yeast extract and allow the females to lay eggs for 2–3 h at 25°C. To obtain enough progeny, repeat this step and mark the time for each egg collection. To gather an appropriate amount of embryos in 2–3 h, adjust the number of MARCM flies.

7. Allow the offspring to develop until specific developmental times and then apply heat shock by incubating in a water bath set at 37°C for an appropriate period of time. (For example, try incubating for 30 min, then adjust the length of heat shock according to the clone induction efficiency.)

8. After heat shock, return the developing animals to the fly incubator set at to 25°C. Wait until the desired developmental stage for mosaic analysis, and proceed to Step 9.

   See Troubleshooting.

Dissection and Tissue Staining

9. Dissect the animals with the right genotype at the developmental stage of interest and perform immunostaining for detection of UAS-marker expression using various fluorophore-conjugated secondary antibodies.

Image Collection and Analysis

10. Analyze the MARCM samples using a regular fluorescence microscope or a confocal microscope. Follow the instructions provided by the microscope manufacturer to collect images of MARCM clones.

    See Troubleshooting.

DISCUSSION

Patterns of wild-type clones reflect patterns of neurogenesis. MARCM has thus become a very powerful tool for studying brain development. In particular, a multicellular NB clone reveals lineage information (e.g., Figs. 1B and 2B). One can further label single neurons derived from the same progenitor for detailed analysis of neuron types and their projections (e.g., Figs. 1B and 2C–F). Most informative MARCM studies involve use of a lineage-specific GAL4 driver plus induction of mitotic recombination at specific times of development to target specific neurons based on parentage as well as birth timing (e.g., Figs. 2C–F). Note that mitotic recombination needs to be induced in the neuronal precursors rather than the postmitotic neurons. The frequency of obtaining clones of particular types may vary drastically depending on the genetic components used and the nature of the lineages involved. Efficiency of mitotic recombination differs significantly among the available hs-FLPs as well as FRT sites. For example, *hs-FLP22* on the X chromosome is more active than *hs-FLP1* on the same chromosome. Also, *FRT40A*, *FRTG13*, and *FRT82B* respond more reliably than *FRT19A* to temporal induction of FLP. Moreover, the efficiency of targeting the same neural progenitor for mitotic recombination may vary at different times of development. For example, the frequency of obtaining NB clones of the MBs drops sharply on rapid production of MB neurons.

TROUBLESHOOTING

Problem (Step 1): MARCM-applicable parental flies are sick.
Solution: Redistribute the MARCM genetic elements between the MARCM-applicable male and
female flies.

Problem (Step 3): MARCM cross yields little or no progeny.
Solution: Arrange the MARCM genetic elements in different combinations, or simply switch the
sexes for parental flies and use more young virgin females in the cross. Try various balancer
chromosomes and avoid coexistence of multiple balancers. Note some *tubP-GAL80* homozy-
gotes are sick and may not be as fertile as *tubP-GAL80* over a balancer chromosome.

Problem (Step 8): Few MARCM progeny survive after heat shock.
Solution: Check the temperature of the water bath. Reduce the length of heat shock. If efficiency
becomes an issue, try multiple rounds of tolerable heat shock.

Problem (Step 10): No MARCM clone is seen.
Solution: Check the MARCM genetic elements individually and ensure they are properly arranged.
First, determine GAL4 and UAS-reporter expression patterns at the stage of phenotypic analysis. Second, examine the efficiency of FLP/FRT using a ubiquitous driver (e.g., *tubP-GAL4*) or
other MARCM-applicable lines known to work efficiently. Third, determine if *tubP-GAL80*
resides distal to FRT and not on the same chromosome arm as GAL4 or UAS-reporter. Finally,
ensure induction of FLP activity in the progenitors rather than their postmitotic neurons. If not
sure, try to induce recombination at different developmental stages to determine an appropriate condition.

Problem (Step 10): Low efficiency of clone induction.
Solution: Synchronize organism development and determine the timing for efficient clone induction. Ensure use of the same FRT insertion in a given set of MARCM-applicable lines. Try more
potent FLP transgenes, and intensify the heat shock for induction of *hs-FLP*. If the efficiency
remains low, consider rebuilding the MARCM-applicable lines to ensure use of intact FRT sites.
After all, note that the efficiency in obtaining different types of clone can differ drastically
despite optimizing all the conditions.

Problem (Step 10): Too many background clones.
Solution: Use a tissue-specific GAL4 driver to selectively mark the clones of interest. Control FLP
activity spatially and/or temporally to selectively target specific progenitors. Try a weaker and
less leaky *hs-FLP*, or reduce the degree of heat shock.

Problem (Step 10): Weak labeling of MARCM clones.
Solution: Try a stronger GAL4 driver, a brighter reporter line, or multiple copies of UAS-reporter.

Problem (Step 10): Weak GAL4-dependent background in MARCM samples.
Solution: One copy of *tubP-GAL80* might not be sufficient to suppress a very strong GAL4 driver. In
addition, depending on insertion sites, *tubP-GAL80* might not be uniformly expressed through
all tissues. To eliminate the residual GAL4 pattern, try multiple copies of *tubP-GAL80* or generate a GAL80 counterpart of the GAL4.

Reverse Genetics by Loss-of-Function Mosaic Analysis

One major application of MARCM is to study cell-autonomous function(s) of a gene within single cells or a group of cells in otherwise unperturbed organisms (e.g., Lee et al. 2000a). In this application, a mutation of interest situated distal to one FRT site is put *in trans* to a *tubP-GAL80*-containing chromosome arm that carries the same FRT. The resulting MARCM clones, which are negative for *tubP-GAL80* and thus specifically marked, will become homozygous for the mutation in otherwise heterozygous organisms. Moreover, by including a UAS-transgene, one can perform rescue experiments in the mutant MARCM clones. Conversely, if the mutation is placed on the same chromosome arm as *tubP-GAL80*, MARCM-labeled cells will be homozygous wild-type and may lie adjacent to sister cells that are homozygous mutant. This MARCM variant, called reverse MARCM, allows one to determine non-cell-autonomous effects of a mutation.

MATERIALS

See Protocol 1.

METHOD

Genetic Requirements and Strategies

1. Prepare MARCM-applicable flies. A desired mutant allele must be inserted behind an FRT site and on the opposite homologous chromosome arm from *tubP-GAL80* (Fig. 1A). This will allow creation of GAL80-minus clones that are homozygous for the mutation of interest. If possible, check the presence of the mutant allele on the FRT recombinant chromosome by a complementation test using another null allele. Ascribing any phenotype of interest to loss of a particular gene further requires MARCM analysis with multiple independent alleles or rescue of the phenotype by a wild-type transgene. For the rescue experiments, UAS-transgene can be placed on any chromosome arm except the one that carries *tubP-GAL80*.

2. Proceed as in Protocol 1, Steps 2–10.

   See Troubleshooting.

Genetic Requirements for Reverse MARCM

3. Prepare MARCM-applicable flies as above, except place the mutation of interest on the same chromosome arm as *tubP-GAL80*.

4. Proceed as in Protocol 1, Steps 2–10.

   See Troubleshooting.

DISCUSSION

The potential for mRNA/protein perdurance means that mutant phenotypes may not be obvious in mutant single-cell clones because the wild-type transcripts or proteins inherited from their heterozygous precursors could persist through initial neuron development. In this case, two-cell or NB clones are more effective for showing mutant phenotypes. Similar situations may occur in the rescue experiments: GAL80 perdurance may delay the expression of wild-type UAS-transgene, thus preventing rescue of early neuron development. If GAL80 perdurance is not a concern, one can also manipulate gene function specifically in GAL80-minus clones using dominant-negative UAS-transgenes or by GAL4-dependent targeted RNA interference. In addition, qualitatively distinct phenotypes may be observed in the clones of different sizes owing to differential requirement of the same gene in the NB versus GMCs or at different stages of the lineage development.

TROUBLESHOOTING

Problem (Steps 2 and 4): No mutant MARCM clone is seen.

Solution: The mutant clones might die or lose proper cell fate and become negative for the GAL4 driver. To distinguish these possibilities and rule out any effect of a background mutation, repeat the MARCM experiments using additional alleles, possibly some hypomorphic mutations, and mark the clones with a ubiquitous GAL4 driver. One should also try to locate clones shortly after the induction. General troubleshooting advice for MARCM clone induction is provided in Protocol 1.

Problem (Steps 2 and 4): Marked clones appear in GAL4-negative lineages.

Solution: Mutant clones could lose cell memory and ectopically express inappropriate cell-fate markers, including diverse cell type–specific GAL4 drivers. Repeat the MARCM experiments using a ubiquitous GAL4 driver and determine if the patterns of clones make sense now.

Problem (Steps 2 and 4): Mutant clones show variable phenotypes.

Solution: Diverse factors may contribute to variable mutant phenotypes in the MARCM clones, such as the perdurance issues or a temporal requirement of the target gene. Normally, perdurance is less in two-cell or NB clones, which should show more severe phenotypes. However, a pleiotropic gene may play distinct roles in NBs versus GMCs and even be reused in postmitotic neurons. Further, it may be differentially required at different stages through development of a protracted lineage. Systematic analysis of diverse MARCM clones generated at different developmental stages would be necessary to resolve such complex situations.

Problem (Steps 2 and 4): Mutant phenotypes cannot be rescued by wild-type transgene.

Solution: UAS-transgene may not be effectively expressed during early neuron development owing to perdurance of GAL80. In addition, GAL4-dependent induction of the rescuing transgene may elicit gain-of-function phenotypes even in wild-type clones. A complete rescue may ultimately require supplement of a genomic fragment or the complementary DNA directly driven by its endogenous promoter.

Genetic Mosaic Screens in the MBs

In neurons whose progenitors can be efficiently targeted for mitotic recombination, genetic mosaic screens can be used to systematically uncover cell-autonomous genes that are required for development or function. This technique involves the generation of numerous FRT lines carrying various independent mutations, followed by derivation and phenotypic analysis of MARCM clones using these mutant FRT lines in combination with a MARCM-enabling stock that carries all the other genetic elements required for MARCM. Mutants of interest are recovered based on the MARCM phenotypes, which are imaged live using diverse fluorescent markers. Mutant genes that underlie the phenotypes of interest can then be identified by conventional genetics including derivation and analysis of series of recombinant chromosomes. Besides chemical mutagenesis, genes on a particular FRT chromosome may be randomly disrupted by P element insertion. A collection of FRT chromosomes carrying various *piggyBac* insertions has been generated by Dr. Liqun Luo's laboratory and these have recently been made available through the *Drosophila* Genetic Resource Center (DGRC) in Kyoto, Japan (Schuldiner et al. 2008). These lines are ready for MARCM experiments and the genomic location of each *piggyBac* transposon has been validated, making forward genetic mosaic screens much less involved. Here we describe procedures specifically used for genetic mosaic screens in the MBs.

MATERIALS

CAUTION: See Appendix for proper handling of materials marked with <!>.

Reagents

Ethyl methanesulfonate (EMS) <!>
See also Protocol 1.

Equipment

See Protocol 1.

METHOD

Prepare MARCM Flies Carrying EMS-Mutagenized FRT Chromosomes

1. Subject about 100 UAS-mCD8::GFP, FRT male flies per vial to 25 mM EMS according to standard procedures (see Fig. 3 for genetic scheme; Lewis and Bacher 1968).

   The concentration of EMS is adjustable and can be monitored by scoring for X-linked lethals. Thirty percent carrying X-linked lethals is approximately equal to one hit per autosome.

2. Cross EMS-treated males with virgin females carrying dominant visible markers on the corresponding homologous chromosomes. Discard parental flies after a certain time to avoid redundantly screening the same mutations.

   See Troubleshooting.

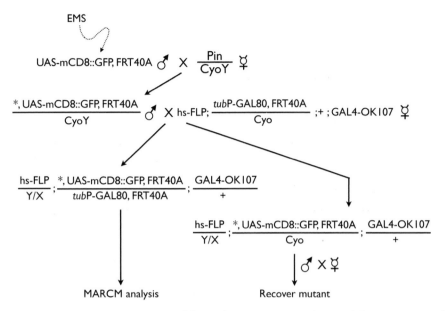

FIGURE 3. Genetic scheme for an EMS-mediated forward genetic screening for MB-defective mutants. The asterisk represents a mutagenized chromosome.

3. Cross individual male progeny that carry the mutagenized FRT chromosome over a balancer or any dominant visible marker with MARCM-applicable females (e.g., hs-FLP; tubP-GAL80, FRT40A; GAL4-OK107 for MB screens on the FRT40A chromosome arm).

Induction of MARCM Clones

4. Follow Protocol 1, Steps 3–8.

Dissection and Analysis

5. Dissect the mosaic animals with the right genotype at the derived developmental stage. Mount live brains in proper orientation for direct analysis of fluorescently labeled MARCM clones under a fluorescent microscope.

   See Troubleshooting.

6. For the lines that consistently yield MB clones showing specific phenotypes of interest, proceed to recover the mutagenized FRT chromosome by conventional genetics.

   See Troubleshooting.

DISCUSSION

Productive genetic mosaic screens require optimized genetics, adequate mutagenesis, efficient production of stereotyped clones, unambiguous detection of mutant clones, and easy in-depth analysis of diverse phenotypes of interest. The five genetic elements required for MARCM should be thoughtfully arranged between the mutagenized male flies and their female partners so that enough well-synchronized MARCM progeny can be readily obtained from individual single-male crosses and that one can promptly recover the mutant chromosomes of interest afterward. An ideal level of mutagenesis is to induce one lethal hit per chromosome arm, to balance the cost-effectiveness of the screen and the challenges in handling multiple lethal hits simultaneously. To effectively generate the desired clones, one should trigger mitotic recombination in various controlled manners to target specific progenitors at right stages. To simplify phenotypic analysis, one should selectively label the clones of particular lineages using a lineage-specific GAL4 driver. A strong GAL4 driver plus multiple copies of UAS-fluorescent reporters might be needed for live imaging of single-cell clones. Productive primary screens ultimately provide the ability to identify your dream mutants, which requires informative phenotypic analysis as to their cell fates, morphologies or subcellular organization, and development. Genetic mosaic screens based on MB gross morphology have permitted identification of genes controlling various aspects of MB development (Lee et al. 2000a,b; Wang et al. 2002; Reuter et al. 2003; Zheng et al. 2003; Zhu et al. 2003, 2006; Schuldiner et al. 2008). Use of a dendritic marker in an analogous screen further allows one to selectively uncover genes governing neuron polarity (Yang et al. 2008).

TROUBLESHOOTING

Problem (Step 2): EMS-treated flies are not fertile.
Solution: Male flies may be sterilized by excessive EMS treatment. Check the EMS concentration and shorten their exposure to EMS if necessary.

Problem (Step 5): MARCM clones are hardly seen in live samples.
Solution: Confirm the presence of MARCM clones by immunostaining. Boost marker expression by increasing the copy number of UAS-reporter or resort to a stronger GAL4 driver or a brighter fluorescent marker.

Problem (Step 6): Abnormal MARCM clones are rarely observed.
Solution: EMS efficiency may vary from batch to batch and decay over time. Ensure flies are adequately mutagenized by determining their average lethal hits.

REFERENCES

Awasaki T, Lai S-L, Ito K, Lee T. 2008. Organization and postembryonic development of glial cells in the adult central brain of *Drosophila*. *J Neurosci* **28:** 13742–13753.

Jefferis GS, Marin EC, Stocker RF, Luo L. 2001. Target neuron prespecification in the olfactory map of *Drosophila*. *Nature* **414:** 204–208.

Knoblich JA. 2008. Mechanisms of asymmetric stem cell division. *Cell* **132:** 583–597.

Lai S-L, Lee T. 2006. Genetic mosaic with dual binary transcriptional systems in *Drosophila*. *Nat Neurosci* **9:** 703–709.

Lai S-L, Awasaki T, Ito K, Lee T. 2008. Clonal analysis of *Drosophila* antennal lobe neurons: Diverse neuronal architectures in the lateral neuroblast lineage. *Development* **135:** 2883–2893.

Lee T, Luo L. 1999. Mosaic analysis with a repressible cell marker for studies of gene function in neuronal morphogenesis. *Neuron* **22:** 451–461.

Lee T, Luo L. 2001. Mosaic analysis with a repressible cell marker (MARCM) for *Drosophila* neural development. *Trends Neurosci* **24:** 251–254.

Lee T, Lee A, Lu L. 1999. Development of the *Drosophila* mushroom bodies: Sequential generation of three distinct types of neurons from a neuroblast. *Development* **126:** 4065–4076.

Lee T, Winter C, Marticke SS, Lee A, Luo L. 2000a. Essential roles of *Drosophila* RhoA in the regulation of neuroblast proliferation and dendritic but not axonal morphogenesis. *Neuron* **25:** 307–316.

Lee T, Marticke S, Sung C, Robinow S, Luo L. 2000b. Cell-autonomous requirement of the USP/EcR-B ecdysone receptor for mushroom body neuronal remodeling in *Drosophila*. *Neuron* **28:** 807–818.

Lewis EB, Bacher F. 1968. Method of feeding ethylmethane sulfonate (EMS) to *Drosophila* males. *Dros Info Ser* **43:** 193.

Reuter JE, Nardine TM, Penton A, Billuart P, Scott EK, Usui T, Uemura T, Luo L. 2003. A mosaic genetic screen for genes necessary for *Drosophila* mushroom body neuronal morphogenesis. *Development* **130:** 1203–1213.

Schuldiner O, Berdnik D, Ma Levy J, Wu JS, Luginbuhl D, Gontang AC, Luo L. 2008. piggyBac-based mosaic screen identifies a postmitotic function for cohesin in regulating developmental axon pruning. *Dev Cell* **14:** 227–238.

Wang J, Zugates CT, Liang IL, Lee C-HJ, Lee T. 2002. *Drosophila* Dscam is required for divergent segregation of sister branches and suppresses ectopic bifurcation of axons. *Neuron* **33:** 559–571.

Wu JS, Luo L. 2006. A protocol for mosaic analysis with a repressible cell marker (MARCM) in *Drosophila*. *Nat Protoc* **1:** 2583–2589.

Yang JS-J, Bai J-M, Lee T. 2008. Dynein-dynactin complex is essential for dendritic restriction of TM1-containing *Drosophila* Dscam. *PLoS ONE* **3:** e3504. doi: 10.1371/journal.pone.0003504.

Yu H-H, Lee T. 2007. Neuronal temporal identity in post-embryonic *Drosophila* brain. *Trends Neurosci* **30:** 520–526.

Zheng X, Wang J, Haerry TE, Wu AY-H, Martin J, B O'Connor M, Lee C-HJ, Lee T. 2003. TGF-β signaling activates steroid hormone receptor expression during neuronal remodeling in the *Drosophila* brain. *Cell* **112:** 303–315.

Zhu S, Perez R, Pan M, Lee T. 2003. Requirement of Cul3 for axonal arborization and dendritic elaboration in *Drosophila* mushroom body neurons. *J Neurosci* **25:** 4189–4197.

Zhu S, Lin S, Kao C-F, Awasaki T, Chiang A-S, Lee T. 2006. Gradients of the *Drosophila* Chinmo BTB-zinc finger protein govern neuronal temporal identity. *Cell* **127:** 409–422.

9 Studying Olfactory Development and Neuroanatomy with Clonal Analysis

Aaron Ostrovsky, Sebastian Cachero, and Gregory Jefferis
*Division of Neurobiology, Medical Research Council Laboratory of Molecular Biology,
Cambridge CB2 0QH, United Kingdom*

ABSTRACT

We describe the use of clonal analysis with the MARCM (mosaic analysis with a repressible cell marker) system for studying cell lineage, development, and anatomy in the fly olfactory system. We provide detailed protocols for generating flies with mosaic labeling (Protocol 1); dissecting, staining, and imaging brains (Protocol 2); and using image registration software to merge images from different brains into a common frame of reference (Protocol 3). These techniques can be applied to the study of development and anatomy of any part of the fly brain.

ANATOMY OF THE OLFACTORY SYSTEM

There are two main parts to the insect olfactory system: the peripheral sensory organs and the central processing areas in the brain. In the periphery, olfactory receptor neurons (ORNs) express one of approximately 60 odorant receptor proteins and each ORN is housed in a sensory bristle or sensillum, generally with one or more other ORNs. This arrangement is stereotyped in the fly, so that each sensillum contains a prescribed set of ORNs. The axons of ORNs expressing the same receptor all converge on individual subregions of the antennal lobe (AL), called glomeruli (Fig. 1A–C). In an adult fly, each glomerulus contains processes from ORNs, local neurons (LNs), and projection neurons (PNs). LNs are diverse in nature and can be either excitatory or inhibitory; they usually have processes in many glomeruli and may synapse on ORNs, PNs, or other LNs. The majority of PNs have dendrites in one glomerulus and all send a projection along one of three antennocerebral tracks to the lateral protocerebrum. PNs innervating a given glomerulus have stereotyped axon terminal arbors in the lateral protocerebrum (Marin et al. 2002; Wong et al. 2002).

Holometabolic insects such as *Drosophila melanogaster* undergo multiple rounds of neurogenesis during development. The first round occurs during embryogenesis and gives rise to the fully functional nervous system of the larva. A second round of neurogenesis during larval development produces neurons specific to the adult nervous system, which form connections during pupal meta-

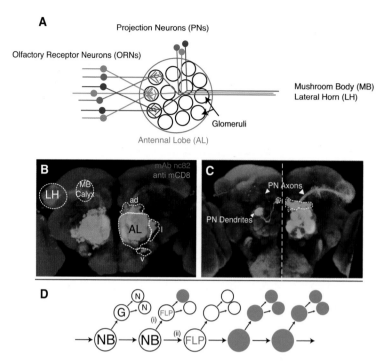

FIGURE 1. Olfactory system organization and neuroblast division patterns. (*A*) Schematic cartoon of the organization of the olfactory receptor neurons and the olfactory projection neurons in the insect antennal lobe. (*B*) Projection of a fly's brain labeled with GH146-Gal4 UAS:mCD8-GFP. On the fly's left side, the antennal lobe (AL) is delineated along with the three cell body clusters (anterodorsal, ad; lateral, l; and ventral, v) each descended from one neuroblast. On the fly's right side, the location of the mushroom body (MB) and lateral horn (LH) are noted. (*C*) Projection of a fly's brain. On the fly's left side, a MARCM clone of the anterodorsal neuroblast clone is outlined. On the fly's right side, a single-cell MARCM clone of the PN innervating glomerulus DL1 is shown. (*D*) A cartoon of the canonical neuroblast division pattern. Asymmetric division gives a ganglion mother cell (G) and another neuroblast (NB). (i) If the FLP activity is driven in the ganglion mother cell, then a single neuron is labeled (green) with mCD8-GFP. (ii) If the FLP activity is driven in a neuroblast, a neuroblast clone can be labeled (green), which includes all the cells expressing Gal4/UAS-mCD8-GFP that descend from that neuroblast.

morphosis. Larvae possess an olfactory circuit that is organizationally similar to, but numerically less complex than, the adult circuit. Embryonically derived ORNs appear to degenerate during this process, but some PNs persist and are then integrated into the adult olfactory system (Marin et al. 2005).

DEVELOPMENT OF THE OLFACTORY SYSTEM

There are two different patterns of neurogenesis in the insect olfactory system. The antennal sensilla containing one to four sensory neurons are each derived from a single progenitor cell that produces both neurons and support cells (Endo et al. 2007). In contrast, central neurons originate from asymmetric division of a long-lived neural stem cell or neuroblast (Fig. 1D). For example, in the antennal lobe three neuroblasts undergo multiple rounds of asymmetric cell division. At each division, the larger daughter cell remains a neuroblast, whereas the smaller daughter is a ganglion mother cell that divides once more, resulting in two postmitotic neurons.

ORNs in the periphery begin sending axons to the central brain 14 h after puparium formation (APF) and a significant volume of processes appear in the developing antennal lobe by 26 h APF. PNs begin developing within 24 h of larval hatching and their axons rapidly grow out to the lateral protocerebrum. However, dendrites and axon terminal arbors do not develop until the pupal stage. PN dendrites first appear at 0 h APF and innervate a newly developing AL that appears to form de novo. By 18 h APF, these PN dendrites form restricted distributions within the developing AL that generally match their positions in the adult AL; a coarse dendritic map therefore forms before ORN axons begin to innervate the developing AL. By 35 h, ORN axons segregate to form protoglomeruli at the periphery of the developing AL and then invade the AL, merging with the dendritic projections of PNs. By 50 h APF, the developing AL has adopted its adult organization (Jefferis et al. 2004).

MOSAIC ANALYSIS WITH A REPRESSIBLE CELL MARKER

A single animal with cells of different genotypes is termed a mosaic. The generation of transgenic mosaic animals is a powerful tool because it allows the creation and analysis of animals with mutations in specific tissues. Mosaic analysis is particularly useful for mutations that are lethal when present in all cells in the animal (Ikeda and Kaplan 1970) or to determine the site of action of a gene. Mosaic strategies can also be very useful for cell lineage tracing. There are a variety of strategies for genetic mosaic analysis. In this chapter, we focus on a particularly powerful technique known as MARCM (Lee and Luo 1999), which requires a marker transgene whose expression is initially prevented by a genetic repressor. The repressor must be present in only one copy (i.e., heterozygous) so that a mitotic recombination event in dividing cells can result in the loss of the repressor from one daughter cell allowing the marker transgene to be expressed (Fig. 2).

The standard form of MARCM requires five different transgenes. A Gal4-UAS pair is selected to define a particular expression pattern of interest. Gal80, a repressor of the binary Gal4-UAS-expression system, is present on one chromosome and is usually driven by a strong ubiquitous promoter (such as tubulin promoter) so that Gal4 is repressed in all tissues. The last two essential transgenes are the flippase (FLP) recombinase enzyme and its FLP recognition target (FRT) DNA sequences, which determine when and where mitotic recombination can take place (Xu and Rubin 1993). An FRT site is placed between the centromere and the Gal80 transgene; the homologous chromosome, which lacks Gal80, must also contain the same FRT site. FLP can catalyze exchange between homologous FRT sites on sister chromatids, resulting in exchange of the distal part of the sister chromatids. If this happens during G2/M phase when chromatids are duplicated (i.e., four copies of each chromosome are present), then normal segregation of these two pairs can result in one daughter cell with two copies of the repressor gene and the other daughter with none.

As mature neurons are generally postmitotic, MARCM is effective only when used during the development of the nervous system when neural precursors are dividing. Most central neuroblasts undergo asymmetric cell division to yield a ganglion mother cell and another neuroblast that can continue this pattern or stop dividing (but see Bello et al. 2008 and Bowman et al. 2008 for exceptions) (Fig. 1D). If the recombination event occurs in the neuroblast, the Gal80 can either be passed to the ganglion mother cell or to the daughter neuroblast. If the neuroblast loses Gal80, then all of

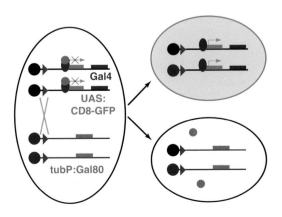

FIGURE 2. MARCM depends on two main elements, the binary expression system Gal4 (blue rectangle and oval) UAS (green rectangle) with its repressor Gal80 (red rectangle and circle) and the FLP-FRT (purple triangles) system needed to induce targeted recombination between homologous chromosomes. On heat shock, cells undergo recombination between FRT sites. If the cell is in G2 phase, the recombination can occur between any of the four replicated sister chromatids. When such a cell divides, one of the daughter cells can inherit two chromosomes with Gal80 transgenes and the other cell would then have no Gal80 transgenes. The result is one cell that is now uniquely labeled (because of the loss of the Gal80 repressor) and one that remains unlabeled.

its progeny that express the Gal4 line of interest will be labeled. Alternatively, if the ganglion mother cell loses Gal80, then its two daughter neurons will be labeled contingent on their expression of Gal4. Finally, if the recombination occurs in the ganglion mother cell, then one of the two daughter neurons will be without Gal80 repression and will be labeled if it expresses the Gal4.

The FLP enzyme is typically driven by a heat-shock promoter, which results in stochastic recombination events after heat shock. This can be advantageous because the duration of heat shock can be titrated to cause recombination in only one or a few dividing cells in each animal. This can be used to simplify a complex Gal4 expression pattern into clusters of neurons made by a single neuroblast or to reveal the morphology of single cells.

A number of possible variants of the basic MARCM scheme have been devised. Perhaps the most basic one is the incorporation of a mutant allele distal to the FRT on the chromosome without the Gal80 transgene; this results in positively labeled clones that are homozygous mutant for the gene of interest. When MARCM is used in this way to generate mutant clones, it is vital to remember that cells that do not express the Gal4 transgene can never be labeled. There may therefore be many more mutant cells in the animal than initially apparent from the Gal4 expression.

An additional UAS transgene can be introduced to misexpress a gene of interest or express a rescue construct in a mutant clone (e.g., Komiyama et al. 2003). Instead of heat-shock FLP, it is also possible to use a cell type–specific FLP to restrict the generation of clones to a specific cell type (e.g., Lee et al. 2001). Finally, a more complex variation is the use of a second binary expression system such as the LexA/LexAOp system in addition to the Gal4/UAS system (Lai and Lee 2006). Using two expression systems allows the expression of two different fluorescent protein markers driven by two different transgenes. This can be used to test whether two groups of neurons, labeled by different enhancer traps, originate from the same neuroblast. The LexA system can also be Gal80 independent. The LexA system can then be used to manipulate one neuronal population whereas the MARCM system is used to visualize a second neuronal population.

MARCM FOR STUDYING DEVELOPMENT AND BIRTH ORDER OF NEURONS

MARCM has now been applied to a large variety of cell populations in the fly. It was first used to study developmental patterning of mushroom body (MB) neurons (Lee et al. 1999), showing that each of the four MB neuroblasts sequentially generates three neuronal types. MARCM has also been used to study the production of glia during development (Awasaki et al. 2008) and examine the full morphology and development of the three antennal lobe neuroblast clones (Lai et al. 2008).

Because the set of cells labeled in a MARCM experiment is under both genetic and temporal control, it can be used to study the birth order of cells. Unlike ORNs, which each express a particular OR that can be used to predict their glomerular target, there is no class of genetic marker that unambiguously identifies each different class of PN. In one study in which Gal4 was expressed in a large subset of PNs, early heat shocks that produced single-cell clones always labeled a specific PN subtype that innervates the DL1 glomerulus (Jefferis et al. 2001). Heat shocks applied progressively later labeled single-cell clones that innervated different glomeruli, establishing a relation between PN birth order and neuronal identity.

Protocol 1

Generation of Flies with Mosaic Labeling

This protocol describes how to establish a mating cage for MARCM in PNs of the fly antennal lobe and then select appropriate flies for dissection and staining using immunohistochemistry. As discussed above, the protocol can be adapted to determine the birth order of neuroblast lineages or individual cells. Alternatively, it can be used to dissect a complicated Gal4 line into its component neuroblast lineages to help elucidate projection patterns and connectivity. Collecting newly hatched larvae during a short time window allows for precise control of the stage during development at which the heat shock is applied.

MATERIALS

See the end of the chapter for recipes for reagents marked with <R>.

Reagents

Agar plates for embryo collection cages (usually made with grape juice or apple juice) <R>
Fly lines (Bloomington *Drosophila* Stock Center, Indiana University)
Food vials (VWR; cat. no 89092-728)
Yeast paste <R>

Equipment

Embryo collection (or mating) cages (e.g., Flystuff; cat. no. 59-100)
Timer/stopwatch
Water bath or incubator (set at 37°C)

METHOD

Generating Larvae

To carry out a MARCM experiment you must mate flies of the appropriate genotypes including the five transgenes discussed above. Different applications will require different parental stocks, but a common starter stock is one containing hs-FLP, UAS-CD8-GFP, FRT, and Gal80 transgenes. This can then be mated with a stock containing an experiment-specific Gal4-enhancer trap and the same FRT element to make labeled clones. In the experiment described here, both stocks contain the FRT insertion G13, which is located on the right arm of the second chromosome, and one contains the GH146-Gal4 element that labels the majority of projection neurons (Stocker et al. 1997). The only restriction on where transgenes can be located for MARCM is that the Gal4 transgene (GH146 in this case) cannot be distal to the FRT site on the same chromosome arm as the tubP-Gal80, otherwise they will cosegregate and the repressor will always turn off Gal4.

1. Expand the fly stocks to be used in the experiment. Collect y,w, hs-FLP; UAS-mCD8-GFP; FRTG13 tubP-Gal80 virgin females and FRTG13 Gal4-GH146 males. For a productive cage, use at least 30 virgin females and an equal or greater number of males.

2. Place these flies in a mating cage with a "dollop" of yeast paste on the agar food plate.

3. Remove the collection plate from the mating cage at least once in the morning, early afternoon, and evening. Replace with a fresh plate containing yeast paste.

> Plates can be swapped on and off the cage during the day and discarded after being left overnight on the cage. This approach is advantageous when you would like to heat-shock larvae at tightly controlled time points, although there may be fewer larvae available for each collection. More larvae are available for collection if the plate is left on longer (e.g., overnight) and multiple rounds of larvae can be collected from each plate, but care must be taken to ensure that larvae are the right age. To ensure this, the plate must be completely cleared after each collection.

> *See Troubleshooting.*

4. Clear the yeast from the old plate.

> If you leave the yeast on the plate, you will find it difficult to collect the hatched larvae because they will have crawled into the yeast.

Heat Shock of Larvae to Induce Mitotic Recombination

5. Using forceps to clear them from the plate, remove the first larvae to hatch and discard them. Note the time. From this time point, any further larvae collected will have a known window in which they hatched.

> Larvae are typically collected over a 2-h window. Calculate when your first collection time will be. At 25°C, it takes ~20 h from the time the collection plate was put on the mating cage for the first larva to hatch, and at 22°C it takes ~22 h. For example, if you place the plate on the cage at 4 p.m. on day 1 and store the cage at 25°C, you should be prepared to clear the plate at noon on day 2.

6. Collect 70–90 larvae and place them in a food vial. Note the time of collection.

> Larvae remain in the food vial until they eclose.

7. Set a timer for 20 min and submerge the food vials in a water bath set at 37°C.

> You will have to decide on a heat-shock regime based on the type of clone desired and any other experimental constraints. For example, for larval shocks, heat shock right after larval hatching to bias for full neuroblast clones, or shock later for more single- or double-cell labeling events. Ensure that the water level in the bath is sufficient to cover the vial up to the cotton plug because larvae can crawl out of the food during the heat shock.

8. After 20 min, remove the vials from the water bath and place them at 25°C.

> It is important to leave the larvae enough time after heat shock to undergo mitotic recombination and begin expression of mCD8-GFP before dissecting them. You should wait at least 24 h after heat shock; 48 h will ensure that all of the Gal80 has been degraded and there is sufficient GFP (green fluorescent protein) fluorescence for accurate imaging (Lee and Luo 1999).

9. Leave the larvae at 25°C until they eclose.

10. Repeat Steps 6–9 until the plate stops yielding larvae, then discard the plate.

11. Collect any adults within 48 h after eclosion. Sort the flies to ensure that those dissected have the appropriate genotype.

> If you anticipate any sexual dimorphisms, it is best to separate males from females before you begin dissections.

> *See Troubleshooting.*

TROUBLESHOOTING

Problem (Step 3): Low numbers of eggs are laid.
Solution: Add more flies to the mating cage. The flies can also be kept at a slightly higher temperature such as 25°C to increase the number of eggs laid.

Problem (Step 11): Too many or too few clones per fly.

Solution: Adjust the time and temperature of heat shock. The length of heat shock will greatly affect clone frequency and we have used times in the range of 10 min to 2 h in different situations. Temperature increases of the order of 1°C–2°C will also make a difference.

Protocol 2

Immunochemistry and Imaging

This protocol describes how to obtain well-stained whole brain images that can be used to examine neuronal morphology and are of sufficient quality to be used for image registration.

MATERIALS

CAUTION: See Appendix for proper handling of materials marked with <!>.
See the end of the chapter for recipes for reagents marked with <R>.

Reagents

Anti-CD8a (rat monoclonal antibody) (Invitrogen; cat. no. MCD0800)
Coverslips (glass; #1 thickness)
Electrostatically charged slides (SuperFrost Plus from Thermo Scientific; cat. no. J1800AMNZ)
Ethanol (100%) <!>
Goat anti-mouse Alexa Fluor 568 (Invitrogen; cat. no. A11031)
    Secondary antibodies are prepared at a concentration of 50% (v/v) in glycerol for storage at −20°C. They are used at a final dilution of 1:400.
Goat anti-rat Alexa Fluor 488 (Invitrogen; cat. no. A11006)
Mounting medium (SlowFade [cat. no. S-2828] from Invitrogen or Vectashield [cat. no. H-1000] from Vector Laboratories)
Nail polish
nc82 (mouse monoclonal antibody that recognizes the presynaptic protein Bruchpilot) (Developmental Studies Hybridoma Bank, University of Iowa)
Normal goat serum (Sigma)
Paraformaldehyde <!> (4% in PB) <R>
PB (phosphate buffer) <R>
PBT (phosphate buffer containing 0.1% Triton X-100 <!>) <R>

Equipment

Dissecting microscope with light source
Dissection wells (www.emsdiasum.com; cat. no.71561)
Forceps (#5 Biologie; Dumont)

METHOD

1. Anesthetize the flies that are to be dissected using CO_2 or ether or by cooling on ice.

2. Transfer the flies to a container with 100% ethanol. Leave them there for 30–60 sec.
    This step makes the cuticle less hydrophobic.

3. Transfer the flies to a prechilled dissection well containing PB.
    The choice of dissection medium is a matter of personal preference. Although phosphate-buffered saline and PBT can be used, PB gives the lowest background in fluorescence immunostainings and the use of Triton X-100 (in PBT) before or during fixation may disturb lipid membranes.

4. Dissect the brain out of the head capsule.

   A video from Kei Ito showing how to dissect a fly brain can be found at http://jfly.iam.u-tokyo.ac.jp/html/movie/index.html. Besides the one shown, there are several different techniques: Some remove the head and then perform the dissection, whereas others dissect the brain from the intact animal.

5. Clean the brain, removing fat and the trachea, especially the trachea on the anterior side of the brain, which can interfere with imaging.

   If preferred, the brain can cleaned after fixation. Fixed tissue is more rigid, making it easier to clean.

6. Transfer the brain to a chilled dissection well containing 500 µL of 4% paraformaldehyde in PB.

7. Keep dissecting for up to 1 h.

   An hour is usually enough time for dissecting 10–30 brains, if they are cleaned before fixation, and 20–50 brains, if they are cleaned after fixation.

8. Transfer the dissection well containing the brains in fixative from the ice to an orbital shaker at room temperature. Incubate with gentle agitation for 25 min.

9. Wash the brains with PBT and incubate in the orbital shaker for 15 min at room temperature.

10. Transfer the brains to a microfuge tube and repeat Step 9 five times.

11. Block by adding 400 µL of PBT containing 5% normal goat serum (NGS). Incubate for at least 1 h.

    The type of serum used depends on the origin of the secondary antibodies; if the secondary antibodies are donkey antibodies then normal donkey serum should be used instead of NGS.

12. Remove the blocking solution and add 400 µL of the primary antibodies diluted in PBT containing 5% NGS.

    In the case of MARCM using CD8-GFP, use anti-CD8 diluted 1:200 and nc82 diluted 1:20.

13. Incubate with gentle agitation for 48 h at 4°C.

    It is important to incubate the nc82 antibody for 48 h to get a high-quality, even staining of the neuropil that can be used for registration of the brain.

14. Wash as described in Steps 9 and 10.

15. Add 400 µL of the secondary antibodies diluted in PBT containing 5% NGS.

    Always use secondary antibodies conjugated to green fluorophores for detecting GFP. This avoids the problem of GFP intrinsic fluorescence interfering with the green channel. In this case, use red secondary antibodies for detecting nc82.

16. Incubate with gentle agitation for 48 h at 4°C.

17. Wash as described in Steps 9 and 10.

18. Remove as much liquid as possible and add 200 µL of mounting medium. Incubate for no less than 1 h to allow equilibration.

    It is important to add no less than 200 µL of mounting medium to avoid diluting it. This is critical for the imaging of the brains where larger mismatches between the refractive index of the mounting medium and the immersion oil will lead to increased spherical aberration at deeper optical sections.

19. Using a blunted pipette tip, transfer the brains onto electrostatically charged slides.

    The brains stick better to positively charged slides and do not move around when mounting medium is added in Step 23.

20. Using a pipette with a yellow tip, remove as much of the mounting medium from the slide as possible. Using forceps, orient the brains with their anterior side facing up.

    It can be difficult to distinguish between the anterior and posterior sides of the brain. The most characteristic features of the anterior side are the antennal lobes and the shape of the subesophageal ganglion. Before proceeding, check that the brains are properly oriented using a fluorescence dissecting microscope.

21. Arrange the brains in a grid-like pattern on the slide.

    Maintaining a well-organized grid will make it easier to find brains for reimaging should this be required.

22. Break a #1 coverslip and put two pieces of it by the sides of the brains. Gently drop a coverslip on top of the brains.

    The two pieces of broken glass will prevent the coverslip from crushing the brains.

23. Using a pipette, add mounting medium to the slide on the edge of the coverslip. The medium should spread by capillarity underneath the coverslip. Keep adding medium until the entire space underneath the coverslip is filled.

24. Apply one layer of nail polish to the edges of the coverslip to seal it. Allow it to dry at room temperature. Apply a second layer of nail polish and a third if mounting medium leaks out.

    The slides can now be kept at 4°C until they are imaged or at −20°C for medium-term storage or at −80°C for long-term storage.

25. Imaging is usually performed with a confocal microscope. A few considerations to bear in mind when choosing the imaging parameters are as follows.

    i. On average, a *Drosophila* central brain sitting on its posterior (without including the optic lobes) is 380 µm wide, 300 µm along the dorsoventral axis, and 165 µm deep (anteroposterior axis).

    ii. Depending on the optics of the instrument used, 40x, 25x, and 20x objectives can be used for imaging the whole brain. If the aim is to include only the olfactory circuit (antennal lobe, mushroom body, and lateral horn), then a 40x would be the objective of choice.

    iii. Laser power, photomultiplier gain, and offset will depend on the microscope and on each particular brain. Normally pixel dwell time can be constant for all specimens.

    iv. For most applications, useful resolutions for imaging whole brains are 768 x 768 and 1024 x 1024 pixels and the slice spacing is ~0.5–1 µm. Slice spacing of 0.5–1 µm is optimal for 40x objectives. If lower power objectives are used, larger slice spacing (i.e., >1 µm) might be optimal.

    v. It is a good practice to include the name of the slide and the number of the brain in the grid in the name of the image file. This makes it easy to revisit the brain when necessary.

    See Troubleshooting.

TROUBLESHOOTING

Problem (Step 25): No fluorescent staining of the brain is observed, especially in the nc82 channel.
Solution: Overfixing of the brains disrupts the binding of antibodies to their antigen and nc82 is particularly prone to this. Make sure that fixation does not exceed 25 min at room temperature.

Problem (Step 25): High background fluorescence or nonspecific staining occurs.
Solution: There are three major factors that will contribute to high background or nonspecific staining. First, ensure that you block with the appropriate serum. Second, ensure that you wash the brains for at least 1 h and make a minimum of five to six changes of the PBT to ensure that fixed and unbound antibodies have a chance to diffuse out of the brain. Finally, you can adjust antibody dilutions as necessary to improve signal to noise; this applies to both primary and secondary antibodies.

Problem (Step 25): Damage to brains has occurred.
Solution: Damage can be done to the brains at any stage of the protocol and you must always be careful when dissecting the brains and handling them. Quality is more important than quantity, and

care should be taken to obtain the least damaged brains when dissecting. When processing through washes and addition of antibodies, do not suck brains up into the pipette tip as this can shear even well-fixed brains. You should also avoid muddling the brains with the pipette tip. When transferring brains from one place to another, it is recommended to use a pipette tip that has had the first 0.5–1 cm removed (blunted); do not pull the brains too far up into the pipette tip.

Image Registration

To compare confocal images of labeled neurons in different brains, it may be desirable to register them to a template or standard brain. There are various image registration approaches available. Some depend on manually specifying landmarks on the brains to be registered. Others depend only on the grayscale intensity value of one of the channels in the confocal image. Another important difference between registration approaches is whether they apply linear or nonlinear (warping) transformations. Linear transformations typically include translation, rotation, and scaling along each axis. Nonlinear transformations are much more computationally intensive, but are required to register brains with different shapes. Here we describe the practical steps required for an intensity-based nonlinear registration that has been used to map the higher olfactory centers of the fly brain using the staining for the presynaptic marker Bruchpilot (nc82). This registration is in fact a two-step process. The first step is a linear transformation that roughly aligns the two brains, followed by a second nonlinear step that allows different parts of the brain to move in slightly different directions.

Before starting a registration project, it is worth thinking carefully about the choice of template brain. Ideally, you would use a publicly accessible template—that way you will be able to compare your results directly with those of other groups. However, you may find that such a template is not available or that your brains do not register efficiently to a template generated by another laboratory (perhaps because of differences in staining protocols or genetic background). In this case, you will have to choose/make your own template. It is possible to use a single brain as a template. However, we have found that using an average of 10 or more brains usually results in a template image with significantly higher signal to noise and consequently higher registration success. The construction of an average template brain is out of the scope of this protocol, but see, for example, Jefferis et al. (2007) and Kurylas et al. (2008) for information.

MATERIALS

You will need to install ImageJ from http://rsbweb.nih.gov/ij/ along with the Biorad Writer and Nrrd Reader plug-ins available at http://flybrain.stanford.edu/imageJplugins. The easiest way to get all of this in a single package is to download the Fiji distribution of ImageJ at http://pacific.mpi-cbg.de. All of these tools are available for Windows, Mac, and Linux; follow the installation procedures detailed on those sites.

The core registration software is also cross platform and is available from http://www.nitrc.org/projects/cmtk. However, for simplicity we strongly recommend that you use the basic graphical user interface (GUI) available at http://flybrain.stanford.edu/GUI. Unfortunately, this GUI is presently only available for Mac OS X. Download the zip file and uncompress it in your /Applications folder. A test data set to confirm the procedure is available at http://flybrain.stanford.edu/warpbench.

METHOD

1. Start with a set of confocal images acquired in Protocol 2; nc82 staining is the basis for the registration procedure.

2. Make a folder for your work and make a subfolder called "images."

   The name "images" (all lowercase) is important.

3. Save all images into the images subfolder in Biorad PIC format using the Biorad Writer plug-in. Save each confocal channel separately with filenames in the form myfirstbrain01.pic, myfirstbrain02.pic, etc. Make sure that the nc82 channel is saved with the suffix 01.pic and do not use spaces or underscores in your filenames.

   You may use an underscore to separate the channel number (e.g., brain1_01.pic, brain1_02.pic), but they must not be present anywhere else in the image file name.

4. Choose a well-dissected and stained specimen as the template.

5. Double click on RegistrationRunner inside /Applications/IGSRegistrationTools and follow the prompts.

6. Choose the images folder containing your images

   If you choose a single image in this folder, then only that image will be registered.

7. Answer "No" to the question "Do you want to use the default reference brain?".

8. Choose your template brain.

9. Edit the command line options in the next window to remove the -t test option.

10. Run the registrations. A whole brain will take several hours on an eight-processor Apple Mac Pro.

11. Use the Nrrd Reader plug-in to open images that will appear in the reformatted subfolder. It is easiest to do this by dragging and dropping them onto the Fiji toolbar.

12. Score registrations by generating a multicolor image merging the reformatted image with the template image. Use the ImageJ/Fiji Image … Color … Merge Channels function for this. Look for good alignment of the edges of neuropil, axon tracts (which will appear largely nc82-negative), and areas of more intense neuropil staining. To be able to turn each channel on and off easily, further convert the merged stack with the Image5D plug-in's RGB to Image5D option.

13. Examine the other channels.

    See Troubleshooting.

TROUBLESHOOTING

Problem (Step 13): Only one of my images is registered.

Solution: In Step 6, make sure that you select the images folder and not a single image inside the folder.

Problem (Step 13): My images register, but some or all of the images are not reformatted.

Solution: This usually happens if the image name does not specify the channel correctly (see Step 3); the exact form of the image names is critical. They must end in 01.pic, 02.pic, etc.

Problem (Step 13): Registrations fail completely.

Solution: If you see an almost completely dark image, it is likely that the first step linear registration is failing. You can check this by generating a reformatted image based only on this linear registration step. To do this, add the command line option "-l a" in Step 9. If the brains are poorly aligned after the affine step, it may be useful to use a few manual landmarks to initialize your registration. There is a plug-in in the latest version of Fiji to do this. For further details, see the Jefferis Laboratory wiki at flybrain.mrc-lmb.cam.ac.uk.

Problem (Step 13): Registrations rarely work perfectly.

Solution: Consistent specimen preparation and imaging is vital for good quality image registration. Make sure that your brains have even nc82 staining throughout the brain from inside to out—the long incubation times in Protocol 2 are critical for this. If the upper slices in the confocal image come out much brighter than the lower slices, consider (1) changing the mounting medium so that its refractive index more closely matches the oil of your immersion lens; (2) changing the photomultiplier tube (PMT) gain/laser power through the specimen if your software allows this; or (3) using image processing to normalize image intensity through the stack. In general choose one set of imaging parameters (lens, Z-slice spacing, pixel size, pixel dwell time, etc.) and use them consistently.

Problem (Step 13): Brains show excessive warping in certain specific regions.

Solution: Your template brain may have an unusual shape in this region. Consider using shape-based averaging to generate a template brain that is more representative of the population you want to register (Kurylas et al. 2008). One specific example in which this can happen is in sexually dimorphic regions in which you may consider using an intersex template.

RECIPES

CAUTION: See Appendix for proper handling of materials marked with <!>.
Recipes for reagents marked with <R> are included in this list.

Agar Plates for Embryo Collection Cages (Enriched with Fruit Juice)

Reagent	Quantity (for 1 L)
Agar	32 g
Sucrose	16.6 g
Fruit juice (apple or grape)	300 mL

Add distilled H_2O to 1 L. Boil the ingredients while stirring on a hot plate at high setting (the solution will foam and rise and it may be necessary to take it off the heat and put it back a couple of times before fully dissolving the agar). Cool to touch. Add 5 mL of propionic acid <!>/phosphoric acid <!>, 9:1, and stir. Pour into 5-cm Petri dishes. Keep the lids off the plates for 10 min. Cover and store the plates at 4°C.

Paraformaldehyde (4% in PB)

Add 100 µL of 20% EM grade paraformaldehyde (www.emsdiasum.com; cat. no. 15713) <!> to 400 µL of PB <R>.

PB (Phosphate Buffer)

Reagent	Quantity (for 1 L)	Final concentration
$NaH_2PO_4 \cdot H_2O$	4.363 g	0.032 M
Na_2HPO_4	9.705 g	0.068 M

Dissolve reagents in 1 L of H_2O.

PBT (Phosphate Buffer Containing 0.1% Triton X-100)

Reagent	Quantity (for 1 L)	Final concentration
$NaH_2PO_4 \cdot H_2O$	0.4363 g	0.0032 M
Na_2HPO_4	0.9705 g	0.0068 M
NaCl	7.953 g	0.136 M
Triton X-100 <!>	1 mL	0.1%

Dissolve reagents in 1 L of H_2O.

Yeast Paste

Take packaged dry baker's yeast and use small volumes of either water or 0.5% propionic acid <!> to make a paste. Propionic acid inhibits the growth of microbes on the yeast and encourages egg-laying in the female flies.

ACKNOWLEDGMENTS

We thank Nicolas Masse for comments on this manuscript. Work in our laboratory is supported by the Medical Research Council and the European Research Council.

REFERENCES

Awasaki T, Lai SL, Ito K, Lee T. 2008. Organization and postembryonic development of glial cells in the adult central brain of *Drosophila*. *J Neurosci* **28:** 13742–13753.

Bello BC, Izergina N, Caussinus E, Reichert H. 2008. Amplification of neural stem cell proliferation by intermediate progenitor cells in *Drosophila* brain development. *Neural Dev* **3:** 5.

Bowman SK, Rolland V, Betschinger J, Kinsey KA, Emery G, Knoblich JA. 2008. The tumor suppressors Brat and Numb regulate transit-amplifying neuroblast lineages in *Drosophila*. *Dev Cell* **14:** 535–546.

Endo K, Aoki T, Yoda Y, Kimura K, Hama C. 2007. Notch signal organizes the *Drosophila* olfactory circuitry by diversifying the sensory neuronal lineages. *Nat Neurosci* **10:** 153–160.

Ikeda K, Kaplan WD. 1970. Unilaterally patterned neural activity of gynandromorphs, mosaic for a neurological mutant of *Drosophila melanogaster*. *Proc Natl Acad Sci* **67:** 1480–1487.

Jefferis GSXE, Marin EC, Stocker RF, Luo L. 2001. Target neuron prespecification in the olfactory map of *Drosophila*. *Nature* **414:** 204–208.

Jefferis GSXE, Vyas RM, Berdnik D, Ramaekers A, Stocker RF, Tanaka NK, Ito K, Luo L. 2004. Developmental origin of wiring specificity in the olfactory system of *Drosophila*. *Development* **131:** 117–130.

Jefferis GSXE, Potter CJ, Chan AM, Marin EC, Rohlfing T, Maurer CR Jr, Luo L. 2007. Comprehensive maps of *Drosophila* higher olfactory centers: Spatially segregated fruit and pheromone representation. *Cell* **128:** 1187–1203.

Komiyama T, Johnson WA, Luo L, Jefferis GSXE. 2003. From lineage to wiring specificity: POU domain transcription factors control precise connections of *Drosophila* olfactory projection neurons. *Cell* **112:** 157–167.

Kurylas AE, Rohlfing T, Krofczik S, Jenett A, Homberg U. 2008. Standardized atlas of the brain of the desert locust, *Schistocerca gregaria*. *Cell Tissue Res* **333:** 125–145.

Lai SL, Lee T. 2006. Genetic mosaic with dual binary transcriptional systems in *Drosophila*. *Nat Neurosci* **9:** 703–709.

Lai SL, Awasaki T, Ito K, Lee T. 2008. Clonal analysis of *Drosophila* antennal lobe neurons: Diverse neuronal architectures in the lateral neuroblast lineage. *Development* **135:** 2883–2893.

Lee T, Luo L. 1999. Mosaic analysis with a repressible cell marker for studies of gene function in neuronal morphogenesis. *Neuron* **22:** 451–461.

Lee T, Lee A, Luo L. 1999. Development of the *Drosophila* mushroom bodies: Sequential generation of three distinct types of neurons from a neuroblast. *Development* **126:** 4065–4076.

Lee CH, Herman T, Clandinin TR, Lee R, Zipursky SL. 2001. N-cadherin regulates target specificity in the *Drosophila* visual system. *Neuron* **30:** 437–450.

Marin EC, Jefferis GSXE, Komiyama T, Zhu H, Luo L. 2002. Representation of the glomerular olfactory map in the Drosophila brain. *Cell* **109:** 243–255.

Marin EC, Watts RJ, Tanaka NK, Ito K, Luo L. 2005. Developmentally programmed remodeling of the *Drosophila* olfactory circuit. *Development* **132:** 725–737.

Wong AM, Wang JW, Axel R. 2002. Spatial representation of the glomerular map in the *Drosophila* protocerebrum. *Cell* **109:** 229–241.

Xu T, Rubin GM. 1993. Analysis of genetic mosaics in developing and adult *Drosophila* tissues. *Development* **117:** 1223–1237.

10 | Dissecting and Staining *Drosophila* Optic Lobes

Javier Morante and Claude Desplan

Department of Biology, New York University, New York, New York 10003

ABSTRACT

The *Drosophila* visual system is composed of the retina and the optic lobes, which are the ganglia where photoreceptors project and initial processing of visual inputs occurs. This chapter briefly describes the different ganglia that form the visual system and provides a protocol for dissecting and staining *Drosophila* optic lobes at different stages of development (larva, pupa, and adult).

THE *DROSOPHILA* VISUAL SYSTEM

In recent years, the *Drosophila* brain has emerged as a model for studying neural circuits and the different sensory inputs that these circuits process to direct behavior (Luo et al. 2008). The *Drosophila* visual system consists of two main structures, the retina and the underlying optic lobe, that process visual information received from the photoreceptors (Tang and Guo 2001; Rister et al. 2007; Gao et al. 2008; Katsov and Clandinin 2008; Yamaguchi et al. 2008). A number of distinct optic lobe cell types have been described as a result of studies using Gal4 drivers and single neuron labeling techniques (Fischbach and Dittrich 1989; Otsuna and Ito 2006; Rister et al. 2007; Gao et al. 2008; Katsov and Clandinin 2008; Morante and Desplan 2008). In a few instances, the behavior directed by these cells has also been analyzed (Rister et al. 2007; Gao et al. 2008; Katsov and Clandinin 2008).

The enormous cell diversity observed in the visual system offers a powerful system for neurobiologists to study neurogenesis, neuronal polarity, axon guidance, and the correlation of specific cell populations with their physiological functions in driving behavior. The Protocol describes methods for the dissection and staining of *Drosophila* optic lobes that can be used to analyze the anatomy and generation of these neural circuits at any stage of development.

The *Drosophila* Retina

The *Drosophila* compound eye is composed of about 800 individual eyes called ommatidia (Wolff and Ready 1993). Each adult ommatidium contains eight photoreceptor cells (PRs) that can be classified into two groups based on their morphology and the visual input for which they specialize: two inner PRs (R7 and R8) surrounded by six outer PRs (R1–R6). R1–R6 express the wide-spectrum photopigment rhodopsin 1 (Rh1) and are important for motion detection. They can be considered homologous to the vertebrate rod cells. In contrast, R7 and R8 express different rhodopsins (Wernet and

157

Desplan 2004) and have a topographic organization with R7 placed on top of R8. These characteristics strongly suggest that R7 and R8 are involved in color vision, thus resembling vertebrate cone cells.

Ommatidia can be divided into three subtypes based on the rhodopsin content of their inner PRs (Wernet and Desplan 2004). The pale subtype (30%) contains ultraviolet (UV)-sensitive Rh3 in R7 and blue-sensitive Rh5 in R8. The yellow subtype (70%) contains UV-sensitive Rh4 in R7 and green-sensitive Rh6 in R8. Pale and yellow ommatidia are distributed stochastically in the main part of the retina and specialize in the discrimination of short or long wavelengths, respectively. A third subtype of ommatidia, located in the dorsal rim area, contains UV-sensitive Rh3 in both R7 and R8 and is specialized in detecting the vector of polarized light (Wernet et al. 2003).

The spatial distribution and expression of rhodopsin genes in the outer and inner PRs reflects the widely different functions of R1–R6 (motion detection) versus R7 and R8 (color vision). Indeed, the inputs transmitted by the outer and inner PRs are processed in two different targets in the optic lobes (Morante and Desplan 2004). Outer PRs project to the first optic neuropil, called the lamina, where they connect to five lamina neuron subtypes, L1–L5. Processing of information coming from R7 and R8 occurs in two distinct layers of the medulla, the second optic ganglion. R8 synapses in the superficial M3 layer, whereas R7 terminates in the deeper M6 layer. The medulla also receives inputs from L1–L5 lamina cells and thus merges motion detection and color vision.

The Optic Lobe Neuropils

The *Drosophila* optic lobes are formed by several structures that mediate different behaviors and represent different levels of processing: the lamina, the medulla, and the lobula complex, which is formed by the lobula and the lobula plate (Fig. 1). The optic lobes develop from two larval neuroepithelia: the inner proliferation center (IPC) and the outer proliferation center (OPC) (White and Kankel 1978) (Fig. 2A), where neuroepithelial cells increase in number through symmetric cell divisions with a plane of cell division parallel to the epithelial surface. In the "transition zones" (Egger et al. 2007; Yasugi et al. 2008), epithelial cells give rise to neuroblasts that divide asymmetrically with a plane of cell division that is now perpendicular to the epithelium to self-renew and produce differentiated optic lobe neurons (Egger et al. 2007; Yasugi et al. 2008). As a result, the IPC gives rise to the lobula complex, and distinct regions of the OPC generate the cells of the medulla cortex, medulla rim, and lamina (Meinertzhagen and Hanson 1993; J Morante and C Desplan, unpubl.) (Fig. 2).

The optic lobes represent the largest structure in the fly brain, and each lobe is formed by an estimated 60,000 neurons (Hofbauer and Campos-Ortega 1990). The optic lobes contain a very large number of cell types; for example, the medulla alone has been estimated to contain more than 60 different cell types (Cajal and Sanchez 1915; Strausfeld 1976; Fischbach and Dittrich 1989; Meinertzhagen and Sorra 2001; Otsuna and Ito 2006; Morante and Desplan 2008). How this cell diversity is generated remains mostly unknown.

Adult optic lobes

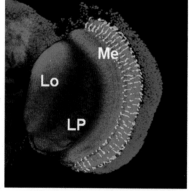

GMR-lacZ **DN-Cad Elav**

FIGURE 1. Organization of the optic lobe in adult *Drosophila*. Optic lobe cell bodies are visualized with Elav (blue) and the neuropil is labeled with *DN*-cadherin (red). Photoreceptor projections in the medulla are visualized with *GMR-lacZ* transgene (green). (Lo) lobula; (LP) lobula plate; (Me) medulla. Lamina is missing in this picture.

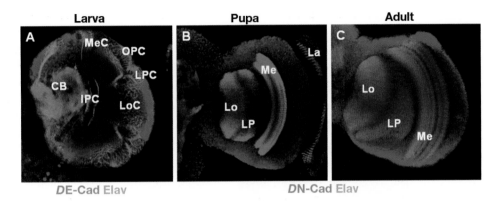

FIGURE 2. Development of the *Drosophila* optic lobe. Organization of the optic lobe at larva (*A*), pupa (*B*), and adult (*C*) stages. (CB) central brain; (IPC) inner proliferation center; (La) lamina; (Lo) lobula; (LoC) lobula cells; (LP) lobula plate; (LPC) lamina precursor cells; (Me) medulla; (MeC) medulla cells; (OPC) outer proliferation center. Neurons are visualized with Elav in all panels. Neuropil and neuroepithelial cells are labeled with *DE*-cadherin in *A*. Neuropils are labeled with *DN*-cadherin in *B* and *C*.

The First Optic Neuropil: The Lamina

In the region of the OPC neuroepithelium that gives rise to the lamina, the release of Hedgehog (Hh) by photoreceptors serves as an inductive signal for proliferation of lamina precursor cells (Huang and Kunes 1996). However, Hh alone is not sufficient for assembly of the lamina cartridge. Hh induces lamina precursor cells to express the epidermal growth factor receptor (EGFR). The EGFR ligand Spitz (Spi) is transported from photoreceptors through their axons and activation of EGFR in lamina precursor cells is necessary for differentiation of the lamina cartridge (Huang and Kunes 1996; Huang et al. 1998).

The lamina is the best-characterized optic neuropil. There are an estimated 6000 cells organized in lamina cartridges (Meinertzhagen and Sorra 2001), which are the functional units in the lamina (Hofbauer and Campos-Ortega 1990). Recently, very elegant work has shown that L1 and L2 lamina neurons, which are postsynaptic targets for R1–R6, are necessary and sufficient for motion-dependent behavior (Rister et al. 2007). In contrast, L3 seems to contribute to orientation behavior but not to motion detection (Rister et al. 2007). Thus, the lamina cartridges represent the first step in visual processing as they are the direct targets for outer PRs and are involved in the initiation of the motion detection pathway.

The Second Optic Neuropil: The Medulla

The medulla, with an estimated 40,000 neurons, is the most complex optic neuropil in *Drosophila* (Fischbach and Dittrich 1989; Morante and Desplan 2008). Medulla cell bodies are located either in the medulla cortex (the region between the lamina and the medulla neuropil) or in the medulla rim (the region between the medulla and the lobula plate) (Fig. 1). The medulla neuropil is stratified into 10 layers (M1–M10) (Fig. 3) that can be grouped into two domains: M1–M6, where R7 and R8 send their projections, and M7–M10 in the lower medulla, which is devoid of photoreceptor projections (Fischbach and Dittrich 1989; Morante and Desplan 2008).

Photoreceptor innervation is not necessary for the development of medulla neurons. Instead, neuroepithelial cells generate neuroblasts that differentiate into the different types of medulla neurons. Expression of Lethal of scute (L[1]sc) signals the transition of OPC neuroepithelial cells to neuroblasts in a proneural wave that sweeps through the surface of the brain. This wave appears to be at least in part regulated by Janus kinase (JAK)-signal transducers and activators of transcription (STAT) signaling; JAK-STAT negatively controls the progression of this wave and thus the final number of OPC neuroblasts that will generate the medulla (Yasugi et al. 2008).

Layering of optic lobes

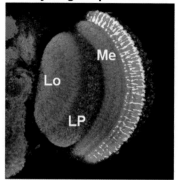

GMR-lacZ DV-Glut **ChAT**

FIGURE 3. Layering of the adult medulla and lobula complex. Layering in the adult optic lobes is revealed with anti-choline acetyltransferase (blue) (ChAT4B1; Developmental Studies Hybridoma Bank) and anti-*Drosophila* vesicular glutamate transporter (red) (*DV*-Glut; Daniels et al. 2004). Photoreceptor projections in the medulla are visualized with a *GMR-lacZ* transgene (green). (Lo) lobula; (LP) lobula plate; (Me) medulla.

The medulla is the target for inner PRs (Takemura et al. 2008) and is thus involved in color vision (Gao et al. 2008). One prerequisite to discriminate colors is to be able to compare the inputs of at least two PRs focusing on the same point in space that are sensitive to different wavelengths of light. *Drosophila* contains the machinery to perform color vision, because R7 and R8 look in the same direction and contain photopigments sensitive to different wavelengths (Salcedo et al. 1999; Yamaguchi et al. 2008). Therefore, the fly color vision system likely relies on comparing the inputs of R7 and R8 in the medulla neuropil.

To understand how color vision is processed, we have described the target neurons of R7 and R8 in the medulla (Morante and Desplan 2008) in a study that builds on the pioneering work in *Calliphora* using Golgi impregnations (Cajal and Sanchez 1915). Using the same technique, several groups have identified many cell types in the medulla of *Drosophila* and *Musca* (Strausfeld 1976; Fischbach and Dittrich 1989). Recently, using mosaic analysis of single cells with a repressible cell marker (MARCM) (Lee and Luo 1999) and a series of enhancer-trap Gal4 lines reporting expression of several transcription factors, we have estimated that more than 60 cells participate in a single medulla column (the functional unit of the medulla) (Morante and Desplan 2008). The enormous number of cell types forming part of a medulla column contrasts with the relatively small number of elements that provide inputs into the medulla (R7 and R8 and R1–R6 through L1–L5 neurons). Thus, the divergence in the flow of information between PRs and medulla neurons argues that much local processing occurs in the medulla.

This analysis has allowed differentiation between two main classes of neurons (Morante and Desplan 2008): (1) projection neurons, both columnar and noncolumnar, that connect the medulla with the lobula only, with both the lobula and lobula plate, or with the lamina; (2) local columnar and noncolumnar interneurons, whose ramifications reside within the limits of the medulla. Although columnar medulla cells maintain the retinotopic map and receive visual inputs from only one ommatidium, noncolumnar cells are able to receive inputs from several ommatidia.

The medulla also receives inputs from the lamina L1–L5 neurons, the primary target neurons for outer PRs (Takemura et al. 2008), which suggests that it is an intermediate target in the motion detection pathway before higher processing in the lobula plate (Morante and Desplan 2008).

The Lobula Complex: Lobula and Lobula Plate

The last step in the visual pathway, at least in the optic lobe, is the lobula complex, which is formed by the lobula plate and the lobula. The lobula complex represents a high-order processing structure for visual information and has been studied in detail in large flies such as *Musca domestica* (Pierantoni 1976) and recently also in *Drosophila* (Scott et al. 2002; Otsuna and Ito 2006; Katsov and Clandinin 2008). Interestingly, lobula complex cells possess wide-field ramifications and thus may be where visual inputs received from the medulla converge.

Dissection and Staining of *Drosophila* Optic Lobes at Different Stages of Development

We outline procedures for dissecting the optic lobes from *Drosophila* larvae, pupae, and adults. We also describe methods for visualizing the anatomy of brain neural circuits by staining with fluorescent secondary antibodies and primary antibodies specific for various neuronal populations and architectural features.

MATERIALS

CAUTION: See Appendix for proper handling of materials marked with <!>.
See the end of the chapter for recipes for reagents marked with <R>.

Reagents

CO_2
Glycerol (50%)
Normal goat serum (5%, v/v) <R>
Paraformaldehyde <!> (16%) (Electron Microscopy Sciences; cat. no. 15710)
PBST (PBS containing 0.3% Triton X-100 <!>) <R>
Phosphate-buffered saline (PBS) (1x), pH 7.4 <R>
Primary antibodies (see Table 1)
Secondary antibodies (see Table 2)
Vectashield mounting medium (for preserving fluorescence) (Vector Laboratories; cat. no. H-1000)

TABLE 1. Primary antibodies used to label *Drosophila* brains

Primary antibody	Source	Dilution
To detect expression of GFP- or lacZ-based constructs:		
Rabbit anti-GFP	Molecular Probes; A11122	1:1000
Sheep anti-GFP	Biogenesis; 4745-1051	1:1000
Rabbit anti-β-Gal	Cappel; 855978	1:20,000
Mouse anti-β-Gal	Promega; Z378A	1:500
To label relevant architectural features:		
Mouse anti-nc82 (to label general larval, pupal, and adult neuropil)	Developmental Studies Hybridoma Bank	1:50
Rat anti-*DE*-cadherin (DCAD2) (to label the larval neuroepithilium and neuropil in larvae and early pupae)	Developmental Studies Hybridoma Bank	1:50
Rat anti-*DN*-cadherin (DN-EX#8) (to label general neuropil in larvae, pupae, and adult)	Developmental Studies Hybridoma Bank	1:50
To label neuronal populations in larvae:		
Mouse anti-24B10 (to label photoreceptors and their projections)	Developmental Studies Hybridoma Bank	1:50
Mouse anti-Dachshund (mAbdac 2-3) (to label lamina precursors and lobula cells)	Developmental Studies Hybridoma Bank	1:100
Mouse anti-Elav (9F8A9) (to label neurons)	Developmental Studies Hybridoma Bank	1:25
Rat anti-Elav (7E8A10) (to label photoreceptors and their projections)	Developmental Studies Hybridoma Bank	1:25

TABLE 2. Secondary antibodies used to label *Drosophila* brains

Secondary antibody	Source	Dilution
Donkey anti-sheep Alexa488	Molecular Probes; A-11015	1:1000
Donkey anti-mouse Alexa488	Molecular Probes; A-21202	1:1000
Goat anti-rabbit Alexa488	Molecular Probes; A-21429	1:1000
Donkey anti-mouse Alexa555	Molecular Probes; A-31570	1:1000
Donkey anti-rabbit Alexa555	Molecular Probes; A-31572	1:1000
Goat anti-guinea pig Alexa555	Molecular Probes; A-21435	1:500
Goat anti-rat Alexa555	Molecular Probes; A-21434	1:500
Donkey anti-mouse Alexa647	Molecular Probes; A-31571	1:500
Donkey anti-rabbit Alexa647	Molecular Probes; A-31573	1:500
Donkey anti-rat Cy5	Jackson ImmunoResearch Laboratories; 712-175-153	1:400
Goat anti-guinea pig Alexa647	Molecular Probes; A-21450	1:500

Equipment

Cover glasses (18 × 18-mm and 24 × 50-mm) (Fisher Scientific; cat. no. 12-542A and 12-545F)
Dissecting microscope
Dissection dishes (three-well, glass) (Fisher Scientific; cat. no. 21-379)
Forceps (Dumont #55; Fine Science Tools)
Imaging microscope and software
Microscope slides (Fisher Scientific; cat. no. 12-550-15)
Minutien pins (0.1-mm-diameter) (Fine Science Tools; cat. no. 26002-10)
Mounted pins (fine)
Nail polish (clear)
Orbital shaker (Bellco Biotechnology)
Paintbrush (small)
Pin holders (12-cm) (Fine Science Tools; cat. no. 26016-12)
Sylgard 184 Silicone Elastomer Kit (Dow Corning) (Stern 1999)

METHOD

Dissection

The dissection procedure should take ~10–25 min per experiment. It is critical to minimize dissection time to avoid tissue degradation.

1. Dissect *Drosophila* brains as described below according to the developmental stage desired.

Larval Brains

i. Collect larvae at the stage of interest either from the walls of the vial using forceps or with a PBS-soaked paintbrush for younger larvae. Alternatively, fill the vial with 50% glycerol and larvae will come up to the surface.

ii. Place larvae in a Petri dish layered with Sylgard transparent resin containing drops of PBS.

iii. Remove the brain lobes by holding the larval body with one pair of forceps and pulling from the larval mouth hook with a second pair. Discard the larval body. Remove the excess tissue surrounding the brain lobes using fine mounted pins.

iv. Place the clean brain lobes in a glass dissection dish well containing 180 µL of PBS. Keep the dish on ice.

A typical experiment requires 10–20 larval brains.

Pupal Brains

    i. Collect pupae at the appropriate developmental stage in a Petri dish layered with Sylgard transparent resin containing drops of PBS.

> Pupae can be distinguished from slow-moving third-instar larvae because they stop crawling and remain attached to the wall of the vial. White pupae are considered the earliest stage in pupal development, whereas older pupae are dark brown. To study specific pupal development stages, wait the appropriate number of hours before collecting.

    ii. Submerge pupae in PBS and hold the pupal case through the abdominal part with one pair of forceps. With another pair, open the pupal case and pull the pupae out. Gently open the pupae and push the brain out. Remove excess tissue surrounding the brain lobes using fine mounted pins.

    iii. Using a pair of sharp forceps held closed, place the clean brain lobes in a glass dissection dish well containing 180 μL of PBS. Keep the dish on ice.

> A typical experiment requires 6–10 pupal brains.

    iv. Using fine pins, remove or gently separate the pupal eye to avoid interference with visualization of the pupal optic lobe.

> This step is critical for visualizing the medulla neuropil.

Adult Brains

    i. Anesthetize adult flies of the appropriate genotype with CO_2. Decapitate and place the heads in a glass dissection dish well containing PBS. Keep the dish on ice.

    ii. Perform the dissection in a Petri dish layered with Sylgard transparent resin containing drops of PBS, or alternatively, in the dissection dish well containing PBS.

    iii. Holding the head through the maxillary cavity, submerge it in PBS, and remove the maxillary palps with forceps. Hold both sides of the maxillary cavity and gently pull the two pairs of forceps away from each other to open the head cuticle. Remove the brain from the head cuticle and remove any excess surrounding tissue using fine mounted pins.

    iv. Using a pair of sharp forceps held closed, place the clean adult brain lobes in a glass dissection dish well containing 180 μL of PBS. Keep the dish on ice.

> A typical experiment requires 6–10 adult brains.

    v. Remove the retina and lamina to avoid interference with the visualization of the medulla/lobula complex and poor antibody penetration.

> This step is critical.

Fixation

The fixation and staining of brains should take ~1.5 d.

2. Add 20 μL of 37% paraformaldehyde to the glass wells containing 180 μL of PBS and the dissected tissue. Fix the larval, pupal, or adult brains by incubating on an orbital shaker with gentle agitation for 20 min at room temperature.

3. Remove the paraformaldehyde and wash with agitation three times in fresh PBS for 5 min each.

Staining

Day 1

4. Add 200 μL of blocking solution (5% normal goat serum) and incubate the fixed brains on an orbital shaker with gentle agitation for at least 30 min at room temperature in darkness.

5. Dilute primary antibodies in PBT in 1.5-mL tubes (you will need 200 μL per well).

6. Remove the blocking solution and add primary antibody. Incubate with agitation overnight at room temperature in darkness.

> For the majority of commercial antibodies, it is possible to omit the blocking step and incubate the dissected brains directly in primary antibodies overnight.

Day 2

7. Remove primary antibody and wash three times in PBS for 20 min each at room temperature.
   > Primary antibodies can be reused three times or more.

8. Dilute secondary antibodies in PBT in 1.5-mL tubes (you will need 200 μL per well). Add secondary antibodies to brain preparations. Incubate on an orbital shaker with agitation for 3 h at room temperature in darkness.
   > Prepare the secondary antibodies fresh and discard after use.

9. Remove secondary antibodies and wash three times in PBS for 20 min each at room temperature.

Mounting

Use the bridge method illustrated in Figure 4 to mount *Drosophila* brains and preserve their three-dimensional configuration. This procedure should take ~5–10 min per experiment.

10. Add two 2-μL drops of 50% glycerol to each end of a microscope slide. Cover these drops with 18 × 18-mm cover glasses.

11. Add a 5-μL drop of Vectashield mounting medium in between the two cover glasses.

12. Using a pair of sharp forceps held closed, place the brains in the mounting medium. Using fine mounted pins, align all the brains in the same orientation for ease of imaging.

13. Place a 24 × 50-mm cover glass on top of the bridge to cover the brains.

14. Seal the edges of the cover glass with nail polish and store the samples at 4°C in a dark slide holder.

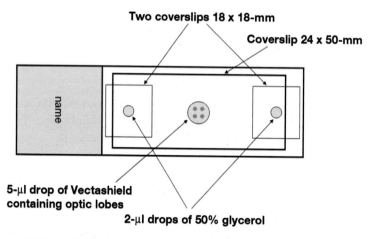

FIGURE 4. Bridge method for mounting *Drosophila* optic lobes.

RECIPES

CAUTION: See Appendix for proper handling of materials marked with <!>.
Recipes for reagents marked with <R> are included in this list.

Normal Goat Serum (5%, v/v)

Add 50 µL of normal goat serum (Lampire Biological Laboratories; cat. no. S2-0609) to 950 µL of PBST <R>. Aliquot the serum in very small volumes and store at −20°C.

PBST (0.3%)

Add 1.5 mL Triton X-100 <!> to 498.5 mL PBS. Store this nonhazardous buffer at room temperature.

Phosphate-Buffered Saline (PBS) (1x)

Reagent	Quantity
NaCl	137 mM
KCl <!>	2.7 mM
Na_2HPO_4	10 mM
KH_2PO_4	2 mM

Dissolve 8 g of NaCl, 0.2 g of KCl, 1.44 g of Na_2HPO_4, and 0.24 g of KH_2PO_4 in 800 ml of distilled H_2O, and mix well. Adjust the pH to 7.4 with HCl <!> and add H_2O to 1 L. Dispense the solution into aliquots and sterilize them by autoclaving for 20 min. Store the buffer at room temperature.

REFERENCES

Cajal SR, Sanchez D. 1915. Contribucion al conocimiento de los centros nerviosos de los insectos. *Trab Lab Invest Biol* **XIII:** 1–167.

Daniels RW, Collins CA, Gelfand MV, Dant J, Brooks ES, Krantz DE, DiAntonio A. 2004. Increased expression of the *Drosophila* vesicular glutamate transporter leads to excess glutamate release and a compensatory decrease in quantal content. *J Neurosci* **24:** 10466–10474.

Egger B, Boone JQ, Stevens NR, Brand AH, Doe CQ. 2007. Regulation of spindle orientation and neural stem cell fate in the *Drosophila* optic lobe. *Neural Dev* **2:** 1.

Fischbach KF, Dittrich APM. 1989. The optic lobe of *Drosophila melanogaster*. I. A Golgi analysis of wild-type structure. *Cell Tissue Res* **258:** 441-475.

Gao S, Takemura SY, Ting CY, Huang S, Lu Z, Luan H, Rister J, Thum AS, Yang M, Hong ST, et al. 2008. The neural substrate of spectral preference in *Drosophila*. *Neuron* **60:** 328–342.

Hofbauer A, Campos-Ortega JA. 1990. Proliferation pattern and early differentiation of the optic lobes in *Drosophila melanogaster*. *Roux's Arch Dev Biol* **198:** 264–274.

Huang Z, Kunes S. 1996. Hedgehog, transmitted along retinal axons, triggers neurogenesis in the developing visual centers of the *Drosophila* brain. *Cell* **86:** 411–422.

Huang Z, Shilo BZ, Kunes S. 1998. A retinal axon fascicle uses spitz, an EGF receptor ligand, to construct a synaptic cartridge in the brain of *Drosophila*. *Cell* **95:** 693–703.

Katsov AY, Clandinin TR. 2008. Motion processing streams in *Drosophila* are behaviorally specialized. *Neuron* **59:** 322–335.

Lee T, Luo L. 1999. Mosaic analysis with a repressible cell marker for studies of gene function in neuronal morphogenesis. *Neuron* **22:** 451–461.

Luo L, Callaway EM, Svoboda K. 2008. Genetic dissection of neural circuits. *Neuron* **57:** 634–660.

Meinertzhagen IA, Hanson TE. 1993. *The development of the optic lobe.* Cold Spring Harbor Laboratory Press, Cold Spring Harbor, NY.

Meinertzhagen IA, Sorra KE. 2001. Synaptic organization in the fly's optic lamina: Few cells, many synapses and divergent microcircuits. *Prog Brain Res* **131:** 53–69.

Morante J, Desplan C. 2004. Building a projection map for photoreceptor neurons in the *Drosophila* optic lobes. *Semin Cell Dev Biol* **15:** 137–143.

Morante J, Desplan C. 2008. The color-vision circuit in the medulla of *Drosophila*. *Curr Biol* **18:** 553–565.

Otsuna H, Ito K. 2006. Systematic analysis of the visual projection neurons of *Drosophila melanogaster*. I. Lobula-specific pathways. *J Comp Neurol* **497:** 928–958.

Pierantoni R. 1976. A look into the cock-pit of the fly. The architecture of the lobular plate. *Cell Tissue Res* **171:** 101–122.

Rister J, Pauls D, Schnell B, Ting CY, Lee CH, Sinakevitch I, Morante J, Strausfeld NJ, Ito K, Heisenberg M. 2007. Dissection of the peripheral motion channel in the visual system of *Drosophila melanogaster*. *Neuron* **56:** 155–170.

Salcedo E, Huber A, Henrich S, Chadwell LV, Chou WH, Paulsen R, Britt SG. 1999. Blue- and green-absorbing visual pigments of *Drosophila*: Ectopic expression and physiological characterization of the R8 photoreceptor cell-specific Rh5 and Rh6 rhodopsins. *J Neurosci* **19:** 10716–10726.

Scott EK, Raabe T, Luo L. 2002. Structure of the vertical and horizontal system neurons of the lobula plate in *Drosophila*. *J Comp Neurol* **454:** 470–481.

Stern CD. 1999. Grafting of somites. *Methods Mol Biol* **97:** 255–264.

Strausfeld NJ. 1976. *Atlas of an insect brain.* Springer-Verlag, Berlin.

Takemura SY, Lu Z, Meinertzhagen IA. 2008. Synaptic circuits of the *Drosophila* optic lobe: The input terminals to the medulla. *J Comp Neurol* **509:** 493–513.

Tang S, Guo A. 2001. Choice behavior of *Drosophila* facing contradictory visual cues. *Science* **294:** 1543–1547.

Wernet MF, Desplan C. 2004. Building a retinal mosaic: Cell-fate decision in the fly eye. *Trends Cell Biol* **14:** 576–584.

Wernet MF, Labhart T, Baumann F, Mazzoni EO, Pichaud F, Desplan C. 2003. Homothorax switches function of *Drosophila* photoreceptors from color to polarized light sensors. *Cell* **115:** 267–279.

White K, Kankel DR. 1978. Patterns of cell division and cell movement in the formation of the imaginal nervous system in *Drosophila melanogaster*. *Dev Biol* **65:** 296–321.

Wolff T, Ready DF. 1993. Pattern formation in the *Drosophila* retina. In *The development of* Drosophila melanogaster (ed M Bate and A Martinez Arias), pp. 1277–1325. Cold Spring Harbor Laboratory Press, Cold Spring Harbor, NY.

Yamaguchi S, Wolf R, Desplan C, Heisenberg M. 2008. Motion vision is independent of color in *Drosophila*. *Proc Natl Acad Sci* **105:** 4910–4915.

Yasugi T, Umetsu D, Murakami S, Sato M, Tabata T. 2008. *Drosophila* optic lobe neuroblasts triggered by a wave of proneural gene expression that is negatively regulated by JAK/STAT. *Development* **135:** 1471–1480.

11 Recording and Imaging in the *Drosophila* Nervous System

An Introduction to Section 2

Bing Zhang, Phillip A. Vanlandingham, and Hong Bao

Department of Zoology, University of Oklahoma, Norman, Oklahoma 73019

NEUROSCIENTISTS HAVE A LONG-STANDING INTEREST IN directly monitoring neuronal and synaptic activity. Monitoring is accomplished by electrically recording membrane potentials with sharp electrodes or optically imaging variations in intracellular ion concentrations and second messenger dynamics. Both of these techniques have enabled landmark discoveries in the elucidation of the cellular mechanisms of action potentials, ion channels, synaptic transmission, synaptic plasticity, and neuronal networks. These techniques remain powerful today primarily because they detect signals at high resolutions, reveal live information on neuronal activity, and allow for manipulation of the electrical excitability of living cells.

Electrophysiology has been a primary domain for fly neurobiologists. Most of the electrophysiological techniques taught in the *Drosophila* Neurobiology Summer Course at Cold Spring Harbor Laboratory were developed in the 1970s by several neurophysiologists who took advantage of the *Drosophila* large muscle fibers for intracellular recordings. Wyman and colleagues used the indirect flight muscle preparation to study the giant fiber pathway (Levine and Wyman 1973; Tanouye and Wyman 1980), while Jan and Jan developed the larval body wall muscle preparation to study synaptic transmission at the neuromuscular junction (NMJ) (Jan and Jan 1976a,b). Building on these seminal studies, Broadie and Bate first applied the patch-clamp technique to the small body wall muscles in fly embryos (Broadie and Bate 1993). These NMJ preparations have played pivotal roles in phenotypic analyses of mutations affecting ion channels, synaptic development and transmission, axonal transport, and neurodegeneration. Together with behavioral analyses, electrophysiological studies laid the intellectual foundation for the drive to clone *Shaker* (Jan et al. 1977; Salkoff 1983; Wu et al. 1983), *slowpoke* (Elkins et al. 1986), *eag* (Wu et al. 1983) and *seizure* potassium channels (Elkins and Ganetzky 1990), and *paralytic* sodium channels (Wu et al. 1978). NMJ recordings were also critical to understanding the cellular and molecular mechanisms by which exocytotic and endocytotic proteins function in synaptic transmission (e.g., Koenig et al. 1983; Littleton et al. 1993; Broadie et al. 1994; DiAntonio and Schwarz 1994; Umbach et al. 1994; Schulze et al. 1995; Stimson et al. 1998; Zhang et al. 1998; Fergestad et al. 1999; Delgado et al. 2000).

Two chapters in this section of the manual highlight the application of NMJ electrophysiology in *Drosophila*. In Chapter 12, Zhang and Stewart describe the basic theory and practice of electrophysiology as it is applied to third-instar larval NMJs. In Chapter 13, Allen and Godenschwege focus on the electrophysiology of the giant fiber pathway in adult flies.

Another major electrophysiological technique is extracellular recording, which monitors the firing of action potentials with a sharp electrode placed outside but near somata or axons. This method is often used to study sensory systems, such as audition (Eberl et al. 2000), gustation (Rodrigues and Siddiqi 1981), olfaction (Alcorta 1991; Ayer and Carlson 1991), phototransduction (Alawi and Pak

1971), and touch (Kernan et al. 1994). Although not suitable for detecting subthreshold synaptic potentials, the extracellular recording configuration is advantageous in that it is technically simpler and less damaging to neurons. Hence, it is ideal for obtaining relatively long-term and stable recordings. Following the NMJ chapters in this section, we include three chapters on select sensory systems: Dolph, Nair, and Raghu on visual physiology (Chapter 14); Eberl and Kernan on mechanosensation (Chapter 15); and Benton and Dahanukar on chemosensory coding (Chapter 16).

Many basic physiological questions about the brain require directly studying neurons in the central nervous system. The use of more advanced patch-clamp techniques makes it possible to record from neurons for a prolonged period in dissected embryonic and larval ventral nerve cords (Baines and Bate 1998) or in the brain of live adult flies (Wilson et al. 2004). The power of in situ recording should be obvious: It offers the unique opportunity to decipher the cellular mechanisms by which central neurons or neural circuits process information. In this section, Marley and Baines describe patch-clamp analysis of neurons in fly embryos and larvae (Chapter 17), and Murthy and Turner cover patch clamping of neurons in the adult fly brain (Chapter 18).

For years, *Drosophila* physiology was almost synonymous with electrophysiology. In recent years, optophysiological techniques, applied to the study of the *Drosophila* nervous system, have advanced and matured. As a result, optical imaging is now exploited to further our understanding of the fly nervous system (Karunanithi et al. 1997; Umbach and Gundersen 1997; Wang et al. 2003; Yu et al. 2004). Three chapters describe optical imaging techniques for studying synaptic function and synapse formation at the larval NMJ. In Chapter 19, Macleod describes calcium imaging of motor neuron terminals. In Chapter 20, Levitan and Shakiryanova demonstrate optical detection of neuropeptide release. In Chapter 21, Andlauer and Sigrist describe the imaging of synapse formation in respect to active zones and postsynaptic receptor assembly. Although not covered in this manual, readers are highly encouraged to follow the new development of optophysiology, which allows one to control neuronal activity as well as fly and mammalian behavior using light and light-sensitive ion channels (Boyden et al. 2005; Lima and Miesenböck 2005; Schroll et al. 2006).

The extensive use of *Drosophila* as a model organism for molecular genetic analyses provides a distinct advantage in studies of the nervous system. Hence, we also include an overview of molecular tools relevant to the study of synaptic physiology. The final two chapters in this section cover molecular genetic methods for manipulating neuronal excitability and synaptic plasticity (Olsen and Keshishian; Chapter 22) and for acute inactivation of synaptic proteins (Habets and Verstreken; Chapter 23).

In this "everything microarray and stem cell" age, it is satisfying that physiology remains a valuable approach in *Drosophila* neurobiology research and in neuroscience at large. One must wonder, however, what the new challenges are for the future of physiology in *Drosophila*. The relatively short history of *Drosophila* neurobiology has encompassed a multitude of technical and intellectual developments. Still, many outstanding questions require new advances. We hope, in particular, to see progress in two challenging areas in the near future. One is the use of electrophysiology for monitoring neuronal activities in real-time fly behavior. In the following section, van Swinderen demonstrates the power of in situ extracellular recording of field potentials in studying fly attention in otherwise immobilized live flies (van Swinderen and Greenspan 2003; Chapter 25). Physiological recording of freely behaving flies, however, has not yet been carried out. One valuable approach is wireless recording, which has been used successfully in freely moving large animals (Pinkwart and Borchers 1987; Hampson et al. 2009). We anticipate with great excitement the day when microelectrodes in conjunction with an ultra-miniature wireless device can be glued on the head or thorax of freely running or flying flies.

The second major challenge is to monitor the activity of a large population of neurons simultaneously or even across an entire brain. Currently, electrophysiology is used, at its best, to record from two or three neurons at one time. Optical imaging is superior in its ability to monitor a large population of neurons simultaneously, but it is too slow to follow high-frequency neuronal firing. To enable us to see both "the trees and the forest," the development of a multiarray of electrodes and high-speed imaging and analysis methods will be of particular importance. Considering these and imagining other, potentially unforeseen, future technical advances, we believe that the most productive days of physiology in *Drosophila* neurobiology are yet to come.

REFERENCES

Alawi AA, Pak WL. 1971. On-transient of insect electroretinogram: Its cellular origin. *Science* **172:** 1055–1057.

Alcorta E. 1991. Characterization of the electroantennogram in *Drosophila melanogaster* and its use for identifying olfactory capture and transduction mutants. *J Neurophysiol* **65:** 702–714.

Ayer RK Jr, Carlson J. 1991. *acj6:* A gene affecting olfactory physiology and behavior in *Drosophila. Proc Natl Acad Sci* **88:** 5467–5471.

Baines RA, Bate M. 1998. Electrophysiological development of central neurons in the *Drosophila* embryo. *J Neurosci* **18:** 4673–4683.

Boyden ES, Zhang F, Bamberg E, Nagel G, Deisseroth K. 2005. Millisecond-timescale, genetically targeted optical control of neural activity. *Nat Neurosci* **8:** 1263–1268.

Broadie K, Bellen HJ, DiAntonio A, Littleton JT, Schwarz TL. 1994. Absence of synaptotagmin disrupts excitation-secretion coupling during synaptic transmission. *Proc Natl Acad Sci* **91:** 10727–10731.

Broadie KS, Bate M. 1993. Development of the embryonic neuromuscular synapse of *Drosophila melanogaster. J Neurosci* **13:** 144–166.

Delgado R, Maureira C, Oliva C, Kidokoro Y, Labarca P. 2000. Size of vesicle pools, rates of mobilization, and recycling at neuromuscular synapses of a *Drosophila* mutant, *shibire. Neuron* **28:** 941–953.

DiAntonio A, Schwarz TL. 1994. The effect on synaptic physiology of synaptotagmin mutations in *Drosophila. Neuron* **12:** 909–920.

Eberl DF, Hardy RW, Kernan MJ. 2000. Genetically similar transduction mechanisms for touch and hearing in *Drosophila. J Neurosci* **20:** 5981–5988

Elkins T, Ganetzky B, Wu CF. 1986. A *Drosophila* mutation that eliminates a calcium-dependent potassium current. *Proc Natl Acad Sci* **83:** 8415–8419.

Elkins T, Ganetzky B. 1990. Conduction in the giant nerve fiber pathway in temperature-sensitive paralytic mutants of *Drosophila. J Neurogenet* **6:** 207–219.

Fergestad T, Davis WS, Broadie K. 1999. The stoned proteins regulate synaptic vesicle recycling in the presynaptic terminal. *J Neurosci* **19:** 5847–5860.

Hampson RE, Collins V, Deadwyler SA. 2009. A wireless recording system that utilizes Bluetooth technology to transmit neural activity in freely moving animals. *J Neurosci Methods* **182:** 195–204.

Jan LY, Jan YN. 1976a. Properties of the larval neuromuscular junction in *Drosophila melanogaster. J Physiol* **262:** 189–214.

Jan LY, Jan YN. 1976b. L-glutamate as an excitatory transmitter at the *Drosophila* larval neuromuscular junction. *J Physiol* **262:** 215–236.

Jan YN, Jan LY, Dennis MJ. 1977. Two mutations of synaptic transmission in *Drosophila. Proc R Soc Lond B Biol Sci* **198:** 987–108.

Karunanithi S, Georgiou J, Charlton MP, Atwood HL. 1997. Imaging of calcium in *Drosophila* larval motor nerve terminals. *J Neurophysiol* **78:** 3465–3467.

Kernan M, Cowan D, Zuker C. 1994. Genetic dissection of mechanosensory transduction: mechanoreception-defective mutations of *Drosophila. Neuron* **12:** 1195–1206.

Koenig JH, Saito K, Ikeda K. 1983. Reversible control of synaptic transmission in a single gene mutant of *Drosophila melanogaster. J Cell Biol* **96:** 1517–1522.

Levine JD, Wyman RJ. 1973. Neurophysiology of flight in wild-type and a mutant *Drosophila. Proc Natl Acad Sci* **70:** 1050–1054.

Lima SQ, Miesenböck G. 2005. Remote control of behavior through genetically targeted photostimulation of neurons. *Cell* **121:** 141–152.

Littleton JT, Stern M, Schulze K, Perin M, Bellen HJ. 1993. Mutational analysis of *Drosophila* synaptotagmin demonstrates its essential role in Ca^{2+}-activated neurotransmitter release. *Cell* **74:** 1125–1134.

Pinkwart C, Borchers HW. 1987. Miniature three-function transmitting system for single neuron recording, wireless brain stimulation and marking. *J Neurosci Methods* **20:** 341–352.

Rodrigues V, Siddiqi O. 1981. A gustatory mutant of *Drosophila* defective in pyranose receptors. *Mol Gen Genet* **181:** 406–408.

Salkoff L. 1983. Genetic and voltage-clamp analysis of a *Drosophila* potassium channel. *Cold Spring Harbor Symp*

Quant Biol **48:** 221–231

Schroll C, Riemensperger T, Bucher D, Ehmer J, Völler T, Erbguth K, Gerber B, Hendel T, Nagel G, Buchner E, Fiala A. 2006. Light-induced activation of distinct modulatory neurons triggers appetitive or aversive learning in *Drosophila* larvae. *Curr Biol* **16:** 1741–1747.

Schulze KL, Broadie K, Perin MS, Bellen HJ. 1995. Genetic and electrophysiological studies of *Drosophila* syntaxin-1A demonstrate its role in nonneuronal secretion and neurotransmission. *Cell* **80:** 311–320.

Stimson DT, Estes PS, Smith M, Kelly LE, Ramaswami M. 1998. A product of the *Drosophila stoned* locus regulates neurotransmitter release. *J Neurosci* **18:** 9638–4969.

Tanouye MA, Wyman RJ. 1980. Motor outputs of giant nerve fiber in *Drosophila*. *J Neurophysiol* **44:** 405–421.

Umbach JA, Gundersen CB. 1997. Evidence that cysteine string proteins regulate an early step in the Ca^{2+}-dependent secretion of neurotransmitter at *Drosophila* neuromuscular junctions. *J Neurosci* **17:** 7203–7209.

Umbach JA, Zinsmaier KE, Eberle KK, Buchner E, Benzer S, Gundersen CB. 1994. Presynaptic dysfunction in *Drosophila csp* mutants. *Neuron* **13:** 899–907.

van Swinderen B, Greenspan RJ. 2003. Salience modulates 20–30 Hz brain activity in *Drosophila*. *Nat Neurosci* **6:** 579–586.

Wang JW, Wong AM, Flores J, Vosshall LB, Axel R. 2003. Two-photon calcium imaging reveals an odor-evoked map of activity in the fly brain. *Cell* **112:** 271–282.

Wilson RI, Turner GC, Laurent G. 2004. Transformation of olfactory representations in the *Drosophila* antennal lobe. *Science* **303:** 366–370.

Wu CF, Ganetzky B, Haugland FN, Liu AX. 1983. Potassium currents in *Drosophila*: Different components affected by mutations of two genes. *Science* **220:** 1076–1078.

Wu CF, Ganetzky B, Jan LY, Jan YN, Benzer S. 1978. A *Drosophila* mutant with a temperature-sensitive block in nerve conduction. *Proc Natl Acad Sci* **75:** 4047–4051.

Yu D, Ponomarev A, Davis RL. 2004. Altered representation of the spatial code for odors after olfactory classical conditioning; memory trace formation by synaptic recruitment. *Neuron* **42:** 437–449.

Zhang B, Koh YH, Beckstead RB, Budnik V, Ganetzky B, Bellen HJ. 1998. Synaptic vesicle size and number are regulated by a clathrin adaptor protein required for endocytosis. *Neuron* **21:** 1465–1475.

12 Synaptic Electrophysiology of the *Drosophila* Neuromuscular Junction

Bing Zhang[1] and Bryan Stewart[2]

[1]*Department of Zoology, University of Oklahoma, Norman, Oklahoma 73019;* [2]*Department of Biology, University of Toronto at Mississauga, Mississauga, Ontario L5L 1C6, Canada*

ABSTRACT

Chemical synaptic transmission is an important means of neuronal communication in the nervous system. On the arrival of an action potential, the nerve terminal experiences an influx of calcium ions, which in turn trigger the exocytosis of synaptic vesicles (SVs) and the release of neurotransmitters into the synaptic cleft. Transmitters elicit synaptic responses in the postsynaptic cell by binding to and activating specific receptors. This is followed by the recycling of SVs at presynaptic terminals. The *Drosophila* larval neuromuscular junction (NMJ) shares many structural and functional similarities to synapses in other animals, including humans. These include the basic feature of synaptic transmission, as well as the molecular mechanisms regulating the synaptic vesicle cycle. Because of its large size, easy accessibility, and the well-characterized genetics, the fly NMJ remains an excellent model system for dissecting the cellular and molecular mechanisms of synaptic transmission. In this chapter, we describe the theory and practice of electrophysiology as applied to the *Drosophila* larval NMJ preparation. First, the basics of membrane potentials are introduced, with an emphasis on the resting potential and synaptic potential. Second, the equipment and methods required to set up an electrophysiology rig are presented. Third, protocols are provided that explain how to use the rig to record from muscles, determine the passive membrane properties of the muscle (i.e., input resistance and time constant), record synaptic potentials both intracellularly and extracellularly, detect synaptic currents by two-electrode voltage clamp, perform quantal analysis, and study short-term synaptic plasticity.

INTRODUCTION

Neurons use electrical and chemical signals to receive, process, and communicate information. The basic currency of signaling by excitable cells (neurons and muscles) is electrical current. Accordingly, electrophysiology is the study of the electrical properties of cells, and we draw on the theoretical background of electrical phenomenon to help understand the behavior of the nervous system. By analogy, the cell membrane is similar to a simple resistor-capacitor (RC) circuit (see Boxes 1 and 2). In an electronic circuit, a battery (voltage source) is connected via conductors (wires) to a resistor and a capacitor arranged in parallel, with a return wire to the battery that completes the circuit. Relatively simple physical relationships describe the flow of electrons through the wires. The cellular equivalents to an electronic circuit are the resting membrane potential (see below), which acts as the battery; the extracellular and intracellular fluids, which act as the low-resistance conductors; ion channels within cell membranes, which act as the resistors; and the phospholipid bilayer, which acts as a capacitor. An important difference between electronic and cellular circuits is that electrons are the charged particles that flow through an electronic circuit, whereas in a biological circuit a variety of cations and anions (e.g., Na^+, K^+, Ca^{2+}, Cl^-) generate membrane currents.

Drosophila has been used to great advantage as a model system for elucidating the molecular and genetic mechanisms that underlie bioelectric and biochemical signaling. The introduction of the larval NMJ as an experimental preparation established a system that permits relatively easy access to large, electrically excitable cells (Jan and Jan 1976a,b). Other electrical techniques applied to *Drosophila* include electroretinograms (Hotta and Benzer 1969; see Chapter 14), extracellular recording from the

---

BOX 1. ELECTRICAL TERMINOLOGY

Charge (Q): Electric charge is a fundamental property of subatomic particles. By convention an electron has a charge of –1 and a proton has a charge of +1. In general, particles with the same charge repel each other, and particles with opposite charges attract each other. The unit of charge is known as the coulomb (C) and 1 C is equal to 6.25×10^{18} charges. In biological cells, the charge is carried by ions such as Na^+, K^+, Ca^{2+}, and Cl^-.

Current (I): The directional movement of charged particles is electric current, such as electrons moving through a metal conductor. Current is measured in amperes (A); 1 A is the movement of 1 C of charge per second.

Voltage (V): Also known as electrical potential or potential difference, voltage refers to the difference in electric potential between two points within an electric field, such as a circuit. By definition, potential difference is the amount of work required to move a unit of charge. A 1-V potential difference requires 1 J of work to separate 1 C of charge.

Conductance (g) or Resistance (R or Ω): The flow of current between two points is called conductance; the greater the conductance, the easier it is for current to flow. Resistance is the inverse of conductance; the greater the resistance, the harder it is for current to flow. Conductance is measured in siemens (S) and resistance is measured in ohms (Ω). The biological equivalent of conductance is the permeability of ion channels. The higher the conductance of a membrane, the greater the number of open ion channels.

Capacitance (C): If two conducting materials are separated by an insulating layer, a capacitor is formed. This could be two metal plates separated by a nonconducting material; in the cell, the lipid bilayer of the membrane separates the conducting intracellular and extracellular fluids. If there is an excess of positive charge on one of the conductors and an excess of negative charge on the other, a potential difference develops between the two conductors. Capacitance is measured in farads (F): A 1-F capacitor will have an electrical potential of 1 V when +1 C of charge is on one conductor and –1 C is on the other conductor. An electric current will change the electric potential of a capacitor by moving charge from one conductor to the other. This is called a capacitive current (I_C). Lipid bilayers serve as biological capacitors.

BOX 2. BASIC EQUATIONS AND RC CIRCUITS

Ohm's law: Ohm's law describes the relationship between current, voltage, and conductance (resistance):

$$I = gV.$$

It states that electric current is a product of conductance and the voltage difference applied to the conductor. Ohm's law can also be written as

$$V = IR,$$

which is the inverse relationship (and the one most often used in our discussions) and states that the voltage between two points is equal to the product of the current and resistance.

Capacitance: The definition of capacitance is given by

$$C = Q/V,$$

which states that capacitance is the ratio of the charges on the conductors to the voltage between the conductors.

An RC circuit:

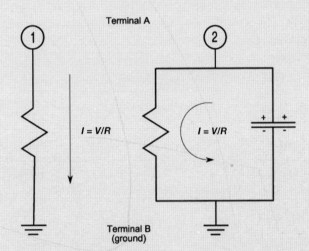

In circuit (1), there is only a resistor in the conductive path between the two terminals. If an electric potential exists between the two terminals, then current will flow according to Ohm's law. In circuit (2) there is a capacitor in parallel to the resistor. In this case, if there is an electric potential between the two terminals, current will also flow, but the capacitor must be charged or discharged before current will flow through the resistor.

We therefore need to know the *rate* of change of the electric potential of the capacitor,

$$C = \frac{Q}{V}.$$

The rate of change of the voltage for a capacitive current is found from the time derivative,

$$\frac{\partial V}{\partial t} = \frac{I_C}{C},$$

and the capacitive current is therefore

$$I_C = C\frac{\partial V}{\partial t}.$$

(Continued on following page.)

BOX 2. (*CONTINUED*)

Therefore, to determine the total amount of current flowing in circuit (2), we can add the current through the resistor and the capacitor:

$$I_{total} = I_C + I_i,$$

$$I_{total} = C\frac{\partial V}{\partial t} + \frac{V}{R}.$$

From this relationship we can see that if the voltage is constant, then $I_C = 0$; when the voltage changes, capacitance is important.

How does capacitance affect the behavior of a circuit? This can be seen easily by measuring voltage differences across the two terminals in the above circuits when the voltage changes:

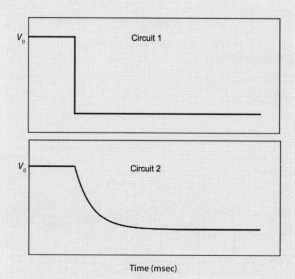

Initially, we will start with an arbitrary voltage (V_0) at time 0. If the voltage is instantly changed, we will see for circuit (1) that the voltage we measure also changes instantly, going in a downward direction. However, for circuit (2) we see that the change is not instant but rather takes some period of time to settle down to the new level. This is the property the capacitance adds—it slows down the response of the circuit to changes in voltage.

The relationship between voltage and time in an RC circuit is given by

$$V = V_0\, e^{-t/R},$$

in which t is time, R is the resistance, and C is the capacitance of the circuit. The component of the equation RC is also known as the time constant or tau (τ) and this is an important parameter for neurophysiology. The time constant is equal to $1/e$ or ~36% of V_0.

Functionally, the membrane time constant reflects how fast a neuron or muscle responds to a synaptic signal. For a given cell, membrane capacitance (C) does not change, and therefore membrane resistance determines the time constant. The more channels open (low R), the faster the response becomes.

giant fiber pathway (Tanouye and Wyman 1980; see Chapter 13), and the use of dissociated cell cultures (Wu et al. 1983) to gain access to the relatively small neurons. More recently, in situ electrical recordings from the central nervous system of the embryo, larvae, and adults (Baines and Bate 1998; Rohrbough and Broadie 2002; Wilson et al. 2004; see Chapters 17 and 18), as well as adoption of optophysiology techniques (Karunanithi et al. 1997; Macleod et al. 2002; Marek and Davis 2002; Wang et al. 2004; Lima and Miesenbock 2005; Rasse et al. 2005; see Chapters 19–21), have expanded the repertoire of techniques available to challenge and measure physiological processes used by neurons.

THE RESTING MEMBRANE POTENTIAL

The physical basis of electrical signaling in cells is the resting membrane potential. Simply stated, there is a charge difference across cell membranes because the inside of a cell is electrically negative when compared with the outside. Understanding the origins of this electrical potential is the foundation for learning how changes in the membrane potential constitute signaling.

How do we know that cells have a resting membrane potential? Julius Bernstein in 1902 was among the first to *predict* that a cell would have a negative resting potential. Direct evidence of cell resting potentials, however, was not available until electrophysiologists measured it in squid giant axons in the 1940s (Hodgkin and Huxley 1945; Hodgkin and Katz 1949). Today, using well-honed techniques, it is relatively straightforward to gain electrical access to the inside of most cells. Glass micropipettes, having a tip diameter on the order of several hundred nanometers (for typical intracellular recording pipettes) are carefully inserted through the plasma membrane and into the cell lumen. Using the appropriate electronic hardware connected to the micropipette, cell resting potentials can be accurately measured and are typically ~50–80 mV less than the outside.

How Is the Membrane Potential Generated?

Two basic factors determine a cell's resting potential: unequal distribution of ions between the inside and outside of a cell and the selective permeability of cell membranes.

Unequal Distribution of Ions between the Inside and Outside of a Cell

The ionic composition of the extracellular fluid is different from that in the intracellular fluid. The extracellular fluid usually contains Na^+ as the major cation, which is electrically balanced by a combination of Cl^- and HCO_3^- ions to maintain electrical neutrality. Smaller contributions to the extracellular charge come from K^+, Ca^{2+}, and Mg^{++}, which again are balanced by Cl^- and HCO_3^- ions.

The major cation of the intracellular fluid is K^+, with a smaller contribution from Na^+. Intracellular free $[Ca^{2+}]$ is usually near 0, because Ca^{2+} are bound to buffers, pumped into intracellular stores, or pumped out of the cell. Intracellular fluid has a greater abundance (compared with the outside of the cell) of large impermeable anions, chiefly proteins, DNA, and RNA. These negative charges are balanced by intracellular K^+, and it is for this reason that there is an imbalance of K^+ across the cell membrane.

Selective Membrane Permeability

The cell membrane is a phospholipid bilayer with associated peripheral and transmembrane proteins. The lipid bilayer portion is impermeable to most things, noteworthy exceptions being O_2, CO_2, and steroid compounds. Charged ions such as K^+ and Na^+ do not readily cross the bilayer. They move in and out of the cell with the aid of transmembrane protein channels and transporters. Among these proteins is a family of ion channels, known as leakage channels, that primarily transport K^+ (Goldstein et al. 1996), making the cell membrane relatively permeable to potassium. The membrane, however, remains relatively impermeable to other ions when it is in the resting state.

The Physical Basis for Membrane Potential

We see now that the resting membrane potential is a function of the ionic composition of the intracellular and extracellular milieus, the charge difference that results, and the selective movement of ions across the membrane with the assistance of ion-selective protein channels and transporters. To

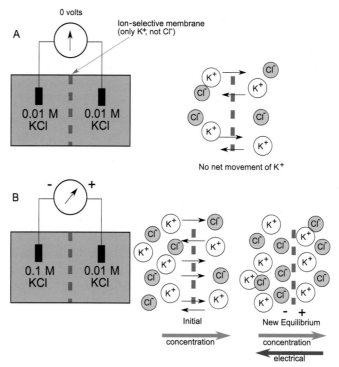

FIGURE 1. Physical basis of the membrane potential. (*A*) In a tank two KCl solutions are separated by a membrane that is only permeable to K$^+$. A voltmeter measures the electrical difference between the two solutions. If the solutions on each side of the membrane are equal, there is no net movement of ions, so the voltmeter reads 0 V. (*B*) If the solution on the left side is increased to 0.1 M KCl, then there is an initial concentration gradient driving K$^+$ and Cl$^-$ from left to right. Because the membrane is permeable only to K$^+$, some K$^+$ ions will be separated from counterbalancing Cl$^-$ ions at the membrane. This charge separation instantly creates an electrical force, which is equal to but in the opposite direction from the concentration gradient, establishing a new equilibrium. The voltmeter now indicates there is an electrical potential difference between the two solutions.

help us visualize how these factors generate a potential, consider a tank divided into two compartments by a membrane that is only permeable to K$^+$ (Fig. 1). Each compartment is filled with 0.01 M KCl, and a voltmeter is used to measure the electrical difference across the membrane (Fig. 1A). The meter indicates that there is no potential difference—that is, there is zero voltage across the membrane. This illustrates that selective permeability without a chemical gradient is insufficient to generate a voltage difference across the membrane.

Next replace the solution in one of the compartments with a 0.1 M KCl solution. Now there is an unequal distribution of ions across the membrane and a concentration gradient exists that could drive K$^+$ and Cl$^-$ from high concentration to low. If the membrane was equally permeable to K$^+$ and Cl$^-$ (not shown in Fig. 1), then all of the ions in the two compartments would redistribute and eventually reach equal concentrations on both sides of the membrane. At this point, the voltage difference across the membrane would again be 0. Thus, having a concentration difference and unrestricted permeability will not generate a potential difference.

Let us now examine the situation when *the membrane is only permeable to K$^+$* and when there is a concentration gradient for KCl across the membrane (Fig. 1B). Following the second law of thermodynamics, K$^+$ will start to diffuse across the membrane, but the counterbalancing negatively charged Cl$^-$ ions cannot follow these positive charges. A *separation of charge* therefore occurs at the membrane interface. There are now two forces at play: the K$^+$ concentration gradient moving ions in one direction and a newly developed electrical gradient, which opposes the concentration gradient, because the negatively charged ions attract the positively charged ions. A new equilibrium condition is quickly established in which the force of the concentration gradient is exactly balanced by an opposing electrical force. This electrical force is the membrane potential of this artificial cell. Therefore, both unequal distribution of ions and selective permeability are required to generate a membrane potential.

Biological systems develop a similar separation of charge across the plasma membrane and this is what we measure as the membrane potential. The K^+ leakage channels are the major contributor to the resting permeability of the plasma membrane, but the ionic and cellular environment within and around a cell is obviously more complicated than the above illustration. Many more species of ions are involved, and, importantly, the plasma membrane is permeable to K^+ and at smaller permeability values to other ions as well.

Equilibrium Potentials of Individual Ions

The simplified experiment in the previous section illustrates that having unequal concentrations of ions across a semipermeable membrane gives rise to an electrical force. If the concentrations of ions are known, then the electrical force that is required to balance the force of the concentration gradient can be predicted using the Nernst equation:

$$E_{ion} = \frac{RT}{zF} \ln \frac{[\text{ion outside cell}]}{[\text{ion inside cell}]},$$

in which E_{ion} is the equilibrium potential for the ion, R is the universal gas constant, T is absolute temperature, z is the valence of the ion, and F is Faraday's constant. This equation is often simplified for monovalent cations when the cell is at 20°C:

$$E_{ion} \text{ (mV)} = 58 \log \frac{[\text{ion outside cell}]}{[\text{ion inside cell}]}.$$

Note the change from natural logarithm to logarithm base-10. In our example above, if the 0.1 M KCl solution is considered to be inside the cell and the 0.01 M KCl solution to be outside the cell, then $E_{K^+} = -58$ mV. This relationship can be used to calculate the equilibrium potential for any ion. The reader will see that this value is independent of permeability. Cations that are at a higher concentration inside the cell will have a negative equilibrium potential, whereas cations that are at a higher concentration outside the cell will have a positive equilibrium potential. For example, typically $[Na^+]_{outside} = 100$ mM, whereas $[Na^+]_{inside} = 5$ mM, so the equilibrium potential for sodium is $E_{Na^+} = +75$ mV.

The Resting Potential for an Entire Cell

Cells are in the presence of many different ions, both inside and outside the cell. To understand the resting potential for an entire cell, we need to consider the concentrations of those ions and their relative permeability. The major contributing cations are Na^+ and K^+, and Cl^- is the major anion. The Goldman–Hodgkin–Katz (GHK) equation describes the relationship between the voltage of the cell membrane (V_m), the permeability (P), and the concentrations of ions as

$$V_m = \frac{RT}{zF} \ln \frac{P_K[K^+]_{out} + P_{Na}[Na^+]_{out} + P_{Cl}[Cl^-]_{in}}{P_K[K^+]_{in} + P_{Na}[Na^+]_{in} + P_{Cl}[Cl^-]_{out}}.$$

Note the reversal of terms for Cl^-, which reflects the negative valence of the chloride ion. Once again converting to logarithm base-10 and assuming a temperature of 20°C, the relationship becomes

$$V_m = 58 \log \frac{P_K[K+]_{out} + P_{Na}[Na+]_{out} + P_{Cl}[Cl^-]_{in}}{P_K[K^+]_{in} + P_{Na}[Na^+]_{in} + P_{Cl}[Cl^-]_{out}}.$$

Much of the work to derive this relationship was performed on the squid axon, in which it was found that at rest the permeabilities of K^+:Na^+:Cl^- are 1:0.03:0.1. That is, in the resting squid axon, the permeability of K^+ is 10 times greater than that of Cl^- and 33.3 times greater than that of Na^+. If $[K^+]_{out} = 10$ mM, $[K^+]_{in} = 400$ mM, $[Na^+]_{out} = 460$ mM, $[Na^+]_{in} = 50$ mM, $[Cl^-]_{out} = 540$ mM, $[Cl^-]_{in} = 40$ mM, then

$$V_m = 58 \log \frac{1(10) + 0.03(460) + 0.1(40)}{1(400) + 0.03(50) + 0.1(540)}$$

$$= -70 \text{ mV}.$$

Comparing the simplified version of the GHK equation with that of the Nernst equation, one sees that the resting membrane potential of the cell will always be close to the equilibrium potential of the most permeable ion. In this example, because K^+ is the most permeable ion, V_m approaches E_{K^+} with smaller contributions from the other ions.

The Na⁺/K⁺ ATPase

When the cell membrane is selectively permeable to a single ion (e.g., K^+), passive diffusion of K^+ down its chemical gradient first generates a negative potential inside. This electrical gradient works in favor of keeping K^+ inside and will eventually become strong enough to completely stop the net efflux of K^+. To reach this equilibrium state, very few ions are required near the membrane to maintain a separation of charges compared with the number of ions that are present in solution, such that the $[K^+]$ on both sides of the membrane remains unaltered. At equilibrium, the net flow of K^+ is 0 and no energy is required to maintain equilibrium. If, on the other hand, the cell is permeable only to Na^+, then the equilibrium potential will be completely different. Unlike K^+, which sets the membrane potential to E_{K^+} (~ −90 mV), Na^+ will bring the membrane potential to its own equilibrium potential, E_{Na^+}, which is typically more depolarized (~+50 to +75 mV).

A more complex and dynamic situation arises, however, when a cell membrane is permeable to more than one ion. In a cell whose membrane is predominantly permeable to K^+, a slight influx of Na^+ will disturb the equilibrium state of K^+. This creates a severe problem for maintaining Na^+ and K^+ distributions across the membrane. One can imagine that each Na^+ influx will lead to a K^+ efflux, and this vicious tug of war is impossible to stop without introducing a "peacemaker." In biological cells, this peacemaker is the Na^+/K^+ pump, which pumps Na^+ out and K^+ in, such that $[Na^+]$ and $[K^+]$ do not change over time. There is a price, however, for maintaining the ion concentrations because the Na^+/K^+ pump has to accomplish its task by going against each ion's electrochemical gradient. A significant amount of cellular energy, in the form of adenosine triphosphate (ATP), is consumed in this process. The Na^+/K^+ pump is also an ATPase. In most cells, it pumps in two K^+ ions in exchange for three Na^+ ions out, thereby also making a direct contribution to the generation of the resting potential.

Membrane Permeability and Ion Channels

If you understand the fundamental principle that the membrane potential of the cell is primarily a function of ion permeability (because ion concentrations are relatively stable), then understanding neural signaling is made much easier. What controls ion permeability? The K^+ leakage channel can be regarded as an unregulated K^+ pore that allows K^+ to flow in or out of the cell. However, there are also a host of ion channels that alter their permeabilities to allow ions to cross the cell membrane. There are two broad categories of such "regulated" ion channels, defined by what regulates their opening and closing: voltage-gated channels (see Chapter 17) and ligand-gated channels.

As the name implies, it is the membrane voltage that controls the opening and closing of voltage-gated channels. Most of these channels are closed at the resting membrane voltage (V_m), but they open up when V_m is depolarized (becomes more positive). These channels usually allow only one specific ion to cross through the channel. As such, there are voltage-gated Na^+ channels, voltage-gated K^+ channels, and voltage-gated Ca^{2+} channels. Within each category there are a variety of channel types, which confer specific properties on the cells that express them.

Ligand-gated channels open when a ligand binds to a receptor site on the channel protein. At the fly NMJ, these are neurotransmitter (glutamate) receptors (indeed the phrases ligand-gated channel and neurotransmitter receptors can be used interchangeably). Unlike voltage-gated channels, ligand-gated channels are often permeable to more than one ion. For example, Na^+, K^+, and Ca^{2+} can pass through the larval NMJ glutamate receptor once it is opened by the transmitter glutamate (Jan and Jan 1976a).

Action Potentials, Synaptic Potentials, and Driving Force

Knowing that there are a variety of regulated ion channels on the plasma membrane of a cell, let us return to our general relationship between membrane permeability and membrane potential. At rest, we know that because K^+ is the most permeable ion, V_m is set slightly more positive than E_{K^+}. What would happen if suddenly the membrane permeability of Na^+ became substantially greater than that of K^+ (i.e., $P_{Na^+} >> P_{K^+}$)? Because our rule of thumb is that the membrane potential is close to the equilibrium potential of the most permeable ion, then we would predict that $V_m \approx E_{Na^+}$. Using the values from the example above, $[Na^+]_{out}$ = 460 mM, $[Na^+]_{in}$ = 50 mM, and using the Nernst equation, E_{Na^+} = +56 mV. If under the new condition of high Na^+ permeability, it is assumed that $P_{K^+}:P_{Na^+}:P_{Cl^-}$ is 1:15:0.1, then, using the GHK equation, the cell's membrane potential would be +44 mV.

This is exactly what happens in a neuron during the rising phase of an action potential. An external stimulus depolarizes the neural membrane, and if the stimulus is strong enough, voltage-gated Na^+ channels open to greatly increase P_{Na^+}, which shifts the membrane potential to a positive value. The voltage-gated Na^+ channels do not stay open long (a few milliseconds), as P_{Na^+} returns to the prestimulus level and the membrane potential also returns to the resting level. With a brief delay, the stimulus also activates voltage-gated K^+ channels. This results in an efflux of K^+ ions and a rapid repolarization of the membrane. Therefore, during the falling phase of an action potential, as the Na^+ channels inactivate and the K^+ channels open, the membrane is once again most permeable to K^+ and the membrane potential once again returns to a voltage near E_{K^+}.

A similar change in ion permeability and membrane potential occurs at the synapse, except that the initiating event is the binding of neurotransmitter molecules to the neurotransmitter receptor. At the *Drosophila* larval NMJ, the excitatory neurotransmitter, glutamate, is released into the synaptic cleft and binds to glutamate receptors (GluRs) in the postsynaptic membrane of the muscle, causing an ion channel to open. As these ligand-gated channels are not specific to one ion, it is more difficult to predict what V_m the membrane approaches on channel opening. By analogy to the voltage-gated channels, the V_m will change according to the *mixture* of equilibrium potentials of the permeable ions.

In synaptic physiology this equilibrium potential is given the name "reversal potential" (V_{rev}), which refers to the membrane potential at which the net receptor/channel current is 0. At this potential, Na^+ influx and K^+ efflux are equal but in opposite directions. The current will flow either inward or outward if the membrane potential deviates away from the reversal potential. For glutamate receptors at the fly NMJ, the reversal potential is ~–10 mV. Practically this means that during synaptic transmission the V_m of the muscle cell will move from its resting value toward V_{rev}. Thus, Na^+ influx dominates K^+ efflux when the membrane potential is hyperpolarized below –10 mV. Under voltage-clamp conditions, it is possible to reveal that K^+ efflux overpowers Na^+ influx when the membrane is depolarized beyond –10 mV. One important practical application of the concept of the reversal potential in synaptic studies is that the peak amplitude of the excitatory junction potential (EJP; the evoked postsynaptic potential in the fly muscle) will always be below but never reach its reversal potential, no matter how much transmitter is released. This phenomenon is known as a "ceiling" effect. A detailed explanation of V_{rev} can be found in specialized neurophysiology textbooks.

Finally, we introduce the concept of an ionic electromotive force (EMF_{ion}) (also called driving force). Essentially, this is the difference between the equilibrium potential for an ion (E_{ion}) and the membrane potential (V_m) at any given point in time:

$$EMF_{ion} = V_m - E_{ion}.$$

According to Ohm's law, an ion current (I_{ion}) with a conductance G_{ion} can be expressed as

$$I_{ion} = G_{ion}(V_m - E_{ion}).$$

The importance of this relationship can be seen by considering the situation in which $V_m = E_{ion}$; that is, $EMF_{ion} = 0$. Under such conditions even if an ion channel for that ion is open there will be no net movement of the ion (I_{ion} =0), because there is no driving force.

On the other hand, if V_m is far from E_{ion}, then there will be a large driving force for that ion. This is the common situation for Na^+ at the resting membrane potential of most cells. V_{rest} for most cells is a negative value (e.g., –70 mV), whereas E_{Na^+} is usually a positive value (e.g., +50 mV). Therefore, in this example the driving force for Na^+ is 120 mV. If Na^+ channels are fully open under these conditions, then there is a large force that will push Na^+ through the channel. In contrast, Na^+ influx at the peak of an action potential (e.g., +48 mV) will be minimal even if all of the Na^+ channels are open. This is because the driving force (2 mV) is very small when the peak of the action potentials closely approaches E_{Na^+}. Hence, ion permeability, as well as the driving force, determines the impact of an ion on the membrane potential.

The concepts of relative permeability and driving force and their dynamics can be illustrated by a "spring" model, as shown in Figure 2. At rest, the thick and thin springs represent a high permeability for K^+ and a low permeability for Na^+, respectively. Because the resting potential is close to E_{K^+}, the driving force (the length of the spring) for K^+ is much smaller (shorter) than the driving force for Na^+. Although the equilibrium potentials for Na^+ and K^+ do not usually change for a given cell, both the permeability and driving force can be highly dynamic. For example, the Na^+ spring becomes much stronger relative to the K^+ spring at the peak of an action potential. At this point, the driving force for Na^+ reaches a minimal level, whereas it is at its largest for K^+. Readers should keep in mind that after a brief delay the K^+ spring strengthens dramatically, whereas the Na^+ spring weakens during repolarization of the action potential (not shown). This spring model also applies to synaptic potentials. The example shown here is an excitatory postsynaptic potential (EPSP) mediated by nonselective cation channels, such as the nicotinic acetylcholine receptor at mammalian NMJs and the glutamate receptor at fly NMJs. At the reversal potential, when the peak of the EPSP approaches –10 mV, the springs for Na^+ and K^+ are represented as equally strong. Hence, it is the relative strength of the springs—that is, the permeability of permeable ions—that determines the membrane potential.

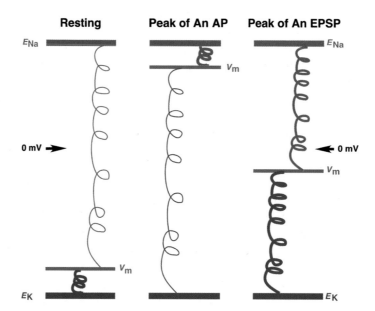

FIGURE 2. A spring model of relative permeability, driving force, and their dynamics. This cartoon models the relative permeability, the driving force, and membrane potentials of a typical cell at three different states. The thickness of the spring correlates positively with the permeability of an ion channel, whereas the length of the spring represents the driving force. At rest, the dominant strength of the K^+ spring (P_{K^+}) brings the membrane potential (V_m) toward E_{K^+}. But the driving force for Na^+ is much higher than that for K^+. The equilibrium potentials for Na^+ (E_{Na^+}) and K^+ (E_{K^+}) rarely change for a given cell. However, both the permeability and driving force can be highly dynamic. During the peak of an action potential, the spring strength and length reverse for K^+ and Na^+ such that P_{Na^+} overpowers P_{K^+}, thereby bringing the membrane potential near E_{Na^+}. In a synaptic potential mediated by nonselective cation channels such as the glutamate receptor at the fly NMJ, the spring is usually equally strong at the reversal potential. Hence, the peak of the EPSP approaches –10 mV. The fundamental take-home message is that it is the strength of the spring (i.e., relative permeability) of permeable ions that determines the membrane potential.

SYNAPTIC TRANSMISSION

Synaptic transmission is the process by which neurons secrete neurotransmitter molecules from the nerve terminal onto target cells. Synaptic physiology has been studied for a long time and remains a favorite topic among neuroscientists. Although a complete review of synaptic transmission is beyond the scope of this chapter, highlighting some of the achievements and contemporary advancements is worthwhile. Synaptic physiology began when Bernard Katz and his colleagues recorded the first synaptic response at the frog NMJ in the 1950s using an intracellular microelectrode. Their important studies, summarized in an elegant book (Katz 1966), provided the essential elements for our understanding of synaptic transmission. Their major findings were as follows.

1. There appeared to be a spontaneous release of neurotransmitter.

2. Action potentials arriving at the nerve terminal caused the release of neurotransmitter.

3. Transmitter release required that calcium be in the extracellular fluid at the time the action potential arrived.

4. Neurotransmitter is released in defined units, which Katz called quantal units.

5. The units appeared to be the same as the spontaneously released units (Fig. 3).

Indeed in subsequent years much of the effort in synaptic physiology has been to test, retest, extend, and expand on these early ideas. A recent study appears to challenge the basic tenet of Katz's quantal theory (Fredj and Burrone 2009).

Nearly simultaneous with the development of Katz's data on synaptic physiology, the first glimpses of the nerve terminal ultrastructure were obtained with an electron microscope. These initial views of the synapse revealed the presence of small clear vesicular structures in the nerve terminal. This observation naturally gave rise to the notion that Katz's quantal units were in fact the synaptic vesicles. This remained a controversial notion, however, until the important work of Heuser, Reese, and others, who showed a correspondence between quantal release as measured by electrophysiology and vesicular release as observed with the electron microscope (Heuser et al. 1979).

In the late 1980s studies began to reveal the molecular nature of synaptic transmission. Synaptophysin, VAMP (vesicle-associated membrane protein)/synaptobrevin, and SNAP-25 (synaptosomal-associated protein 25) were among the first synaptic proteins to be identified and characterized (Trimble et al. 1988; Oyler et al. 1989; Sudhof et al. 1989). This was followed by the identification of synaptotagmin (Perin et al. 1990), syntaxin (Bennett et al. 1992), and Rab3 (Fischer et al. 1990) in the early 1990s (also see reviews by Jahn and Sudhof 1994; Hay and Scheller 1997; Chen and Scheller 2001; Jahn et al. 2003). Research into the molecular nature of cellular processes was (of course) not limited to neuroscience, and studies on neural proteins soon became greatly influenced by molecular studies in related vesicle trafficking disciplines within the broader framework of cell biology. Particularly noteworthy was the identification of endocytic protein dynamin (Chen et al. 1991; van der Bliek and Meyerowitz 1991) following earlier discoveries of clathrin (Pearse 1976) and AP2 (Keen 1987), as well as seminal studies emerging from the vesicle trafficking field (see reviews by Cremona and De Camilli 1997; Brodsky et al. 2001; Conner and Schmid 2003). Most noteworthy are the studies that first identified *N*-ethylmaleimide-sensitive factor (NSF) and the other SNARE (soluble NSF attachment receptor) proteins and the realization that VAMP, syntaxin, and SNAP-25 were members of large protein families involved in a host of vesicular transport processes throughout the cell (Sollner and Rothman 1994). It had long been known that tetanus toxin and the botulinum toxins have profound effects on synaptic transmission. The discovery that these toxins were proteases that specifically cleaved VAMP, syntaxin, and SNAP-25 led to studies using them as probes that helped explain synaptic processes at the molecular level (Jahn and Niemann 1994). In particular, identifying the target molecules of these toxins quickly assigned a function to these recently discovered synaptic molecules.

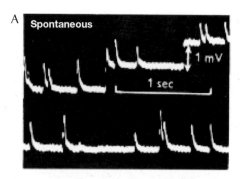

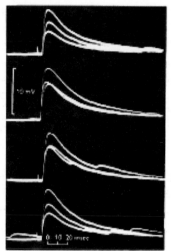

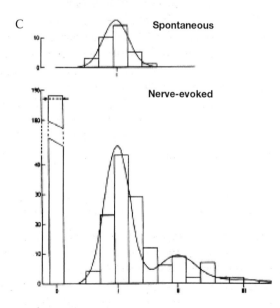

FIGURE 3. Quantal units of transmitter release. (*A*) A constant feature of synaptic transmission is the spontaneous (i.e., in the absence of action potentials) release of neurotransmitter. These miniature synaptic events appear similar to those caused by action potentials at low extracellular [Ca^{2+}]. (*B*) Nerve-evoked EJPs at medium extracellular [Ca^{2+}]. (*C*) If one collects EJP amplitude data of spontaneous events and nerve-stimulated events (under low-release conditions), two observations can be made. First, the amplitude distribution of the spontaneous events is the same as for the smallest nerve-stimulated events; second, the larger nerve-stimulated EJPs fluctuate in a manner that suggests the larger EJPs are multiples of the smallest EJP. These observations gave rise to the quantal hypothesis, which is that neurotransmitter is released from the nerve terminal in small packets (or quanta). (Reprinted, with permission, from del Castillo and Katz 1954.)

A Brief History of Neurogenetic and Electrophysiological Studies of Synaptic Transmission

Neurogenetic studies in *Drosophila* began with the premise that mutations of single molecules would alter the function of the nervous system, which would cause detectable changes in the behaviors of flies (Benzer 1973). With advances in molecular cloning techniques, it became possible to identify the genes whose mutations affected neural function. Touchstone publications include the report of temperature-sensitive alleles of *shibire*, which cause paralysis by blocking endocytosis of synaptic vesicles. It was discovered that *shibire* encodes the dynamin protein (Poodry and Edgar 1979; Koenig and Ikeda 1989; van der Bliek and Meyerowitz 1991). A temperature-sensitive allele of another *Drosophila* gene, *comatose*, also causes temperature-dependent paralysis. *comatose* encodes NSF (Ordway et al. 1994; Pallanck et al. 1995). Following these early studies, a number of other mutations affecting synaptic transmission have been discovered using forward genetics (e.g., *stoned*, Stimson et al. 1998; Fergestad et al. 1999; *awd*, Krishnan et al. 2001).

Using reverse genetics, researchers within the *Drosophila* community began looking for fly homologs of the proteins being characterized in other systems. The general strategy was to perform searches, such as screening cDNA libraries, to identify the genetic regions that encode known synaptic proteins, and then to make new mutants of these genes to study their effects on synaptic function. In this way synaptotagmin, syntaxin, SNAP-25, VAMP/synaptobrevin, and complexin, among others, have been analyzed in *Drosophila* (DiAntonio et al. 1993; Littleton et al. 1993; Broadie et al. 1994; Schulze et al. 1995; Deitcher et al. 1998; Rao et al. 2001; Huntwork and Littleton 2007). Using reverse genetics, a number of endocytotic mutants (such as AP2, AP180, and endophilin) have also been generated and studied to understand their roles in synaptic vesicle recycling (Gonzalez-Gaitan and Jackle 1997; Zhang et al. 1998; Verstreken et al. 2002).

The creation of transgenics has also been a valuable method in understanding synaptic transmission in *Drosophila* (Spradling and Rubin 1982; Brand and Perrimon 1993). These studies include expression of genes bearing engineered site-directed mutants, expression of exogenous tetanus toxin, gene replacement strategies, acute inactivation of synaptic proteins, and tissue-specific and temporally controlled gene expression (e.g., Sweeney et al. 1995; Osterwalder et al. 2001; Roman et al. 2001; Marek and Davis 2002; Venken and Bellen 2007; also see Chapters 22 and 23). A noteworthy example is research into the Ca^{2+}-binding properties of synaptotagmin I (Syt I) (Mackler et al. 2002; Robinson et al. 2002). Syt I was found to have two Ca^{2+}-binding C2 domains that interact with phospholipid and SNAREs, suggesting that it triggers transmitter release (Perin et al. 1990). Direct testing of Syt I's ability to perform this function was only achieved by constructing site-directed mutants that alter the Ca^{2+}-coordinating amino acids and introducing the transgene into *syt I*-null flies (Mackler et al. 2002; Robinson et al. 2002). These studies underscore the importance of combining molecular genetics with electrophysiology in understanding synaptic transmission.

USING *DROSOPHILA* LARVAL NMJ TO STUDY SYNAPTIC TRANSMISSION

The fly larval body wall muscles are ideal for studying synaptic transmission. They do not express voltage-gated Na^+ channels (Hong and Ganetzky 1994) and thus do not produce rapid action potentials. Neither do they exhibit Ca^{2+}-dependent regenerate potentials unless the tracheal system remains intact and is perfused with air (Yamaoka and Ikeda 1988). Hence, all recordings from the body wall muscle are passive membrane properties or pure synaptic responses. Furthermore, the muscle is isopotential, making it possible to record synaptic potentials anywhere in the muscle (in contrast to the frog NMJ in which minis [spontaneous miniature synaptic potentials] can only be detected when the electrode is inserted near the synapse). For these reasons the fly larval NMJ preparation has become the gold standard in the *Drosophila* neurobiology field ever since it was first developed for electrophysiological studies (Jan and Jan 1976a,b).

The key to understanding the electrophysiology of synaptic transmission lies in understanding the concepts of equilibrium potentials, driving force, selective permeability, and the dynamic interplay of different channels open at a specific time. The intracellular recording technique detailed in the following protocols is relatively simple, yet powerful, allowing one to address all of the basic questions concerning passive membrane properties and synaptic transmission. The technique also lays the foundation for more advanced techniques, such as voltage clamp and patch clamp. All of these experimental methods are based on similar biophysical principles and have methodological and operational features in common. Hence, we highly recommend that the reader first master the skills of intracellular recording (Protocols 1 and 2) before moving on to the focal loose patch and the two-electrode voltage-clamp (TEVC) techniques (Protocols 3 and 4).

In the CSHL course, students were provided with a number of "mystery" mutant strains so that they could use their newly acquired skills to characterize the mutant phenotypes and compare them with the wild-type larvae. These mutants included Syx^{3-69}, $snap$-25^{ts}, lap, K_{ir} mutant, and eag, Sh double mutants. Syx^{3-69} and $snap$-25^{ts} mutants were chosen because they have higher rates of minis and produce larger EJPs compared with the wild type (Rao et al. 2001; Lagow et al. 2007). Once a student was more experienced, mutant flies were provided that had reduced synaptic transmission (such as syt, DiAntonio et al. 1993; Littleton et al. 1993; Mackler et al. 2002; lap, Zhang et al. 1998; Bao et al. 2005). The eag, Sh double mutant is unique because the motor axons are hyperexcitable, making it possible to observe spontaneous EJPs without nerve stimulation (Feng et al. 2004). The K_{ir} mutant is ideal for studying the effect of leak potassium channels on resting potential and on passive membrane properties (Paradis et al. 2001). A bonus of the K_{ir} mutant is that it gives students the opportunity to appreciate the power of retrograde compensation in maintaining synaptic homeostasis (Paradis et al. 2001).

Electrophysiology Equipment

The electrophysiology setup is commonly referred to as a rig. The rig used in the following protocols is based on equipment and software used in the Cold Spring Harbor Laboratory (CSHL) course, Neurobiology of *Drosophila*. Manufacturers regularly update their products and periodically replace old products with new models. Equipment from reputable commercial suppliers other than those mentioned here should work equally well. Hence, we strongly recommend that you contact the vendors and work with them closely before purchasing a piece of equipment for a rig. Contact information for the major suppliers of electrophysiology equipment may be found at the end of the chapter under the section WWW Resources. Here we describe briefly the components of an electrophysiology rig. Figure 4 shows examples of the major rig components using either an upright compound microscope (Fig. 4A–C) or a stereomicroscope (Fig. 4D).

Vibration Isolation Table: The table is intended to reduce vibration and minimize mechanical damage to cells. Suggested suppliers are Newport Corporation and Technical Manufacturing Corporation.

Microscope: A microscope is used to visualize the preparation and electrodes. Any stereomicroscope with a long working distance is sufficient for most of the intracellular recording and TEVC experiments routinely conducted with fly larval NMJ preparations. Examples of stereomicroscopes include the Leica M50, Nikon SMZ645, and Olympus SZX7. In fact, any dissection microscope will do the job, provided you can clearly visualize the body wall muscle. Some investigators prefer an upright compound microscope equipped with 5× and 10× air objectives and a 60× water-immersion objective with long working distance (e.g., Olympus BX51WI with LUMPFL 60xW WD3.3 mm objective). A compound microscope is more costly, but experiments can be conducted that are not possible with a stereomicroscope. For example, with a compound microscope, single-bouton recordings can be made within the NMJ, the morphology of a single synaptic bouton can be explored, and optical imaging of the NMJ (using green fluorescent protein [GFP], red fluorescent protein [RFP], or calcium indicators) can be performed at the level of a single synaptic bouton.

Amplifier: It amplifies small synaptic or membrane electrical signals above noise levels so those physiological signals can be visualized and recorded. In the CSHL course, an AxoClamp2B or AxoPatch 200B (Molecular Devices) amplifier was used. However, any good amplifier from a reputable supplier will work well. The amplifier should come with holders for microelectrode and head

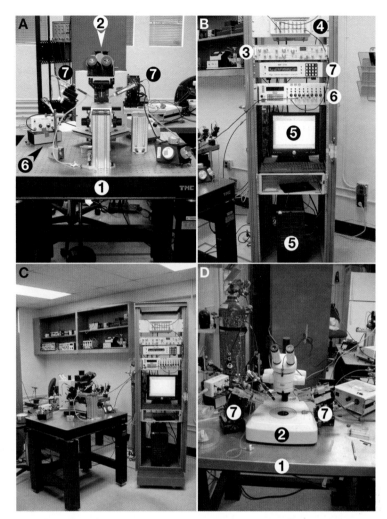

FIGURE 4. Examples of electrophysiology rigs. These pictures illustrate the major components of an electrophysiology setup commonly used in fly neurophysiology laboratories. (1) Vibration isolation (reduction) table; (2) microscopes (compound and stereomicroscope); (3) amplifier; (4) analog-to-digital (AD) and digital-to-analog (DA) interface board; (5) computer, monitor, and acquisition and analysis software (latter two not shown); (6) stimulator and stimulus isolator; (7) micromanipulators (motorized and manual). (*A–C*) A rig using an upright compound microscope. This configuration can be used for a variety of both basic and sophisticated experiments, including patch clamping of neurons or muscles (in embryos, larvae, and adults), two-electrode voltage clamping (TEVC), intracellular recording, single-bouton focal recording, and optical imaging. (*D*) A rig using a stereomicroscope. This simple setup is highly functional for intracellular recordings and TEVC in larvae or adult muscles. It is, however, not suitable for recordings from embryos or single boutons.

stages. Make sure that the outer diameter of the capillary glass to be used for manufacturing microelectrodes matches the size of the electrode holder. Suggested suppliers are Molecular Devices, HEKA, and Warner Instruments.

AD/DA Interface Board: This device changes analog (A) signals to digital (D) form so that a computer can record electrical traces. It also converts digital signals into analog ones, such as when your computer sends command signals to stimulate the nerve. A suggested product is Digidata 1440A (Molecular Devices).

Computer and Software: The software used in the course is called pClamp 10.0, which is compatible with the Digidata 1440A AD/DA interface board. The pClamp package has two separate programs: The Clampex program makes it possible to observe and acquire data and the Clampfit program is used for data analysis. In addition, the Mini Analysis Program (Synaptosoft Inc.) is used to analyze spontaneous and miniature synaptic potentials.

Stimulator: To excite an axon to fire action potentials, a brief electrical shock must be delivered to it. A stimulator and isolator are used to program the stimulation paradigms (such as stimulation frequency, duration, and strength). Suggested products are the S48 square pulse stimulator and SIU5 isolator (Grass Technologies/Astro-Med, Inc.) or the Master-8 and ISO-Flex stimulus isolator (A.M.P.I., Israel). If you are only delivering simple pulses (such as a single stimulus or two pulses), a stimulator is not required. Instead, use the built-in software with the Clampex program as a "stimulator" (for safety reasons a stimulus isolator must be used).

Micromanipulators: Micromanipulators help position the electrode(s) and assist in impaling the cell membrane with minimal physical damage. Larval NMJ recordings require one "coarse" manual manipulator (e.g., Narishige MM-3; Siskiyou MX110), which is usually placed on the left side and is used to position the stimulating ("suction") electrode. Two "fine" manipulators are needed to help position the recording electrode and impale muscle fibers. A manipulator with fine movement can be manual (e.g., Narishige NMN-25), hydraulic (e.g., Narishige MHW-3), or motorized (e.g., Sutter Instruments MP-225 or MP-285; Burleigh PCS-5000). The fine manipulator is a critically important piece of equipment because it is used to direct the electrode to impale muscles and to hold the electrode in a stable position. It is prudent to obtain the best micromanipulator possible.

Microelectrode puller: This device is used to pull the high-quality microelectrodes needed for impaling cell membranes. As one of the most important (but often overlooked) pieces of electrophysiology equipment, an electrode puller will determine how reliable your electrodes will be over time. Suggested products are the P-97 or P-1000 Flaming/Brown Micropipette Puller (Sutter Instruments) or the PC-10 Puller (Narishige).

Other Items:
BNC cables
Glass capillaries for manufacturing microelectrodes
Grounding wires and silver wires
Light source
MicroFil (World Precision Instruments)
Syringe (1-mL), filled with 3 M KCl and to be used with MicroFil to backfill a microelectrode

Video Monitor System: This is optional, but for teaching purposes it is extremely useful, because it allows up to five students to simultaneously visualize NMJ preparations and to watch their own actions (such as electrode position/movement) on a large monitor.

Oscilloscope: This is optional, although an oscilloscope has more functions than does a computer.

No two electrophysiology rigs are identical, because each laboratory configures theirs to suit their needs. In addition, different experiments require minor modifications to the rig, such as varying the number of micromanipulators needed depending on the number of electrodes being used. Historically, the system for signal observation and data storage has gone through the most dramatic change (and improvement) within the last 20 years. Instead of the traditional oscilloscope, chart recorder, and magnetic tapes, electrophysiological data are now visualized, acquired, and stored in digital formats in a computer using specially designed software.

Using Axoclamp 2B and the Digidata data acquisition systems as examples, a diagram of wiring connections for an electrophysiology rig that can be used for both intracellular recording and TEVC is provided in Figure 5.

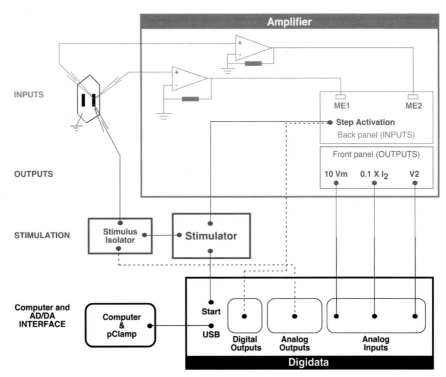

FIGURE 5. A wiring diagram for an electrophysiology rig. A rig can be divided into four major components based on the flow and processing of electrical signals: inputs, outputs, stimulation, and interface and data acquisition. Once connected through a BNC cable as illustrated here, these components will become a functional electrophysiology rig. The wiring diagram shown here is based on the AxoClamp 2B amplifier, Digidata, and pClamp software. However, this diagram can be used as a reference for any other similar products. Membrane potentials recorded by microelectrodes are detected by specific probes and then sent into an amplifier through corresponding channels (such as ME1 and ME2). Once amplified by the amplifier, the electrical signals travel to the AD interface board to be converted into digital forms and then acquired by the computer using specially designed software. The interface board is also capable of producing signals to command either the stimulator or the amplifier. A voltage or current delivered through the stimulus isolator and a suction electrode is used to stimulate the motor axons into firing action potentials. A stimulator is used to program different stimulus parameters so that one can control the stimulus duration and frequency. The software and DA interface board can also be used to control the stimulus pulse, but they do not have great flexibility for users if one wants to have more complex stimulus paradigms.

Electrophysiological Recording from a "Model" Cell

The muscle or neuron membrane is functionally equivalent to an RC circuit with the membrane resistance and capacitor in parallel (see Box 2). Once inserted inside the membrane, an electrode introduces a serial resistance and small capacitance to the RC circuit (Fig. 6). Through a narrow opening at its tip (~0.1 µm), current can pass through the electrode, into the cell, and back to the outside (ground) across the membrane to complete the circuit. This arrangement enables a voltage difference between the outside and inside of the cell membrane to be recorded. To determine cell membrane properties, a current can be injected into the cell through the electrode. One complication with this approach, however, is that the voltage difference measured with the electrode includes the voltage drop across the cell membrane and that across the electrode. Furthermore, a small amount of current is drawn by the electrode capacitor, thereby slowing the current flow across the membrane. Fortunately, most amplifiers are equipped with bridge balance and capacitance compensation functions so that the effects of the electrode on cell membrane properties can be canceled out. This protocol describes the basics of setting up and conducting electrophysiological experiments using a model cell. For the novice, a model cell provides a way to learn the operation of electrophysiology equipment and software without the anxiety of damaging living cells. This protocol also illustrates passive membrane properties such as the input resistance, capacitance, and time constant.

MATERIALS

Equipment

CLAMP-1U
Electrophysiology rig
Model cell (CLAMP-1U; Molecular Devices Inc.)

METHOD

Setting Up to Record from a Model Cell

1. Turn on the electrophysiology rig (including the amplifier, A/D board, and computer) and let them warm up.

2. On the front panel of the AxoClamp amplifier, make sure that the capacitance compensation knob and bridge balance dial on panel ME1 are turned fully counterclockwise to 0.

3. Insert the ME1 side of the model cell (CLAMP-1U) into the head stage (HS-2A-0.1 LU) of the amplifier.

4. Insert the gold pin on the other end of the black wire into the yellow socket (the ground) located on the back of the head stage.

5. Adjust the "Input Offset" knob so that the reading on the digital meter is 0.

6. Open the Clampex software and leave it on the "View" mode; there should be a flat line on or near 0 mV. To record and save data, click on the red button on the top panel of Clampex.

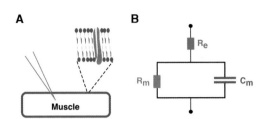

FIGURE 6. Illustration of intracellular recording and the equivalent electrical circuit. (A) The membrane potential of a muscle (or a neuron) can be recorded and monitored using a sharp intracellular recording electrode. The lipid bilayer and an ion channel of the membrane are shown in the *inset*. (B) In electrical terms, the ion channel and the lipid bilayer of the cell membrane can be regarded as a resistor (R_m) and capacitor (C_m), respectively. They are arranged as a parallel RC circuit. The electrode resistor (R_e) and capacitor (C_e, not shown) are in serial with the cell membrane RC circuit. For a given cell, C_m is fixed and thus alterations of ion channels (R_m) determine the passive membrane properties (membrane resistance and time constant) and the rate and magnitude of synaptic responses.

Adjusting Capacitance Compensation in the Bath Mode

7. Flip the toggle switch on the model cell to the "bath" mode. Set the current amplitude to +0.5 nA using the thumb wheel on the "Step Command" panel of the amplifier. Deliver a brief current pulse (5 msec, +0.5 nA) to the model cell. Note the "square" waveform generated by the current pulse (Fig. 7, top panel).

8. Increase the pulse delivery frequency (e.g., to 1 Hz) and gradually turn the capacitance compensation knob clockwise. The rise time of the rising phase should become shorter (Fig. 7, middle panel). You should also notice that the background noise of the trace gets larger (this is normal). Keep turning the capacitance compensation knob until you start to see oscillations. The optimal setting should be slightly lower, so that these oscillations just disappear. An example of mild oscillations is shown in Figure 7 (bottom panel).

   An increase in compensation beyond the start of the oscillations will lead to uncontrolled large-amplitude oscillations, which often result in the death of recorded cells. Thus, it is a good idea to perform capacitance compensation in the bath mode when the microelectrode is in the bath instead of inside a cell.

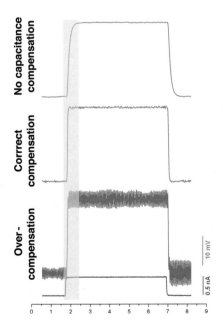

FIGURE 7. Capacitance compensation in a model cell. In many electrophysiological experiments an experimenter needs to inject currents into cells to manipulate the membrane potential or to learn the kinetics of membrane potential responses. A common problem is that the capacitor of the microelectrode draws current away from the cell, slowing down the response time of the cell membrane. Using capacitance compensation one can reduce this electrode capacitance effect. A square waveform generated here in the *top panel* has a slower rise and fall phase. After capacitance compensation, the rise and fall phase becomes much faster (*middle panel*). However, the waveform becomes noisy with high-frequency oscillations if it is slightly overcompensated (*bottom panel*). A further increase in capacitance compensation will result in large amplitude and uncontrollable oscillations (not shown) sufficient to kill a living cell. The *x*-axis is in milliseconds.

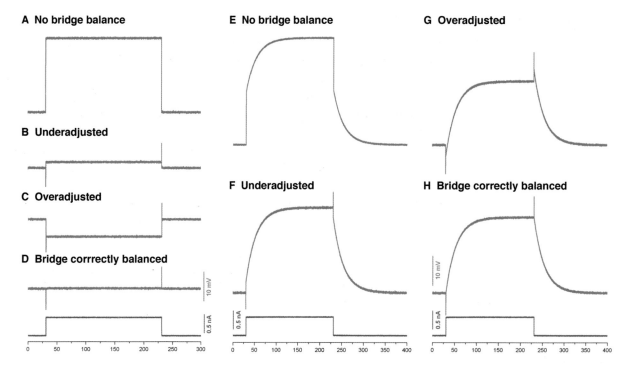

A No bridge balance

B Underadjusted

C Overadjusted

D Bridge corrrectly balanced

E No bridge balance

F Underadjusted

G Overadjusted

H Bridge correctly balanced

FIGURE 8. Bridge balance in a model cell. Another problem one encounters during current injection is the serial resistance from the microelectrode, which will mask the voltage response from the cell membrane following current injection. Bridge balance allows one to determine the input resistance of the cell membrane by canceling out the serial resistance of the microelectrode. (A–D) Performing bridge balance in the "bath mode" of a model cell. Following current injection the square waveform represents the voltage drop across the electrode resistor. The peak amplitude (i.e., the plateau portion) allows one to determine the input resistance of the electrode (A). Examples of incomplete and overadjustments of bridge balance are shown in B and C, respectively. A perfect bridge balance can be achieved by carefully canceling out the electrode response (A). (E–H) Performing bridge balance in the "cell mode" of a model cell. Before any bridge balance, current injection results in a complex waveform, which has sharp rise and fall phases as well as slowly charging and discharging components (E). The sharp components, which are essentially identical to the square wave in A, come from the electrode resistor. The slow components are responses from the cell membrane. Examples of under- and overadjustments of the bridge are shown in F and G. Once the bridge is perfectly adjusted, the only remaining part is response from the cell membrane resulting from a parallel resistor-capacitor (RC) circuit (H) (see Box 2). Membrane input resistance as well as time constant can be readily determined from this waveform. Note the initial and tail sharp "spikes" (in B–D and F–H), which result from capacitance currents of the electrode. The x-axis is in msec.

Balancing the Bridge in the Bath Mode

9. In the bath mode, deliver a +0.5 nA current pulse of a longer duration (200 msec) through the model cell, observe the waveform, and acquire data. You should expect to obtain a square wave, similar to the one shown in Figure 8A. The peak voltage should be near 25 mV if the model cell 1U is in use. These results indicate that the input resistance of the model electrode is 50 MΩ.

10. Increase the pulse frequency (e.g., at 1 Hz) and continuously monitor the voltage traces. Gradually turn the "bridge balance" knob clockwise and observe the change of the waveform. At the point where there is no voltage change in response to the current injection, the bridge is considered balanced, meaning that the voltage drop across the electrode is 0 (Fig. 8D). Now the electrode resistance has been effectively canceled out of the circuit. Above or below this balanced point the bridge is "under-" or "over-" adjusted, respectively (Figs. 8B,C).

Balancing the Bridge in the Cell Mode

11. Reset the bridge balance to 0, flip the toggle switch on the model cell to "cell" mode, deliver a +0.5 nA current pulse through the model cell, and acquire data (Fig. 8E). Note the compound waveform with the initial sharp rising and falling phases followed by the slow components (compare with Fig. 8A). Also note the complex waveform in which the sharp component arises from the model electrode, whereas the exponential component arises from the model cell membrane.

12. Deliver a repeating +0.5 nA current (e.g., at 1 Hz) and continuously monitor the voltage traces. Gradually turn the "bridge balance" knob clockwise and observe the change of the waveform until the sharp components have completely disappeared (Fig. 8H). This cancels out the model electrode resistance, so that the remaining waveform is from the model cell membrane only. The input resistance of the model cell membrane is 50 MΩ as well. By curve filling, you can also determine the time constant of the model cell membrane.

    Note the examples of traces when the bridge is "under-" or "over-" adjusted, respectively (Fig. 8F,G). When using a glass electrode, this step is performed to cancel out the glass electrode resistance and determine the input resistance and time constant of the muscle membrane.

13. Repeat Steps 7 and 8 and Steps 11 and 12 using –0.5 nA current.

DATA ANALYSIS

14. Open Clampfit and open the dataset of interest.

15. Place one cursor on the flat line and another cursor on the square wave. The difference between these two cursors is the amplitude of the square wave (i.e., 25 mV). Hence, the model electrode resistance can be determined using Ohm's law ($V = IR$): $R_e = 25 \times 10^{-3}$ V/0.5×10^{-9} A = 50×10^6 Ω = 50 MΩ.

16. With a perfect bridge balance, the peak amplitude of the waveform in the "cell" mode is also close to 25 mV, indicating that the input resistance for the model cell membrane is 50 MΩ.

17. To obtain the model cell membrane time constant, place one cursor on the beginning part of the charging curve and another cursor at the end of the curve. Under "Analysis," select "Fit," which will lead you to a table of various functions and mathematical equations. Select "Exponential, standard" and perform the curve fitting.

18. A second line will appear on the waveform (often in different colors; not shown). These two lines should overlap with each other. Go to "Windows," select "Results1," which opens a table of numbers that includes the membrane time constant (τ = 24 msec). This means that it takes ~24 msec to charge up the model cell membrane to two-thirds of the final amplitude.

TROUBLESHOOTING

Problem: The recording is too noisy.

Solution: Noise is a common and constant problem for electrophysiology. It may never be fully eliminated, but there are ways to reduce it. First, identify the source of the noise. Noise can be intrinsic to the amplifier, for which little can be done, although you could contact the commercial supplier if its amplifier has unusually high intrinsic noise. Common sources of noise are radiative electrical pickup (line frequency noise, typically oscillating at 60 cycles, from lights, power strips, or outlets, and high-frequency noise from computers and monitors) and ground-loop noise. Light sources and the power supply to the microscope are often the main culprits.

Shielding with a sheet of tinfoil is usually effective in reducing or completely eliminating line frequency noise. Positioning the computer and monitor away from the head stage may also reduce noise. Some monitors are noisier than others. Replace a particularly noisy monitor if necessary. Finally, ground-loop noise can be avoided by using a common ground.

Problem: Voltage is unstable and undergoes uncontrolled, large-amplitude oscillations.
Solution: The capacitance compensation dial may be turned beyond its ideal value. Turn the knob counterclockwise slightly until the oscillation stops. Although it does not matter to the model cell, this type of oscillation usually kills cells when it occurs in a live recording!

Problem: When a current pulse (150 msec, +0.5 nA) is delivered, the model cell membrane potential initially drops below the baseline.
Solution: The bridge is overadjusted. Turn the "bridge balance" counterclockwise as far as it will go and rebalance the bridge (see Fig. 8G).

Problem: The voltmeter reading is unrealistic (e.g., –896). In addition, the reading does not change when the "Input offset" dial is adjusted.
Solution: The circuit is not complete. Verify that the ground wire is in the bath solution. If the ground wire is not the problem, look for a break or bad connection among the BNC cables or ground wires. A voltmeter can be used to diagnose connection problems of the circuit or the wires.

Protocol 2

Recording from *Drosophila* Larval Body Wall Muscles: Passive Membrane Properties and Basic Features of Synaptic Transmission

This protocol first describes the steps for performing intracellular recording from fly larval body wall muscles and then explains how to record and analyze spontaneous and evoked synaptic potentials. The *Drosophila* larval body wall muscle preparation was first used for electrophysiological analysis in the 1970s (Jan and Jan 1976a,b). This preparation has become the "gold standard" for studying neuronal excitability as well as synaptic transmission.

Many aspects of the exercise described here are similar to those you have experienced with the model cell. However, there are a number of new challenges, including larval dissection (filleting), identification of muscle fibers and their innervating nerves, the use of micromanipulator and microelectrode in penetrating the muscle membrane, and nerve stimulation to evoke synaptic potentials.

MATERIALS

CAUTION: See Appendix for proper handling of materials marked with <!>.
See the end of the chapter for recipes for reagents marked with <R>.

Reagents

Drosophila melanogaster synaptic transmission mutants (e.g., *Syx*[3-69], *snap-25*[ts], *syt I*, and *lap*), neuronal excitability mutants (e.g., *eag, Sh* double mutant), or muscle excitability mutants (e.g., larvae that overexpress the inward rectifier potassium channel $K_{ir}2.1$ in body wall muscles [referred to as the K_{ir} "mutant"])

D. melanogaster wild-type strains (Canton-S or Oregon-R)
    All flies (mutants and wild type) are maintained and raised on cornmeal-based fly food at room temperature. Avoid crowding in the vials, so that larvae can grow healthy and large.

HL-3 saline <R>
HL-3 saline without Ca^{2+}, ice cold
KCl <!>, 3 M

Equipment

Dissection instruments
Electrophysiology rig
Microelectrodes, intracellular (as prepared in Protocol 5)
Microscope
Suction electrode (as prepared in Protocol 5)
Syringe, 1-mL or 5-mL

METHOD

Dissecting Third-Instar Larvae

1. Dissect third-instar larvae as described in Chapter 7, Protocol 1. Use either a Sylgard-coated Petri dish and small insect pins or a magnetic dish with paper clip pins. Perform the dissection in ice-cold, Ca^{2+}-free HL-3 saline (Stewart et al. 1994).

2. Using a pair of fine scissors, transect the segmental nerves loose from the base of the ventral nerve cord so that the nerve end can be picked up with a suction electrode (Step 20).

3. Immediately following dissection, rinse the NMJ preparation at least three times with room temperature HL-3 saline containing the desired level of Ca^{2+}. The purpose of these washes is to ensure that the bathing solution is free of any material that could clog the suction electrode. We recommend using 0.8 mM Ca^{2+} in the HL-3 saline.

4. Position the NMJ dish on the center of the microscope stage. Bring the preparation into focus and take note of the orientation, segments, and muscle patterns. Place the preparation so that the posterior end of the larvae points toward you. This gives enough space on the two lateral sides so that the electrodes can access the NMJ without bumping into the insect pins.

Recording Membrane Potentials

5. Backfill a microelectrode with 3 M KCl solution. Fill the electrode approximately two-thirds full and add one drop of the KCl saline at the end of the glass pipette so that the solution can be drawn into the fine tip of the electrode through the built-in capillary (see Protocol 5). Use a piece of soft tissue (e.g., Kimwipe) to remove any liquids outside of the electrode so that salt crystals do not accumulate inside the electrode holder.

6. Carefully insert the microelectrode into the electrode holder and tighten the screw cap slightly so that the microelectrode is held firmly in place. Make sure that the silver wire from the electrode holder is in contact with the KCl saline inside the electrode lumen. Do not screw the cap on too tightly or the glass electrode may break. Screw the electrode holder onto the head stage.

7. Place the reference (ground) wire into the bath saline.

8. Adjust the focal plane above your preparation and lower the microelectrode into contact with the saline. The first sign that the electrode has entered the saline is that the digital voltmeter begins to display a steady voltage.

9. Turn the "Input offset" knob in the ME1 panel until V_m reads 0 mV. Lock the knob if necessary to avoid accidentally changing the setting.

10. Visually inspect the tip of the electrode for any obvious damage under the microscope. Replace it if the electrode is broken or contains an excessive amount of debris or dirt.

11. Check the input resistance of the electrode by turning the toggle switch to the "on" position on the "Step Command" panel and manually injecting 1 nA current.

    A good electrode for intracellular recording usually has an input resistance ranging from 15 to 25 MΩ. Record the electrode input resistance in a notebook to serve as a reference on which electrode to use in the future.

12. Perform bridge balance to cancel out the resistance of the electrode until the square wave disappears (similar to those conducted in Fig. 8A–D). Use the "Capacitance compensation" knob (with a yellow top) to increase the rise time. Avoid overcompensation (i.e., when the trace will be ringing with high-frequency oscillations).

13. Lower the focal plane first and then the electrode so that the tip of the electrode is always within the visual field. Repeat the "focus down first and lower electrode next" motion until the electrode tip is just above the preparation.

    A common mistake that most novices make is to lower the electrode too fast and too far, with the result that the electrode is "driven" through the preparation!

14. Once you are close to the muscle surface, switch the manipulator to the fine control movement mode to position the microelectrode above a muscle fiber (e.g., muscle 6 in abdominal segment 4). Carefully lower the electrode above the preparation while adjusting the focal plane toward the preparation.

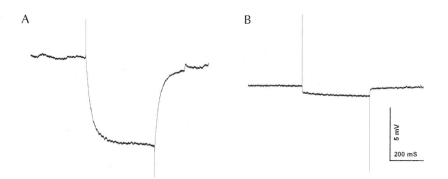

FIGURE 9. Bridge balance in muscle cells. The examples shown here are intracellular recorded waveforms in response to injection of –1 nA current square wave (400 msec duration) in body wall muscles of a wild-type larva (*A*) and a larva whose muscles express the human inward rectifying potassium channel $K_{ir}2.1$ (K_{ir} "mutant") (*B*). Note the dramatic reduction in both membrane input resistance and time constant in the K_{ir} "mutant." The electrode capacitance "spikes" at the beginning and end of the current pulse are evident in these examples.

15. Penetrate the muscle membrane slowly and gently. If you are using the Sutter P-250 motorized manipulator, switch to the "diagonal'" mode so that the electrode moves diagonally toward the muscle, which makes membrane penetration easier. Two hints that your electrode is touching or pressing on the muscle surface are that (1) the digital voltmeter begins to show negative potentials up to –4 to –5 mV and (2) you may notice a dent on the muscle surface at the tip of the electrode. The potential readings are a more reliable and sensitive indicator, especially if it is difficult to clearly the electrode tip. At this point, a slight advance or vibration of the electrode is sufficient to enter the cell. There should be a rapid voltage drop to at least –45 mV.

16. Wait for the membrane potential to become more hyperpolarized. It should stabilize at –65 to –75 mV. If the resting potential remains at –45 mV or is even more depolarized, withdraw the electrode and try inserting it into the adjacent muscle or try a muscle cell in a different segment. Continue this trial-and-error approach until a reasonably good resting potential is obtained.

    Before inserting the electrode into another muscle cell, check its input resistance again, and replace the electrode if its input resistance is dramatically altered.

17. Once a suitable and stable resting potential is obtained, cancel out the electrode resistance by adjusting the bridge balance again. Specifically, inject a –1 nA current pulse for 400 msec duration at 1 Hz (Fig. 9A; Fig. 10A,B). Readjusting the bridge balance is necessary because muscle membrane debris may alter the input resistance of the electrode. Determine and record the muscle-input resistance (R_m) and the membrane time constant τ. Record the R_m and τ values in the notebook. Fly muscle R_m varies between 5 and 12 MΩ, and τ is ~30 msec. Try a new muscle fiber if the input resistance of the recorded muscle is <5 MΩ.

Synaptic Potentials

This section explains how to record synaptic potentials. Many of the previous steps are applicable here, but in addition a suction electrode must be properly positioned and used to stimulate motor nerves. Typically, the manipulator for the suction electrode is placed on the side of the microscope opposite the recording electrode.

18. Lower the suction electrode just below the surface of the bath saline and use a 1- or 5-mL syringe to gently apply negative pressure (suction) so that the saline fills one-half of the length of the suction electrode. This ensures that the stimulating silver wire inside the electrode is deeply immersed in the saline. Place the silver wire near the shank of the suction electrode so that the wire remains in touch with the saline.

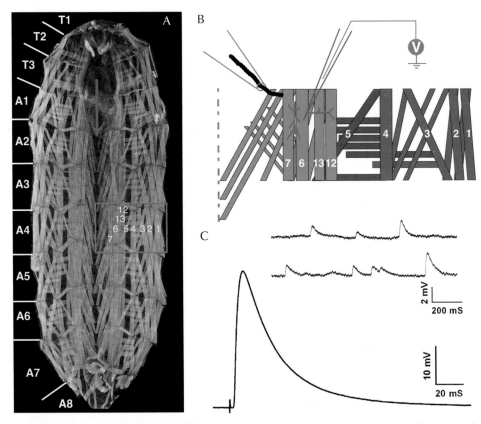

FIGURE 10. Intracellular recording of synaptic potentials from larval body wall muscles. (*A*) An illustration of the thoracic (T) and abdominal (A) segments and the repeated pattern of musculature arrangements (the brain and the ventral nerve cord are removed in this image) in a dissected, flattened, and fixed third-instar larva. (Adapted, with permission, from Bellen and Budnik 2000.) Major ventral, lateral, and dorsal muscles are labeled with numbers. The ventral midline is indicated by the blue arrow. (*B*) One hemisegment of the body wall muscles is illustrated here, in which ventral, lateral, and dorsal muscles are marked with green, black, and blue, respectively (Hoang and Chiba 2001). Major surface muscles are identified by their numbers. The ventral midline is shown as the dashed red line. A suction electrode is shown to pick up a segmental nerve and used to stimulate the motor axons within the nerve. A microelectrode is used to record minis and evoked synaptic potentials from muscle 6. Segments 4 and 5 are often used for synaptic transmission studies because the nerve leading to these muscles is long and easier to pick up by the suction electrode. Ventral longitudinal muscles such as muscles 6, 7, 12, and 13 are typically used for intracellular recordings largely because of their large sizes and ventral positions. (*C*) Without nerve stimulation the muscle constantly displays spontaneous miniature synaptic potentials (called minis or mEJPs). Following a single stimulus, the whole muscle's response to transmitter released from the nerve terminal is recorded as a compound excitatory junction potential (EJP), which is more generally known as an excitatory postsynaptic potential (EPSP).

Avoid using a larger (e.g., 10-mL) syringe because the suction it generates is too strong and difficult to control.

19. Stop suctioning by applying a small amount of positive pressure into the suction electrode. This prevents the suction electrode from continuing to suck and picking up nerves randomly along its path. Avoid pushing the saline out of the electrode. It is possible to maintain a "steady state" by gently balancing the positive and negative pressures inside the electrode.

20. Lower the suction electrode to the segment where recordings from muscles will occur (e.g., abdominal segment 4, muscles 6 and 7). Keep in mind that the nerve enters each segment from the anterior end, near the denticle belt area. Once the open tip of the suction electrode is positioned near the nerve entry point, gently apply a negative pressure until the nerve is pulled into the electrode lumen.

Avoid applying too much pressure into the suction electrode. Excessive pressure will cause the nerve to stretch and pull away from the muscle, resulting in a partial or a complete failure to evoke synaptic responses.

21. Once the suction electrode is in place, position the intracellular electrode and impale a muscle of interest (e.g., muscle 6) (Fig. 10A,B). Perform a routine check of the input resistance and time constant, balance the bridge, and perform capacitance compensation. Record spontaneous minis, which occur randomly at 2–5 Hz with average amplitude ranging from 0.6 to 1 mV in wild-type larvae (Fig. 10C). Record minis for 45 sec to 1 min. From these recordings you can determine the average amplitude of the minis as well as the average mini frequency.

22. Stimulate the motor nerve by delivering a current pulse of 150–200 μsec duration, gradually increasing the stimulus strength until a synaptic potential (commonly called an EJP) is evoked. Once one EJP has been observed, increase the stimulus strength to excite another axon (Kurdyak et al. 1994) because muscles 6 and 7 are dually innervated. The threshold voltage needed to evoke EJPs is ~0.6 V. An example of a compound EJP on muscle 6 on segment A4 is shown in Figure 10C.

23. Stimulate the motor nerve at a low rate (0.1–0.2 Hz) and record at least five different EJPs. Knowing the average EJP amplitude and mini amplitude, you can estimate the quantal content by determining the ratio of the average EJP amplitude to the average mini amplitude.

> The average amplitude of the EJPs will not be determined accurately if the nerve is stimulated at too high a frequency (≥1 Hz), because frequent stimulation depletes the releasable pool of synaptic vesicles, causing the EJP amplitude to gradually reduce with consecutive stimuli. This will lead to an underestimate of quantal content.

> To gain more experience recording synaptic potentials and performing quantal analysis, take recordings from other well-established mutant flies. Compare your observations of wild-type larvae with the mutants and with data from the published literature.

[Ca^{2+}]-Dependent Short-Term Synaptic Plasticity

Once intracellular recording and nerve stimulation skills have been mastered, there are many other exciting areas of synaptic transmission to explore. These include testing the dependence of transmitter release on extracellular [Ca^{2+}] (Jan and Jan 1976b), investigating the effect of repetitive nerve stimulation on the rate of synaptic vesicle pool depletion (Delgado et al. 2002), and studying activity- and [Ca^{2+}]-dependent short-term plasticity (Zhong and Wu 1991; Rohrbough et al. 1999; Bao et al. 2005).

24. Design a stimulation paradigm that can deliver two pulses separated by 100 msec (10 Hz), 50 msec (20 Hz), 20 msec (50 Hz), and 10 msec (100 Hz), respectively. Observe the relative amplitude of two evoked EJPs in HL-3 saline containing 0.6 mM Ca^{2+}. This is a classic exercise that is often referred to as "twin-pulse" or "two-pulse" facilitation (TPF).

25. Repeat the above experiments in NMJ preparations based in HL-3 saline containing 1, 1.5, and 2 mM Ca^{2+}, respectively. Examples of short-term synaptic plasticity at 30 Hz at 0.6 and 1.5 mM Ca^{2+} are shown in Figure 11, A and B, respectively.

Electrode Cleanup

26. At the end of the day, rinse the suction electrode thoroughly with distilled water and leave the electrode to air dry to prevent bacterial growth. When rinsing the suction electrode, always suck water in and do not push the liquid out, as small bits of debris may clog the tip of the electrode, rendering it useless. A suction electrode properly cared for can be used repeatedly for as long as it works (up to 6 mo or longer).

27. Discard the intracellular microelectrode, and turn off the electronics.

A   B

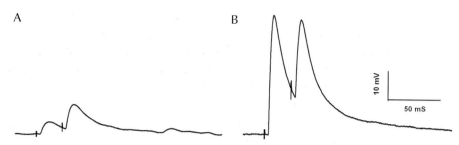

FIGURE 11. Short-term synaptic plasticity at the larval NMJ. Synaptic release of transmitter is not fixed but rather plastic depending on the amount of extracellular $[Ca^{2+}]$ and nerve activity patterns. (*A*) At low $[Ca^{2+}]$ (0.6 mM), the basal amplitude of EJPs is small. In response to twin-pulse stimulation at 30 Hz, the amplitude of the second EJP is larger compared with the amplitude of the first one, revealing the ability of the synapse to facilitate. (*B*) At high $[Ca^{2+}]$ (1.5 mM), the basal amplitude of EJPs is large. Instead of facilitation, now the same synapse displays short-term synaptic depression following twin-pulse nerve stimulation.

Data Analysis

28. Using Clampfit, analyze the input resistance and time constant of the muscle membrane, and the amplitude of EJPs. Use Mini Analysis to analyze the mini frequency and amplitude. Quantal content of a synapse can be estimated by calculating the ratio of the average EJP amplitude and the average mini amplitude provided that the resting potential of different recorded muscles is similar and that the $[Ca^{2+}]$ is relatively low. Otherwise, nonlinear summation should be taken into consideration in determining quantal content (Martin 1976; Stevens 1978; Feeney et al. 1998).

TROUBLESHOOTING

Problem: No voltage can be detected when the microelectrode is placed in the bath saline.
Solution: There may be an "open" circuit. Consider the following.

1. Look for broken wires, and replace any that are found.

2. Confirm that the ground wire is in the bath.

3. Look for a large air bubble inside the electrode tip that got there when the microelectrode was backfilled.

4. Determine whether the silver wire inside the electrode holder is contacting the golden pin.

Problem: Resting potentials of the impaled muscle cell never hyperpolarize beyond –35 to –45 mV.
Solution: Wild-type larval body wall muscles have a resting potential from –65 to –75 mV. A poor resting potential usually means that the muscle has been damaged during dissection or penetration with the microelectrode. Consider the following.

1. Healthy muscles have a translucent appearance, whereas degenerating muscles are brownish or visibly tearing apart from the muscle attachment points in the denticle belt area. This tends to occur if the preparation is left at warm temperatures. Keep and work with muscle preps in a cooler room or cooled microscope stage (18°C –20°C).

2. Physical tearing or damage to muscles should be avoided at all costs during dissection or electrode impalement. However, electrode-caused damages are difficult to detect by eye. One way to detect any damage is to measure the muscle input resistance and time constant. Both of these parameters will reduce dramatically in a damaged muscle. Thus, it is always a good idea to routinely monitor muscle input resistance and time constant in all of your experiments.

Problem: Although the resting potential of the muscle and minis were normal, no evoked synaptic potentials were detected when the segmental nerve was stimulated.

Solution: Consider the following when evoking synaptic potentials.

1. The stimulator must be turned on and the voltage output level must be high enough to evoke action potentials. Gradually increase the amplitude of the stimulus strength until an evoked potential is seen.

2. Check to see if the polarity of the stimulus was reversed. Switch the polarity of the stimulus isolator and test again.

3. Determine if the suction electrode contains enough saline so that the stimulating wire is in contact with the saline. (If the wire is not in contact with the saline, then no current can pass through the suction electrode.) Gently push the nerve ending off the suction electrode, raise the suction electrode away and above the NMJ, and refill it with saline. It is important to raise the suction electrode so that no nerves are picked up during refilling. Lower the electrode and pick up the nerve ending again.

4. Inspect the nerve for damage that may have occurred during dissection. If the forceps have pinched the nerve, you will not get normal EJPs. Start over with a new dissection.

5. Too much pressure was applied to the suction electrode when picking up the nerve causing disruption of the synaptic contact. Switch the suction electrode to a different segment and try to gently pick up the nerve.

6. Verify that the saline contains Ca^{2+}, which is essential to trigger exocytosis. Prepare saline containing Ca^{2+} and repeat the experiment.

Problem: Instead of an EJP with peak amplitude of 25 mV, an "odd"-looking EJP was obtained having much smaller amplitude and unusually slow rising and falling rates.

Solution: Most likely the motor nerve from a neighboring segment was stimulated. Larval body wall muscles are electrically coupled by gap junctions with homologous muscles in the neighboring segments (Gho 1994). EJPs as well as minis can passively invade the neighboring muscles resulting in EJPs or minis with much slower kinetics. The minis with slow rising and falling phases should be excluded in determining mini frequency and amplitude.

Protocol 3

Voltage-Clamp Analysis of Synaptic Transmission at the *Drosophila* Larval NMJ

Although intracellular recording is particularly valuable in revealing membrane potential changes, it has a number of limitations. Primarily, it does not offer information on the kinetics of membrane currents associated with ion channels or synaptic receptors responsible for the potential change. Furthermore, the resting potential of the *Drosophila* body wall muscle varies naturally such that the driving force also varies considerably, making it difficult to accurately compare the amplitude of minis or evoked EJPs. Finally, accurate determination of quantal content based on minis and EJPs is possible only at low release conditions when nonlinear summation is not a major issue. As the EJP amplitude increases, it creates a "ceiling effect," because the same amount of transmitter will be less effective in depolarizing the membrane when the potential is approaching the reversal potential of glutamate receptors/channels.

To overcome these limitations, the voltage-clamp technique can be used, which uses negative feedback mechanisms to keep the cell membrane potential steady at any reasonable set points. In voltage-clamp mode, the amplitude and kinetics of membrane currents can be determined. In the large larval muscle cells of *Drosophila*, the TEVC method is used, in which one electrode monitors the cell membrane potential while the other electrode passes electric currents. This protocol introduces the application of TEVC in analysis of synaptic currents using the larval NMJ preparation.

MATERIALS

CAUTION: See Appendix for proper handling of materials marked with <!>.
See the end of the chapter for recipes for reagents marked with <R>.

Reagents

Drosophila melanogaster synaptic transmission mutants (e.g., Syx^{3-69}, $snap-25^{ts}$, $syt I$, and lap), neuronal excitability mutants (e.g., eag, Sh double mutant), or muscle excitability mutants (e.g., larvae that overexpress the inward rectifier potassium channel $K_{ir}2.1$ in body wall muscles [referred to as the K_{ir} "mutant"])

D. melanogaster wild-type strains (Canton-S or Oregon-R)
  All flies (mutants and wild type) are maintained and raised on cornmeal-based fly food at room temperature. Avoid crowding in the vials, so that larvae can grow healthy and large.
HL-3 saline <R>
HL-3 saline without Ca^{2+}, ice cold
KCl <!>, 3 M

Equipment

Dissection instruments
Electrophysiology rig
Microelectrode, current-passing (as prepared in Protocol 5)
  Prepare the current-passing electrode with an input resistance of 5–8 MΩ so that it will be easier to inject current into muscles.
Microelectrode, voltage-monitoring (as prepared in Protocol 5)
  Prepare the voltage-monitoring electrode with an input resistance of 15–18 MΩ.

200

BOX 3. A FEW WORDS OF CAUTION CONCERNING VOLTAGE-CLAMP EXPERIMENTS

- When using a water-immersion lens (such as 60x) to visualize the NMJ for electrophysiology, a long working distance is required to provide room for positioning three electrodes (voltage, current, and suction).

- For safety reasons, an "electrophysiology-ready" specialized water-immersion lens must be used. The tip of the lens is coated with electrical insulators, such as ceramics or special resins (e.g., LUMPLFL 60 W/IR objective from Olympus). Voltage-clamp methods generate high voltages through the current-passing electrode. A metal-coated lens dipped in saline will short-circuit the amplifier and could be hazardous to you and the equipment.

- Avoid touching the current injection head stage while it is in the TEVC mode. Switch back to the bridge mode if you need to change the electrode.

Microscope
Suction electrode (as prepared in Protocol 5)
Syringe, 1-mL or 5-mL

METHOD

1. The head stage (HS-2A-x0.1 LU) and connections to the AD board remain the same as in the intracellular (bridge) mode (see Fig. 6). Connect the current-passing and low-resistance head stage (e.g., HS-2A-x1LU) to the ME2 channel on the back of the amplifier.

2. Connect V_2 and I_2 outputs to defined input channels (such as channels 2 and 3, respectively) on the AD board and set up the pClampex software accordingly to record signals from these channels.

   Use a model cell to practice TEVC if you are unfamiliar with the operation of the amplifier and the software used for voltage-clamp experiments (see Protocol 1).

3. Pull voltage and current microelectrodes (as in Protocol 5), fill them with 3 M KCl, and place them in the electrode holders.

4. Dissect and prepare a *Drosophila* larval body wall preparation as described in Chapter 7.

Performing TEVC

5. To practice using TEVC, measure spontaneous minis without stimulating the nerve.

6. Remain in the "bridge mode." Set the electrode potentials to 0, check the input resistance, and balance the bridge after the electrode is in the bath solution.

7. Impale the target muscle fiber with the current-passing electrode. Measure the resting potential.

   Because it has a larger tip (hence a lower resistance) than an intracellular electrode, the current-passing electrode can cause more damage to the muscle membrane and consequent deterioration in the initial resting potential.

8. Once the resting potential of the muscle has recovered, impale the same muscle fiber with the voltage electrode. Wait 1–2 min until the readings from both electrodes become similar. If these readings differ by more than a few millivolts, or if the resting potential is poor, reject this muscle and try again with a new muscle fiber.

9. Tune the clamp conditions to ensure that the muscle fiber can be clamped at a set voltage and that the quality of the current response is fast and not too noisy. Set the clamp "Gain" to the minimum and turn the "Phase Lag" and "Anti-alias" filter off completely. Adjust the "Holding

Position" dial until the "RMP Balance" LEDs reach an equal intensity (which should be dim). This allows the holding potential to be close to the resting potential of the muscle. If the holding potential is far from the resting potential, the muscle could be killed because of excessive current shock.

10. Switch from the "Bridge" mode to "TEVC" mode by pressing the large blue TEVC button.

11. Set the holding potential to –80 mV. Set the step command to +20 mV. Repeatedly deliver a 5-Hz square pulse (500-msec duration), and observe the shape of the command voltage from the voltage electrode (the V_m channel), and the current (the I_2 channel) and voltage responses (the V_2 channel) from the current electrode. The initial response will not look like a perfect square wave and the peak amplitude of the command voltage may not reach 20 mV.

12. Gradually increase the gain to achieve the best square wave allowed under the experimental conditions. The voltage trace should appear sharper and its rise and fall time should be faster.

> Avoid turning the gain too high, which will lead to oscillations and likely kill the clamped cell. The current trace will have two capacitance components (an initial and a tail current), which initially have a slower time course and should also become sharper and faster when the gain is properly set.

13. To increase further the speed of the voltage and current responses, careful apply capacitance neutralization of the voltage-monitoring electrode on ME1 (no need to adjust ME2).

Recording Miniature Synaptic Currents (mEJC)

14. Once the holding potential is steady and the noise level of the current trace is low, begin recording mEJCs. Alter the holding potentials from –80 mV to –100, –90, –60, and –40 mV, and observe the change of mEJC amplitude.

15. Once you are fully comfortable with TEVC, use the suction electrode to stimulate the motor nerve and record evoked EJCs (Fig. 12). Vary the extracellular $[Ca^{2+}]$, determine the EJC amplitude at each concentration, and compare them at the same holding potential. Also, vary the holding potential and observe the change of EJC amplitude at a given $[Ca^{2+}]$.

16. Examine short-term synaptic plasticity by measuring TPF or depression. Monitor long-term depletion of the synaptic vesicle pool by prolonged and repetitive stimulation. Use the same stimulation parameters that were used for EJP recordings in Protocol 2.

TROUBLESHOOTING

Problem: Both electrodes recorded very good resting potentials, but the voltage electrode did not respond at all to current injection via the current-passing electrode.

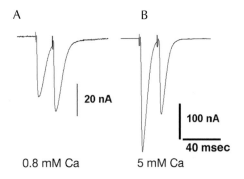

A   B

20 nA

100 nA

40 msec

0.8 mM Ca   5 mM Ca

FIGURE 12. Two-electrode voltage-clamp (TEVC) analysis of synaptic transmission at the larval NMJ. Short-term synaptic facilitation (*A*) and depression (*B*) at a wild-type larval NMJ are revealed by TEVC recording of synaptic currents (EJCs) at different $[Ca^{2+}]$ in response to twin-pulse stimuli (50 Hz).

Solution: The body wall muscles have three layers. It is likely that the two electrodes were not in the same muscle fiber. One of the electrodes likely penetrated a muscle fiber below the target muscle. Replace the current-passing electrode and try again.

Problem: The baseline of the clamped current was not steady.
Solution: Small variations of the baseline over time are expected. However, large variations indicate poor "space clamp." Start with another muscle with new electrodes.

Problem: After inserting the current-passing electrode, the muscle resting potential was very poor (−20 mV) and never recovered to the normal range.
Solution: The large-tipped current-passing electrode can easily injure muscle cells. The poor resting potential results from an unregulated influx of Na^+ ions into the cell through the damaged membrane. Consider fabricating sharper and more efficient current-passing electrodes. Bevel the tip of the electrode on a grinder at a 45° angle to sharpen the bottom portion of the tip. This creates a more pointed tip with a larger opening. In addition, the muscle will be better "space clamped" because of the increased efficiency of current passing.

Problem: The TEVC currents are too noisy.
Solution: Reduce noise using common methods, such as shielding, common grounding, etc. In addition, lower bath saline levels. Keeping the recording chamber dry also helps to reduce noise.

Problem: Muscle contraction disrupted what was a perfect clamp.
Solution: Losing a good clamp to muscle contraction, especially when the nerve is stimulated at a high frequency in high $[Ca^{2+}]$, is expected. Two things can be done that will reduce (although not eliminate) the chance of muscle contractions for TEVC and for all other recordings. When pinning the larvae, try to stretch them out as much as possible, without tearing the muscle or the preparation. Also, perform the experiments in a cooler room (18°C–20°C) or on a cooled microscope stage to reduce the chance of muscle contractions.

Protocol 4

Focal Recording of Synaptic Currents from Single Boutons at the Larval NMJ

Focal recording is an extracellular method designed for the study of synaptic activity of one or a few synaptic boutons rather than the ensemble activity of all the boutons (Kurdyak et al. 1994), as occurs with intracellular recording or TEVC. This is a useful technique for investigating the properties of different motor neurons that innervate the same muscle, applying statistical analysis to discrete synaptic events, and investigating the heterogeneity of synaptic release properties among boutons. A compound microscope with epifluourescent imaging capability is very helpful but not essential; any GFP *Drosophila* strain that labels the nerve terminal or synaptic boutons can be used to locate the boutons. A particularly useful strain is Mhc-CD8-Sh-GFP, containing a GFP molecule that is expressed in muscle, localizes to the postsynaptic apparatus, and outlines boutons (Zito et al. 1999). Vital fluorescent dyes (such as 4-Di-2-Asp) may also be applied to the dissected preparation to help locate boutons. The microscope should be equipped for differential interference contrast (DIC or Nomarski) optics, if fluorescence is not used.

MATERIALS

CAUTION: See Appendix for proper handling of materials marked with <!>.
See the end of the chapter for recipes for reagents marked with <R>.

Reagents

Drosophila melanogaster synaptic transmission mutants (e.g., Syx^{3-69}, $snap\text{-}25^{ts}$, $syt\ I$, and lap), neuronal excitability mutants (e.g., *eag, Sh* double mutant), or muscle excitability mutants (e.g., larvae that overexpress the inward rectifier potassium channel $K_{ir}2.1$ in body wall muscles [referred to as the K_{ir} "mutant"])
D. melanogaster wild-type strains (Canton-S or Oregon-R)
   All flies (mutants and wild type) are maintained and raised on cornmeal-based fly food at room temperature. Avoid crowding in the vials, so that larvae can grow healthy and large.
HL-3 saline <R>
HL-3 saline without Ca^{2+}, ice cold
KCl <!>, 3 M

Equipment

Dissection instruments
Electrophysiology rig
Microelectrode, focal extracellular (as prepared in Protocol 5)
Microscope, with a water-immersion objective (40x or 60x)
   Select an objective with a long working distance for focal recording experiments. Working distance for most brands is 1.5–3 mm.
Suction electrode (as prepared in Protocol 5)
Syringe, 1-mL or 5-mL

METHOD

1. Dissect third-instar *Drosophila* larvae to expose the musculature and nervous system (as per Chapter 7). If the motor neurons will be stimulated, cut the segmental nerves. Bathe the preparations in physiological saline.

2. Mount the larvae, in its dissection chamber, onto the microscope.

3. Place the stimulating and focal recording electrodes as close as possible to a target muscle, under a low-power air objective. Center the electrodes in the field of view. If stimulating the nerve, suck the nerve into the stimulating pipette.

4. Defocus the microscope if necessary and move the water-immersion lens into the light path. Be careful not to touch the pipettes with the lens. Add saline if necessary to make contact between the lens and the bathing solution. Carefully focus the lens, attempting to locate the pipettes somewhere above the muscle.

5. Once the pipettes are in focus, carefully manipulate the recording electrode down toward the muscle, while continuing to focus the lens onto the muscles.

6. Locate the nerve terminal boutons. If using transmitted light and DIC, visible boutons will appear as "fried egg" or "doughnut" structures on the muscle surface. Fluorescent labels, if used, will reveal the location of boutons.

7. Once a target bouton has been selected, manipulate the recording pipette toward the bouton and begin stimulation of the motor neuron. If the bouton is being recruited, a deflection of the amplifier output should be detected just after the stimulus artifact.

8. Continue to make small positional adjustments to the recording electrode, moving it in all three planes of motion until an optimal signal is found.

9. Proceed with the planned experiment.

TROUBLESHOOTING

Problem: The recording electrodes are hard to locate under the water-immersion lens.
Solution: Good quality lenses have a relatively high numerical aperture (NA) so their depth of field is narrow. Use the low-power air lenses to position the recording electrode in the center of the field and as close to the muscles as possible.

Problem: The recording pipette is making contact with the lens.
Solution: There are two potential problems. The pipette is too long on the distal side of the bend or the micromanipulators are not angled properly. Learn how to make shorter bent pipettes through trial and error. The micromanipulator must be adjusted so that the pipette enters the recording chamber at a relatively flat angle.

Problem: The nerve terminal boutons cannot be located with DIC optics.
Solution: If it is not possible to switch to a fluorescence-based system, then ensure that the microscope illumination system is properly set and aligned for Köhler illumination. Adjust the condenser diaphragm and light intensity to change the contrast and brightness, and adjust the polarizer. Not all the boutons of the NMJ will be easily visible; usually a short string of them will be visible as they cross over the surface of a muscle.

Problem: There is little or no synaptic response.
Solution: When first learning this technique, start with 1.5 mM extracellular Ca^{2+} to ensure that a significant synaptic current is present. The responses from individual boutons will be much smaller than those recorded with an intracellular technique. Ensure that the segmental nerve is in the stimulating pipette.

Fabrication of Microelectrodes, Suction Electrodes, and Focal Electrodes

This protocol details some basic methods for manufacturing microelectrodes used for intracellular recording and TEVC and loose patch electrodes used for focal recording. In addition, a method is provided for manufacturing homemade suction electrodes used for nerve stimulation.

MATERIALS

Equipment

Capillaries, borosilicate glass for producing microeletrodes (e.g., 1B100F-4, 100-mm-long; 1.5-mm-OD, 0.84-mm-ID; inner filament, and fire-polished; or TW150F-4, 100-mm-long, 1.5-mm-OD, 1.12-mm-ID, inner filament; WPI)

Connectors, gold pin

Connectors, male (red and black)

Luer valve, three-way

Microelectrode puller (e.g., P-97 or P-1000; Sutter Instruments)

MicroFil (34-gauge, 100-μm-ID, 164-μm-OD, 67-mm-long, MF-34g-5; WPI)

Microforge (e.g., MF-830 Narishige; MF200-1 WPI; de Fonbrune microforge, Warner Instruments)

Microgrinder (e.g., EG-400; Narishige; or MicroBeveler, SYS-48000; WPI)

Pipettes, 5-mL plastic (or a similar sized plastic pipettes)

Soldering iron and solder

Suction electrode holder (e.g., microslectrode holder, straight, 1.5-mm-OD, MEH2SW; WPI)

Syringes, 1-mL or 5-mL

Tubing for connecting the suction electrode holder with the syringe (e.g., Masterflex peroxide-cured silicone tubing, 1.6-mm-ID, EW 96400-14; Cole-Parmer Instrument Co)

Tubing, heat-shrinkable

Wire, silver (e.g., 280-μm-OD Teflon-coated silver wire, 786500; A-M Systems)

Wires, electrical (red and black)

METHOD

Manufacturing Microelectrodes

Microelectrode quality is critically important to the success of intracellular recording. Microelectrodes are pulled from glass capillaries or pipettes using an electrode puller. Glass capillaries are typically 4-in (100-mm) long, and from each piece of glass capillary two nearly identical microelectrodes can be fashioned. Obtaining ideal microelectrodes for your recordings is an empirical process. Begin by following the instructions that accompany the electrode puller as you test pulling conditions, such as heat levels, pulling strength, and pulling speed.

For fly larval body wall recordings, a microelectrode works best when its input resistance is 15–25 MΩ. An electrode with lower input resistance tends to damage muscles during impalement, whereas an electrode with higher input resistance introduces more noise. For the current-passing electrode in the TEVC, the input resistance is typically much lower, ranging from 5 to 8 MΩ.

When pulling an electrode keep the following points in mind.

- Review and follow the electrode puller instructions.

- Avoid touching the middle section of the glass capillary with bare hands, as oil and sweat from hands could affect the quality of the microelectrode.

- Never touch the box filament on the puller either by hand or with the glass capillary.

- With the exception of patch-clamp electrodes, microelectrodes can be prepared fresh or stored in a dust-proof container (e.g., a large culture dish with a strip of clay to hold the electrodes).

- Backfill a microelectrode with 3 M KCl solution just before using it.

When selecting a glass capillary, consider the following.

- Select a glass capillary with a single barrel, with an OD that matches the size of the electrode holder and an ID that is larger than the thickness of the silver wire on the electrode holder.

- Select glass capillaries containing an inner filament, which facilitates rapid and automatic filling of the fine tip of the electrode (e.g., TW150F-4 or 1B100F-4; WPI).

- Make sure that the glass type and size match the heating filament on the electrode puller.

Manufacturing Suction Electrodes

Suction electrodes are often homemade and thus each laboratory has its favorite manufacturing method. In general, a suction electrode setup has five components: the electrode holder with a side port, the electrode itself, electric wires, a rod (for mounting on a manipulator), and a pressure application system. Here, we offer a simple design for making a working suction electrode (Fig. 13).

1. Cut off the pointed end of a 5-ml plastic pipette leaving a 20-cm-long tube. Reshape the inner lumen of one end of the tube so that it fits well with the end of the electrode holder (e.g., MEH2SW; WPI).

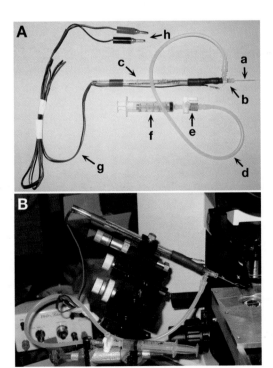

FIGURE 13. A homemade suction electrode system. (*A*) An example of a homemade suction electrode. (a) Suction electrode glass pipette; (b) electrode holder (with a side port); (c) a rod and holder made from a 5-mL plastic pipette; (d) air tubing; (e) a three-way Luer valve; (f) a 5-ml syringe; (g) stimulation (red) and ground wire (black); (h) male connectors to a stimulus isolator. See the text for detailed instructions on the assembly of this suction electrode system. (*B*) A view of the suction electrode system mounted on a micromanipulator in an electrophysiology rig.

2. Solder a 3.5-ft (~105-cm) piece of insulated electric wire (use the red wire) to a female gold connector. Pull this wire through the plastic tube (from Step 1) so that the female connector is outside the reshaped end of the tube. Connect the female connector firmly onto the gold pin on the electrode holder. This will serve as the stimulating wire of the suction electrode.

3. Push the electrode holder into the reshaped end of the plastic tube. Use a 4-cm-long piece of heat-shrinkable tubing to anchor and reinforce the connection between the plastic tube and the electrode holder.

4. Solder another piece of similarly insulated electric wire (wire coated in black) to a gold pin connector (male and female together).

5. Solder a 25-cm-long piece of soft Teflon-coated silver wire (e.g., A-M Systems) to the female connector. Use a razor blade to gently remove the Teflon coating on the two ends of the silver wire, so that it can be soldered onto the gold connector, and to access the bath solution. This will serve as a ground wire for the suction electrode.

6. Position the silver ground wire so that it will be long enough to wrap around a glass suction electrode. Use two pieces of heat-shrinkable tubing (~1.5-cm-long), one near each end, to anchor the black lead onto the outside of the plastic tube.

7. Solder male connectors onto the other end of the electric wire (use a red connector to match the red lead, and a black connector to match the black lead).

8. Cut a 3-ft-long piece of Masterflex tubing. On one end, put a female connector (provided by WPI) and connect it with the male side port of the electrode holder. On the other end, connect the tubing with a three-way Luer valve and a 1-ml syringe.

9. Pull at least 10 microelectrodes using the same glass capillaries and puller settings as you would to make intracellular microelectrodes. Use a piece of soft tissue (such as a Kimwipe) to briefly touch the tip of the microelectrode. Examine the microelectrode under a microscope to ensure that the break is small and clean (i.e., the broken tip is not jagged). Fire-polish the tip of the broken microelectrode in a microforge so that the inner diameter of the electrode is 8–10 µm, slightly larger than the segmental nerve of a third-instar larvae. This is the "functional part" of the suction electrode.

10. Insert one suction electrode into the electrode holder and test for leakiness of the system using distilled water. You should be able to suck water in or push it out with the syringe (the water may flow slowly because of the small opening of the electrode tip).

11. The suction electrode can be tested and used with an NMJ preparation (see Protocol 2).

> After each use, remember to rinse the electrode with distilled water and let it dry thoroughly in the air to prevent salt buildup and bacterial growth.

Manufacturing Focal Electrodes

12. Pull a glass capillary into two micropipettes that are somewhat longer than suction or intracellular pipettes. Prepare the micropipette tips for recording using either Steps 2 and 3 or Step 4.

13. Gently break the tip of the pipette and examine it with a microforge. If the tip is cleanly broken, then heat and polish the tip to an internal diameter of 2–5 µm.

14. Bend the micropipette tip, so that when the pipette is mounted in the electrode holder and on the micromanipulator the opening of the pipette will be almost at a right angle to the muscle surface (Fig. 14). To bend the glass, proceed as follows.

    i. Use a micromanipulator to move the pipette into position above the heating filament of a microforge.

    > The design of the microforge must permit the glass pipette to be held horizontally and allow a heating filament to be brought under the pipette while the process is viewed through a

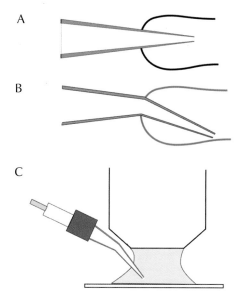

FIGURE 14. Constructing focal micropipettes for single-bouton recordings. A pipette is pulled in a similar manner to making a suction electrode and the tip is broken by gently flexing the tip against another pipette or sharp tweezer. The tip is fire-polished to a final diameter of ~5 μm. (*A*) The pipette is held horizontally over the heating filament of the microforge. (*B*) As current is applied through the filament it heats and begins to melt the pipette glass. The glass will bend around the filament and when it reaches an angle of ~30° from horizontal the filament heat should be turned off. (*C*) On mounting the pipette in the holder and micromanipulator, the final angle of the pipette must be adjusted to fit under the water-immersion objective, and then lowered onto the preparation.

low-power lens of a compound microscope. There are commercial microforges available that can do this, such as the de Fonbrune microforge. Alternatively, such a microforge can be constructed with relatively inexpensive components.

ii. Turn on the filament.

iii. As the filament gently heats up, the glass will warm and the tip will start to fall down around the filament. The lumen of the pipette normally will become smaller within the angle, but it must not seal closed. Likewise, heating the glass at this stage should not close the opening of the pipette tip. It is a matter of trial and error to get the right starting pipette plus the right amount of filament heating to bend the glass without closing the pipette.

15. As an alternative to Steps 2 and 3, use a microgrinder to grind the tip opening to match the size of a single synaptic bouton.

   i. Position your microelectrode at a 30° angle to the grinding wheel.

   ii. Gently grind the tip, keeping the surface of the wheel moist with water.

   iii. Check the size of electrode tip frequently under a high-powered microscope.

   iv. Stop grinding when the diameter of the inner lumen of the tip reaches 8–10 μm, the size of a single bouton.

   > There is no need to bend the tip of the electrode if this method is used, because the open tip of the electrode is now parallel with the surface of the muscle and synaptic boutons.

TROUBLESHOOTING

Problem: The input resistance of the microelectrode is either too low or too high, rendering the electrode unsuitable for use in an experiment.

Solution: If the input resistance is too high, reduce the heat used while pulling the electrode. If the input resistance is too low, then increase the heat. Be sure to follow suggestions provided in the puller operations manual.

Problem: The silver wire from the electrode holder is bent, making it difficult to attach the suction electrode.

Solution: Similar to the microelectrode holder, the silver wire that goes inside the microelectrode works best when it is straight. Thus, avoid bending it. If the wire becomes bent or twisted, gently straighten it out by hand.

Problem: The suction electrode failed to deliver any electrical pulses to the nerve.
Solution: The suction electrode may have an open circuit. When assembling a new suction electrode system, use a voltmeter to verify that the electrical connections are good. If there is a problem, inspect the Teflon coating on the silver wire and confirm that it was removed at the solder points. Also, make sure that the stimulating silver wire is in contact with the saline inside the suction electrode. Inspect electrical connections regularly.

Problem: The suction electrode does not suck well or does not hold the saline at a steady level inside the electrode.
Solution: There may be a leak within the pressure system. Inspect the tubing and the connectors for a leak. A common problem is that the rubber washer inside the electrode holder gets old. Replace it with a new one.

RECIPE

CAUTION: See Appendix for proper handling of materials marked with <!>.
Recipes for reagents marked with <R> are included in this list.

HL-3 Saline

Component	Amount per liter	Final concentration
NaCl	4.09 g	70 mM
$MgCl_2 \cdot 6H_2O$ <!>	4.07 g	20 mM
$NaHCO_3$	0.84 g	10 mM
Trehalose	1.89 g	5 mM
Sucrose	39.36 g	115 mM
HEPES	1.19 g	5 mM
KCl <!>	5 mL	5 mM
$CaCl_2$ <!>	1 mL	1 mM

1. Add all of the dry components to ~750 ml ddH_2O. Stir to dissolve.
2. Add 5 ml of 1 M KCl solution.
3. Add 1 ml of 1 M $CaCl_2$ solution.
   $[Ca^{2+}]$ can be varied from 0.1 to 2 mM if needed.
4. Adjust the pH to 7.2 with NaOH <!> or HCl <!>.
   Do not use KOH because it will alter $[K^+]$ in the saline.
5. Adjust the final volume to 1000 mL with H_2O.

ACKNOWLEDGMENTS

We thank the members of our laboratories for their important contributions to our research programs. Bing Zhang especially wishes to thank Dr. Hong Bao for her expert assistance with the Neurobiology of *Drosophila* course and for providing some of the figures for this chapter. Research in the Zhang laboratory was supported by grants from National Science Foundation (NSF), National Institutes of Health (NIH), Oklahoma Center for the Advancement of Science and Technology (OCAST), and research funds from the University of Oklahoma. Bing Zhang also thanks Drs. Phillip

Vanlandingham, Rudhof Bohm, and Hong Bao for their constructive comments. Bryan Stewart's research was supported by a Canada Research Chair and grants from the Canadian Institute for Health Research, the Natural Science and Engineering Research Council of Canada, and the Canadian Foundation for Innovation. An extensive citation of all publications related to the topics covered in this chapter was not possible because of space limitations. We therefore focused on select initial publications and wish to apologize to our colleagues whose publications were not cited.

REFERENCES

Baines RA, Bate M. 1998. Electrophysiological development of central neurons in the *Drosophila* embryo. *J Neurosci* **18:** 4673–4683.

Bao H, Daniels RW, MacLeod GT, Charlton MP, Atwood HL, Zhang B. 2005. AP180 maintains the distribution of synaptic and vesicle proteins in the nerve terminal and indirectly regulates the efficacy of Ca^{2+}-triggered exocytosis. *J Neurophysiol* **94:** 1888–1903.

Bellen HJ, Budnik V. 2000. The neuromuscular junction. In Drosophila *protocols* (ed. W Sullivan et al.), pp. 175–199. Cold Spring Harbor Laboratory Press, Cold Spring Harbor, NY.

Bennett MK, Calakos N, Scheller RH. 1992. Syntaxin: A synaptic protein implicated in docking of synaptic vesicles at presynaptic active zones. *Science* **257:** 255–259.

Benzer S. 1973. Genetic dissection of behavior. *Sci Am* **229:** 24–37.

Brand AH, Perrimon N. 1993. Targeted gene expression as a means of altering cell fates and generating dominant phenotypes. *Development* **118:** 401–415.

Broadie K, Bellen HJ, DiAntonio A, Littleton JT, Schwarz TL. 1994. Absence of synaptotagmin disrupts excitation-secretion coupling during synaptic transmission. *Proc Natl Acad Sci* **91:** 10727–10731.

Brodsky FM, Chen CY, Knuehl C, Towler MC, Wakeham DE. 2001. Biological basket weaving: Formation and function of clathrin-coated vesicles. *Annu Rev Cell Dev Biol.* **17:** 517–568.

Chen MS, Obar RA, Schroeder CC, Austin TW, Poodry CA, Wadsworth SC, Vallee RB. 1991. Multiple forms of dynamin are encoded by *shibire*, a *Drosophila* gene involved in endocytosis. *Nature* **351:** 583–586.

Chen YA, Scheller RH. 2001. SNARE-mediated membrane fusion. *Nat Rev Mol Cell Biol* **2:** 98–106.

Cremona O, De Camilli P. 1997. Synaptic vesicle endocytosis. *Curr Opin Neurobiol* **7:** 323–330.

Conner SD, Schmid SL. 2003. Regulated portals of entry into the cell. *Nature* **422:** 37–44.

Del Castillo J, Katz B. 1954. Quantal components of the end-plate potential. *J Physiol* **124:** 560–573.

Delgado R, Maureira C, Oliva C, Kidokoro Y, Labarca P. 2000. Size of vesicle pools, rates of mobilization, and recycling at neuromuscular synapses of a *Drosophila* mutant, *shibire*. *Neuron* **28:** 941–953.

Deitcher DL, Ueda A, Stewart BA, Burgess RW, Kidokoro Y, Schwarz TL. 1998. Distinct requirements for evoked and spontaneous release of neurotransmitter are revealed by mutations in the *Drosophila* gene neuronal-synaptobrevin. *J Neurosci* **18:** 2028–2039.

DiAntonio A, Parfitt KD, Schwarz TL. 1993. Synaptic transmission persists in synaptotagmin mutants of *Drosophila*. *Cell* **73:** 1281–1290.

Feeney CJ, Karunanithi S, Pearce J, Govind CK, Atwood HL. 1998. Motor nerve terminals on abdominal muscles in larval flesh flies, *Sarcophaga bullata*: Comparisons with *Drosophila*. *J Comp Neurol* **402:** 197–209.

Feng Y, Ueda A, Wu CF. 2004. A modified minimal hemolymph-like solution, HL3.1, for physiological recordings at the neuromuscular junctions of normal and mutant *Drosophila* larvae. *J Neurogenet* **18:** 377–402.

Fergestad T, Davis WS, Broadie K. 1999. The stoned proteins regulate synaptic vesicle recycling in the presynaptic terminal. *J Neurosci* **19:** 5847–5860.

Fischer von Mollard G, Mignery GA, Baumert M, Perin MS, Hanson TJ, Burger PM, Jahn R, Südhof TC. 1990. rab3 is a small GTP-binding protein exclusively localized to synaptic vesicles. *Proc Natl Acad Sci* **87:** 1988–1992.

Fredj NB, Burrone J. 2009. A resting pool of vesicles is responsible for spontaneous vesicle fusion at the synapse. *Nat Neurosci* **12:** 751–758.

Goldstein SA, Price LA, Rosenthal DN, Pausch MH. 1996. ORK1, a potassium-selective leak channel with two pore domains cloned from *Drosophila melanogaster* by expression in *Saccharomyces cerevisiae*. *Proc Natl Acad Sci* **93:** 13256–13261.

González-Gaitán M, Jäckle H. 1997. Role of *Drosophila* α-adaptin in presynaptic vesicle recycling. *Cell* **88:** 767–776.

Gho M. 1994. Voltage-clamp analysis of gap junctions between embryonic muscles in *Drosophila*. *J Physiol.* (Pt 2) **481:** 371–383

Hay JC, Scheller RH. 1997. SNAREs and NSF in targeted membrane fusion. *Curr Opin Cell Biol* **9:** 505–512.

Heuser JE, Reese TS, Dennis MJ, Jan Y, Jan L, Evans L. 1979. Synaptic vesicle exocytosis captured by quick freezing and correlated with quantal transmitter release. *J Cell Biol* **81:** 275–300.

Hoang B, Chiba A. 2001. Single-cell analysis of *Drosophila* larval neuromuscular synapses. *Dev Biol* **229:** 55–70.

Hodgkin AL, Huxley AF. 1945. Resting and action potentials in single nerve fibres. *J Physiol* **104:** 176–195.

Hodgkin AL, Katz B. 1949. The effect of sodium ions on the electrical activity of the giant axon of the squid. *J Physiol* **108:** 37–77.

Hong CS, Ganetzky B. 1994. Spatial and temporal expression patterns of two sodium channel genes in *Drosophila*. *J Neurosci* **14:** 5160–5169.

Hotta Y, Benzer, S. 1969. Abnormal electroretinograms in visual mutants of *Drosophila*. *Nature* **222:** 354–356.

Huntwork S, Littleton JT. 2007. A complexin fusion clamp regulates spontaneous neurotransmitter release and synaptic growth. *Nat Neurosci* **10:** 1235–1237.

Jahn R, Niemann H. 1994. Molecular mechanisms of clostridial neurotoxins. *Ann NY Acad Sci* **733:** 245–255.

Jahn R, Sudhof TC. 1994. Synaptic vesicles and exocytosis. *Annu Rev Neurosci* **17:** 219–246.

Jahn R, Lang T, Sudhof TC. 2003. Membrane fusion. *Cell* **112:** 519–533.

Jan LY, Jan YN. 1976a. L-glutamate as an excitatory transmitter at the *Drosophila* larval neuromuscular junction. *J Physiol (Lond)* **262:** 215–236.

Jan LY, Jan YN. 1976b. Properties of the larval neuromuscular junction in *Drosophila melanogaster*. *J Physiol (Lond)* **262:** 189–214.

Karunanithi S, Georgiou J, Charlton MP, Atwood HL. 1997. Imaging of calcium in *Drosophila* larval motor nerve terminals. *J Neurophysiol* **78:** 3465–3467.

Katz B. 1966. *Nerve, muscle and synapse*. McGraw-Hill, New York.

Keen JH. 1987. Clathrin assembly proteins: Affinity purification and a model for coat assembly. *J Cell Biol* **105:** 1989–1998.

Koenig JH, Ikeda K. 1989. Disappearance and reformation of synaptic vesicle membrane upon transmitter release observed under reversible blockage of membrane retrieval. *J Neurosci* **9:** 3844–3860.

Krishnan KS, Rikhy R, Rao S, Shivalkar M, Mosko M, Narayanan R, Etter P, Estes PS, Ramaswami M. 2001. Nucleoside diphosphate kinase, a source of GTP, is required for dynamin-dependent synaptic vesicle recycling. *Neuron* **30:** 197–210.

Kurdyak P, Atwood HL, Stewart BA, Wu CF. 1994. Differential physiology and morphology of motor axons to ventral longitudinal muscles in larval *Drosophila*. *J Comp Neurol* **350:** 463–472.

Lagow RD, Bao H, Cohen EN, Daniels RW, Zuzek A, Williams WH, Macleod GT, Sutton RB, Zhang B. 2007. Modification of a hydrophobic layer by a point mutation in syntaxin 1A regulates the rate of synaptic vesicle fusion. *PLoS Biol* **5:** e72. Erratum in *PLoS Biol* 2007 **5:** e175.

Lima SQ, Miesenbock G. 2005. Remote control of behavior through genetically targeted photostimulation of neurons. *Cell* **121:** 141–152.

Littleton JT, Stern M, Schulze K, Perin M, Bellen HJ. 1993. Mutational analysis of *Drosophila* synaptotagmin demonstrates its essential role in Ca^{2+}-activated neurotransmitter release. *Cell* **74:** 1125–1134.

Macleod GT, Hegström-Wojtowicz M, Charlton MP, Atwood HL. 2002. Fast calcium signals in *Drosophila* motor neuron terminals. *J Neurophysiol* **88:** 2659–2663.

Mackler JM, Drummond JA, Loewen CA, Robinson IM, Reist NE. 2002. The C^{2}B Ca^{2+}-binding motif of synaptotagmin is required for synaptic transmission in vivo. *Nature* **418:** 340–344.

Marek KW, Davis GW. 2002. Transgenically encoded protein photoinactivation (FlAsH-FALI): acute inactivation of synaptotagmin I. *Neuron* **36:** 805–813.

Martin AR. 1976. The effect of membrane capacitance on non-linear summation of synaptic potentials. *J Theor Biol* **59:** 179–187.

Ordway RW, Pallanck L, Ganetzky B. 1994. Neurally expressed *Drosophila* genes encoding homologs of the NSF and SNAP secretory proteins. *Proc Natl Acad Sci* **91:** 5715–5719.

Osterwalder T, Yoon KS, White BH, Keshishian H. 2001. A conditional tissue-specific transgene expression system using inducible GAL4. *Proc Natl Acad Sci* **98:** 12596–125601.

Oyler GA, Higgins GA, Hart RA, Battenberg E, Billingsley M, Bloom FE, Wilson MC. 1989. The identification of a novel synaptosomal-associated protein, SNAP-25, differentially expressed by neuronal subpopulations. *J Cell Biol* **109:** 3039–3052.

Pallanck L, Ordway RW, Ganetzky B. 1995. A *Drosophila* NSF mutant. *Nature* **376:** 25.

Paradis S, Sweeney ST, Davis GW. 2001. Homeostatic control of presynaptic release is triggered by postsynaptic membrane depolarization. *Neuron* **30:** 737–749. Erratum in *Neuron* 2001 **31:** 167.

Pearse BM. 1976. Clathrin: A unique protein associated with intracellular transfer of membrane by coated vesicles. *Proc Natl Acad Sci* **73:** 1255–1259.

Perin MS, Fried VA, Mignery GA, Jahn R, Südhof TC. 1990. Phospholipid binding by a synaptic vesicle protein homologous to the regulatory region of protein kinase C. *Nature* **345:** 260–263.

Poodry CA, Edgar L. 1979. Reversible alteration in the neuromuscular junctions of *Drosophila melanogaster* bearing a temperature-sensitive mutation, *shibire*. *J Cell Biol* **81:** 520–527.

Rao SS, Stewart BA, Rivlin PK, Vilinsky I, Watson BO, Lang C, Boulianne G, Salpeter MM, Deitcher DL. 2001. Two distinct effects on neurotransmission in a temperature-sensitive SNAP-25 mutant. *EMBO J* **20:** 6761–6771.

Rasse TM, Fouquet W, Schmid A, Kittel RJ, Mertel S, Sigrist CB, Schmidt M, Guzman A, Merino C, Qin G, et al. 2005. Glutamate receptor dynamics organizing synapse formation in vivo. *Nat Neurosci.* **8:** 898–905.

Robinson IM, Ranjan R, Schwarz TL. 2002. Synaptotagmins I and IV promote transmitter release independently of Ca^{2+} binding in the C(2)A domain. *Nature* **418:** 336–340.

Rohrbough J, Broadie K. 2002. Electrophysiological analysis of synaptic transmission in central neurons of *Drosophila* larvae. *J Neurophysiol* **88:** 847–860.

Rohrbough J, Pinto S, Mihalek RM, Tully T, Broadie,K. 1999. *latheo*, a *Drosophila* gene involved in learning, regulates functional synaptic plasticity. *Neuron* **23:** 55–70.

Roman G, Endo K, Zong L, Davis RL. 2001. P{Switch}, a system for spatial and temporal control of gene expression in *Drosophila melanogaster*. *Proc Natl Acad Sci* **98:** 12602–12607.

Schulze KL, Broadie K, Perin MS, Bellen HJ. 1995. Genetic and electrophysiological studies of *Drosophila* syntaxin-1A demonstrate its role in nonneuronal secretion and neurotransmission. *Cell* **80:** 311–320.

Sollner T, Rothman JE. 1994. Neurotransmission: Harnessing fusion machinery at the synapse. *Trends Neurosci* **17:** 344–348.

Spradling AC, Rubin GM. 1982. Transposition of cloned P elements into *Drosophila* germ line chromosomes. *Science* **218:** 341–347.

Stevens CF. 1976. A comment on Martin's relation. *Biophys J* **16:** 891–895.

Stewart BA, Atwood HL, Renger JJ, Wang J, Wu CF. 1994. Improved stability of *Drosophila* larval neuromuscular preparations in haemolymph-like physiological solutions. *J Comp Physiol (A)* **175:** 179–191.

Stimson DT, Estes PS, Smith M, Kelly LE, Ramaswami M. 1998. A product of the *Drosophila* stoned locus regulates neurotransmitter release. *J Neurosci* **18:** 9638–9649.

Südhof TC, Lottspeich F, Greengard P, Mehl E, Jahn R. 1987. A synaptic vesicle protein with a novel cytoplasmic domain and four transmembrane regions. *Science* **238:** 1142–1144.

Sweeney ST, Broadie K, Keane J, Niemann H, O'Kane CJ. 1995. Targeted expression of tetanus toxin light chain in *Drosophila* specifically eliminates synaptic transmission and causes behavioral defects. *Neuron* **14:** 341–351.

Tanouye MA, Wyman RJ. 1980. Motor outputs of giant nerve fiber in *Drosophila*. *J Neurophysiol* **44:** 405–421.

Trimble WS, Cowan DM, Scheller RH. 1988. VAMP-1: A synaptic vesicle-associated integral membrane protein. *Proc Natl Acad Sci* **85:** 4538–4542.

van der Bliek AM, Meyerowitz EM. 1991. Dynamin-like protein encoded by the *Drosophila* shibire gene associated with vesicular traffic. *Nature* **351:** 411–414.

Venken KJ, Bellen HJ. 2007. Transgenesis upgrades for *Drosophila melanogaster*. *Development* **134:** 3571–3584.

Verstreken P, Kjaerulff O, Lloyd TE, Atkinson R, Zhou Y, Meinertzhagen IA, Bellen HJ. 2002. Endophilin mutations block clathrin-mediated endocytosis but not neurotransmitter release. *Cell* **109:** 101–112.

Wang Y, Guo HF, Pologruto TA, Hannan F, Hakker I, Svoboda K, Zhong Y. 2004. Stereotyped odor-evoked activity in the mushroom body of *Drosophila* revealed by green fluorescent protein-based Ca^{2+} imaging. *J Neurosci* **24:** 6507–6514.

Wilson RI, Turner GC, Laurent G. 2004. Transformation of olfactory representations in the *Drosophila* antennal lobe. *Science* **303:** 366–370.

Wu CF, Suzuki N, Poo MM. 1983. Dissociated neurons from normal and mutant *Drosophila* larval central nervous system in cell culture. *J Neurosci* **3:** 1888–1899.

Yamaoka K, Ikeda K. 1988. Electrogenic responses elicited by transmembrane depolarizing current in aerated body wall muscles of *Drosophila melanogaster* larvae. *J Comp Physiol A* **163:** 705–714.

Zhang B, Koh YH, Beckstead RB, Budnik V, Ganetzky B, Bellen HJ. 1998. Synaptic vesicle size and number are regulated by a clathrin adaptor protein required for endocytosis. *Neuron* **21:** 1465–1475.

Zhong Y, Wu CF. 1991. Altered synaptic plasticity in *Drosophila* memory mutants with a defective cyclic AMP cascade. *Science* **251:** 198–201.

Zito K, Parnas D, Fetter RD, Isacoff EY, Goodman CS. 1999. Watching a synapse grow: Noninvasive confocal imaging of synaptic growth in *Drosophila*. *Neuron* **22:** 719–729.

WWW RESOURCES: SUPPLIERS OF MAJOR ELECTROPHYSIOLOGY EQUIPMENT

http://www.a-msystems.com/ A-M Systems, Inc., PO Box 850, Sequim, WA 98324

http://www.coleparmer.com Cole-Parmer, 625 East Bunker Court, Vernon Hills, IL 60061

http://www.moleculardevices.com/home.html Molecular Devices, 1311 Orleans Drive, Sunnyvale, CA 94089-1136

http://narishige-group.com/ Narishige International USA, Inc., 1710 Hempstead Turnpike, East Meadow, NY 11554

http://www.siskiyou.com Siskiyou Corporation, 110 S.W. Booth Street, Grants Pass, OR 97526
http://www.sutter.com/ Sutter Instruments Company, One Digital Drive, Novato, CA 94949
http://www.warneronline.com Warner Instruments, LLC1125 Dixwell Avenue, Hamden, CT 06514
http://www.wpiinc.com/ World Precision Instruments, Inc., Toll Free 866-606-1974

13 | Electrophysiological Recordings from the *Drosophila* Giant Fiber System

Marcus J. Allen[1] and Tanja A. Godenschwege[2]

[1]*University of Kent, School of Biosciences, Kent CT2 7NJ, United Kingdom;* [2]*Florida Atlantic University, Biological Sciences, Boca Raton, Florida 33431*

ABSTRACT

The giant fiber system (GFS) of *Drosophila* is a well-characterized neuronal circuit that mediates the escape response in the fly. It is one of the few adult neural circuits from which electrophysiological recordings can be made routinely. This chapter describes a simple procedure for stimulating the giant fiber neurons directly in the brain of the adult fly and obtaining recordings from the output muscles of the GFS.

INTRODUCTION

The GFS mediates a fast escape behavior in adult flies (Allen et al. 2006). Behaviorally, it is characterized by an initial extension of the mesothoracic leg, to propel the flies off the substrate, followed by a wing downbeat to initiate flight. The efferent (output) pathways of the GFS have been well defined (Fig. 1) for the most part by work from Wyman and others in the 1980s using a combination of dye injection, electron microscopy, and electrophysiological techniques (Ikeda et al. 1980; King and Wyman 1980; Koto et al. 1981). The two largest interneurons in the fly, the aptly named giant fibers (GFs), relay the signal from the brain to the mesothoracic neuromere where each makes two identified synapses. The first is to a large motorneuron (TTMn) that drives the tergotrochanteral "jump" muscle (TTM), which is also referred to in the literature as the tergal depressor of trochanter or TDT. This GF–TTMn synapse, which is the largest central synapse in the fly, is a mixed synapse with the electrical gap-junction component encoded by the *shaking-B* (*shakB*) gene and the chemical component using acetylcholine its neurotransmitter (Blagburn et al. 1999; Allen and Murphey 2007; Phelan et al. 2008). The second identified synapse of the GF is to another interneuron, the peripherally synapsing interneuron (PSI), which exits the ganglion via the posterior dorsal medial nerve (PDMN) and synapses with dorsal longitudinal motorneurons (DLMns) within the PDMN. The DLMns drive the large indirect flight muscles (DLMs). Electrophysiological recordings can be made from the GFS in a simple noninvasive manner to determine the function of the central synapses within the circuit. Using combinations of adult viable mutants and/or GAL4 lines that express in its neurons, the GFS has provided a useful model circuit to investigate the role of several molecules in the formation of

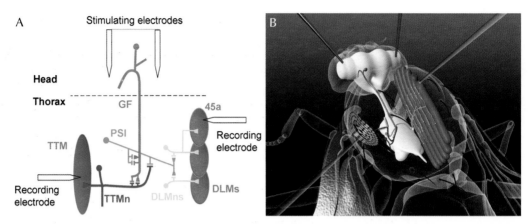

FIGURE 1. The giant fiber system (GFS): neurons and muscles. (A) Schematic indicating the neurons and connections of the GFS. For clarity, only one-half of the bilateral circuit is shown. The giant fiber (GF; red) relays information from the brain to the thoracic ganglia where it makes an electrochemical synapse to the tergotrochanteral motor neuron (TTMn; blue), which innervates the tergotrochanteral muscle (TTM). It also makes an electrochemical synapse to the peripherally synapsing interneuron (PSI; green), which, in turn, makes chemical synapses to the dorsal longitudinal motorneurons (DLMns; yellow) that innervate the dorsal longitudinal muscles (DLMs). The relative positions of the stimulating and recording electrodes are indicated. (Adapted, with permission of Elsevier, from Allen et al. 2006.) (B) Artist's impression of the GFS showing the central nervous system within the fly's body. The neurons and muscles of the GFS are shown in their approximate positions and the best positions for the stimulating and recording electrodes are indicated.

central synapses including Glued, Rac1, Robo, Semaphorin 1a, and Neuroglian (Allen et al. 1999, 2000; Godenschwege et al. 2002a,b, 2006). The GFS has also been used to investigate the effects of aging, sensitivity to anesthetics, the effects of neurodegeneration, and the molecular basis of habituation (Engel and Wu 1996, 1998; Lin and Nash 1996; Martinez et al. 2007; Watson et al. 2008).

Stimulating and Recording from the GFS

The GFs can be activated directly with brain stimulation, and the two output pathways can be monitored by recording simultaneously from the TTM and DLMs. The original rationale was that by placing the stimulating electrodes into the brain and slowly increasing the stimulation voltage, a point would be reached where only the GF interneurons would propagate an action potential because their large size would mean they have the least resistance and thus the lowest threshold. Although this may theoretically be true, in practice, accurate positioning of the electrodes is hard to achieve, so the stimulation voltage given is much above threshold. This ensures that the GFs are activated directly and not by upstream neurons (unless that is desired, see below). Although many neurons in the brain may be activated, the only route to the TTMs and DLMs from the brain activated by this procedure seems to be via the GFs. This is supported by findings that genetic ablation of the GFs, or abrogation of the electrochemical synapses between the GF and the TTMn and PSI, results in total loss of TTM and DLM responses on brain stimulation (Allen et al. 2000; Allen and Murphey 2007). However, both TTMn and the DLMns have other unidentified inputs, one of which is triggered by looming stimuli (Fotowat et al. 2009). Once direct activation of the GFs is achieved, recordings from TTM monitor the function of the GF–TTMn central synapse along with the neuromuscular junction (NMJ) and recordings from DLM monitor the function of the GF–PSI and PSI–DLMns synapses as well as the NMJ.

Standard Tests of Synaptic Function

The most commonly used tests for the GFS are response latency, the refractory period, and the ability to follow high-frequency stimulation. These will be described in turn.

Response Latency: This is the time taken for the output muscle to respond to a single stimulus activating the GFs. In the TTM of wild-type flies this is ~0.8 msec after GF activation and is via the monosynaptic pathway through the large electrochemical GF–TTMn synapse. The response in a DLM, through the disynaptic pathway, is seen ~1.2 msec after GF activation. These latencies correspond to the escape behavior in which the jump always occurs before the wing downbeat. This robust short-latency (SL) response is a good indicator of synaptic function, and any abnormalities in the synapses of the GFS will result in an increase in the latency or a loss of the response—for example, loss of gap junctions or structural malformations of the synapse that alter its shape or size (Thomas and Wyman 1984; Oh et al. 1994; Allen et al. 1999, 2000; Godenschwege et al. 2002a,b, 2006; Allen and Murphey 2007; Uthaman et al. 2008).

In addition to SL responses, intermediate-latency (IL) responses (TTM ~ 1.8 msec, DLM ~ 2.2 msec), and long-latency (LL) responses (TTM ~ 3.9 msec, DLM ~ 4.3 msec) can be elicited by simply reducing the voltage during brain stimulation or providing a light-off stimulus to a tethered fly. All these responses are still conducted through the GF; note the delay between the TTM and DLM response is always ~0.4 msec, indicating the disynaptic pathway from GF to DLM via the PSI and DLMn. The longer IL and LL responses, during low-voltage electrical stimulation or a light-off stimulus, are attributed to indirect activation of the GF by the afferent neurons in the brain. These neurons still remain unidentified but have interesting properties as they show both sensitivity to anesthetics and habituation to repeated stimuli (Engel and Wu 1996, 1998; Lin and Nash 1996).

Refractory Period: In this test, twin stimuli are given, initially 10 msec apart, and the responses from both TTM and DLM are recorded. The interval between the two stimuli is then gradually reduced until the second stimulus fails to elicit a response. The shortest time between two stimuli that still produces two responses is defined as the refractory period. For TTM this is ~3 msec and for DLM it is ~5 msec because of the greater time needed for the PSI–DLMn chemical synapses to replenish their synaptic vesicles. This test is less common than the other two as similar information can be gleaned if you observe the responses to the first two stimuli in the "following at high frequencies" test.

Following at High Frequencies: In this test a train of 10 stimuli are given to the preparation at high frequency and the number of responses is recorded. These trains of stimuli are usually given at 100, 200, and either 250 or 300 Hz. At 100 Hz (stimuli 10 msec apart) both TTM and DLM should respond 1:1 and give 10 responses. At the higher frequencies—for example, 250 Hz (stimuli 4 msec apart)—TTM will still respond 1:1 because of the robust GF–TTMn electrochemical synapse; however, DLM recordings will start to show failures as the time between stimuli is less than the refractory period of the PSI–DLMns synapses. An alternative way of performing the test is to gradually increase the frequency of the stimuli until the response rates fall below 50% (5 out of 10). This is described as the Following Frequency$_{50}$ (FF$_{50}$) (Gorczyca and Hall 1984). This test will often reveal an abnormality in synaptic function that does not cause an abnormal response latency (Allen et al. 1999), although it usually confirms an aberrant response latency.

Recording from TTM and DLM: The Outputs of the GFS

This protocol is a standard method for recording from the GFS of *Drosophila*. It is a relatively non-invasive method that allows the investigator to stimulate the giant fibers in the brain and assay the function of several central synapses within this neural circuit by recording from the thoracic musculature.

MATERIALS

CAUTION: See Appendix for proper handling of materials marked with <!>.
See the end of the chapter for recipes for reagents marked with <R>.

Reagents

CO_2 or ice (Step 1)
Dental wax, soft (available from most dental product suppliers)
Drosophila melanogaster wild-type/control flies (e.g., Oregon-R, w^{1118}, *bendless/+; shakB²/+*) and
  mutant strains (e.g., *bendless, shakB²*)
Forceps
KCl <!>, 3 M, or GFS saline <R>
Slide or mounting tray
  These can be made from a small Petri dish filled with tooth carding wax (shown in Fig. 2B), from a
  piece of Plexiglas or a coin, or from a small piece of wood.

Equipment

Electrodes, recording (glass with a resistance of 40–60 MΩ; need two of these)
  These are fabricated using a good glass microelectrode puller (e.g., a Sutter P-95). Again preformed
  microelectrodes can be purchased if desired.
Electrodes, tungsten and sharpened (one ground and two stimulation electrodes)
  These can be fabricated from 0.005-in-diameter tungsten wire sharpened electrolytically using 4 M
  NaOH. Alternatively, commercially available tungsten electrodes can be used. The electrophysiology
  rig is shown in Figure 2A,B (the figure legend contains equipment source information).
Faraday cage (optional)

METHOD

Mounting Flies

1. Anesthetize the fly on ice or with CO_2.

   The fly should be left for 20–30 min after mounting if CO_2 is used, because occasionally it can
   affect recordings. This is not a problem when using ice; however, the fly must be secured in the
   wax more quickly as recovery from cooling can be quite rapid.

2. Using forceps, transfer the anesthetized fly to the wax by its legs, and mount it into soft wax on
   a slide or tray with the ventral side down, pushing the legs into the wax to secure.

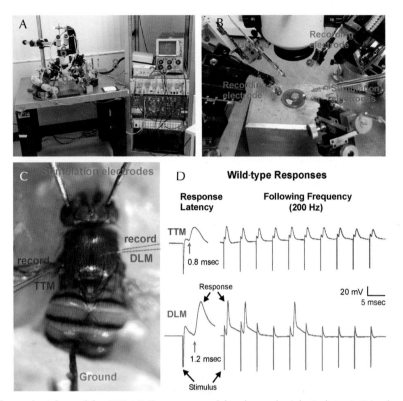

FIGURE 2. Electrophysiology of the GFS. (*A*) Components of the electrophysiological rig. 1: Stimulator (S48 Square Pulse Stimulator, Grass Instruments); 2: stimulation isolation unit (SIU5 RF Transformer Isolation Unit, Grass Instruments); 3: two-channel intracellular amplifier (Model 5A Microelectrode Amplifier, Getting Instruments); 4: data acquisition system (Digidata 1440A, Molecular Devices) and computer with software (not shown); 5: storage oscilloscope 5111A (Tektronix); 6: stereomicroscope (Wild M5) on a boom stand; 7: vibration isolation table (TMC); 8: light source (Fostec); 9: recording platform with five manual multi-axis micromanipulators (Narashigi, Sutter Instrument Company, and World Precision Instruments). (*B*) Magnification of the recording platform. Around the recording tray are arranged two stimulation electrodes (sharp tungsten electrodes), two recording electrodes (glass electrodes filled with saline), and one ground electrode (sharp tungsten electrodes). (*C*) *Drosophila melanogaster* impaled with stimulation electrodes through the eyes in the brain and a ground in the abdomen. Two glass electrodes are placed in the thorax for recording responses from the TTM and DLM. (*D*) Sample electrophysiological traces from recordings of the TTM and DLM on brain stimulation of a wild-type fly. The response latency of the GF-TTM pathway is 0.8 msec and it can follow stimuli 1:1 at 200 Hz. In contrast, the response latency of the GF-DLM pathway is 1.2 msec, and responses are not seen after every stimulus when given 10 stimuli at 200 Hz.

3. Pull the proboscis outward and push into the wax so that the head lies slightly forward and down on the surface.

   This step is important because the head needs to be secure and not move when the stimulating electrodes are inserted (Step 6). Keeping the head slightly stretched in front of the thorax will also help prevent inadvertent stimulation of the ventral nerve cord.

4. Pull the wings outward, away from the thorax, and secure. Ensure that the fly cannot move its thorax and that the areas of the DLM and TTM (Fig. 3, dotted areas) are visible and accessible.

   If the fly is mounted incorrectly or not securely, it becomes very difficult to obtain recordings, so it is advisable to practice these steps several times before proceeding with the protocol.

Placement of Electrodes

Successful recording from the GFS depends on being able to arrange the five micromanipulators so that the electrodes can be placed within several millimeters of each other. It is worth spending some time moving and adjusting these before a preparation is introduced so that minimal adjustment is required when recordings are needed.

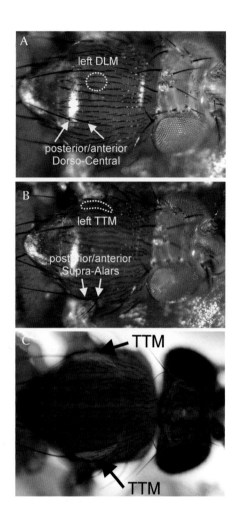

FIGURE 3. Localization of the TTM and DLM using bristles and illumination. (*A*) There are six pairs of indirect flight muscles, but only the dorsal longitudinal muscle pair 45a (also called the dorsal median muscle or muscle number 6) are innervated by the DLM motor neurons that receive input via the PSI from the contralateral GF (Demerec 1994). The attachment site of the DLM 45a muscles are under the cuticle between the anterior dorsocentral setae (yellow arrows) and the midline of the animal (yellow circle indicates site of left DLM). (*B*) The TTM is underneath the cuticle, just dorsal of the anterior and posterior supra-alar setae (yellow arrows) as indicated by the area circled by a dotted line (Demerec 1994). (*C*) As the TTM fibers are running along the dorsoventral axis they can be nicely visualized when a light source is placed underneath the fly (black arrows).

5. Place the ground electrode into the posterior end of the abdomen (Fig. 2C).

6. Place the stimulating electrodes through the eyes into the brain (Fig. 2C).

   The brain sits at the back of the head capsule, but electrodes pushed in too far may traverse the head capsule and enter the thorax where they may stimulate the ventral nerve cord directly.

7. Give single pulses of 30–60 V for 0.03 msec and check for successful activation of the GFS by looking for movement of the wings and/or TTM muscle on stimulation.

   See Troubleshooting.

8. Place the GFS saline (or 3 M KCl)-filled glass electrode for intracellular recordings into the left (or right) DLM muscle fiber 45a, which is immediately below the cuticle (see Figs. 2C and 3A).

   See Troubleshooting.

Stimulation and Recording

9. Give single stimuli as in Step 7 and modulate the stimulus strength by varying the voltage to determine the threshold for eliciting a response.

   The response of a good DLM recording is ~50–70 mV and has a latency of ~1.2–1.4 msec (Fig. 2D). Set the voltage 5–10 V above the determined threshold for the remainder of the experiment.

   See Troubleshooting.

10. Place the second intracellular recording electrode in the right (or left) TTM muscle on the contralateral side with respect to the recording electrode for the DLM (see Figs. 2C and 3B).

The TTM muscle fibers are much smaller than the DLM muscle fibers and hence it is more difficult to obtain and maintain a good recording. The response of a good recording from the TTM is ~30–50 mV and has a latency of ~0.8 msec (Fig. 2D). Protocols can be programmed in software such as pCLAMP to capture 10-msec sweeps to collect data.

See Troubleshooting.

11. Once good recordings have been obtained from the TTM and DLM, give 10 single stimuli with an interval of ~5 sec between the stimuli and determine the average response latency for both GF outputs.

    For this step a separate software protocol that captures 120-msec sweeps can be used to collect the data.

12. Finally, determine the following frequency by giving 10 trains of 10 stimuli at 100 Hz, with an intermittence of ~2 sec between the trains. Calculate the percentage of the total responses. Perform the same assay for trains of stimuli given at 200 Hz and 300 Hz.

13. Compare the TTM and DLM response latencies as well as the following frequencies at 100, 200, and 300 Hz between wild-type and mutant flies.

TROUBLESHOOTING

Problem (Steps 8 and 10): The recording electrodes are sliding on the cuticle and are unable to pierce it to impale the correct muscle.

Solution: The more perpendicular the electrode is to the cuticle, the easier it is for the electrode to get through the cuticle. Possible changes to make the electrode more perpendicular to the cuticle are to move the electrode to a slightly different location within the target area, change the angle of the micromanipulator, try for the muscle on the contralateral side, or remount the fly in a differently angled position.

Problem (Steps 8 and 10): The recording electrodes are indenting the cuticle or the recording electrodes are bending without piercing the cuticle.

Solution: Confirm that the electrode is not broken and has the appropriate shape. The tip of your electrode should have the approximate shape and size similar to the posterior supra-alar setae (Fig. 3B). If the electrode is not broken and has the appropriate shape, try gently tapping on the back of the forward-moving knob of the micromanipulator (once there is slight indentation) to encourage penetration through the cuticle.

Problem (Steps 7, 9, and 10): There is no stimulation artifact and no response.

Solution: Check whether all equipment is turned on. Double-check whether the fly is responding on stimulation (Step 7). If it does not, there is something wrong with your stimulation (check stimulation electrodes, ground and stimulator settings, etc.). If the fly does respond, then there is something wrong with your recording (check recording electrodes and amplifier settings).

Problem: The muscle response has an unusual shape with multiple peaks.

Solution: The microelectrode is not recording from a single muscle cell. This can occur in recordings from either muscle but is more common in TTM recordings, because this muscle is composed of many small fibers and maintaining the position of the electrode after several muscle contractions is problematic. An unusual shaped or multipeaked response trace does not affect the data, because response latencies and followings will still be preserved.

Problem: There is a very large stimulation artifact obscuring the muscle response and/or recordings of multiple stimuli are drifting on the recording monitor.

Solution: Consider the following.
1. Confirm that the ground electrode is properly in the fly.
2. Double-check the voltage and duration of the stimuli given.
3. Also, when the hemolymph dries up around the ground wire it results in a loss of conductance. This can be prevented and restored with a small drop of GFS saline on the fly where the ground electrode enters the abdomen.

Problem: There are long latencies or no responses in wild-type flies.
Solution: Consider the following.
1. Double-check whether the electrode is in the correct target area for the appropriate muscle.
2. The electrode might have pierced through the correct muscle. Both muscles are just underneath the cuticle. The cuticle is approximately no thicker than two to three times the thickness of a posterior supra-alar setae at it thickest visible point (Fig. 3B).
3. Stimulation is below threshold. Try increasing the voltage (duration).
4. If CO_2 was used to anesthetize the fly, either leave the fly to recover from CO_2 longer before testing or anesthetize flies using ice.
5. The wild-type fly may be a mutant.

Problem: Very short latencies are obtained for both TTM (<0.7 msec) and DLM (<1 msec).
Solution: This occurs if the ventral nerve cord, and thus the TTMn and DLMn motorneurons, are being activated directly. Check the position of the stimulating electrodes and replace them in the brain if necessary.

DISCUSSION

In wild-type flies average response latencies to a single stimulus are in the range 0.8 msec ± 0.1 msec for the GF-TTM pathway and 1.4 msec ± 0.3 msec for the GF-DLM pathway, depending on genotype and genetic background. Similarly, with respect to following frequencies the GF-TTM path is able to follow 10 stimuli 1:1 up to 300 Hz and the GF-DLM pathway is able to follow 10 stimuli up to 100 Hz, but variability between individual flies of different genotypes and genetic background has been observed. Hence, it is important to choose carefully the appropriate control flies when analyzing the electrophysiological phenotypes of mutants or targeted disruptions in the GFS. Two classic mutants that do affect the function of the GFS dramatically are *shakB*[2] and *bendless* (Thomas and Wyman 1984; Blagburn et al. 1999; Allen and Murphey 2007; Phelan et al. 2008; Uthaman et al. 2008). In *shakB*[2] flies, the GF–TTMn synapse lacks the gap junctions, but the chemical component is still present. The average response latency for the TTM in these flies is consistently increased to an average of 1.5 msec and it is not able to follow stimuli given at either 100, 200, or 300 Hz because of the weak labile nature of the resultant GF–TTMn synapse. In addition, no responses are obtained from the DLM when the GF is stimulated in the brain. Proof that the lack of responses is not due to a defect at the NMJ comes from the ability to record responses from the DLM muscle when the motorneurons are stimulated directly by stimulation electrodes placed in the thorax (Thomas and Wyman 1984).

In contrast, in *bendless* flies the GF-DLM pathway remains unaffected when compared to wild-type control flies. However, the GF-TTM connection is consistently increased to an average of >2 msec and is not able to following stimuli given at either 100, 200, or 300 Hz.

The reason that these indirect electrophysiological tests of these central synapses of the GFS are successful is that the NMJs at both TTM and the DLMs are large and extensive with many synaptic boutons. They rarely fail; the motorneurons can be stimulated directly at frequencies up to 500 Hz and the muscles will still show 1:1 responses to stimuli (MJ Allen and TA Godenschwege, unpubl.). Thus any effects seen on transmission through the pathways from the GF can be attributed to central synaptic defects. If defects are seen when testing it is always prudent to stimulate the motoneurons directly to confirm that the NMJs are functioning correctly in at least a few flies of the same genotype, because some mutants do affect the adult NMJ (Huang et al. 2006).

RECIPE

CAUTION: See Appendix for proper handling of materials marked with <!>.
Recipes for reagents marked with <R> are included in this list.

GFS Saline

NaCl	101 mM
CaCl$_2$ <!>	1 mM
MgCl$_2$ <!>	4 mM
KCl <!>	3 mM
Glucose	5 mM
NaH$_2$PO$_4$	1.25 mM
NaHCO$_3$	20.7 mM

Adjust the pH to 7.2. From Gu and O'Dowd 2006.

ACKNOWLEDGMENTS

Work in the M.J.A. laboratory has been supported by the Wellcome Trust and the Leverhulme Trust. T.A.G. is supported by R01 HD050725. Thanks to Robin Konieczny for the artwork in Figure 1. We also owe much to R.K. Murphey for his enthusiasm and encouragement regarding the GFS.

REFERENCES

Allen MJ, Murphey RK. 2007. The chemical component of the mixed GF-TTMn synapse in *Drosophila melanogaster* uses acetylcholine as its neurotransmitter. *Eur J Neurosci* **26:** 439–445.

Allen MJ, Shan X, Caruccio P, Froggett SJ, Moffat KG, Murphey RK. 1999. Targeted expression of truncated *Glued* disrupts giant fiber synapse formation in *Drosophila. J Neurosci* **19:** 9374–9384.

Allen MJ, Shan X, Murphey RK. 2000. A role for *Drosophila* Drac1 in neurite outgrowth and synaptogenesis in the giant fiber system. *Mol Cell Neurosci* **16:** 754–765.

Allen MJ, Godenschwege TA, Tanouye MA, Phelan P. 2006. Making an escape: Development and function of the *Drosophila* giant fibre system. *Semin Cell Dev Biol* **17:** 31–41.

Blagburn JM, Alexopoulos H, Davies JA, Bacon JP. 1999. Null mutation in *shaking-B* eliminates electrical, but not chemical, synapses in the *Drosophila* giant fiber system: A structural study. *J Comp Neurol* **404:** 449–458.

Demerec, M, ed. 1994. *Biology of* Drosophila. Cold Spring Harbor Laboratory Press, Cold Spring Harbor, NY.

Engel JE, Wu CF. 1996. Altered habituation of an identified escape circuit in *Drosophila* memory mutants. *J Neurosci* **16:** 3486–3499.

Engel JE, Wu CF. 1998. Genetic dissection of functional contributions of specific potassium channel subunits in habituation of an escape circuit in *Drosophila. J Neurosci* **18:** 2254–2267.

Fotowat H, Fayyazuddin A, Bellen HJ, Gabbiani F. 2009. A novel neuronal pathway for visually guided escape in *Drosophila melanogaster. J Neurophysiol* **102:** 875–885.

Godenschwege TA, Hu H, Shan-Crofts X, Goodman CS, Murphey RK. 2002a. Bi-directional signaling by Semaphorin 1a during central synapse formation in *Drosophila. Nat Neurosci* **5:** 1294–1301.

Godenschwege TA, Simpson JH, Shan X, Bashaw GJ, Goodman CS, Murphey RK. 2002b. Ectopic expression in the giant fiber system of *Drosophila* reveals distinct roles for Roundabout (Robo), Robo2, and Robo3 in dendritic guidance and synaptic connectivity. *J Neurosci* **22:** 3117–3129.

Godenschwege TA, Kristiansen LV, Uthaman SB, Hortsch M, Murphey RK. 2006. A conserved role for *Drosophila* Neuroglian and human L1-CAM in central-synapse formation. *Curr Biol* **16:** 12–23.

Gorczyca M, Hall JC. 1984. Identification of a cholinergic synapse in the giant fiber pathway of *Drosophila* using conditional mutations of acetylcholine synthesis. *J Neurogenet* **1:** 289–313.

Gu H, O'Dowd DK. 2006. Cholinergic synaptic transmission in adult *Drosophila* Kenyon cells in situ. *J Neurosci* **26:** 265–272.

Huang FD, Woodruff E, Mohrmann R, Broadie K. 2006. Rolling blackout is required for synaptic vesicle exocytosis. *J Neurosci* **26:** 2369–2379.

Ikeda K, Koenig JH, Tsuruhara T. 1980. Organization of identified axons innervating the dorsal longitudinal flight muscle of *Drosophila melanogaster*. *J Neurocytol* **9:** 799–823.

King DG, Wyman RJ. 1980. Anatomy of the giant fibre pathway in *Drosophila*. I. Three thoracic components of the pathway. *J Neurocytol* **9:** 753–770.

Koto M, Tanouye MA, Ferrus A, Thomas JB, Wyman RJ. 1981. The morphology of the cervical giant fiber neuron of *Drosophila*. *Brain Res* **221:** 213–217.

Lin MQ, Nash HA. 1996. Influence of general anesthetics on a specific neural pathway in *Drosophila melanogaster*. *Proc Natl Acad Sci* **93:** 10446–10451.

Martinez VG, Javadi CS, Ngo E, Ngo L, Lagow RD, Zhang B. 2007. Age-related changes in climbing behavior and neural circuit physiology in *Drosophila*. *Dev Neurobiol* **67:** 778–791.

Oh CE, McMahon R, Benzer S, Tanouye MA. 1994. *bendless*, a *Drosophila* gene affecting neuronal connectivity, encodes a ubiquitin-conjugating enzyme homolog. *J Neurosci* **14:** 3166–3179.

Phelan P, Goulding LA, Tam JL, Allen,MJ, Dawber RJ, Davies JA, Bacon JP. 2008. Molecular mechanism of rectification at identified electrical synapses in the *Drosophila* giant fiber system. *Curr Biol* **18:** 1955–1960.

Thomas JB, Wyman RJ. 1984. Mutations altering synaptic connectivity between identified neurons in *Drosophila*. *J Neurosci* **4:** 530–538.

Uthaman SB, Godenschwege TA, Murphey RK. 2008. A mechanism distinct from highwire for the *Drosophila* ubiquitin conjugase bendless in synaptic growth and maturation. *J Neurosci* **28:** 8615–8623.

Watson MR, Lagow RD, Xu K, Zhang B, Bonini NM. 2008. A *Drosophila* model for amyotrophic lateral sclerosis reveals motor neuron damage by human SOD1. *J Biol Chem* **283:** 24972–24981.

14 Analysis of Visual Physiology in the Adult *Drosophila* Eye

Patrick Dolph,[1] Amit Nair,[2] and Padinjat Raghu[2]

[1] Department of Biology, Dartmouth College, Hanover, New Hampshire 03755; [2] Inositide Laboratory, Babraham Institute, Babraham Research Campus, Cambridge CB22 3AT, United Kingdom

ABSTRACT

Historically, the electrophysiological analysis of responses to light has been a powerful tool for characterizing and understanding visual transduction in *Drosophila* photoreceptors. Initially, analysis of visual transduction in *Drosophila* focused on extracellular recordings (electroretinograms [ERGs]) in live animals. Combined with genetic screens, ERGs were used as an assay to isolate mutants in several components of the visual transduction pathway. More recently, a preparation of isolated ommatidia has been developed that allows more detailed analysis of the electrical activity of photoreceptors. Using this approach, voltage-clamp recordings of currents from photoreceptors can be performed allowing detailed analysis of ion channels, such as the TRP channels, and potassium channels in photoreceptors. This chapter presents the methodology for both ERGs and voltage-clamp recordings from *Drosophila* photoreceptors.

INTRODUCTION

Adult *Drosophila* have a compound eye composed of approximately 750 repeating units, called ommatidia, packed together to form the retina. Each ommatidium is a precise arrangement of 19 cells; eight photoreceptors (primary sensory neurons) and 11 accessory cells (Wolff and Ready 1993). The structure and organization of the adult eye is generated through a series of complex and coordinated developmental events that begin in the larval eye imaginal disc and are completed at eclosion. *Drosophila* photoreceptors become competent to respond to light from ~70% pupal development (Hardie et al. 1993). The principal function of the eye is phototransduction, the detection and transduction of photons of light into an electrical signal that is then transmitted to the brain. To do this, photoreceptors are designed to achieve two conflicting goals: (1) to detect single photons of light, and (2) to avoid saturation while working in a linear range during the high photon fluxes characteristic of daylight intensities (Hardie and Raghu 2001).

Sensory transduction in *Drosophila* photoreceptors is initiated by the absorption of a photon of light by the seven-transmembrane G-protein-coupled receptor (GPCR) rhodopsin (Zuker 1996;

Montell 1999). Thus phototransduction is a paradigm for understanding questions in sensory transduction and adaptation. In *Drosophila* photoreceptors, the photoisomerization of rhodopsin triggers G-protein-coupled phospholipase C activity and ends with the activation of two classes of Ca^{2+} and cation selective ion channels, TRP and TRPL, founding members of the TRP superfamily of ion channels. Analysis of visual physiology in *Drosophila* photoreceptors has been central to understanding a number of important areas of modern biology including the GPCR cycle, phosphoinositide signaling, and calcium signaling. In addition, analysis of photoreceptor performance (Juusola and Hardie 2001; Vahasoyrinki et al. 2006) and synaptic transmission (Stowers and Schwarz 1999) are areas of neurobiology that have been studied using *Drosophila* photoreceptors as a model system.

The use of physiological approaches in the *Drosophila* eye has been complemented by the ability to adapt and exploit modern molecular genetic tools. For example, the *Drosophila* eye is not required for viability or reproduction, and thus a homozygous mutation can be generated and studied in the eye without any detrimental effect on the whole animal.

From an experimental point of view, the repeating structure of the fly compound eye offers a large sensory organ from which electrophysiological analysis can be performed with relative ease. Given that transduction is essentially similar across the approximately 750 ommatidia (except for the complement of rhodopsins expressed) the use of ERGs has allowed genetic screens to be performed, isolating mutants with defective visual physiology (Pak 1995). In addition, the development of methods that allow whole-cell voltage clamp experiments has allowed sensory transduction to be studied at an unprecedented level of resolution (Hardie 1991b). Finally, the use of intracellular recording methods has allowed the study of aspects of neurobiology such as information processing.

Electroretinogram Recordings of *Drosophila*

Extracellular recordings or ERGs have long been used as a physiological assay in the *Drosophila* visual system. In these recordings a microelectrode is placed on the eye and a reference electrode is placed elsewhere on the animal (typically in the thorax). Upon light stimulation, the voltage difference between these two electrodes is measured and displayed in real time. It has the advantage that the recording is performed in live animals so that all photoreceptor cells are intact and therefore surrounded by fluid containing endogenous concentrations of ions. Moreover, synaptic transmission between the photoreceptor and downstream laminar neurons can be detected.

MATERIALS

CAUTION: See Appendix for proper handling of materials marked with <!>.

Reagents

CO_2 <!>
Drosophila melanogaster, adults
Myristic acid <!>
Saline (0.7% [w/v] NaCl) or electrical conductive gel (Step 5)

Equipment

Amplifier (DAM50; World Precision Instruments [WPI])
Capillaries, glass
Chart and scope software (ADInstruments)
Collimating beam probe, fused silica (Oriel 77644)
Data acquisition hardware (PowerLab 4/30)
Dissecting microscope
Electrode adapters (5447; WPI)
Fiber optic bundle, fused silica (Oriel 77564)
Fiber optic input assembly (Oriel 77800)
Filter wheels (Oriel 77370)
Fly wheel
> We use a 4-in diameter black plastic disc, 1/8-in thick with 1/16-in diameter holes drilled around the edge.

Fly wheel holder
> This holds the fly wheel such that it can be secured under the dissecting microscope and only a single fly will be exposed to light at a time.

Forceps, fine
Interference filter, 480 nm, 10-nm bandwidth (Oriel 53850)
Interference filter, 530 nm, 10-nm bandwidth (Oriel 53875)
Interference filter, 580 nm, 10-nm bandwidth (Oriel 53910)
Liquid filter and recirculating cooler to cool light path (Oriel 61945, Oriel 60200)
Microelectrodes and microelectrode holders (1.0 mm OD; WPI IB100-4, WPI MEH3S)

Micromanipulators
Needle puller
Neutral density filter set, fused silica (Oriel 50430)
Shutter, 25-mm aperture (#2 case, AlMgF blade finish, solid state; Vincent Associates VS25)
Shutter controller (Vincent Associates VMM-T1)
Socket adapter for 300-W xenon lamp (Oriel 66160)
Soldering iron and rheostat
Xenon 300-W arc lamp, ozone-free (Oriel 6258)
Xenon power supply, xenon lamp housing, and fused silica condensor (Oriel 66084)

METHOD

Mounting Flies

1. Anesthetize flies with CO_2.

2. Using fine forceps, insert a fly headfirst into a hole in the fly wheel with proboscis facing away from the center of the wheel.

3. Gently push on the abdomen of the fly until the head and about one-third of the thorax is visible on the other side of the wheel.

4. Secure the fly to the wheel with melted myristic acid using a soldering iron.

   Do not get any myristic acid on the eye. If the legs are free, it may be necessary to remove them or immobilize them with myristic acid, as they may interfere with the recording.

5. Pull glass capillaries on a needle puller and backfill them with saline or electrical conductive gel.

   The tips of the pipettes may need to be broken off to allow the electrolyte fluid to flow.

   It is not necessary to carefully control the pipette tip diameter, although it is useful to have a larger diameter tip to place on the cornea and a smaller diameter tip to pierce the thorax.

Performing ERG

6. Using the left-hand micromanipulator, gently press the reference electrode into the thorax.

7. Using the right-hand micromanipulator, place the second electrode on the corneal surface of the eye.

   Do not pierce the eye or indent the cornea with the electrode.

8. Dark-adapt the animal (~60 sec) before recording.

Testing for a Prolonged Depolarizing Afterpotential

A prolonged depolarizing afterpotential (PDA) occurs when saturating light stimulus activates rhodopsin in excess of the regulatory protein arrestin. Under such conditions the photoreceptors remain depolarized after the cessation of light stimulus until the rhodopsin is inactivated by orange (580 nm) light (Hillman et al. 1972).

9. Stimulate flies with high-intensity 480 nm light. Depending on the setup, this can be achieved by a 1–5-sec stimulation in the absence of an inline neutral density filter.

   The trace should show a depolarization that will continue after the light stimulus has terminated. During this prolonged depolarization the fly should be refractory to further light stimulation and can remain depolarized for up to several hours.

10. Stimulate flies with 1–5 sec of high-intensity 580 nm light (no inline filter)

    The voltage response should return to baseline and respond normally to subsequent stimuli.

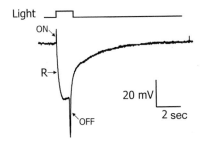

FIGURE 1. An ERG recorded from the eye of a wild-type (Oregon-R) fly within 6 h of eclosion. The duration of the light stimulus is shown (Light). The three major components of the ERG are marked: the receptor potential (R), the on-transient (ON), and the off-transient (OFF).

TROUBLESHOOTING

Problem: There is electrical noise.
Solution: Electrical noise can be a problem, but a Faraday cage is not usually necessary. Many people perform ERG recordings on a bench top with well-grounded components.

Problem: Because of the high gain of the phototransduction cascade, defects in the ERG signal are often not evident when high-intensity illumination is used to stimulate the eye.
Solution: Using neutral density filters to attenuate the stimulating light, construct a light intensity ERG response curve. Then choose an intensity on the linear part of this curve for subsequent experiments.

DISCUSSION

Wild-type flies show a stereotypical ERG response (Fig. 1). Upon light stimulation, there is an upward voltage spike, referred to as an on-transient, which is elicited from the laminar neurons, followed by a depolarization elicited from the photoreceptor cell. After cessation of the light stimulus there is a downward spike, referred to as an off-transient, and the voltage trace rapidly returns to baseline. Defects in this stereotypical trace have been used to identify components of the visual transduction cascade. For example, animals that fail to show on-transients or off-transients are defective in synaptic signaling between the photoreceptor and the laminar neurons (Hotta and Benzer 1970; Pak 1970), and flies that show voltage responses in which the trace returns slowly to baseline after lights off are defective in regulating or inactivating a component of the phototransduction cascade (Smith et al. 1991; Dolph et al. 1993). ERG responses also show a phenomenon referred to as a PDA, which occurs upon saturating light stimulus where rhodopsin is activated in excess of the regulatory protein arrestin. Under such conditions the photoreceptors remain depolarized after the cessation of light stimulus until the rhodopsin is inactivated by orange (580-nm) light (Hillman et al. 1972). Flies that fail to show a PDA typically express less rhodopsin than wild-type controls, and the loss of the PDA has been used as the basis for genetic screens to isolate many mutations that affect visual transduction (Pak 1979).

Preparation of Dissociated Ommatidia from *Drosophila*

Mechanical dissociation of ommatidia is typically performed in flies that are <4 h posteclosion. In animals older than this, the ommatidia become progressively more firmly attached to each other and cannot be dissociated mechanically without causing substantial damage to the constituent photoreceptors. Like the rest of the fly, the eye is surrounded by a cuticle. Thus, before the dissociation process, the eye is dissected from the rest of the head and the retina is scooped out of the cuticular covering.

MATERIALS

See the end of the chapter for recipes for reagents marked with <R>.

Reagents

Drosophila melanogaster, adults
Drosophila Ringer's solution <R>
Fetal calf serum (FCS)
> FCS is used to coat the trituration pipettes to prevent the ommatidia from sticking to them.

Ice (Step 1)
Trituration solution <R>

Equipment

Capillaries, thin wall borosilicate without filament (GC150T-10; Harvard Apparatus)
Dissection instruments (blade holder, needle holder, blades, and insect pins)
Dissection microscope with a red light filter
> Any standard stereomicroscope equipped with a red filter will suffice. The microscope manufacturer can usually supply this filter.

Dissection stage
> Make from a plastic Petri dissection dish with a cured Sylgard base.

Electrode puller (e.g., P-97 or P-1000; Sutter Instruments)
Minutien pins, stainless steel (10 × 0.1-mm; Fine Science Tools)
> Use these to fashion the retinal scoop.

Petri dishes
Tygon Tubing to fit GC150T-10 capillaries (Harvard Apparatus)

METHOD

Perform the following procedure in a dark room under red light illumination. The microscope used for dissection should have a red filter.

1. Pull the glass capillaries to fashion a series of fire-polished trituration pipettes with tip diameters between 100 and 250 μm. Coat the pipettes with FCS.

2. Anesthetize flies by cooling on ice. Decapitate using a sharp razor blade and fix on the Sylgard dissection stage using an insect pin inserted through the head between the two eyes.

3. Slice off the eyes using a sharp razor blade and carefully dissect away bits of attached brain and trachea. In a red-eyed fly these can be clearly seen as white material on an otherwise red background.

4. Using a minutien pin held in a needle holder, scoop out the retina from the overlying cuticle.

5. Using the largest of the trituration pipettes, transfer the retina into a drop of Ringers with 10% FCS. Begin dissociating cells by gently passing the retina through the tip of the trituration pipette in a Petri dish.

6. Transfer the retina into a drop of trituration solution and use progressively smaller pipettes to dissociate ommatidia, repeating the transfer into a fresh drop of trituration solution each time. Retinae can be maneuvered by sucking (by mouth) them into the trituration pipette using the Tygon tubing attached to one end of the glass capillary. Then the retinae can be released from the pipette into the trituration solution by blowing into the Tygon tubing.

7. Once ommatidia are fully dissociated, proceed with Protocol 3.

TROUBLESHOOTING

Problem: It is important to obtain a high-quality gigaseal with no leak current so that low-amplitude electrical events such as quantum bumps (see below) can be recorded.

Solution: Preparation of high-quality dissociated ommatidia is the critical factor. Carefully dissect the retina and gently dissociate the cells.

Whole-Cell Patch-Clamp Analysis of Adult *Drosophila* Photoreceptors

The individual units of the *Drosophila* compound eye, the ommatidia, can be dissociated to generate healthy cells under a limited set of conditions (Hardie 1991a,b). The basolateral plasma membrane of the photoreceptors then becomes accessible to a patch pipette and light responses can be recorded in the whole-cell configuration. This approach has principally been used to study the activity of the light-activated channels in voltage-clamped photoreceptors. Both macroscopic responses (Fig. 2A) and unitary responses to single photons of light (the quantum bumps labeled Q, Fig. 2B) can be recorded using this approach. However, because the light-sensitive channels are localized to the rhabdomeral membrane, which is not accessible to the patch pipette, analysis of single-channel activity using inside-out patches is currently not feasible.

MATERIALS

See the end of the chapter for recipes for reagents marked with <R>.

Reagents

Bath solution <R>
Intracellular solution <R>
Ommatidia, dissociated from *Drosophila* adult eyes (Protocol 2)

Equipment

AD/DA interface board (e.g., Digidata 1440A; Molecular Devices)
Amplifier, patch-clamp (e.g., Axopatch 200B; Axon Instruments)
Capillaries, thick-walled siliconized glass with filament center (GC100-F10; Harvard Apparatus)
Computer with software for data acquisition and analysis
> The software needs to be compatible with the digital interface chosen (pCLAMP 10.0 from Axon Instruments is compatible with Digidata 1440A).

Faraday cage for electrical noise insulation
Light box controlling red and green LED light source
Micromanipulator, low-drift for patch electrode (e.g., Burleigh PCS-5000 series)
Micropipette puller
Microscope, inverted (e.g., Olympus IX-71)
> This allows visualization of dissociated ommatidia under red light illumination for patching. It also provides the optical system to stimulate the photoreceptors with light. The traditional bulb can be replaced with a dual red/green LED that is available from most electronics suppliers.

Patch pipettes
> Pull from borosilicate glass capillaries with internal filament (GC100F-10; Harvard Apparatus). Pipettes should have a resistance of 5–10 MΩ.

Recording chamber, Perspex
Vibration isolation table
Wire, silver

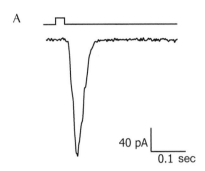

A

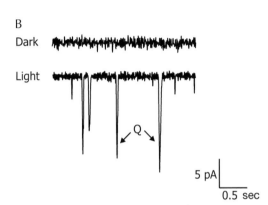

B
Dark

Light

Q

5 pA

0.5 sec

40 pA

0.1 sec

FIGURE 2. (A) Whole-cell voltage-clamp recording from a wild-type (Oregon-R) photoreceptor. The response to a flash of light (several hundred photons of light) from a green LED is shown. (B) Recording from a wild-type photoreceptor with no illumination (Dark) and in constant low-intensity red light (Light) are shown. The unitary responses to light, called quantum bumps, are shown (Q).

METHOD

Perform all recordings in a dark room.

1. Place dissociated ommatidia in a Perspex recording chamber on the stage of an inverted microscope. The light source can be replaced with twin LEDs: (i) a red LED for viewing and selecting ommatidia for recording and (ii) a green LED for stimulating photoreceptors. Calibrate the light source using a combination of photon counting and a light meter.

2. View the ommatidia using a red light source. Assess the quality of the cells at this stage. Healthy ommatidia will look smooth and show no evidence of blebbing.

3. Perform patch-clamp recording using standard procedures (see Chapter 12). Analyze light-sensitive channels in whole-cell mode using voltage-clamp conditions. Stimulate the photoreceptors using the green LED controlled by an external light box and triggered using a stimulus generator or using elements of the patch-clamp software.

TROUBLESHOOTING

Problem: Access resistance may increase over the period of the recording.
Solution: This can be overcome by gentle suction to unblock the tip of the patch pipette.

DISCUSSION

Semi-Intact Preparation

Hevers and Hardie (1995) have described a preparation in which the *Drosophila* retina remains intact with the first optic ganglion, the lamina. The dissection procedure frees the distal parts of the photoreceptors from adjacent pigment cells. Photoreceptors are accessible to whole-cell or cell-attached patch-clamp recordings. Recordings in this setting can potentially be used to study the effects of secondary neurons on the physiology of photoreceptors.

Intracellular Recordings

Sharp glass electrodes with high resistance can be used to perform intracellular recordings from *Drosophila* photoreceptors. A small hole the size of a few ommatidia can be cut in the cornea with a sharp razor blade and used to insert the intracellular electrode (Juusola and Hardie 2001). Intracellular recordings can be used to make recordings of photoreceptor responses. Such recordings can be used to study questions such as light adaptation and information processing by photoreceptors.

RECIPES

CAUTION: See Appendix for proper handling of materials marked with <!>.
Recipes for reagents marked with <R> are listed here.

Bath Solution

NaCl	120 mM
KCl <!>	5 mM
TES (2-[(2-hydroxy-1, 1-bis (hydroxymethyl)ethyl)amino] ethanesulfonic acid)	10 mM
Proline	25 mM
Alanine	5 mM
MgCl$_2$ <!>	4 mM
CaCl$_2$ <!>	4 mM

Adjust the pH to 7.15 and the osmolarity to 82 mOsm.

Drosophila Ringer's Solution

Tris-Cl	10 mM
KCl <!>	182 mM
NaCl	46 mM
CaCl$_2$ <!>	3 mM

Adjust the pH to 7.2.

Intracellular Solution

Potassium gluconate <!>	140 mM
TES <!>	10 mM
MgCl$_2$ <!>	2 mM
Mg-ATP (adenosine triphosphate)	4 mM
GTP (guanosine triphosphate)	0.4 mM

Adjust the osmolarity to 276 mOsm.

Trituration Solution

NaCl	120 mM
KCl <!>	5 mM
TES <!>	10 mM
Sucrose	40 mM
MgCl$_2$ <!>	4 mM
FCS	10%

Adjust the pH to 7.15.

REFERENCES

Dolph PJ, Ranganathan R, Colley NJ, Hard, RW, Socolich M, Zuker CS. 1993. Arrestin function in inactivation of G protein-coupled receptor rhodopsin in vivo. *Science* **260:** 1910–1916.

Hardie RC. 1991a. Voltage-sensitive potassium channels In *Drosophila* photoreceptors. *J Neurosci* **11:** 3079–3095.

Hardie RC. 1991b. Whole-cell recordings of the light-induced current In dissociated *Drosophila* photoreceptors— Evidence for feedback by calcium permeating the light-sensitive channels. *Proc R Soc Lond Series B-Biol Sci* **245:** 203–210.

Hardie RC, Raghu P. 2001. Visual transduction in *Drosophila*. *Nature* **413:** 186–193.

Hardie RC, Peretz A, Pollock JA, Minke B. 1993. Ca^{2+} limits the development of the light response in *Drosophila* photoreceptors. *Proc R Soc Lond Series B-Biol Sci* **252:** 223–229.

Hevers W, Hardie RC. 1995. Serotonin modulates the voltage-dependence of delayed rectifier and shaker potassium channels in *Drosophila* photoreceptors. *Neuron* **14:** 845–856.

Hillman P, Hochstein S, Minke B. 1972. A visual pigment with two physiologically active stable states. *Science* **175:** 1486–1488.

Hotta Y, Benzer S. 1970. Genetic dissection of the *Drosophila* nervous system by means of mosaics. *Proc Natl Acad Sci* **67:** 1156–1163.

Juusola M, Hardie RC. 2001. Light adaptation in *Drosophila* photoreceptors: I. Response dynamics and signaling efficiency at 25°C. *J Gen Physiol* **117:** 3–25.

Montell C. 1999. Visual transduction in *Drosophila*. *Annu Rev Cell Dev Biol* **15:** 231–268.

Pak WL. 1979. Study of photoreceptor function using *Drosophila* mutants. In *Neurogenetics: Genetic approaches to the nervous system* (ed. XO Breakfield), pp. 67–99. Elsevier, New York.

Pak WL. 1995. *Drosophila* in vision research—The Friedenwald-Lecture. *Investig Ophthalmol Vis Sci* **36:** 2340–2357.

Pak WL, Grossfield J, Arnold KS. 1970. Mutants in the visual pathway of *Drosophila melanogaster*. *Nature* **227:** 518–520.

Smith DP, Ranganathan R, Hardy RW, Marx J, Tsuchida T, Zuker CS. 1991. Photoreceptor deactivation and retinal degeneration mediated by a photoreceptor-specific protein kinase C. *Science* **254:** 1478–1484.

Stowers RS, Schwarz TL. 1999. A genetic method for generating *Drosophila* eyes composed exclusively of mitotic clones of a single genotype. *Genetics* **152:** 1631–1639.

Vahasoyrinki M, Niven JE, Hardie RC, Weckstrom M, Juusola M. 2006. Robustness of neural coding in *Drosophila* photoreceptors in the absence of slow delayed rectifier K$^+$ channels. *J Neurosci* **26:** 2652–2660.

Wolff T, Ready DF. 1993. Pattern formation in the *Drosophila* retina. In *The development of* Drosophila melanogaster (ed. Bate M, Martinez Arias A), pp. 1277–1325. Cold Spring Harbor Laboratory Press, Cold Spring Harbor, NY.

Zuker CS. 1996. The biology of vision in *Drosophila*. *J Neurosci* **93:** 571–576.

15 | Measuring Sound-Evoked Potentials from *Drosophila* Johnston's Organ

Daniel F. Eberl[1] and Maurice J. Kernan[2]

[1]*Department of Biology, University of Iowa, Iowa City, Iowa 52242;* [2]*Department of Neurobiology and Behavior, State University of New York, Stony Brook, Stony Brook, New York 11974*

ABSTRACT

Courtship song signals are transduced by Johnston's organ (JO), which is the chordotonal organ in the *Drosophila* antennae. In this chapter, we discuss auditory responses in *Drosophila* and describe a protocol for recording potentials in the antennal nerve that occur in response to acoustic signals generated electronically to mimic courtship songs.

INTRODUCTION

Information about an animal's social and ecological environment takes many forms, several of which can be sensed as mechanical signals. The mechanosensory world of a fly such as *Drosophila* is especially complex, and its transduction involves a diverse set of specialized sense organs. Touch is mediated by sensory bristles, dome-shaped campaniform organs, and body-wall chordotonal organs. Proprioception depends primarily on small hair plates at joints, with contributions from chordotonal organs at joints as well as campaniform organ fields in the halteres and along the wings. Near-field sounds, such as the male courtship song, are transduced by sensory units in JO. Low-frequency stimuli such as wind and gravity are also detected by JO, with likely contributions from leg and abdominal proprioceptors. Dethier (1963) provided a comprehensive discussion of most of these sensory modalities, particularly focusing on their physiological aspects rather than on behavior. Since then, enormous progress has been made in *Drosophila* behavioral assays, electrophysiological preparations, mutagenesis screens, and molecular genetic characterization of developmental and functional aspects of multiple mechanosensory modalities. For further discussion of hearing in *Drosophila*, see reviews by Eberl (1999), Caldwell and Eberl (2002), Jarman (2002), Tauber and Eberl (2003), Todi et al. (2004), Eberl and Boekhoff-Falk (2007), and Kernan (2007).

JO AND ANTENNAL AUDITORY RESPONSES

Courtship Songs

In *Drosophila*, hearing is specialized for detecting courtship songs, which are generated by unilateral wing vibration in male flies. These species-specific songs provide one channel through which females can assess whether the courting male is conspecific. The *Drosophila melanogaster* song includes trains of brief pulses and a sine-wave hum phase. The specific interval between pulses (35 msec) may convey the species information, and oscillation of this interpulse interval is preferred. The effect on female receptivity can be measured by comparing latency with copulation in courting male/female pairs in which the male has been muted (wings removed) in the presence and absence of song playback (Kyriacou and Hall 1982, 1984, 1986; Wheeler et al. 1991). Males also respond behaviorally to song playback by increasing their courtship behavior, a response that may be adaptive either by enabling them to detect and to interrupt nearby courting males or by enhancing the male's own courtship endurance and vigor (von Schilcher 1976; Hall 1994; Eberl et al. 1997; Tauber and Eberl 2001, 2002, 2003).

Fine Structure of JO and the Mechanism of Mechanotransduction

The features of scolopidia (chordotonal sensory units) in the *Drosophila* JO that may specialize them for vibratory stimuli have been reviewed recently (Eberl and Boekhoff-Falk 2007). In contrast to embryonic body wall scolopidia, each of which encloses the sensory process from a single neuron, adult scolopidia are doubly innervated, and a subset of JO scolopidia are triply innervated. Furthermore, the dendritic cap of JO scolopidia is tubular and has the potential for relatively uninhibited longitudinal movement through the support cells. The dendritic cap depends on the NompA protein for attachment to the ciliated sensory dendrites of the neurons (Chung et al. 2001) and on the Ck/MyoVIIA motor protein for its apical attachment to the cuticle at the antennal joint (Todi et al. 2005, 2008). Elaboration of the sensory cilium constituting the outer dendritic segment depends on ciliogenesis and intraflagellar transport (IFT) proteins, including the anterograde Kinesin II motors (Sarpal et al. 2003) and the Btv retrograde cytoplasmic dynein motor (Sharma 2004), as well as a growing collection of IFT particle proteins and other ciliogenesis factors (Dubruille et al. 2002; Han et al. 2003; Avidor-Reiss et al. 2004; Baker et al. 2004; Laurençon et al. 2007; Lee et al. 2008). Auditory mechanotransduction depends on at least three members of the transient receptor potential (TRP) superfamily, including the TRPV channel subunits inactive and Nanchung (Kim et al. 2003; Gong et al. 2004) and the TRPN channel NompC (Kernan et al. 1994; Eberl et al. 2000; Göpfert et al. 2006; Sun et al. 2009).

Studies using laser vibrometry approaches to detail the mechanical responses of antennae in the presence and absence of acoustic stimuli (Göpfert and Robert 2003; Göpfert et al. 2005, 2006; Albert et al. 2007; Nadrowski et al. 2008) have led to the intriguing notion that JO scolopidia are not just passive sensors. Rather, by exerting forces on the a2/a3 joint, these scolopidia are, in fact, also motors that set the arista in motion even in the absence of acoustic stimuli. This invested energy can make the fly's auditory system biologically more effective. Oscillations, tuned to ~200 Hz and introduced in the antennae, make the antennae much more sensitive to very low intensity near-field sounds in this frequency range, notably the courtship songs. At the same time, the oscillations can also damp high-intensity sounds. Thereby, the dynamic range of acoustic sensitivity is extended at both ends of the intensity spectrum. Furthermore, the tuning can also act as a filter to favor the frequencies in the courtship songs. Thus, JO is an exquisitely sensitive sensor whose sensitivity is enhanced by motor activity.

As a multimodal sense organ for hearing, gravity, and wind sensation, JO is organized to separate spatially the fast phasic vibrational response required for hearing from the slower tonic responses of gravity and wind sensation. A well-characterized set of Gal4 enhancer trap lines that express in different subsets of JO neurons (Kamikouchi et al. 2006) have been used to drive reporter constructs, optical sensors, and toxins in these different subsets to identify groups of JO neurons that respond differentially to these mechanical modalities and deliver the corresponding signals to divergent projection sites in the antennal mechanosensory and motor center (AMMC) (Kamikouchi et al. 2009; Yorozu et al. 2009).

Recording Sound-Evoked Potentials from the *Drosophila* Antennal Nerve

This protocol was developed to record extracellular potentials in the antennal nerve in response to near-field acoustic signals. These signals represent the combined action potentials of many or perhaps even all of the responding JO sensory neurons whose axons project along the antennal nerve en route to the AMMC.

MATERIALS

CAUTION: See Appendix for proper handling of materials marked with <!>.

Reagents

Potassium hydroxide (10%, w/v) <!>

Equipment

Air table or breadboard (optional for Step 7)
BNC cable
Clamp stands (two: one to hold speaker and the other to hold tubing in speaker cone)
Computer (Mac or PC)
Dental wax or modeling clay (soft)
Digitizer equipment with analog-voltage-out channels as well as analog-in channels (e.g., instruNet i100B network device with i200 PCI controller; GW Instruments, Inc.)
Dissecting microscope (preferably boom-mounted)
Electrode holders (e.g., the Plexiglas electrode holder that comes with the MM33 micromanipulator)
   Modify for holding tungsten rod electrodes (see Step 6 below).
Electrolytic sharpening equipment (9-V battery or low-voltage direct current power supply, copper wire, and alligator clips)
Faraday cage
Forceps (e.g., Dumont #5; Fine Science Tools)
Light source (e.g., fiber-optic)
Loudspeaker (e.g., 8-in woofer from Radio Shack; cat. no. 40-1016A)
Micromanipulators (two or three, with magnetic stands) (e.g., MM33 from Fine Science Tools)
Micropipette tips (200 µL)
Signal amplifier (e.g., DAM50 differential amplifier from World Precision Instruments)
Software (e.g., instruNet World software, which is compatible with LabVIEW [Mac or PC], SuperScope II [Mac], DASYLab [PC], and others)
Speaker wire
Specimen stage (e.g., x–y stage or rotating stage [optional])
Tungsten rod (0.004 x 3-in) (e.g., A-M Systems; cat. no. 719000)
Tygon tubing (e.g., 0.25-in inner diameter)
Wire electrode leads (e.g., the connector cable with telephone connector that comes with DAM50)

METHOD

Fly Preparation

Each fly is mounted in a 200-µL micropipette tip that has been trimmed to fit the size of the fly. It is useful to prepare a series of tips trimmed to different lengths (and, therefore, opening sizes) to accommodate size differences between flies from different strains, genders, or culture conditions.

1. Aspirate a fly into the large end of a 200-µL micropipette tip, and make sure it enters the small end headfirst.

   Anesthesia with ether or CO_2 is not recommended because it may depress the fly's responses for a long time. If sedation is required, chilling the fly in an empty glass vial on ice is preferred.

2. Lodge the fly in the micropipette tip so that its head is at or near the opening as the thorax becomes wedged against the side. Using an untrimmed micropipette tip, push a small cone of cotton behind the fly to wedge it slightly more tightly against the walls of the tip, and make the head protrude from the tip.

3. Secure the head by packing a small amount of dental wax or modeling clay under it. It is a good idea to prevent the forelegs from protruding beside the head or, if they do, to secure them with wax or clay. Be sure not to let the clay interfere with antennal motion or with the movement of air currents around the front and sides of the head.

4. Once the head is secure, relieve any pressure on the fly imposed by the cotton wad—this will reduce the likelihood of losing hemolymph when the electrodes are inserted. Attach the micropipette tip to its holder on the electrophysiology rig (Fig. 1A,E–G).

Electrode Preparation and Insertion

5. Prepare electrodes from the tungsten rod. To sharpen the tungsten electrodes electrolytically, connect the tungsten rod to one terminal of a 9-V battery, connect the other terminal to a copper wire immersed in a small beaker of 10% potassium hydroxide, and dip the electrode repeatedly into the solution.

   To some extent, the taper of the electrode can be controlled by the depth and speed of dipping; a gradual taper, especially on the recording electrode, is most effective.

6. Make an electrode holder. Solder the wire lead onto a thin brass plate, and clamp the sharpened tungsten rod electrode to the brass plate (Fig. 1C). The clamp ensures a good connection with the tungsten; solder connections with tungsten are difficult to achieve.

   There are many other possible designs for electrode holders. For example, one design runs sharpened tungsten wire through a glass micropipette, where it is clamped with a small alligator clip.

7. Fasten the specimen stage and micromanipulators to a solid base, such as a steel plate or the surface of an air table or breadboard, and position the stereo microscope over the preparation. With sharpened electrodes clamped firmly into the electrode holders (Fig. 1D), position the holders using micromanipulators so that the reference electrode penetrates the head cuticle and the recording electrode penetrates the a1/a2 joint (Fig. 1H).

   For consistently reliable recordings, use specific bristles as landmarks to aid in electrode placement. For the reference electrode, always aim, for example, at the base of the ipsilateral vertical bristle. The placement of the recording electrode is very important. The best recordings are achieved by approaching from the dorso-fronto-medial aspect. Again, use the single row of bristles on the a1 segment as a guide for consistent placement. Insertion proximal to the a1 segment results in much less reliable signals. Successful electrode placement in a wild-type antenna will be reflected in an oscilloscope trace as a reduction in 60-Hz noise (if any) and in a spiky baseline because of ambient noise and spontaneous nerve activity.

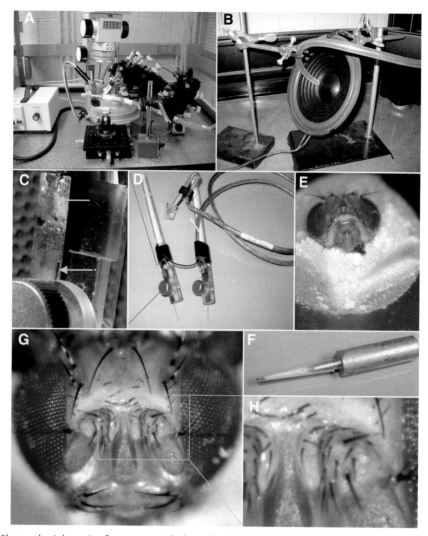

FIGURE 1. Electrophysiology rig, fly mount, and electrode placement. (A) Rig, showing arrangement of fly mount, manipulators, and tubing. (B) Speaker setup. Clamp stands hold the tubing and the speaker, in this case, on the floor under the vibration table. (C) Detail of Plexiglas electrode holder modification. Upper arrow indicates wire connector soldered to the custom-made brass plate. Lower arrow shows the shank of the tungsten rod electrode clamped tightly against the brass plate. (D) Electrode set attached to four-conductor wire connector. The two electrodes are connected to the black and red leads, the green lead (ground) is soldered to the cable shielding, and the second ground lead (white) is not connected. Red lines indicate the approximate view shown in detail in C. Black bands are heat-shrink tubing. (E) Fly head immobilized in the micropipette tip with modeling clay. (F) Pipette tip mount with fly inserted. (G) Fly's head in mount. Yellow box shows region magnified in *H*. (H) Detail of antenna. Yellow star indicates location of electrode placement.

Acoustic Stimulation

The antenna is a very sensitive sound receiver for particle velocity (or near-field sound) but is insensitive to pressure (or far-field sound). This presents some challenges for acoustic stimulation of the fly's antenna using a loudspeaker. If the loudspeaker, which uses a large electromagnet, is placed inside the Faraday cage near enough to the fly to be within near-field range, the actuating electrical signal will cause a large stimulation artifact in the recordings. To circumvent this problem, use the following strategy.

8. Place the speaker outside the Faraday cage. Deliver the near-field component of the sound to the fly using Tygon Tubing with one end placed near, but not touching, the speaker cone, and

the other end placed head-on as close to the fly as possible without touching the electrodes (Fig. 1B). Ensure that the fly's head is within the hemisphere circumscribed by the opening of the tubing so that it remains entirely within the near-field range.

> The time for sound to travel through the tube introduces an extra apparent latency. Try varying tube length, and then extrapolate to a tube of zero length to estimate the actual latency.

9. Electronically generate waveforms that correspond to acoustic signals as described by Eberl et al. (2000) and Tauber and Eberl (2003).

> These can either simulate courtship song stimuli or user-defined waveforms such as tone pulses with various frequencies.

10. To present computer-generated stimuli, route the digital signal from the computer to a digital-to-analog converter such as the instrNet i100B, and amplify the resulting analog signal with an audio amplifier en route to the loudspeaker. Calibration of particle velocity (near-field sound) is beyond the scope of this chapter, but see Crocker (1998) for a starting reference.

> An alternative stimulation method, not described here, is to electrostatically actuate the arista. This method applies the waveform voltage directly to an electrode placed in the air close to the arista. Voltage changes affect the level of electrostatic attraction/repulsion between the electrode and the arista and result in precise control of arista movement. Refer to Albert et al. (2007) for details and a comparison of electrostatic and acoustic actuations.

Data Acquisition

11. Route the leads from the two electrodes to a signal amplifier such as the DAM50 differential amplifier. Typical settings are differential mode (A–B), AC mode, 1000x gain, low-pass filter 10 Hz, and high-pass filter 10 kHz.

12. Connect the amplifier output to a digitizer such as the instruNet i100B. Set the sampling rate at a minimum of 2 kHz, although 10–15 kHz is typically used.

Data Analysis

13. Collect traces individually, or average them, and compute standard measurements such as amplitude (maximum and minimum).

> Data are typically plotted in histograms showing means with error bars that represent standard error of the mean or standard deviation. Sometimes it is useful to visualize the full dispersion of the recording data, for example, by plotting each data point in the bar with distinct symbols for means. Within an individual trace or an averaged trace, power spectrum analysis, for example, using the fast Fourier transform, can be performed to quantify the response energy across the frequency spectrum. The response to a sine-wave stimulus will typically have peaks at double and higher multiples of the stimulus frequency, probably because of populations of JO neurons responding to different phases of the stimulus. Several aspects of recording these extracellular signals can vary between preparations and over time. The greatest source of variance in response amplitude is probably the exact placement of the recording electrode tip relative to the antennal nerve, so a consistent mounting and electrode placement technique is important. Electrical properties of the rig (particularly when changing or sharpening electrodes), circadian effects, and humidity can all contribute to variation in the response amplitudes. Therefore, it is imperative always to alternate recordings from experimental and control flies. Furthermore, there are significant differences between wild-type laboratory strains, so it is also important, where possible, to use controls that arise as siblings of experimental genotypes.

> *See Troubleshooting.*

TROUBLESHOOTING

Problem (Step 13): There is 60-Hz noise in the trace.
Solution: Make sure all equipment is well connected to a common ground (beware of ground loop paths). If necessary, unplug light sources. Try making a shield to cover the front of the prepara-

tion; a piece of cardboard works well if covered in aluminum foil and alligator clipped to the ground. Check that both electrodes are actually penetrating the cuticle; if not, then the effective electrode resistance is very high, and it responds more easily to electrical noise.

Problem (Step 13): There is low response amplitude.

Solution: Check that the a3 segment and the arista are unobstructed and rotate freely; this can be confirmed by gently blowing on the fly while watching for antennal movement under the dissecting microscope. Check that the sound delivery tubing is positioned properly, both at the fly and in the speaker cone. Check for optimal electrode placement. Check for good contact between the tungsten electrode and the brass conductor plate. Periodic removal of oxidation products on the surface of the brass plate (using an emory cloth or a file) may help. Check the electrode tip because waxy cuticular components can accumulate and can serve as an electrical insulator. Clean using a fine paintbrush dipped in a 50% bleach solution, and then rinse with the brush dipped in water.

Problem (Step 13): There is a very flat baseline (no response).

Solution: Check for a short between recording and reference electrodes (if mounted on the same holder).

Problem (Step 13): There are large spontaneous spikes in the trace.

Solution: If the fly is less than 24-h old, it is possible to pick up activity from the ptilinal retractor muscles. These supercontracting muscles, which are of larval origin, degenerate within ~1-d posteclosion.

Problem (Step 13): There is a 2–3-Hz continuous oscillation in the baseline.

Solution: This occurs in young (<1-d old) flies because of activity of the pulsatile organ in the head.

Problem (Step 13): There are response echoes.

Solution: With pulse song stimuli, secondary responses often appear after each primary response to the pulse. These arise from acoustic echoes: Some sound energy reflects from the preparation, returns along the tubing, reflects from the speaker cone, and presents as an echo to the fly. The latency of these echoes depends on the length of the tubing. The amplitude of the echo can be damped somewhat by inserting a loose wad of cotton in the tubing end near the fly (this should be very loose to minimize damping effects on the primary acoustic signal).

ACKNOWLEDGMENTS

Research in the Eberl Laboratory was supported by National Institutes of Health (NIH) grant DC004848 and in the Kernan Laboratory by NIH grant DC002780.

REFERENCES

Albert JT, Nadrowski B, Göpfert MC. 2007. Mechanical signatures of transducer gating in the *Drosophila* ear. *Curr Biol* 17: 1000–1006.

Avidor-Reiss T, Maer AM, Koundakjian E, Polyanovsky A, Keil T, Subramaniam S, Zuker CS. 2004. Decoding cilia function: Defining specialized genes required for compartmentalized cilia biogenesis. *Cell* 117: 527–539.

Baker JD, Adhikarakunnathu S, Kernan MJ. 2004. Mechanosensory-defective, male-sterile *unc* mutants identify a novel basal body protein required for ciliogenesis in *Drosophila*. *Development* 131: 3411–3422.

Caldwell JC, Eberl DF. 2002. Towards a molecular understanding of *Drosophila* hearing. *J Neurobiol* 53: 172–189.

Chung YD, Zhu J, Han Y-G, Kernan MJ. 2001. *nompA* encodes a PNS-specific, ZP domain protein required to connect mechanosensory dendrites to sensory structures. *Neuron* 29: 415–428.

Crocker MJ, ed. 1998. *Handbook of acoustics.* Wiley, New York.

Dethier,VG. 1963. *The physiology of insect senses.* Methuen & Co, London.

Dubruille R, Laurençon A, Vandaele C, Shishido E, Coulon-Bublex M, Swoboda P, Couble P, Kernan M, Durand B. 2002. *Drosophila* regulatory factor X is necessary for ciliated sensory neuron differentiation. *Development* **129:** 5487–5498.

Eberl DF. 1999. Feeling the vibes: Chordotonal mechanisms in insect hearing. *Curr Opin Neurobiol* **9:** 389–393.

Eberl DF, Boekhoff-Falk G. 2007. Development of Johnston's organ in *Drosophila. Int J Dev Biol* **51:** 679–687.

Eberl DF, Duyk GM, Perrimon N. 1997. A genetic screen for mutations that disrupt an auditory response in *Drosophila melanogaster. Proc Natl Acad Sci* **94:** 14837–14842.

Eberl DF, Hardy RW, Kernan M. 2000. Genetically similar transduction mechanisms for touch and hearing in *Drosophila. J Neurosci* **20:** 5981–5988.

Gong Z, Son W, Chung YD, Kim J, Shin DW, McClung CA, Lee Y, Lee HW, Chang D-J, Kaang B-K, et al. 2004. Two interdependent TRPV channel subunits, inactive and Nanchung, mediate hearing in *Drosophila. J Neurosci* **24:** 9059–9066.

Göpfert MC, Robert D. 2003. Motion generation by *Drosophila* mechanosensory neurons. *Proc Natl Acad Sci* **100:** 5514–5519.

Göpfert MC, Humphris ADL, Albert JT, Robert D, Hendrich O. 2005. Power gain exhibited by motile mechanosensory neurons in *Drosophila* ears. *Proc Natl Acad Sci* **102:** 325–330.

Göpfert MC, Albert JT, Nadrowski A, Kamikouchi A. 2006. Specification of auditory sensitivity by *Drosophila* TRP channels. *Nat Neurosci* **9:** 999–1000.

Hall JC 1994. The mating of a fly. *Science* **264:** 1702–1714.

Han Y-G, Kwok BH, Kernan MJ. 2003. Intraflagellar transport is required in *Drosophila* to differentiate sensory cilia but not sperm. *Curr Biol* **13:** 1679–1686.

Jarman AP. 2002. Studies of mechanosensation using the fly. *Hum Mol Genet* **11:** 1215–1218.

Kamikouchi A, Shimada T, Ito K. 2006. Comprehensive classification of auditory sensory projections in the brain of the fruit fly *Drosophila melanogaster. J Comp Neurol* **499:** 317–356.

Kamikouchi A, Inagaki HK, Effertz T, Hendrich O, Fiala A, Göpfert MC, Ito K. 2009. The neural basis of *Drosophila* gravity-sensing and hearing. *Nature* **458:** 165–171.

Kernan M, Cowan D, Zuker C. 1994. Genetic dissection of mechanosensory transduction: Mechanoreception-defective mutations of *Drosophila. Neuron* **12:** 1195–1206.

Kernan MJ. 2007. Mechanotransduction and auditory transduction in *Drosophila. Pflügers Arch* **454:** 703–720.

Kim J, Chung YD, Park D-y, Choi S, Shin DW, Soh H, Lee HW, Son W, Yim J, Park C-S, et al. 2003. A TRPV family ion channel required for hearing in *Drosophila. Nature* **424:** 81–84.

Kyriacou CP, Hall JC. 1982. The function of courtship song rhythms in *Drosophila. Anim Behav* **30:** 794–801.

Kyriacou CP, Hall JC. 1984. Learning and memory mutations impair acoustic priming of mating behaviour in *Drosophila. Nature* **308:** 62–65.

Kyriacou CP, Hall JC. 1986. Interspecific genetic control of courtship song production and reception in *Drosophila. Science* **232:** 494–497.

Laurençon A, Dubruille R, Efimenko E, Grenier G, Bissett R, Cortier E, Rolland V, Swoboda P, Durand B. 2007. Identification of novel regulatory factor X (RFX) target genes by comparative genomics in *Drosophila* species. *Genome Biol* **8:** R195.

Lee E, Sivan-Loukianova E, Eberl DF, Kernan MJ. 2008. An IFT-A protein is required to delimit functionally distinct zones in mechanosensory cilia. *Curr Biol* **18:** 1899–1906.

Nadrowski B, Albert JT, Göpfert MC 2008. Tranducer-based force generation explains active process in *Drosophila* hearing. *Curr Biol* **18:** 1365–1372.

Sarpal R, Todi SV, Sivan-Loukianova E, Shirolikar S, Subramanyan N, Raff EC, Erickson JW, Ray K, Eberl DF. 2003. The *Drosophila* kinesin associated protein (DmKAP) interacts with the Kinesin II motor subunit Klp64D to assemble chordotonal organ sensory cilia but not sperm tails. *Curr Biol* **13:** 1687–1696.

Sharma Y. 2004. "The *Drosophila* deafness gene *beethoven* encodes the dynein heavy chain 1b isoform required for intraflagellar transport." PhD thesis, University of Iowa, Iowa City.

Sun Y, Liu L, Ben-Shahar Y, Jacobs JS, Eberl DF, Welsh MJ. 2009. TRPA channels distinguish gravity sensing from hearing in Johnston's organ. *Proc Natl Acad Sci* **106:** 13606–13611.

Tauber E, Eberl DF. 2001. Song production in auditory mutants of *Drosophila*: The role of sensory feedback. *J Comp Physiol A* **187:** 341–348.

Tauber E, Eberl DF. 2002. The effect of male competition on the courtship song of *Drosophila melanogaster. J Insect Behav* **15:** 109–120.

Tauber E, Eberl DF. 2003. Acoustic communication in *Drosophila. Behav Proc* **64:** 197–210.

Todi SV, Sharma Y, Eberl DF. 2004. Anatomical and molecular design of the *Drosophila* antenna as a flagellar auditory organ. *Microsc Res Tech* **63:** 388–399.

Todi SV, Franke JD, Kiehart DP, Eberl DF. 2005. Myosin VIIA defects, which underlie the Usher 1B syndrome in humans, lead to deafness in *Drosophila. Curr Biol* **15:** 862–868.

Todi SV, Sivan-Loukianova E, Jacobs JS, Kiehart DP, Eberl DF. 2008. Myosin VIIA, important for human auditory function, is necessary for *Drosophila* auditory organ development. *PLoS ONE* **3:** e2115. doi: 10.1371/journal. pone.0002115.

von Schilcher F. 1976. The function of pulse song and sine song in the courtship of *Drosophila melanogaster. Anim Behav* **24:** 622–625.

Wheeler DA, Kyriacou CP, Greenacre ML, Yu Q, Rutila JE, Rosbash M, Hall JC. 1991. Molecular transfer of a species-specific behavior from *Drosophila simulans* to *Drosophila melanogaster. Science* **251:** 1082–1085.

Yorozu S, Wong A, Fischer BJ, Dankert H, Kernan MJ, Kamikouchi A, Ito K, Anderson DJ. 2009. Distinct sensory representations of wind and near-field sound in the *Drosophila* brain. *Nature* **458:** 201–205.

16 | Chemosensory Coding in Single Sensilla

Richard Benton[1] and Anupama Dahanukar[2]

[1]*Center for Integrative Genomics, Faculty of Biology and Medicine, University of Lausanne, CH-1015, Lausanne, Switzerland;* [2]*Department of Entomology, University of California, Riverside, California 92521*

ABSTRACT

The chemical senses, smell and taste, detect and discriminate an enormous diversity of environmental stimuli and provide fascinating but challenging models to investigate how sensory cues are represented in the brain. Important stimulus coding events occur in peripheral sensory neurons, which express specific combinations of chemosensory receptors with defined ligand-response profiles. These receptors convert ligand recognition into spatial and temporal patterns of neural activity that are transmitted to and interpreted in central brain regions. *Drosophila* provides an attractive model to study chemosensory coding because it possesses relatively simple peripheral olfactory and gustatory systems that display many organizational parallels to those of vertebrates. Moreover, virtually all of the peripheral chemosensory neurons are easily accessible for physiological analysis, as they are exposed on the surface of sensory organs in specialized sensory hairs called sensilla. In this chapter, we briefly review anatomical, molecular, and physiological properties of adult *Drosophila* olfactory and gustatory systems and describe protocols for electrophysiological recordings of ligand-evoked activity from different types of chemosensory sensilla.

INTRODUCTION

Drosophila, like most animals, relies significantly on chemosensory cues to drive important behaviors. Environmental chemicals can be detected either as airborne volatiles by the olfactory system, which often underlies food-seeking behaviors, or avoidance of distant dangers. Detection can also occur through direct contact with the source by the gustatory system, which underlies assessment of food quality, mating partner compatibility, or oviposition site. Unlike vertebrates, which use sensory neurons for odor detection and secondary (nonneuronal) sensory cells for gustatory detection, peripheral chemical detection in both olfactory and gustatory systems of insects is mediated by primary sensory neurons, which carry information from specialized structures on the body surface to specific regions of the brain (Wyatt 2003; Vosshall and Stocker 2007). Despite this distinction, and other significant molecular and developmental differences (Benton 2006), many important parallels do exist between the organization and coding mechanisms of insect and vertebrate olfactory and gustatory systems (Ache and Young 2005). These similarities suggest a conserved or convergent logic of how diverse types of chemical cues can be detected and discriminated. Thus *Drosophila*, with its sophisticated genetic tools and relatively straightforward behavioral assays—described elsewhere in this book—has become a premier model system to investigate the molecular and neural basis of chemosensory coding (Benton 2007; Vosshall and Stocker 2007).

The initial steps in chemosensory recognition occur at the periphery, where chemical ligands bind to receptor proteins exposed on the surface of chemosensory neuron dendrites to induce physiological responses in the form of trains of action potentials or "spikes." The location of most chemosensory dendrites in thin, cuticular hairs, or sensilla, which decorate the surface of olfactory and gustatory organs, makes electrophysiological analysis of these responses reasonably accessible. In this chapter, we provide an overview of the anatomical, molecular, and physiological properties of the peripheral olfactory and gustatory systems, and a detailed practical guide to the methods permitting recordings from individual chemosensory sensilla. It should be emphasized that these electrophysiological techniques were established half a century ago by pioneers of the insect chemosensory field in several other insect models (Hodgson et al. 1955; Morita et al. 1957; Schneider 1957; Boeckh 1962). In recent years, improvements in microscopy and instrumentation for electrode manipulation have opened up the much smaller *Drosophila* system to these techniques, powerfully complementing many years of molecular genetic studies. As with most electrophysiological methods, there is probably no substitute for learning this technique directly from a laboratory in which it is already established. Nevertheless, with some practice, recordings from single chemosensory sensilla are fairly straightforward, and we hope that this guide will encourage and support new researchers in applying these approaches to their investigations.

CHEMOSENSORY ORGAN ANATOMY

Olfactory System

Adult *Drosophila* have two distinct olfactory organs, the third segment of the antenna (referred hereafter simply as the "antenna") and the maxillary palp. Both are located on the anterior of the head in bilaterally symmetric pairs (Fig. 1A). These organs are covered in hundreds of porous hairs, or sensilla, which are innervated by the dendrites of olfactory sensory neurons (OSNs) (Stocker 1994; Shanbhag et al. 1999, 2000; Stocker 2001). There are approximately 500 antennal olfactory sensilla, which are classified into several morphological types—club-shaped, large and small basiconic sensilla (approximately 240–280), spine-shaped trichoid sensilla (approximately 140), and small, coned-shaped coeloconic sensilla (approximately 80)—and house between one and four OSNs (Fig. 1B). Maxillary palps have only about 60 basiconic sensilla, each housing two OSNs. The palps also bear large, easily distinguishable mechanosensory bristles (approximately 20), and the antenna has two structures of unknown sensory function, a feather-like projection called the arista (Fig. 1A) and an internal multichambered pocket called the sacculus (Foelix et al. 1989; Shanbhag et al. 1995). In addition, the surface of both organs bears numerous slender, cuticular, noninnervated projections called trichomes (Fig. 1B).

OSNs are bipolar neurons, extending a single axon to the antennal lobe (the primary olfactory center in the brain) and a single dendrite into the hair. The dendrite bears a specialized ciliated ending, which can be highly branched in certain types of sensilla (Shanbhag et al. 2000; Stocker 2001). These cilia are bathed in a lymph fluid rich in proteins and potassium ions that fills the sensillar hair. Surrounding the neurons are several nonneuronal support cells that secrete the proteinaceous components of the lymph, and which physically and electrically isolate OSNs between different sensilla. Thus, each sensillum can be regarded as an isolated microenvironment for odor detection, with distinctive molecular and physiological properties (Benton 2007), as described below.

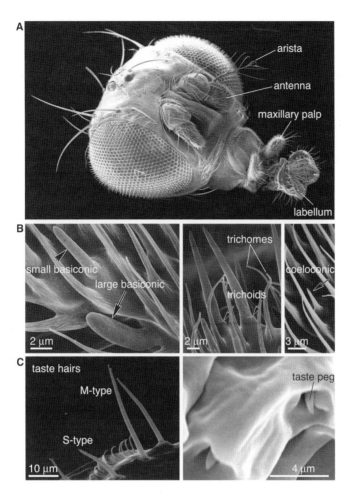

FIGURE 1. Chemosensory organs and sensilla. (*A*) Scanning electron micrograph of the *Drosophila* head showing the major chemosensory organs. (*B*) Scanning electron micrographs of the major types of antennal olfactory sensilla: large (arrow) and small (arrowhead) basiconic (*left*), trichoid (*center*), and coeloconic (arrow, *right*). Trichomes (*center*) are noninnervated hairs of unknown function. (Adapted, with permission, from Riesgo-Escovar et al. 1997.) (*C*) Scanning electron micrographs of labellar taste hairs (*left*) and taste pegs (*right*).

Gustatory System

Contact chemosensory sensilla are distributed in various organs in both external and internal regions of the body (Stocker 1994). A majority of the approximately 350 sensilla are present in the labellum at the distal end of the proboscis (Fig. 1A) and the legs (Nayak and Singh 1983; Shanbhag et al. 2001b). In addition, taste sensilla are located along the anterior wing margins and in internal pharyngeal organs: the labral sense organ, the ventral cibarial sense organ, and the dorsal cibarial sense organ. Sensillar distributions display some sexual dimorphisms: Males have more sensilla on their forelegs, whereas females have a larger number of sensilla in mouthparts and also have a few in the ovipositor.

There are two types of external taste sensilla: taste hairs, which are on the labellum, the legs, and the wings, and taste pegs, which are present in the pseudotracheal folds of the labellum (Fig. 1C). Based on their size, morphology, and position, taste hairs on the labellum have been classified in the literature into either three (L, S, I) or five (L, M, S, I, P) subtypes (Ray et al. 1993; Shanbhag et al. 2001b), and sensilla on the legs have been numbered individually (Meunier et al. 2000). All these subtypes, however, share the same basic anatomical features.

Taste sensilla are innervated by bipolar, primary sensory neurons whose axons project to the thoracic–abdominal ganglia or the subesophageal ganglion in the brain (Stocker 1994). Within each sensillum, several support cells envelop the cell bodies of taste neurons. Taste hairs are long (~15–45 μm) and are characterized by the presence of two independent lumens within a cuticular shaft that bears a single pore at the tip (Falk et al. 1976; Nayak and Singh 1983; Shanbhag et al. 2001b). One lumen, which is present toward the concave side of the hair, contains the unbranched dendrites of two to four chemosensory neurons. The second lumen, which appears crescent-shaped in cross sections, is filled with lymph. Taste pegs are ~3–4 μm in length with a pore or slit at the tip, and a single lumen that contains the dendrite of one chemosensory neuron (Falk et al. 1976). Both types of sensilla are also innervated by a single mechanosensory neuron whose dendrite terminates at the base of the shaft.

Taste sensilla in internal pharyngeal organs do not have a shaft and are innervated by varying numbers of chemosensory neurons whose dendrites have access to taste stimuli via a single pore (Stocker 1994). Unlike external taste sensilla, pharyngeal sensilla are not always accompanied by a mechanosensory neuron. Little is known about the function of these sensilla, in part because they are not as easily accessible for electrophysiological analysis.

MOLECULAR BIOLOGY OF CHEMOSENSATION

Olfactory System

On passing through sensillar pores, odor molecules are likely to associate first with members of the odorant binding protein (OBP) repertoire, a family of small, divergent proteins secreted at high concentration into the lymph by support cells. Stereotyped expression of different combinations of OBPs in different sensillar classes has suggested that they contribute to olfactory specificity (Shanbhag et al. 2001a), but very little is known about the in vivo function of most OBPs. Nevertheless, genetic and structural analysis of the OBP LUSH in recognition of the volatile pheromone *cis*-vaccenyl acetate has implied a direct role for this OBP in odor recognition and neuronal stimulation (Xu et al. 2005; Laughlin et al. 2008). OBPs are also expressed in gustatory sensilla (Galindo and Smith 2001; Shanbhag et al. 2001a), indicating that this family is likely to have roles in both sensory systems.

Odor recognition in OSNs is accomplished by one of three classes of membrane chemosensory receptor repertoires, the odorant receptors (ORs), the gustatory receptors (GRs), and the ionotropic receptors (IRs). ORs and GRs are distantly related, novel, seven-transmembrane domain proteins (Clyne et al. 1999, 2000; Gao and Chess 1999; Vosshall et al. 1999, 2000; Dunipace et al. 2001; Scott et al. 2001; Robertson et al. 2003). Although originally presumed to be G-protein-coupled receptors, ORs adopt a distinct membrane topology and display functional properties of ligand-gated, nonselective, cation channels in heterologous expression systems (Benton et al. 2006; Lundin et al. 2007; Sato et al. 2008; Smart et al. 2008; Wicher et al. 2008). IRs are structurally related to ionotropic glutamate receptors, a widely conserved family of ligand-gated ion channels, but have highly divergent ligand-binding domains (Benton et al. 2009). The use of presumed ligand-gated ion channels as primary sensory receptors obviates the need for complex downstream signaling pathways, although some in vivo analyses have suggested the involvement of various second messenger cascades in mediating or regulating odor-evoked signaling (Kain et al. 2008; Wicher et al. 2008). It is conceivable, however, that odor-evoked current flow through OR and IR ion channels produces a transmembrane potential difference that is sufficient to trigger action potentials in these neurons. The identities of the channels involved in spike generation in OSNs, and where they localize, are unknown.

ORs are expressed in all maxillary palp OSNs, in antennal basiconic and trichoid OSNs, as well as in one population of coeloconic OSNs (Vosshall et al. 2000; Couto et al. 2005; Fishilevich and Vosshall 2005; Yao et al. 2005). OSNs express a single ligand-binding OR (or occasionally more) (Couto et al. 2005; Fishilevich and Vosshall 2005; Goldman et al. 2005)—similar to vertebrate olfactory systems (Mombaerts 2004)—together with a broadly expressed OR coreceptor, OR83b (Larsson et al. 2004; Benton et al. 2006). Most GRs are expressed in taste neurons (see below), but two receptors, GR21a and GR63a, are coexpressed in a population of basiconic OSNs involved in carbon dioxide detection (Jones et al. 2007; Kwon et al. 2007). IRs are expressed in the complementary population of coeloconic OSNs, as well as in sacculus and aristal neurons (Benton et al. 2009). Coeloconic neurons express one to three IRs, with some overlap between different neurons, as well as one or both of the broadly expressed IR8a and IR25a. Some IRs are also expressed in gustatory organs, suggesting that this family may play roles in both types of chemosensory detection.

The spatial distributions of OR, GR, and IR neurons appear to be highly stereotyped across animals, forming a molecular map of OSN identity that, together with functional characterization described below, has defined three types of maxillary palp basiconic sensilla (pb1–pb3), 10 types of antennal basiconic sensilla (ab1–ab10), four types of trichoid sensilla (at1–at4), and four types of coeloconic sensilla (ac1–ac4) (de Bruyne et al. 1999, 2001; Couto et al. 2005; Yao et al. 2005; van der Goes van Naters and Carlson 2007; Benton et al. 2009). Each sensillum type has a characteristic, although overlapping, spatial distribution on the surface of the olfactory organs. Knowledge of these chemosensory receptors has also offered the possibility of generating important genetic tools by using the corresponding receptor gene promoters to drive expression of cellular reporters to visualize, silence, kill, or artificially activate these neurons using the GAL4/UAS system (Brand and Perrimon 1993; Dobritsa et al. 2003; Suh et al. 2007).

Gustatory System

With a few exceptions (Scott et al. 2001; Thorne and Amrein 2008), GRs are expressed in chemosensory neurons in various gustatory organs. Individual GRs have diverse expression patterns: some, such as GR5a and GR66a, are broadly expressed in nonoverlapping populations of taste neurons, whereas others are confined to small numbers of neurons (Thorne et al. 2004; Wang et al. 2004). In contrast to most OSNs, individual taste neurons express a number of GRs (Thorne et al. 2004; Wang et al. 2004; Dahanukar et al. 2007; Jiao et al. 2007; Slone et al. 2007). However, much like chemosensory receptor expression in OSNs, the spatial distribution of GR expression in taste neurons appears highly stereotyped: GR5a neurons also express one or more GR genes that are related to GR5a (Dahanukar et al. 2007), and GR66a neurons may be categorized into molecular classes that express overlapping but distinct subsets of other GRs (Thorne et al. 2004; Wang et al. 2004).

The GR family does not have an OR83b-like counterpart, and evidence suggests that GRs may heteromerize in various combinations (Moon et al. 2006, 2009; Jones et al. 2007; Jiao et al. 2008; Lee et al. 2009). Misexpression of a combination of GRs from one class of taste neurons in another has not yet been shown to be sufficient to confer a novel response (Jiao et al. 2008), suggesting the need for additional cofactors that may be segregated in different classes of taste neurons.

Based on their sequence similarity to each other, GRs are likely to have the same topology as ORs, and perhaps also function as ligand-gated ion channels. However, this has not been shown directly. Genetic evidence suggests a role, at least in part, for G-protein-coupled signaling mechanisms in some taste neurons. Mutations in either $G\alpha s$ or $G\gamma 1$ reduce physiological and behavioral responses to sugars (Ishimoto et al. 2005; Ueno et al. 2006). In both cases, however, sugar responses are only partially reduced, supporting the idea that these are not the only signaling mechanisms.

ELECTROPHYSIOLOGICAL RECORDINGS FROM CHEMOSENSORY SENSILLA

Olfactory System

Stimulus-evoked activity in insect OSNs was first analyzed by the electroantennogram (Schneider 1957; Alcorta 1991), which measures summated responses of many different sensilla. Single-sensillum recordings (SSRs) have largely superseded this technique because of their greater sensitivity and specificity. First developed in moth and beetle olfactory systems (Schneider 1957; Boeckh 1962), the SSR technique involves the insertion of a fine recording electrode into the shaft of a single olfactory hair and a reference (or indifferent) electrode into the head capsule (either through the eye or the proboscis), permitting extracellular recording of voltage differences between the sensillar lymph and the hemolymph (Stengl et al. 1992; Clyne et al. 1997; Ignell and Hansson 2005; de Bruyne 2006). These voltage differences are believed to be because of current flow in the lymph generated by receptor potentials across the dendritic membranes and are visible as spikes of 0.1 to a few millivolt amplitude. In the absence of stimuli, OSNs show spontaneous firing rates (~1–30 spikes/sec depending on the specific neuron), which may reflect basal current leak of chemosensory receptors (Sato et al. 2008).

Two types of electrodes have been used: tungsten and glass. Tungsten electrodes can be etched to a very fine diameter while maintaining strength at the tip (Hubel 1957), which is valuable for penetration of small olfactory hairs and will be the focus of the olfactory recording protocols described here. Glass electrodes were first used in tip recordings from individual olfactory sensilla (Kaissling 1974) in which the electrode is placed over the cut end of a hair, similar to the taste sensilla recording methods described below. Sharpened glass electrodes are now more commonly used to penetrate the walls of olfactory sensilla (Dobritsa et al. 2003), offering the ability to perfuse drugs, dyes, or even proteins into the lymph via this micropipette (Xu et al. 2005), and also permitting detection of slower "sensillar potentials" (also known as "receptor potentials"), thought to be an extracellular derivative of the membrane potentials of OSNs as well as the ion pumping activities of their associated support cells (Kaissling 1986; de Bruyne 2006).

Recording and reference electrodes are connected to a high-impedance, high-gain, differential amplifier (1000x), which is essential to acquire the signals of low-amplitude spikes over the noise level of the amplifier. Signals are also filtered (e.g., low pass 10 kHz, high pass 0.1 kHz) to remove both high- and low-frequency background outside the spike bandwidth and digitally converted for analysis on a computer.

Gustatory System

The development of a technique to measure neural activity of taste neurons in a single sensillum (Hodgson et al. 1955; Hodgson and Roeder 1956; Morita et al. 1957) pioneered the way to study the response specificities of each of the taste neurons in a sensillum, as well as to examine features such as the response threshold, temporal dynamics, and short-term and long-term adaptation to the stimulus. This technique, called the "tip-recording" method, involves the use of a glass micropipette, ~10–20 μm in diameter at the tip, to present a stimulus to the tip of the taste hair. The stimulus pipette also serves as a recording electrode; it is filled with a solution that contains the taste stimulus, as well as an electrolyte that functions as a salt bridge to deliver the neural impulses to a silver wire leading to an amplifier. The reference electrode filled with Ringer's solution is inserted into the proximal region of the proboscis. When a taste neuron is excited to generate impulses, fluctuations in potential between the two electrodes are recorded and amplified through a high input impedance preamplifier circuit. An important feature of the preamplifier circuit that was developed later is that it compensates for the large offset potential between the sensillum and the reference electrode (typically 100 mV) (Marion-Poll and van der Pers 1996). The signal is then filtered (e.g., low pass 3 kHz, high pass 0.1 kHz), further amplified (100x–1000x), and transferred to a computer (sampling rate typically ≥10 kHz) that has the appropriate software for storing and analyzing the data. The amplitudes of impulses recorded from taste neurons are on the order of 0.5–5 mV.

An obvious drawback of the tip-recording method is that the impulses of four taste neurons are recorded simultaneously. If trace amounts of either sodium chloride or potassium chloride are used as recording electrolytes along with a taste compound, the solution also elicits action potentials from the salt and/or water neurons (Dethier 1976). This problem was circumvented by the use of tricholine citrate as the electrolyte (Wieczorek and Wolff 1989), which does not elicit a significant response from any of the taste neurons. Tricholine citrate also stabilizes the sizes of the spike amplitudes of each of the taste neurons (Wieczorek and Wolff 1989), which otherwise increase with an increase in the rate of firing (Fujishiro et al. 1984). The use of this electrolyte, therefore, facilitates unambiguous assignment of impulses to individual taste neurons.

A second disadvantage of the tip-recording method is the need to include salt in the recording electrode. As a result, compounds that are insoluble in water cannot be tested in this fashion. In a modified method called "side-wall recording" (Morita 1959), the recording electrode is brought into contact with the dendrites via a crack in the cuticle of the hair. More recently, recordings from taste sensilla have been obtained by insertion of a sharpened tungsten electrode in the base of the sensillum (Lacaille et al. 2007), which is much like the technique for recording from olfactory sensilla. These methods leave the tip of the sensillum free for application of many different physicochemical categories of stimuli.

CHEMOSENSORY CODING PRINCIPLES

Olfactory System

The *Drosophila* olfactory system has been subject to intensive electrophysiological analysis by single sensilla recordings. The responses of nearly all sensilla types on both antennae and maxillary palps have been examined by using a large panel of odorants (de Bruyne et al. 1999, 2001; Yao et al. 2005; van der Goes van Naters and Carlson 2007). These analyses have revealed that individual odor ligands can stimulate (or inhibit) multiple different classes of OSNs (often at very different thresholds) and that tuning properties of individual classes of OSNs range along a continuum from the very narrowly tuned to the broadly tuned. These observations have suggested a combinatorial coding theory for odor discrimination in which odor identity and intensity are represented by specific combinations of activated sensory neurons. Such a theory, which has also been proposed in vertebrate olfactory systems (Malnic et al. 1999), provides an elegant explanation for how *Drosophila* might be able to detect and discriminate a greater number of odor ligands than there are receptors. Consistent with the stereotyped expression map of chemosensory receptor genes, responses of individual OSNs are highly reproducible and the identity of individual sensillum types, guided by their morphology and their topological position, can be easily determined using a small number of diagnostic odor stimuli.

Although there is some overlap between the ligand specificity of different morphological classes of sensilla, some functional subdivisions are apparent. For example, many antennal basiconic sensilla neurons are broadly tuned to various fruit-derived odors (de Bruyne et al. 2001), whereas trichoid sensilla neurons, like other insect trichoid OSNs, may be tuned specifically to fly-derived (pheromone) ligands, most of which are of unknown molecular identity (van der Goes van Naters and Carlson 2007). Coeloconic OSNs, which express IRs, appear to be much more narrowly tuned than basiconic sensilla, but some of these neurons have not yet been associated with strong ligands (Yao et al. 2005).

In addition to the analysis of receptor responses in endogenous sensilla, much has been learned about the properties of ORs from in vivo receptor misexpression experiments, in which individual ORs are expressed in the "empty neuron" system, a mutant for OR22a/b that lacks the endogenous receptors (Dobritsa et al. 2003; Hallem et al. 2004; Hallem and Carlson 2006). These experiments have allowed electrophysiological matching of ORs to specific sensillum types, and provided strong evidence for the central role of ORs as determinants of the specificity and sensitivity of their corresponding neurons, at least in basiconic sensilla of the antenna and maxillary palp.

In addition to their different ligand specificities, OSN spontaneous firing rates and temporal dynamic properties of odor-evoked responses (e.g., the increase and the termination rate) are often quite distinctive (de Bruyne et al. 1999; Yao et al. 2005). The functional significance of these properties is as yet unknown, but OR misexpression experiments have shown that these properties are also dictated by the specific receptor (Hallem et al. 2004). Finally, spike amplitudes are also characteristic of individual OSNs, although this property varies slightly depending on insertion position of the electrode along the length of the hair shaft and often changes significantly on odor-evoked responses (Clyne et al. 1997). Here, it is the sensory neuron and not the receptor that specifies this property, which may reflect the precise morphology or ion channel composition of the neuronal membranes (Hallem et al. 2004).

Gustatory System

Electrophysiological recordings performed first on larger flies (Hodgson et al. 1955; Hodgson and Roeder 1956; Wolbarsht 1957) and subsequently on *Drosophila* (Fujishiro et al. 1984), established that each of the four taste neurons is tuned to a different class of tastants. These observations, along with more recent studies (Liscia and Solari 2000; Meunier et al. 2003b), established that within each sensillum one neuron responds to sugars (sugar neuron), one responds to salts (salt neuron), one responds to alkaloids and other aversive compounds (bitter neuron), and one responds to pure water and is inhibited by solutions of high osmolarity (water neuron). In sensilla that are innervated by only two taste neurons, one is responsive to sugars as well as low concentrations of salts, whereas the other is responsive to bitter compounds and high concentrations of salts (Hiroi et al. 2004).

GR5a neurons respond to a variety of sugars and correspond to the "sugar" neuron in each sensillum (Thorne et al. 2004; Wang et al. 2004; Marella et al. 2006; Dahanukar et al. 2007). Conversely, GR66a neurons respond to caffeine and other aversive compounds and correspond to the "bitter" neurons (Thorne et al. 2004; Wang et al. 2004; Marella et al. 2006). These two neuronal populations mediate distinct innate behaviors: appetitive for GR5a neurons and aversive for GR66a neurons. This organization is fundamentally similar to that observed in the mammalian gustatory system, in which receptors for sugars and bitter compounds are expressed in dedicated populations of taste cells (Chandrashekar et al. 2006).

The physiological responses of individual sensilla have not been documented as extensively as those of olfactory sensilla, but a few sensilla on the labellum and the legs have been tested with a variety of sugars (Wieczorek and Wolff 1989; Dahanukar et al. 2007) or some bitter compounds (Meunier et al. 2003b). Initial observations suggest that there is some functional heterogeneity among sensilla (Meunier et al. 2003b; Hiroi et al. 2004), and further electrophysiological experiments will likely reveal additional functional classes of taste neurons.

Sugar neurons respond to a number of monosaccharides, disaccharides, and oligosaccharides (Wieczorek and Wolff 1989; Marella et al. 2006; Dahanukar et al. 2007). The strength of the response depends on the identity as well as the concentration of the stimulus, but the temporal response pattern is fairly consistent irrespective of these factors (Dethier 1976; Fujishiro et al. 1984; Dahanukar et al. 2007). In contrast, individual bitter stimuli have been reported to elicit unique patterns of temporal activity from bitter neurons (Meunier et al. 2003b). Typically, responses show some latency that varies with the identity of the stimulus. Such variations in bitter responses are not entirely surprising, given that different classes of bitter neurons express distinct repertoires of GRs.

Recording from Olfactory Sensilla

This protocol describes the basics of setting up the electrophysiology rig and stimulus delivery device, sample preparation, and how to perform and analyze recordings of odor-evoked activity from *Drosophila* olfactory sensilla.

MATERIALS

CAUTION: See Appendix for proper handling of materials marked with <!>.

A comprehensive list of required materials and equipment is provided below. For certain items, specific companies and catalog numbers are proposed but equivalents for almost all items can be obtained from several other suppliers.

Reagents

Drosophila, newly eclosed flies of the desired genotype
Odors (typically use highest purity chemicals from Sigma-Aldrich)
Potassium hydroxide <!>, 1 M solution
> Sodium nitrite or potassium nitrite (each at 10%) can be used instead of potassium hydroxide (see Step 1).

Solvents (paraffin oil or water)
Wax (e.g., HB 196, Dentalfachhandel Joachim Zill catalog)

Equipment

Drosophila aspirator parts:
  Muslin, or similar material (~4-cm^2)
  Pipette tips, 1000-µL, two
  Plastic tubing (~1-m) (e.g., 500PAL, Python Products)
Drosophila preparation materials:
  Double-stick tape
  Glass capillary (used only as a physical support) (e.g., 27-32-0-075, Frederick Haber and Co.)
  Glass coverslip and glass microscope slide
  Needle puller (e.g., Flaming/Brown puller, Sutter Instrument Company)
  Pipette tips, 20-µL and 200-µL
  Razor blade
  Stereomicroscope with illumination (e.g., SMZ645, Nikon)
Electrodes and electrode sharpener, assembled using the following:
  Basic stereomicroscope with illumination (e.g., MTX 3c, Müller Opronic)
  Crocodile clips, two
  Electric cables, two
  Electrode holder (Syntech)
  Manual manipulator (e.g., YOU-1, Narishige)
  Petri dish
  Regulatable power supply (e.g., PS-1152A, Conrad Electronic)
  Retort stand and clamp
  Stainless steel shaft support for electrode holder (Syntech)
  Syringe (20-mL) and syringe needle (>20 mm in length)

Tungsten wire or rods (diameter 0.1 mm) (e.g., M210, Micro Probe Inc.)

Electrophysiological recording rig, assembled using the following:

Antivibration table (1 x 1-m is sufficient) (e.g., TMC63-534, Technical Manufacturing Corporation)

External audio output

Faraday cage (e.g., TMC81-333-03, Technical Manufacturing Corporation)

Grounding wires

Manual manipulator on magnetic stand for reference electrode (e.g., YOU-1, Narishige)

Motorized (or hydraulic) manipulator on magnetic stand for recording electrode (e.g., PCS-6400 with PCS-500-PP11, Burleigh micromanipulator systems)

PC workstation

Signal Acquisition System (e.g., IDAC-4, Syntech)

Software (e.g., AutoSpike, Syntech)

Universal single ended electrode probe (e.g., Syntech)

Upright microscope equipped with 10x and 100x air objectives with long working distance (e.g., Nikon Eclipse FN1 with CFI[LU]plan Epi SLWD 100x/0.70 -WD 6.5 mm)

Odor cartridges, assembled using the following:

Brown glass storage bottles (various sizes), preferably with Teflon caps

Forceps

Pipettes

Small filter paper disks (cut from Whatmann with a clean hole punch)

Solvent dispenser (optional, but highly convenient; e.g., Dispensette III BRAND-4700320, Milian USA)

Tuberculin syringes, 1-mL

Odor delivery device, assembled using the following:

Conical flask with side arm and rubber sealed top

Dental wax

Flow meters (e.g., Thermo Fischer Scientific)

Hand power tool with cutting blade and drill attachments

Magnetic Holding Stand (e.g., M1, World Precision Instruments)

Plastic connectors (kits containing various sizes are available e.g., W2 64-1565, Harvard Apparatus)

Plastic pipette, 10-mL

Plastic tubing (several meters) (e.g., 500PAL, Python Products)

Stimulus Controller (e.g., CS-55, Syntech)

METHOD

Assembly and Use of the Tungsten Electrode Sharpener

See Figure 2.

1. Prepare the syringe.

    i. Trim the end of the syringe tip to leave the opening flush with the body of the syringe.

    ii. Fill with potassium hydroxide solution and clamp with the retort stand under the microscope so that the opening is visible in the field of view (FOV). Place a Petri dish underneath to catch any drips.

    10% sodium nitrite or 10% potassium nitrite can be used instead of potassium hydroxide.

    iii. Push the end of the syringe needle through the top of the syringe wall to contact the liquid but leave the base of the needle projecting from the syringe.

    iv. Attach an electric cable to the base using a crocodile clip, and connect this cable to the power supply anode.

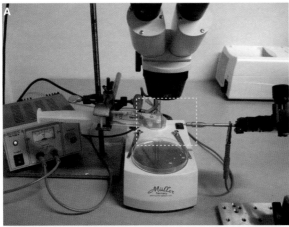

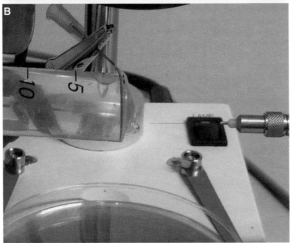

FIGURE 2. A tungsten electrode etcher. (*A*) Overview of the etcher, showing the syringe containing the electrolyte potassium hydroxide on the *left* and the electrode holder on the *right*. (*B*) Close-up image of the boxed region in *A*.

2. On the other side of the stereomicroscope, fix the electrode shaft support in the manual manipulator so that it is in-line with the syringe opening. Attach an electric cable to this support with a crocodile clip and connect this cable to the cathode.

3. Fix the tungsten wire (or rod) into the electrode holder with no more than 4 cm projecting from the end (longer electrodes will tend to vibrate in the airstream on the rig). Insert the electrode holder into the shaft support and adjust its position so that the tip of the electrode is exactly in-line with the syringe opening and is visible in the FOV.

4. Set the power supply to ~6 V. Press the syringe plunger gently so that a small drop of potassium hydroxide appears. Viewing through the microscope, repeatedly dip the end of the tungsten wire into this drop using the manipulator. The flow of current through the tungsten wire in the potassium hydroxide solution will result in removal of metal by electrolysis. To produce a very sharp point (~1 μm diameter), it is best to perform the etching in a stepwise fashion; that is, first insert a longer length of tungsten into the syringe for up to 1 min to thin the wire along this length, then insert half this length to further thin this section, and continue until the tip becomes very narrow. The voltage may be lowered if the electrolysis proceeds too quickly.

> Perfecting electrode sharpening takes practice. Although you can observe the process at a gross level under the stereomicroscope, it is only really possible to judge whether an electrode is sufficiently sharp under the 100x objective on the rig. In general, spear-shaped electrodes, which come to a short, fine point, are sturdier than long tapers. Once sharpened satisfactorily, take great care not to damage the tip when transferring the electrode to the probe or moving it about on the rig as it will bend extremely easily.

Preparation of Odor Cartridges

5. Prepare dilutions of odors in appropriate solvent (typically water or paraffin oil) in small brown glass bottles (many chemicals are light sensitive) with Teflon caps (many odors react with conventional plastic caps). A typical dilution used is 1% v/v, but the appropriate concentration or range of concentrations for a given experiment must be determined for each particular odor and sensillum. A calibrated dispenser is invaluable for measuring equal aliquots of solvent, particularly viscous paraffin oil. Vortex the dilutions well and store in the dark in a ventilated area either at room temperature or at 4°C.

   Odor dilutions should be stable for several weeks or months, depending on the chemical.

6. To prepare an odor cartridge, place a small disk of filter paper into the 1-mL syringe with a clean pair of forceps. Pipette the desired volume of odor dilution on to the paper and insert the plunger. Cap the other end of the syringe with a syringe needle cap or a piece of parafilm to reduce odor evaporation, which is particularly important for highly volatile chemicals.

   In general, fresh odor cartridges should be prepared every day.

Assembly of the Odor Stimulus Device

A commercial stimulus controller from Syntech (CS-55), although expensive, is simple to use, and conveniently integrates stimulus timing with the AutoSpike data analysis software. If other software is used that offers signal pulse outputs, a homemade stimulus device can be constructed using solenoid valves (e.g., from the Lee Co.) (Clyne et al. 1997). An important property of any such device is to avoid pressure changes on addition of the odor stimulus, which is typically achieved by diverting a small percentage (e.g., 10%–25%) of a continuous airstream (typically 2 L/min) through an odor cartridge during the stimulation period.

7. To deliver odors to the preparation from the odor delivery device, trim a 10-mL plastic pipette to create a tube of uniform diameter and drill two small holes, sufficient to fit the 1-mL odor cartridge syringes, halfway along the length. Connect this pipette with plastic tubing to a continuous airflow (port C on the CS-55) with appropriate adaptors or wax.

   For a humidified air source, which helps maintain the viability of the preparation, you can bubble this flow through a conical flask of water. You can also incorporate flow meters into this circuit to measure and precisely regulate the airflow. For certain odors, such as highly hydrophobic pheromones, it is better to use a glass delivery pipette, which adsorbs volatiles less than plastic. There is no "standard" odor delivery method reported in the literature, and it is important to appreciate that odor responses may be significantly affected by minor changes in the mode of delivery (Vetter et al. 2006) and knowledge of the absolute concentration of odor molecules reaching a given sensillum is difficult, if not impossible, to determine. A photoionization detector (PID) may permit quantitative analysis of the concentration of some odors in an airstream (Vetter et al. 2006).

8. To add the odor cartridge, remove the syringe plunger and use appropriate plastic adaptors to connect the cartridge with plastic tubing to the pulse flow output and insert into the hole in the delivery pipette. The CS-55 has two pulse flow ports (A and B), and switches from B to A during odor stimulation to maintain total airflow. Connect a control cartridge (containing filter paper but no odor) to port B and the odor cartridge to port A, and insert these syringes into the two adjacent holes in the delivery pipette.

Assembly of the Recording Rig

See Figure 3.

9. Set up the vibration-controlled table and Faraday cage in an isolated corner of the laboratory or, ideally, a separate room.

   The table is essential because of the small size of the preparation and diameter of the electrodes. A Faraday cage is desirable to shield against background electrical noise. It is particularly impor-

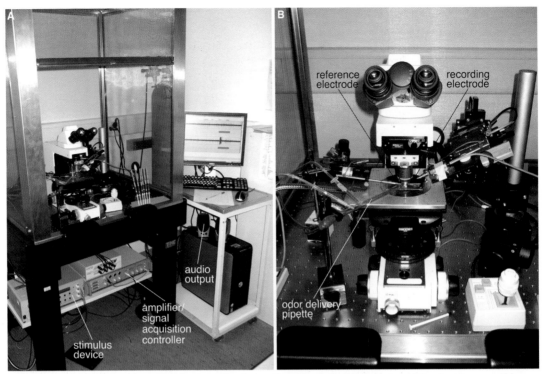

FIGURE 3. An electrophysiological rig for recording from olfactory sensilla. (*A*) Overview of the rig. Many possible configurations of the equipment are possible depending on the available space. (*B*) Close-up view of the rig, revealing the reference electrode mounted on a manual manipulator, the recording electrode mounted on a motorized manipulator, and the odor delivery pipette with odor and control cartridges inserted.

tant to choose a location that is separated from large electrical equipment (e.g., freezers, centrifuges) and that contains electrically isolated outlets. Consult "The Axon Guide" from Molecular Devices (http://www.moleculardevices.com/pages/instruments/axon_guide.html) for extensive background information on building an electrophysiology rig.

10. Place the upright microscope in the center of the table and fix the manual and motorized manipulators at either side (the orientation does not matter, but may depend on your handedness).

    i. Mount the shaft support for the reference electrode on the manual manipulator and the recording electrode probe on the motorized manipulator.

    ii. Fix the odor delivery pipette on a magnetic stand so that the opening is close (10–20 mm) to the center of the microscope stage.

    iii. Connect the electrodes to the signal acquisition system, and connect this system to the odor stimulus controller and the computer (on which the appropriate software has been downloaded).

11. Ground the microscope, manipulators, and odor delivery pipette to a central location within the rig.

    Improper grounding will result in increased noise when the electrode contacts a sensillum. Be prepared to adjust grounding until a good signal-to-noise ratio is reached, but attempt to ground as few locations as possible as too many wires can lead to ground loops.

Drosophila Preparation

See Figure 4.

12. Select newly eclosed flies of the desired genotype and place them in an empty food vial to allow them to reach the appropriate age (typically 2–10 d old).

    Both males and females may be used: No striking molecular or physiological sexual dimorphisms have yet been reported in the peripheral olfactory system in *Drosophila*.

FIGURE 4. A *Drosophila* preparation for recording from antennal olfactory sensilla. (*A*) Overview of the preparation revealing the major support components embedded in small pieces of wax on a microscope slide. The head of the fly is visible protruding from the end of the pipette tip. (*B*) Close-up image of the boxed region in *A*. The right antenna is held against the coverslip by a glass capillary whose tip lies within the joint between the second (II) and third (III) antennal segments.

13. Assemble the aspirator.

    i. Trim one 1000-µL pipette tip ~1 cm from the narrow end and insert into one end of the plastic tubing to form the mouthpiece.

    ii. Place the muslin square over the other end of the tubing and fix in place by inserting the tubing into the broad end of the other pipette tip.

    iii. Trim this pipette tip ~2 cm from the other end to create an opening sufficiently large to aspirate a fly.

14. When you are ready to start recording, draw a single fly from the vial into the aspirator. Withdraw the aspirator and place your finger over the end to trap the animal.

    Anesthesia is therefore avoided, a highly desirable feature for subsequent physiological analysis.

15. Expel the fly into a 200-µL pipette tip. Keeping the end of your aspirator in the base of the pipette tip, hold the tip vertically (narrow end up) to encourage the fly to walk up the tip (*Drosophila* are negatively geotactic). When it is close to the top, blow sharply to trap the fly in the narrow neck of the tip.

16. View the trapped animal under a stereomicroscope to ensure that it is in the correct orientation (head toward the narrow end of the tip) and that no legs or wings are trapped anterior to the head. If they are, suck to dislodge the fly and repeat the trapping procedure.

    It is essential to avoid having any appendages folded against the head, as they will either block access to the antenna or cause movement of the preparation.

17. Trim the pipette tip at each end.

    i. From the narrow end, sufficient plastic should be cut to leave an opening *slightly smaller* than the diameter of the fly head. Typically we make this cut ~2 mm from where the fly's head is located, but the precise position of this cut depends on the size of the fly and the shape of the pipette tip.

ii. From the wide end of the pipette tip, cut ~6 mm from the end of the abdomen of the fly. This distance is not critical, but should leave you with a short piece of pipette tip containing the fly that can be comfortably handled between two fingers.

18. Immobilize the animal within the tip with a small plug of dental wax. Viewing under the stereomicroscope, use the wax to push the head of the fly partially out of the cut pipette tip using a 20-μL pipette tip.

> In an ideal preparation for recording from antennal sensilla (see below for an alternative procedure for maxillary palp sensilla), you want to have about half of the eyes exposed. If the size of the opening is cut correctly, the very slight pressure on the head will cause the antennae to spring forward, whereas the proboscis will remain trapped within the tip. If the opening is too big, the whole head will pass through and move around freely; if it is too small, the head will be crushed. In either of these cases, you will have to discard the preparation and start again!

19. Mount the fly on the glass slide.

    i. Place two small pieces of wax at each side of a glass microscope slide halfway along the length.

    ii. Break a glass coverslip and retrieve a fragment with a short straight edge. Press this fragment into one of the pieces of wax so that it is angled at ~45° with the straight edge at the top.

    iii. Press the pipette tip piece containing the fly into the other piece of wax and, viewing under the stereomicroscope, move it (either by moving the piece of wax on the slide or the angle of the pipette tip in the wax) so that the head of the fly comes into contact with the edge of the coverslip, very close to one end (Fig. 4A).

    > The orientation of the fly will determine which surface of the antenna you can record from: If ventral is up, this will permit recording from the posterior face of the antenna (where many basiconic sensilla are distributed); if dorsal is up, this will permit recording from the anterior face (where, for example, a certain class of coeloconic sensilla are concentrated). The curved surface of the antenna means that the precise angle that you fix it at will affect viewing and access to sensilla located at different positions along its long axis. Alternative mounting procedures have been described that lay the fly against a horizontal coverslip (Clyne et al. 1997), but we find it convenient to angle the coverslip to prevent excessive squashing of the arista against the coverslip for recording from the posterior face of the antenna.

20. Adjust the position of the animal so that the edge of the coverslip presses gently against the head just beneath the antennae. This slight pressure should cause the antennae to project out over the top of the coverslip away from the rest of the head.

21. Set up the capillary to support the fly.

    i. Pull the glass capillary to a fine point using a standard needle puller; the precise shape is unimportant as these capillaries are used only as physical supports. It is convenient to have a box of pulled capillaries prepared, because they can occasionally break during setup of the preparation.

    ii. Fix the capillary on the slide on a small piece of wax on the side of the coverslip where the antennae are closest to the edge.

    iii. While continuing to view under the stereomicroscope, hook the tip of the capillary in the joint between the second and third segments of the antenna closest to the edge of the coverslip (the other antenna can be left free to move). If the capillary tip is too thin and flexible, you can trim it by snipping the tip with a pair of forceps.

    iv. Pull the antenna down and away from the head such that it becomes held gently but firmly between the coverslip and the capillary (Fig. 4B). The long axis of the antenna should be parallel with the slide face to permit optimal viewing of, and access to, sensilla. Variations in these manipulations may be necessary to access different surfaces of the antenna.

    See Troubleshooting.

    > Alternative procedure for recording from sensilla of the maxillary palp: Cut a slightly larger opening in the narrow end of the pipette tip containing the fly to permit pushing of the entire head

out. Cut a small piece of double-stick tape (~3 x 6-mm) and stick it on the pipette tip such that the short edge of the tape is flush with the edge of the tip. Gently bend and push the proboscis until it is glued on to the tape. In a similar manner as for the antenna, the maxillary palps can then be stabilized against a coverslip with a glass capillary.

Performing Recordings

22. Place the preparation slide on the stage with the 10x objective in position, and move the stage until the fixed antenna is in focus in the center of the FOV for both 10x and 100x objectives. Individual sensilla should be clearly visible under the 100x objective.

> If any movement of the antenna is observed, remove the preparation and attempt to restabilize by adjusting the glass capillary support.

See Troubleshooting.

23. Move the electrodes to approximately the right position relative to the preparation. Move the odor stimulus delivery pipette so that the fly is as close as possible to the opening (without being in the way of the electrodes) and is in the center of the airstream.

> It is worthwhile making these adjustments with an old preparation slide in position before starting the recording session.

24. Return to the 10x objective. Adjust the focus so the eye opposite to the restrained antenna is in focus and bring the reference electrode into the same plane. Gently press the electrode against the eye until it enters. Do not insert the electrode too far: You only need contact of the electrode with the hemolymph, and the larger the peripheral hole the more rapidly the preparation will dehydrate.

> In recordings from palp sensilla, the reference electrode can be inserted more proximally into the proboscis.

25. Carefully bring the recording electrode into the FOV with the coarse motorized controller. Avoid rapid movements, because it is very easy to damage the tip by touching the pipette tip or other body parts. If you cannot see the electrode, check its global position by viewing it directly. Adjust its position until the tip is adjacent to the antenna.

26. Insert the electrode tip into the sensillum and prepare to record.

    i. Switch to the 100x objective. The recording electrode tip and antenna should be visible.

    ii. Start the signal recording software and switch on the audio output, if available.

    iii. Select the desired sensillum to record from, and with the fine motorized control, move the electrode tip to near its base. You will probably need to adjust the focus to get a sense of the exact direction that the sensillum points in.

    iv. Gently touch the sensillum with the electrode tip; it should flex in its socket (and produce brief voltage changes). If the electrode tip does not strike the hair face-on, it may slide around the edge, requiring readjustment of its position.

    v. Continue to push the electrode against the sensillum until it pierces the cuticular wall; this may not occur until the hair bends quite significantly, particularly if the tip is not extremely sharp.

    > Successful entry of the electrode into the hair will only be apparent by the appearance of clear spikes on the digital voltage trace. Audio output representation of these spikes is therefore invaluable to judge successful entry while viewing the preparation through the microscope. Although sensillar morphology and spatial position on the antennal surface can provide a guide to finding a specific type of sensillum, it is only from recordings of stimulus-evoked activity using diagnostic odors that sensillar identity can be unambiguously determined. However, with experience, it may be possible to judge sensillar identity from the number of distinct spike amplitudes and the "rhythm" of their combined spontaneous activities.

See Troubleshooting.

27. Once a stable signal has been established, proceed with the desired series of odor stimulations. If the preparation does not move, it should be possible to record from a single sensillum for many minutes.

TROUBLESHOOTING

Problem (Steps 21 and 22): The preparation is unstable.

Solution: If the antenna is moving, readjust the position of the glass capillary support to stabilize it. Many movements are caused by pharyngeal pumping, which is transmitted to the antenna as pulses of movement if the head is not sufficiently constrained in the pipette tip. In such cases, a new fly must be prepared for recording.

Problem (Step 26): You are unable to identify the desired sensillum.

Solution: Consult published maps of the distribution of specific sensilla and ensure that diagnostic odors are being tested at the appropriate concentrations (de Bruyne et al. 2001; van der Goes van Naters and Carlson 2007; Benton et al. 2009). Note that some variations may exist in sensilla numbers between different strains (Stocker 2001). If your microscope can be equipped with fluorescent optics, specific sensillum types can be labeled by using the appropriate chemosensory receptor promoter to drive expression of a membrane targeted mCD8:GFP reporter in the neuron via the GAL4/UAS system (Dobritsa et al. 2003). However, fluorescent signals can be very hard to detect, especially in trichoid and coeloconic sensilla, because of the low surface area of dendritic membranes and the background fluorescence of the cuticle.

Problem (Step 26): There is no electrical contact.

Solution: Check all connections in the circuit, in particular, that the ground electrode is in contact with the hemolymph. It is also possible that the tip of the electrode is covered with a piece of cuticle or membrane, particularly if it has been used extensively for recordings in other sensilla. The electrode can be "cleaned" using the electrode etcher briefly.

Problem (Step 26): The signal-to-noise ratio is low.

Solution: If the noise level is very high, adjust the ground wires to reduce possible sources of noise. Alternatively, the recording electrode may not have fully punctured the sensillar cuticular wall, which can be remedied by applying a little more pressure through movement of the recording electrode into the hair. If neither of these factors are the problem, try a different sensillum. If all sensilla in an antenna give only a weak signal, the preparation may be unhealthy (e.g., dehydrated, visible as "deflation" of the antenna), which requires a new preparation. The best recordings from coeloconic sensilla give only a relatively modest signal-to-noise ratio compared with basiconics, and it has been reported that recording of current changes in the sensilla using glass electrodes and a patch-clamp amplifier in voltage-clamp mode yields a higher signal-to-noise ratio (Yao et al. 2005). This may in part be caused by the smaller conducting area of the glass electrode aperture that is in contact with the sensillar lymph, compared with the entire surface of the tungsten electrode.

DATA ANALYSIS

See Figure 5.

Spike sorting: Differentiation of spikes of different neurons by amplitude is relatively straightforward in large basiconic sensilla but more challenging in small basiconic, trichoid, and coeloconic sensilla. For many experiments, in which it is known that a given ligand stimulates only a single neuron within a sensillum, it may be adequate to sum spikes from all neurons in a sensillum (Yao et al. 2005; Ditzen et al. 2008). Although it may be possible to resolve spikes by shape or other features by applying more sophisticated spike sorting algorithms, there are no published reports of such analyses in the *Drosophila* olfactory system. However, given the knowledge of the molecular markers for most, if not all, OSNs, genetic tools can now be easily generated to selectively silence individual neurons, allowing precise assignment of responses to specific neurons without spike sorting (Schlief and Wilson 2007).

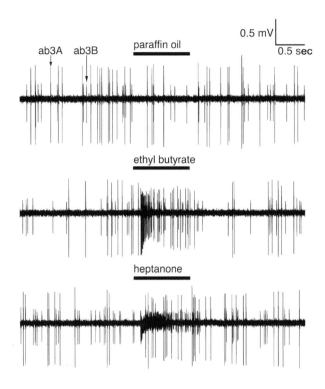

FIGURE 5. Sample recordings from an olfactory sensillum. Electrophysiological traces from the basiconic sensillum ab3, which contains two neurons named ab3A and ab3B with large and small spike amplitudes, respectively. The sensillum was stimulated with a paraffin oil solvent control (*top*), ethyl butyrate, a ligand for the ab3A neuron (*middle*), or heptanone, a ligand for the ab3B neuron (*bottom*). Odors were diluted 0.001% volume/volume in paraffin oil and 10 μL solution added to the odor cartridge. Bars indicate stimulus period (1 sec).

Quantification of spike frequency: Whether the activity of individual OSNs or the summed activity of all OSNs in a sensillum are analyzed, quantification of spike responses to a given stimulus is typically made by counting the number of spikes in a given window (often 0.5 sec for a 1-sec odor pulse) during odor stimulation, subtracting the number of spontaneous spikes before stimulation, and multiplying the difference by 2 to obtain a "corrected response" in spikes per second. There is usually a short delay (100–300 msec) in the onset of the neuronal response after odor stimulation, which is most likely because of the time it takes for the odor stimulus to travel from the cartridge to the preparation. It is therefore sensible to define the start of the period of odor stimulation by determining the time of onset of a response of an OSN to a control stimulus. This approach helps comparison of responses across experiments, even if the position of the odor delivery device or animal changes slightly. Temporal dynamics of spike trains can be analyzed by counting spike numbers in bins before, during, and after odor stimulation.

DISCUSSION

With this protocol—and several weeks to months of practice—it should be possible to perform successful recordings of odor-evoked activity from single olfactory sensilla. Many technical variations are of course possible depending on the specific equipment available and the particular experiment to be performed. Comparison of spike frequencies of odor-evoked responses with published reports is a useful way to assess the functionality of your system, although some differences do exist between different strains of *Drosophila*.

Although conceptually simple, this electrophysiological method is extremely powerful for dissecting the coding mechanisms (de Bruyne et al. 2001; Hallem and Carlson 2006) and molecular genetics (Dobritsa et al. 2003; Larsson et al. 2004; Benton et al. 2007; Jones et al. 2007) of odor detection. This technique is complemented by other physiological approaches in the olfactory (and gustatory) systems, notably optical imaging using genetically encoded calcium indicators (Wang et al. 2003; Marella et al. 2006; Pelz et al. 2006).

Recording from Taste Sensilla

This protocol describes the basics of setting up the electrophysiology rig and stimulus delivery device, sample preparation, and performing and analyzing recordings of stimulus-evoked activity from *Drosophila* taste sensilla.

MATERIALS

CAUTION: See Appendix for proper handling of materials marked with <!>.
See the end of the chapter for recipes for reagents marked with <R>.

Reagents

Beadle-Ephrussi Ringer solution <R>
Drosophila, newly eclosed flies of the desired genotype
Taste compounds (typically use highest purity chemicals from Sigma-Aldrich)

> Dissolve compounds in 30 mM tricholine citrate. Alternatively, 1 mM potassium chloride <!> can also be used as the electrolyte. Dispense 5-mL aliquots in glass vials and store at –20°C. Tastant solutions should be stable for several weeks or months. For experiments, deliver 500-μL aliquots into Eppendorf tubes and store at 4°C for no more than 1 wk.

Taste compounds (typically use highest purity chemicals from Sigma-Aldrich)
Tricholine citrate (T0252, Sigma-Aldrich)

Equipment

Drosophila preparation materials:
  Drosophila aspirator (as described in Protocol 1, Step 13)
  Electrode holder (Syntech, Germany)
  Fine forceps, two (e.g., 11254-20, Fine Science Tools)
  Glass capillary with microfilament (e.g., 1B100F-4, World Precision Instruments)
  Ice bucket
  Modeling clay (e.g., Claytoon, Van Aken International, Inc.)
  Needle puller (e.g., Flaming/Brown puller, Sutter Instrument Company)
  Pipette tip, 200-μl
  Silver wire, 0.3–0.5-mm diameter (e.g., AGW1510, World Precision Instruments)
  Stereomicroscope with illumination (e.g., SZ51, Olympus)
  Syringe, 1-mL
  Syringe needle or pipette tip for loading micropipette (e.g., Microfil MF34G-5, World Precision Instruments, or Microloader tip, Eppendorf)
Electrophysiological recording rig, assembled using the following:
  Aluminum foil
  Amplifier and data acquisition system (e.g., IDAC-4, Syntech, Germany; or CyberAmp380 with Digidata1440A, Molecular Devices)
  Antivibration platform with mechanical pump (e.g., 2210 series BenchMate, Kinetic Systems Inc.)
  Aquarium pump

Conical flask with side arm and rubber sealed top

Electrical connectors: alligator clips (e.g., T115SB and T115SR, Hoffman Products), banana plugs (e.g., SPC15427, Newark), insulated wire, gold wire contacts (19003-01 and 19003-00, Fine Science Tools), soldering iron, wire solder

Faraday cage

Fiber-optic light source

Glass tube (~1-cm inner diameter)

Manual manipulator on magnetic stand for glass tube (e.g., UM-3C with GJ-1, Narishige)

Manual manipulator on magnetic stand for reference electrode (e.g., YOU-1, Narishige)

Manual three-axis manipulator with no drift (e.g., NMN-21 with NMN-C stand, Narishige)

PC workstation

Plastic tubing, various sizes (e.g., R3603, Tygon)

Software (e.g., AutoSpike, Syntech)

Stereo zoom microscope with a table mount stand, equipped with high magnification (100x –150x) and long working distance (≥3 cm) (e.g., Olympus SZX7 zoom 0.8x –5.6x, with 1.5x objective WD 45.5 mm, and 15x or 20x eyepieces)

TasteProbe preamplifier (Syntech)

White card paper

Recording electrodes and micropipette rinsing device:

Electrode holder (Syntech, Germany)

Freezer (–20°C)

Glass capillary with microfilament (e.g., 1B100F-4, World Precision Instruments)

Glass vials with foil-lined screw caps (e.g., 986492, Wheaton Science Products)

Needle puller (e.g., Flaming/Brown puller, Sutter Instrument Company)

Plastic tubing, 8–10-inches (e.g., AAC00001, R3603 tubing, Tygon)

Retort stand and clamp

Silver wire, 0.3–0.5-mm diameter (e.g., AGW1510, World Precision Instruments)

Syringe (10-mL) and syringe needle (18G)

METHOD

Recording Electrodes and Assembly of the Micropipette Washer

1. Use a multistep program of a patch pipette puller to pull electrodes with a tip diameter of ~10–20 μm.

   In the absence of a puller with patch pipette capabilities, carefully break the tip of a standard electrode against a pair of forceps while observing under the low magnification stereomicroscope. Determine whether the tip diameter is sufficiently wide by placing in the recording electrode holder and examining it under high magnification. The exact tip diameter is not critical, although pipettes with smaller diameter are more useful for recording from some smaller sensilla that are closely spaced.

2. Attach an 18G needle to a 10-mL syringe. Slip the needle into one end of the flexible tubing, cut ~8–10 inches in length. Lift the plunger and clamp the plunger on to a stand; the syringe will dangle below the clamp. Push the back end of the recording micropipette into the free end of the tubing. Rinse the electrode with chosen tastant solution by pushing down the body of the syringe (effectively lifting the plunger) and rapidly dipping the electrode tip in the solution 5–10 times. After rinsing, allow the solution to fill halfway lengthwise and the electrode will be ready for use. In the same manner, rinse 5–10 times with water between tastants.

   Fill the electrode just before use or else evaporation at the tip may cause a change in concentration or some precipitation or crystallization of the solute.

Assembly of the Recording Rig

See Figure 6.

3. Choose a bench that is relatively isolated from mechanical or electrical disturbances. Ideally, the area should be far away from chemical hoods or equipment such as freezers, shakers or centrifuges and should have available at least two electrically isolated sockets.

4. Prepare the recording setup.

   i. Set up the vibration-free platform and mount the microscope in the center (Fig. 6A).

   ii. Fix the manipulator for the TasteProbe on one side, and the manipulators for the reference electrode and the glass tube on the other (Fig. 6B).

   iii. Fix the glass tube in its manipulator such that the opening is close to the center of the FOV.

   iv. Using appropriately sized tubing connect the glass tube to the aquarium pump via the filter flask, and connect the TasteProbe to the signal acquisition system.

   v. Install software on the PC workstation and connect it to the data acquisition system.

   vi. Set the fiber optic light source on the bench (not on the vibration-free platform) and align the arms such that the sample can be illuminated with light reflected off white card paper that is placed under the preparation.

   vii. Plug the power supply cord of the TasteProbe into a socket that is isolated from those used for the PC workstation, the fiber optic light source, or the aquarium pump.

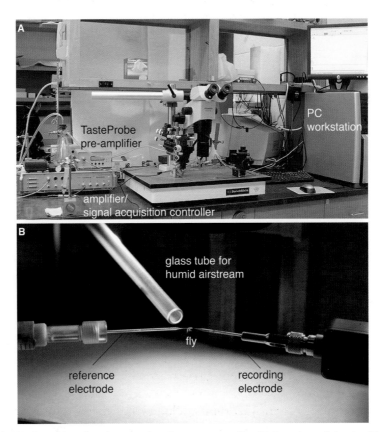

FIGURE 6. An electrophysiological rig for recording from taste sensilla. (*A*) Overview of the rig. (*B*) Close-up view of the rig, revealing the reference and recording electrodes mounted on manipulators and the glass tube used for delivery of humidified air.

viii. Ground the microscope, manipulators, and glass tube.

5. Set up the Faraday cage, if available. A Faraday cage is desirable, but not essential, to shield against background electrical noise.

> Alternatively, we have used a grounded small metal plate (10 x 15-cm), placed between the preparation and the experimenter before recording, which does the job as well.

Drosophila Preparation

See Figure 7.

For recordings from labellar sensilla, follow Steps 6–13; for recordings from leg sensilla, omit Steps 6–13 and follow Steps 14–17.

6. Select newly eclosed flies of the desired genotype and place them in a fresh food vial to allow them to reach the appropriate age (typically 5–10-d old).

7. When you are ready to start recording, fill the shaft of a reference micropipette with Beadle-Ephrussi Ringer solution using a syringe and a microloader tip. Allow a few minutes for the solution to fill to the tip of the pipette by capillary action. Tap to dislodge any air bubbles, particularly near the tip. Place the pipette into the reference electrode holder and, while observing under a stereomicroscope, break the tip very gently against a pair of forceps.

> If the holder allows (e.g., MEH3SW, World Precision Instruments), then tighten till a drop of solution is just expelled from the tip. A small degree of positive pressure ensures that the micropipette does not suction hemolymph/tissue while the fly is being mounted.

8. Set up an ice bucket with some ice next to the workstation. Place the microscope base plate in ice.

9. Using the aspirator draw a single fly from the vial (as described in Steps 14 and 15 in Protocol 1), and blow the fly into a 200-µL pipette tip such that it is trapped in the narrow neck of the tip. Withdraw the aspirator and immediately place the tip in ice for ~30 sec to slow the fly.

> Do not blow too sharply because the fly will have to be removed from the tip for further preparation.

10. Replace the cooled plate in the microscope base. Tap the wide end of pipette tip on the plate until the fly falls on the plate. Using two pairs of forceps, one to stabilize the fly and another to pinch the forelegs, gently remove the fly's forelegs. This is easiest to perform while the fly is lying on its back.

11. Turn the fly over on to its ventral side and insert the tip of the reference electrode into the dorsal side of the thorax (Fig. 7A). The ideal location to insert the electrode for ease of maneuver-

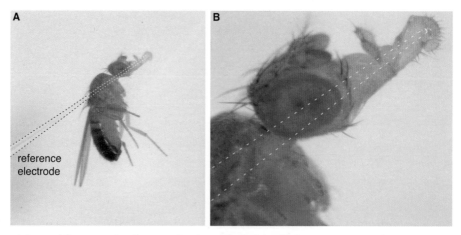

FIGURE 7. A *Drosophila* preparation for recording from labellar taste hairs. (*A*) Overview of the preparation. Dotted lines highlight the position of the reference electrode in the fly. (*B*) Close-up view of the preparation in *A*.

ability is at the center of the ridge that separates the scutellum from the rest of the notum. First pierce through the cuticle while holding the electrode at right angles to the fly, and then slide the electrode at an angle toward the neck.

12. Increase the magnification and nudge the fly so that the electrode slips through the neck and into the head. During this step, you may stabilize the electrode holder on the microscope base using modeling clay. This allows use of forceps in both hands, if necessary, to manipulate the fly. In general, contact with the anterior or lateral aspects of the proboscis is preferable to that with the posterior aspect of the proboscis, which offers more opportunity to damage the esophagus. Avoid contact with taste sensilla at all times.

13. Use the forceps to push the animal toward the posterior until the tip of the electrode is just barely inserted into the labellum. Sometimes, the tip of the electrode will get stuck halfway along the length of the proboscis. If that happens, pinch the maxillary palps with a pair of forceps, lift the anterior fold of cuticle over the tip of the electrode and gently push the proboscis toward the posterior till the tip comes to rest in the labellum (Fig. 7B). Proceed to Step 18 to begin recording.

    If the electrode pierces or pushes against the wall of the labellum, taste neurons may be damaged, and you will have to discard the preparation and start again!

    Alternatively, the proboscis may be "glued" in the retracted position using lanolin (Wako Pure Chemical Industries, Ltd., Japan), in which case the tip of the electrode may rest in a more proximal position.

    See Troubleshooting.

14. Capture and chill a single fly on ice as described in Step 9.

15. Decapitate the fly and secure it to a flat support using insect pins. A microscope slide covered with a thin layer of Sylgard 184 silicone elastomer (Dow Corning Corp.) serves the purpose nicely. Use the pins to secure the legs in a manner such that the distal tarsal segments protrude off the edge of the slide.

16. Secure the slide on a holder, and check by eye that the sensilla of interest are positioned in the center of the FOV and are at a level that can be accessed by the recording electrode. Observe under the microscope and determine whether the preparation is stable.

17. Prepare a reference electrode as described in Step 7 and attach the holder in its manipulator. At low magnification, use the manipulator to bring the electrode close to the preparation and insert it into the abdomen. As described for olfactory recordings, it is worthwhile to make adjustments to the positions of the manipulators using an old preparation slide.

    See Troubleshooting.

Performing Recordings

18. Secure the electrode holder in the appropriate manipulator so that the fly is positioned in the center of the FOV. If the fly is mounted properly, the labellum (or legs) will be completely immobile. If the preparation is not stable, take it back to the smaller microscope and attempt to stabilize by adjusting the position of the reference micropipette. If that fails, start with a new fly.

    See Troubleshooting.

19. Turn on the aquarium pump and adjust the position and the flow rate of the humidified airstream onto the fly.

20. Mount the recording micropipette in the holder connected to the TasteProbe head stage. Center the position of the manipulator along each of the three axes. Bring the tip of the recording electrode to the center of the FOV and focus. Move the fly into the same plane.

21. Set up recordings from the sensilla.

For tip recordings:

   i. Zoom in to view the sensilla of interest. Select a sensillum to record from, focus on its tip, and gently move the recording electrode till it faces the tip head-on. Enable the trigger and bring the electrode in contact with the tip of the hair. The tip should just break the meniscus of the solution in the micropipette. On contact, a beep will be audible and recording will commence. Withdraw contact after ~2–3 sec.

   ii. If the preparation is stable, proceed with the desired series of tastants. Apply consecutive stimuli at least 1 min apart, making sure to wash the electrode thoroughly between applications of different stimuli. It is usually possible to record from a single fly for several hours.

See Troubleshooting.

For side-wall recordings:

   i. Bring a capillary (tip diameter ~10–15 µm) in contact with the wall of the sensillum, roughly at a midpoint along the length.

   ii. Maneuver a second micropipette with a smaller tip diameter that is filled with electrolyte to rest directly opposite the first capillary. Gently press against the sensillum till there is a crack in the wall. This serves as the recording micropipette and should be connected to a universal single ended probe via a silver wire.

   iii. Use a third micropipette with a tip diameter of ~10–20 µm to apply taste stimuli to the tip of a sensillum as described above. In contrast with tip recordings, the stimulus micropipette is not connected to the headstage for these recordings. Therefore, the stimuli need not be dissolved in electrolyte solution.

See Troubleshooting.

Alternative approach for side-wall recordings: Immobilize the fly on a slide as described for olfactory recordings from the antenna, with the exception that the entire head of the animal should be forced out of the narrow end of the pipette tip. Stabilize the labellum between the coverslip and a glass capillary and insert a sharpened tungsten recording electrode in the base of the sensillum as described for olfactory recordings. Deliver taste stimuli to the tip of this sensillum as described in Step 21.iii above.

TROUBLESHOOTING

Problem (Steps 13, 17, 18): The preparation is unstable.
Solution: A factor that should be considered is the shape of the reference micropipette. Both the taper angle and length influence the stability of the preparation, and the ideal form depends on the size and shape of the fly. The electrode should be thin enough toward the tip so that it can be pushed all the way into the proboscis, but wide enough at the base so that the thoracic tissue grips it tightly. If the electrode is too thin the fly tends to spin around on its axis, whereas if it is too wide the fly dehydrates rapidly and may not survive for long.

Problem (Step 21): There is no electrical contact.
Solution: The most common reasons for this are poor conductance through the taste electrode or injury to the taste neurons. Check that all components are connected properly and change the recording micropipette. Start over with a new preparation. Other factors that may affect taste recordings are the age of the fly and the temperature and humidity conditions that the flies are raised at. Advanced age (>20 d), high temperatures, and low humidity often affect the outcome

adversely. Reports in the literature also suggest that the frequency of contact varies with the type of the sensillum under scrutiny (Hiroi et al. 2002).

Problem (Step 21): There is background noise.
Solution: Ensure that all components of the rig are grounded. Test the signal in the "pass through" mode of the TasteProbe and place shields until the baseline appears even. A small foil shield may also be placed on the recording micropipette (this shield should not be grounded). Shorten the length of the recording micropipette using a glass cutter.

DATA ANALYSIS

See Figure 8.

Spike sorting: Differentiation of spikes of different neurons by amplitude is relatively straightforward in recordings from taste sensilla (see Fig. 8 for examples of easily distinguishable spikes from sugar, bitter, and water neurons). With the use of tricholine citrate as recording electrolyte, taste neurons are typically silent in the absence of a stimulus and in many cases a given stimulus activates only one neuron in a sensillum. In some cases in which a single stimulus evokes responses from more than one neuron, algorithms that analyze interspike intervals and spike doublet frequency have been applied for assignment of responses to individual neurons (Meunier et al. 2003a). The identity of responsive neurons may also be determined by using genetic tools to silence individual neurons or by testing binary mixtures in which one component is a known activator of a specific neuron.

Quantification of spike frequency: Count the number of spikes in given window (typically 500 msec, beginning 200 msec after contact), and multiply the number by the appropriate factor (typically 2) to obtain a response in spikes per second. The 200–700-msec period falls within the tonic part of the neuronal response. There is usually a large artifact on contact of the electrode with the sensillum tip, followed by a high frequency of action potentials that rapidly decays over the next 200 msec. For analysis of the temporal structure of the response, spikes may be counted in 50–100-msec bins starting at contact and multiplied by the appropriate factor to obtain a response in spikes per second. Neuronal responses have also been quantified by instantaneous frequencies (=1/interspike

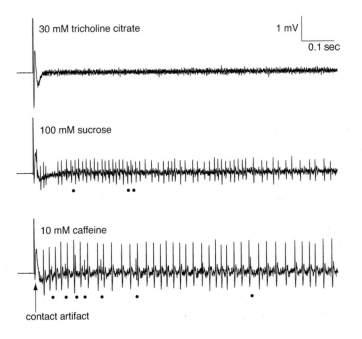

FIGURE 8. Sample recordings from S-type taste sensilla. Sensilla were stimulated with either tricholine citrate (*top*), sucrose, which activates the "sugar" neuron (*middle*), or caffeine, which activates the "bitter" neuron (*bottom*). Occasional spikes of smaller amplitude representing the "water" neuron are also observed (solid circles). Contact of the stimulus/recording micropipette with the tip of the sensillum is evident by an artifact at the beginning of the trace (contact artifact).

interval), which provide a high-resolution view of the temporal dynamics of the response (van der Goes van Naters and den Otter 1998).

DISCUSSION

Successful assembly of a taste physiology rig should make possible a direct analysis of stimulus-evoked activity from taste neurons in various sensilla on the fly body. All the taste neurons in a sensillum should be "visible" in recordings and for a typical sensillum you should be able to identify each neuron as belonging to one of the four described categories. A consideration to keep in mind while surveying responses from "wild-type" strains is that there may be differences in physiological responses to taste stimuli; there is precedent for polymorphisms in GRs (Ueno et al. 2001).

The literature predominantly describes the use of the tip-recording method for *Drosophila*, which has been invaluable for probing responses to various classes of water-soluble stimuli (Meunier et al. 2000, 2003b; Hiroi et al. 2002, 2004; Koseki et al. 2004) and for studying the roles of receptors that underlie these responses (Dahanukar et al. 2001, 2007; Ueno et al. 2001; Moon et al. 2006; Jiao et al. 2007, 2008). Side-wall or tungsten electrode recordings, which are more suitable for testing water-insoluble stimuli such as the cuticular hydrocarbons that are important during courtship (Ferveur 2005), are technically more challenging and are more injurious to the sensillum than tip recordings. A number of factors, therefore, determine the choice of the technique: the nature of the stimulus, the location of the sensillum, and the experimental goal (e.g., whether it is critical to use the same animal over several hours).

Based on the protocols for both olfactory and gustatory sensilla recordings, we hope to have conveyed the principle that if an animal is properly restrained, one can record from virtually any type of sensillum with an appropriate electrode. We have described the use of tungsten and glass electrodes for olfactory and gustatory recordings, respectively, but depending on the objective of the experiment one may easily be substituted for the other.

RECIPE

CAUTION: See Appendix for proper handling of materials marked with <!>.
Recipes for reagents marked with <R> are included in this list.

Beadle-Ephrussi Ringer Solution

Reagent	Amount
NaCl	7.5 g
KCl <!>	0.35 g
CaCl$_2$·2H$_2$O <!>	0.279 g

Dissolve the ingredients in 1 L of H$_2$O. Aliquot and store at –20°C.

ACKNOWLEDGMENTS

We are grateful to Linnea Weiss for the photograph in Figure 1A, Makoto Hiroi and Linnea Weiss for the photographs in Figure 1C, Brice Petit for the photograph in Figure 4A, Linnea Weiss and Woodstock Tom for the photographs in Figure 6, and Linnea Weiss for the traces in Figure 8. We also thank Walt Jones, Matthieu Louis, Takao Nakagawa, Shannon Olsson, Maurizio Pellegrino, Marcus

Stensmyr, and Wynand van der Goes van Naters for discussions and comments on this chapter. Research in R.B.'s laboratory is supported by the University of Lausanne, Swiss National Science Foundation, Roche Research Foundation, and a European Starting Independent Researcher Grant.

REFERENCES

Ache BW, Young JM. 2005. Olfaction: Diverse species, conserved principles. *Neuron* **48:** 417–430.

Alcorta E. 1991. Characterization of the electroantennogram in *Drosophila melanogaster* and its use for identifying olfactory capture and transduction mutants. *J Neurophysiol* **65:** 702–714.

Benton R. 2006. On the origin of smell: Odorant receptors in insects. *Cell Mol Life Sci* **63:** 1579–1585.

Benton R. 2007. Sensitivity and specificity in *Drosophila* pheromone perception. *Trends Neurosci* **30:** 512–519.

Benton R, Sachse S, Michnick SW, Vosshall LB. 2006. Atypical membrane topology and heteromeric function of *Drosophila* odorant receptors in vivo. *PLoS Biol* **4:** e20. doi:10.1371/journal.pbio.0040020.

Benton R, Vannice KS, Vosshall LB. 2007. An essential role for a CD36-related receptor in pheromone detection in *Drosophila. Nature* **450:** 289–293.

Benton R, Vannice KS, Gomez-Diaz C, Vosshall LB. 2009. Variant ionotropic glutamate receptors as chemosensory receptors in *Drosophila. Cell* **136:** 149–162.

Boeckh J. 1962. Elektrophysiologishe Untersuchungen an einzelnen Geruchsrezeptoren auf den Antennen des Totengräbers (Necrophorus, Coleoptera). *Z Vergl Physiol* **46:** 212–248.

Brand AH, Perrimon N. 1993. Targeted gene expression as a means of altering cell fates and generating dominant phenotypes. *Development* **118:** 401–415.

Chandrashekar J, Hoon MA, Ryba NJ, Zuker CS. 2006. The receptors and cells for mammalian taste. *Nature* **444:** 288–294.

Clyne P, Grant A, O'Connell R, Carlson JR. 1997. Odorant response of individual sensilla on the *Drosophila* antenna. *Invert Neurosci* **3:** 127–135.

Clyne PJ, Warr CG, Freeman MR, Lessing D, Kim J, Carlson JR. 1999. A novel family of divergent seven-transmembrane proteins: Candidate odorant receptors in *Drosophila. Neuron* **22:** 327–338.

Clyne PJ, Warr CG, Carlson JR. 2000. Candidate taste receptors in *Drosophila. Science* **287:** 1830–1834.

Couto A, Alenius M, Dickson BJ. 2005. Molecular, anatomical, and functional organization of the *Drosophila* olfactory system. *Curr Biol* **15:** 1535–1547.

Dahanukar A, Foster K, van der Goes van Naters WM, Carlson JR. 2001. A Gr receptor is required for response to the sugar trehalose in taste neurons of *Drosophila. Nat Neurosci* **4:** 1182–1186.

Dahanukar A, Lei YT, Kwon JY, Carlson JR. 2007. Two *Gr* genes underlie sugar reception in *Drosophila. Neuron* **56:** 503–516.

de Bruyne M. 2006. Visualizing a fly's nose. In *Chemical ecology: From gene to ecosystem* (ed. M Dicke, W Takken), pp. 105–125. Springer, Berlin.

de Bruyne M, Clyne PJ, Carlson JR. 1999. Odor coding in a model olfactory organ: The *Drosophila* maxillary palp. *J Neurosci* **19:** 4520–4532.

de Bruyne M, Foster K, Carlson JR. 2001. Odor coding in the *Drosophila* antenna. *Neuron* **30:** 537–552.

Dethier VG. 1976. *The hungry fly.* Harvard University Press, Cambridge.

Ditzen M, Pellegrino M, Vosshall LB. 2008. Insect odorant receptors are molecular targets of the insect repellent DEET. *Science* **319:** 1838–1842.

Dobritsa AA, van der Goes van Naters W, Warr CG, Steinbrecht RA, Carlson JR. 2003. Integrating the molecular and cellular basis of odor coding in the *Drosophila* antenna. *Neuron* **37:** 827–841.

Dunipace L, Meister, S, McNealy C, Amrein H. 2001. Spatially restricted expression of candidate taste receptors in the *Drosophila* gustatory system. *Curr Biol* **11:** 822–835.

Falk R, Bleiser-Avivi N, Atidia J. 1976. Labellar taste organs of *Drosophila melanogaster. J Morphol* **150:** 327–342.

Ferveur JF. 2005. Cuticular hydrocarbons: Their evolution and roles in *Drosophila* pheromonal communication. *Behav Genet* **35:** 279–295.

Fishilevich E, Vosshall LB. 2005. Genetic and functional subdivision of the *Drosophila* antennal lobe. *Curr Biol* **15:** 1548–1553.

Foelix RF, Stocker RF, Steinbrecht RA. 1989. Fine structure of a sensory organ in the arista of *Drosophila melanogaster* and some other dipterans. *Cell Tissue Res* **258:** 277–287.

Fujishiro N, Kijima H, Morita H. 1984. Impulse frequency and action potential amplitude in labellar chemosensory neurons of *Drosophila melanogaster. J Insect Physiol* **30:** 317–325.

Galindo K, Smith DP. 2001. A large family of divergent *Drosophila* odorant-binding proteins expressed in gustatory and olfactory sensilla. *Genetics* **159:** 1059–1072.

Gao Q, Chess A. 1999. Identification of candidate *Drosophila* olfactory receptors from genomic DNA sequence. *Genomics* **60:** 31–39.

Goldman AL, van der Goes van Naters W, Lessing D, Warr CG, Carlson JR. 2005. Coexpression of two functional odor receptors in one neuron. *Neuron* **45:** 661–666.

Hallem EA, Carlson JR. 2006. Coding of odors by a receptor repertoire. *Cell* **125:** 143–160.

Hallem EA, Ho MG, Carlson JR. 2004. The molecular basis of odor coding in the *Drosophila* antenna. *Cell* **117:** 965–979.

Hiroi M, Marion-Poll F, Tanimura T. 2002. Differentiated response to sugars among labellar chemosensilla in *Drosophila*. *Zoolog Sci* **19:** 1009–1018.

Hiroi M, Meunier N, Marion-Poll F, Tanimura T. 2004. Two antagonistic gustatory receptor neurons responding to sweet-salty and bitter taste in *Drosophila*. *J Neurobiol* **61:** 333–342.

Hodgson ES, Roeder KD. 1956. Electrophysiological studies of arthropod chemoreception. I. General properties of the labellar chemoreceptors of Diptera. *J Cell Physiol* **48:** 51–75.

Hodgson E, Lettvin JY, Roeder KD. 1955. Physiology of a primary chemoreceptor unit. *Science* **122:** 417–418.

Hubel DH. 1957. Tungsten microelectrode for recording from single units. *Science* **125:** 549–550.

Ignell R, Hansson BS. 2005. Insect olfactory neuroethology—An electrophysiological perspective. In *Methods in insect sensory neuroscience* (ed. Christensen TA), pp. 319–347. CRC Press, Boca Raton, FL.

Ishimoto H, Takahashi K, Ueda R, Tanimura T. 2005. G-protein γ subunit 1 is required for sugar reception in *Drosophila*. *EMBO J* **24:** 3259–3265.

Jiao Y, Moon SJ, Montell C. 2007. A *Drosophila* gustatory receptor required for the responses to sucrose, glucose, and maltose identified by mRNA tagging. *Proc Natl Acad Sci* **104:** 14110–14115.

Jiao Y, Moon SJ, Wang X, Ren Q, Montell C. 2008. Gr64f is required in combination with other gustatory receptors for sugar detection in *Drosophila*. *Curr Biol* **18:** 1797–1801.

Jones WD, Cayirlioglu P, Grunwald Kadow I, Vosshall LB. 2007. Two chemosensory receptors together mediate carbon dioxide detection in *Drosophila*. *Nature* **445:** 86–90.

Kain P, Chakraborty TS, Sundaram S, Siddiqi O, Rodrigues V, Hasan G. 2008. Reduced odor responses from antennal neurons of $G(q)\alpha$, *phospholipase C*β, and *rdgA* mutants in *Drosophila* support a role for a phospholipid intermediate in insect olfactory transduction. *J Neurosci* **28:** 4745–4755.

Kaissling KE. 1974. Sensory transduction in insect olfactory receptors. In *Biochemistry of sensory functions* (ed D Jaenicke), pp. 243–273. Springer, Berlin.

Kaissling KE. 1986. Chemo-electrical transduction in insect olfactory receptors. *Annu Rev Neurosci* **9:** 121–145.

Koseki T, Koganezawa M, Furuyama A, Isono K, Shimada I. 2004. A specific receptor site for glycerol, a new sweet tastant for *Drosophila*: Structure-taste relationship of glycerol in the labellar sugar receptor cell. *Chem Senses* **29:** 703–711.

Kwon JY, Dahanukar A, Weiss LA, Carlson JR. 2007. The molecular basis of CO_2 reception in *Drosophila*. *Proc Natl Acad Sci* **104:** 3574–3578.

Lacaille F, Hiroi M, Twele R, Inoshita T, Umemoto D, Maniere G, Marion-Poll F, Ozaki M, Francke W, Cobb M, et al. 2007. An inhibitory sex pheromone tastes bitter for *Drosophila* males. *PLoS ONE* **2:** e661. doi:10.1371/journal.pone.0000661.

Larsson MC, Domingos AI, Jones WD, Chiappe ME, Amrei H, Vosshall LB. 2004. Or83b encodes a broadly expressed odorant receptor essential for *Drosophila* olfaction. *Neuron* **43:** 703–714.

Laughlin JD, Ha TS, Jones DN, Smith DP. 2008. Activation of pheromone-sensitive neurons is mediated by conformational activation of pheromone-binding protein. *Cell* **133:** 1255–1265.

Lee Y, Moon SJ, Montell C. 2009. Multiple gustatory receptors required for the caffeine response in *Drosophila*. *Proc Natl Acad Sci* **106:** 4495–4500.

Liscia A, Solari P. 2000. Bitter taste recognition in the blowfly: Electrophysiological and behavioral evidence. *Physiol Behav* **70:** 61–65.

Lundin C, Kall L, Kreher SA, Kapp K, Sonnhammer EL, Carlson JR, Heijne G, Nilsson I. 2007. Membrane topology of the *Drosophila* OR83b odorant receptor. *FEBS Lett* **581:** 5601–5604.

Malnic B, Hirono J, Sato T, Buck L.B. 1999. Combinatorial receptor codes for odors. *Cell* **96:** 713–723.

Marella S, Fischler W, Kong P, Asgarian S, Rueckert E, Scott K. 2006. Imaging taste responses in the fly brain reveals a functional map of taste category and behavior. *Neuron* **49:** 285–295.

Marion-Poll F, van der Pers J. 1996. Un-filtered recordings from insect taste sensilla. *Entomol Exp Appl* **80:** 113–115.

Meunier N, Ferveur JF, Marion-Poll F. 2000. Sex-specific non-pheromonal taste receptors in *Drosophila*. *Curr Biol* **10:** 1583–1586.

Meunier N, Marion-Poll F, Lansky P, Rospars JP. 2003a. Estimation of the individual firing frequencies of two neurons recorded with a single electrode. *Chem Senses* **28:** 671–679.

Meunier N, Marion-Poll F, Rospars JP, Tanimura T. 2003b. Peripheral coding of bitter taste in *Drosophila*. *J Neurobiol* **56:** 139–152.

Mombaerts P. 2004. Odorant receptor gene choice in olfactory sensory neurons: The one receptor–one neuron hypothesis revisited. *Curr Opin Neurobiol* **14:** 31–36.

Moon SJ, Kottgen M, Jiao Y, Xu H, Montell C. 2006. A taste receptor required for the caffeine response in vivo. *Curr Biol* **16:** 1812–1817.

Moon SJ, Lee Y, Jiao Y, Montell C. 2009. A *Drosophila* gustatory receptor essential for aversive taste and inhibiting male-to-male courtship. *Curr Biol* **19:** 1623–1627.

Morita H. 1959. Initiation of spike potentials in contact chemosensory hairs of insects. III. D.C. stimulation and generator potential of labellar chemoreceptor of *Calliphora. J Cell Comp Physiol* **54:** 189–204.

Morita H, Doira S, Takeda K, Kuwabara M. 1957. Electrical response of contact chemoreceptor on tarsus of the butterfly, *Vanessa indica. Mem Fac Sci Kyushu Univ Ser E Biol* **2:** 119–139.

Nayak SV, Singh RN. 1983. Sensilla on the tarsal segments and mouthparts of adult *Drosophila melanogaster* Meigen (Diptera: Drosophilidae). *Int J Insect Morphol Embryol* **12:** 273–291.

Pelz D, Roeske T, Syed Z, de Bruyne M, Galizia CG. 2006. The molecular receptive range of an olfactory receptor in vivo (*Drosophila melanogaster* Or22a). *J Neurobiol* **66:** 1544–1563.

Ray K, Hartenstein V, Rodrigues V. 1993. Development of the taste bristles on the labellum of *Drosophila melanogaster. Dev Biol* **155:** 26–37.

Riesgo-Escovar JR, Piekos WB, Carlson, J.R. 1997. The *Drosophila* antenna: Ultrastructural and physiological studies in wild-type and *lozenge* mutants. *J Comp Physiol A* **180:** 151–160.

Robertson HM, Warr CG, Carlson JR. 2003. Molecular evolution of the insect chemoreceptor gene superfamily in *Drosophila melanogaster. Proc Natl Acad Sci (Suppl 2)* **100:** 14537–14542.

Sato K, Pellegrino M, Nakagawa T, Nakagawa T, Vosshall LB, Touhara K. 2008. Insect olfactory receptors are heteromeric ligand-gated ion channels. *Nature* **452:** 1002–1006.

Schlief ML, Wilson RI. 2007. Olfactory processing and behavior downstream from highly selective receptor neurons. *Nat Neurosci* **10:** 623–630.

Schneider D. 1957. Elektrophysiologische untersuchen con Chemo- und Mechanorezeptoren der Antenne des Seidenspinners *Bombyx mori. Z Vergl Physiol* **40:** 8–41.

Scott K, Brady R Jr, Cravchik A, Morozov P, Rzhetsky A, Zuker C, Axel, R. 2001. A chemosensory gene family encoding candidate gustatory and olfactory receptors in *Drosophila. Cell* **104:** 661–673.

Shanbhag SR, Singh K, Singh RN. 1995. Fine structure and primary sensory projections of sensilla located in the sacculus of the antenna of *Drosophila melanogaster. Cell Tissue Res* **282:** 237–249.

Shanbhag SR, Muller B, Steinbrecht RA. 1999. Atlas of olfactory organs of *Drosophila melanogaster.* 1. Types, external organization, innervation and distribution of olfactory sensilla. *Int J Insect Morphol Embryol* **28:** 377–397.

Shanbhag SR, Muller B, Steinbrecht RA. 2000. Atlas of olfactory organs of *Drosophila melanogaster* 2. Internal organization and cellular architecture of olfactory sensilla. *Arthropod Struct Dev* **29:** 211–229.

Shanbhag SR, Hekmat-Scafe D, Kim MS, Park SK, Carlson JR, Pikielny C, Smith DP, Steinbrecht RA. 2001a. Expression mosaic of odorant-binding proteins in *Drosophila* olfactory organs. *Microsc Res Tech* **55:** 297–306.

Shanbhag SR, Park SK, Pikielny CW, Steinbrecht RA. 2001b. Gustatory organs of *Drosophila melanogaster:* Fine structure and expression of the putative odorant-binding protein PBPRP2. *Cell Tissue Res* **304:** 423–437.

Slone J, Daniels J, Amrein H. 2007. Sugar receptors in *Drosophila. Curr Biol* **17:** 1809–1816.

Smart R, Kiely A, Beale M, Vargas E, Carraher C, Kralicek AV, Christie DL, Chen C, Newcomb RD, Warr CG. 2008. *Drosophila* odorant receptors are novel seven transmembrane domain proteins that can signal independently of heterotrimeric G proteins. *Insect Biochem Mol Biol* **38:** 770–780.

Stengl M, Hatt H, Breer H. 1992. Peripheral processes in insect olfaction. *Annu Rev Physiol* **54:** 665–681.

Stocker RF. 1994. The organization of the chemosensory system in *Drosophila* melanogaster: A review. *Cell Tissue Res* **275:** 3–26.

Stocker RF. 2001. *Drosophila* as a focus in olfactory research: Mapping of olfactory sensilla by fine structure, odor specificity, odorant receptor expression, and central connectivity. *Microsc Res Tech* **55:** 284–296.

Suh GS, Ben-Tabou de Leon S, Tanimoto H, Fiala A, Benzer S, Anderson DJ. 2007. Light activation of an innate olfactory avoidance response in *Drosophila. Curr Biol* **17:** 905–908.

Thorne N, Amrein H. 2008. Atypical expression of *Drosophila* gustatory receptor genes in sensory and central neurons. *J Comp Neurol* **506:** 548–568.

Thorne N, Chromey C, Bray S, Amrein H. 2004. Taste perception and coding in *Drosophila. Curr Biol* **14:** 1065–1079.

Ueno K, Ohta M, Morita H, Mikuni Y, Nakajima S, Yamamoto K, Isono K. 2001. Trehalose sensitivity in *Drosophila* correlates with mutations in and expression of the gustatory receptor gene *Gr5a. Curr Biol* **11:** 1451–1455.

Ueno K, Kohatsu S, Clay C, Forte M, Isono K, Kidokoro Y. 2006. Gsα is involved in sugar perception in *Drosophila melanogaster. J Neurosci* **26:** 6143–6152.

van der Goes van Naters W, Carlson JR. 2007. Receptors and neurons for fly odors in *Drosophila. Curr Biol* **17:** 606–612.

van der Goes van Naters W, den Otter CJ. 1998. Amino acids as taste stimuli for tsetse flies. *Physiol Entomol* **23:** 278–284.

Vetter RS, Sage AE, Justus KA, Carde RT, Galizia CG. 2006. Temporal integrity of an airborne odor stimulus is greatly affected by physical aspects of the odor delivery system. *Chem Senses* **31:** 359–369.

Vosshall LB, Stocker RF. 2007. Molecular architecture of smell and taste in *Drosophila. Annu Rev Neurosci* **30:** 505-533.

Vosshall LB, Amrein H, Morozov PS, Rzhetsky A, Axel R. 1999. A spatial map of olfactory receptor expression in the *Drosophila* antenna. *Cell* **96:** 725–736.

Vosshall LB, Wong AM, Axel R. 2000. An olfactory sensory map in the fly brain. *Cell* **102:** 147–159.

Wang JW, Wong AM, Flores J, Vosshall LB, Axel R. 2003. Two-photon calcium imaging reveals an odor-evoked map of activity in the fly brain. *Cell* **112:** 271–282.

Wang Z, Singhvi A, Kong P, Scott K. 2004. Taste representations in the *Drosophila* brain. *Cell* **117:** 981–991.

Wicher D, Schafer R, Bauernfeind R, Stensmyr MC, Heller R, Heinemann SH, Hansson BS. 2008. *Drosophila* odorant receptors are both ligand-gated and cyclic-nucleotide-activated cation channels. *Nature* **452:** 1007–1011.

Wieczorek H, Wolff G. 1989. The labellar sugar receptor of *Drosophila. J Comp Physiol A* **164:** 825–834.

Wolbarsht ML. 1957. Water taste in *Phormia. Science* **125:** 1248.

Wyatt TD. 2003. *Pheromones and animal behaviour: Communication by smell and taste.* Oxford University Press, Oxford.

Xu P, Atkinson R, Jones DN, Smith DP. 2005. *Drosophila* OBP LUSH is required for activity of pheromone-sensitive neurons. *Neuron* **45:** 193–200.

Yao CA, Ignell R, Carlson JR. 2005. Chemosensory coding by neurons in the coeloconic sensilla of the *Drosophila* antenna. *J Neurosci* **25:** 8359–8367.

17 Electrophysiological Recording from Neurons in *Drosophila* Embryos and Larvae

Richard Marley and Richard A. Baines

Faculty of Life Sciences, University of Manchester, Manchester M13 9PT, United Kingdom

ABSTRACT

The fruit fly *Drosophila melanogaster* has been instrumental in expanding our understanding of early aspects of neural development. For example, the use of this model system has greatly added to our knowledge of neural cell fate determination, axon guidance, and synapse formation. A little more than 10 years ago, it also became possible to access and make electrophysiological recordings directly from neurons in situ in an intact central nervous system (CNS). This advance facilitated studies aimed at understanding both the development and regulation of neuronal signaling. It has now been determined when *Drosophila* motor neurons

first become electrically active, that synaptic activity is required for appropriate electrical development, and that homeostatic and retrograde signaling pathways exist to help shape electrical development. The aim of this chapter is to provide a brief introduction to the basis of electrical excitability in neurons and a basic description of *Drosophila* electrophysiological preparations to enable the reader to make recordings from motor neurons in the ventral nerve cord of embryos and larvae.

INTRODUCTION

Neurons can be considered to be components in an electrical circuit, adhering to standard electrical principles such as Ohm's law and showing properties of resistance, capacitance, conductance, and rectification. Because of this, the discipline of electrophysiology draws from both physics and biology, a blend that provides the tools required to study the electrical activity of dozens of cells functioning in a network as well as the gating of a single ion channel in an isolated patch of membrane. This chapter will discuss the nature of cellular excitability and the applications of electrophysiology to neurons in *Drosophila* larvae.

Neurons are electrically excitable. This is because neurons maintain a potential voltage difference across, and have gated ion-selective ion channels in, their membranes. At rest (i.e., when not firing action potentials) the voltage difference is ~-60 mV (negative inside the cell with respect to outside).

To signal, specific ion channels must first open and, in doing so, they allow an influx of cations (notably Na$^+$ and Ca^{2+}) driven by electrochemical gradients. Numerous neuroscience textbooks have more than adequate explanations of the action potential and the reader is directed to these if needed.

Ion channels are large transmembrane protein molecules comprised of a number of homomeric or heteromeric subunits, which assemble to form a functional pore. Primary subunits, often denoted as α subunits, constitute the pore-forming channel that selectively allows ions to cross the membrane. Additional subunits, often termed β subunits, modulate the gating (opening) properties of the α subunits. Ion channels are gated by various stimuli, including membrane voltage, ligand binding, stretch, and absorption of photons. This review focuses on voltage-gated (VG) ion channels that are involved in the generation of action potentials.

The primary pore forming α subunits of VG channels typically comprise four homologous domains (termed I, II, III, and IV), and each domain contains six transmembrane (TM) segments (Fig. 1). In the case of VG K$^+$ channels, each α subunit contains only one domain and therefore four α subunits are required to form an active channel. In contrast, the α subunits of VG Na$^+$ and Ca^{2+} channels each contain four homologous domains, and so only one α subunit is required to form a functional channel. Although all regions of VG channels are contributory to function and regulation, perhaps the two most significant TM regions are S4 and S5/6. The S4 region contains the primary voltage sensor, which has positively charged amino acid residues (arginine or lysine) at every third position (Noda et al. 1984). This region initiates the conformational change of channel structure that follows membrane depolarization and leads to channel opening. The pore itself is formed by the region linking the S5 and S6 regions of the channel (the imaginatively named S5–S6 linker)

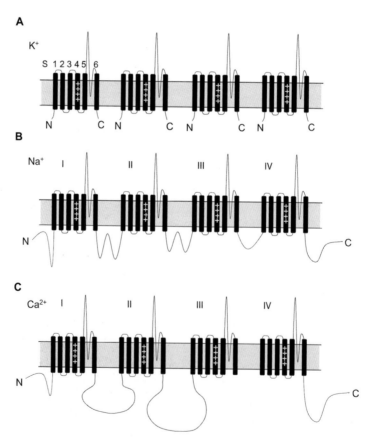

FIGURE 1. Diagram showing the basic structure of the pore-forming α subunits of (A) K$^+$, (B) Na$^+$, and (C) Ca^{2+} voltage-gated ion channels. Each functional channel is composed of four homologous domains (I, II, III, IV), each of which comprises six TM-spanning regions (termed S1–6). The α subunits of Na$^+$ and Ca^{2+} channels are made from one protein, whereas the α subunit of the K$^+$ channel is a tetramer of four separate proteins.

(Taglialatela et al. 1994; Catterall 2000a). The pore, its diameter, and any positively or negatively charged amino acid residues present govern which ion will be allowed to traverse the pore when the channel is open. It is important to be aware that the ion that possesses the greatest permeability for an ion channel is not necessarily the physiological ion. For example, VG Ca^{2+} channels are more permeable to Ba^{++} than Ca^{2+} because Ca^{2+} acts on the cytoplasmic regions of its own channel to inhibit additional Ca^{2+} entry, a feature not shared by Ba^{++} (Hagiwara and Ohmori 1982).

Neurons move ions against their concentration gradients to maintain the resting potential that is a crucial prerequisite for excitability. Once the voltage sensors present in VG Na^+ channels detect a suitable membrane depolarization, the ion channels open to allow Na^+ to flow down its respective electrochemical gradient. This influx is regenerative in the sense that it further depolarizes the membrane causing more channels to open, leading to a full-blown action potential. It is important to note that action potentials are an all-or-nothing event. This is to say that, within a neuron, it is not physiologically possible to generate half an action potential. However, based on the precise assortment of ion channels present, action potentials can vary quite widely in amplitude, duration, and frequency between differing neurons. Action potentials, or spikes as they are usually called, code and transmit information through the nervous system. Because action potentials are all-or-nothing events, information transfer in the CNS is encoded mainly by frequency rather than amplitude.

Na^+, K^+, and Ca^{2+} are the main ion species whose distribution and movement across the cell membrane combine to allow neurons to fire action potentials. Each ion makes its own unique contribution and governs a different aspect of neuronal excitability. Moreover, each ion crosses the membrane through specific ion channels that show greatly increased permeability to that ion relative to others. Experimentally, one can show that channels are permeable to other ions, but in practice the permeability to other ions is so low that it can be effectively discounted. The roles of each ion and its respective ion channels are described below. Briefly, and in simplistic terms, they can be thought of as Na^+ underlying the depolarization of the action potential, K^+ setting both the resting membrane potential and mediating the hyperpolarization phase of the action potential, and Ca^{2+} mediating the coupling of electrical activity with biological processes.

SODIUM CHANNELS

Many biologists are familiar with the role of Na^+ for the depolarization that underlies the action potential. The influx of Na^+ is through a VG Na^+ channel that responds to increases in membrane potential (depolarization) by opening. As stated above, the first channels to open facilitate a further increase in membrane potential resulting in more channel openings until a threshold is reached whereupon a sufficient number of channels are open to allow the neuron to fire an action potential. The action potential is terminated by the inactivation (closure) of Na^+ channels, and subsequent opening of delayed VG K^+ channels that allow the efflux of K^+ out of the cell to bring about a rapid hyperpolarization. To maintain action potential firing, neurons must continually reestablish the differential distribution of both Na^+ and K^+ across the membrane, and this energy-consuming process is achieved through activity of the Na^+/K^+ exchanger (which pumps three Na^+ ions out for every two K^+ ions that come in).

The *Drosophila* genome contains one confirmed gene that encodes a VG Na^+ channel: this gene is termed *paralytic* (*para*) (Suzuki et al. 1971). Other candidate genes have been identified (e.g., *NaCP60E*) but have not yet been electrophysiologically characterized (Littleton and Ganetzky 2000). A typical Na^+ current (termed I_{Na}) from *Drosophila* motor neurons is shown in Figure 2A. The current is comprised of two components: (1) a transient current (I_{NaT}) that rapidly activates and inactivates, and (2) a smaller amplitude, slowly activating and noninactivating, persistent current (I_{NaP}). The transient component underlies the generation of the depolarizing phase of the action potential. In contrast, the persistent fraction of sodium currents is activated in the subthreshold voltage range and is thought to contribute to plateau generation, pacemaker activity, and increased firing frequencies (Li and Bennett 2003; Nikitin et al. 2006; Tazerart et al. 2008). Significantly, this current component is also a principal target for antiepileptic drugs (Lampl et al. 1998).

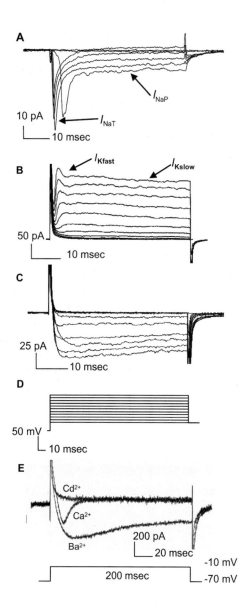

FIGURE 2. Voltage-dependent ion channel characteristics in aCC and RP2 motor neurons. (A) Blocking both outward I_K and inward I_{Ca} isolates the voltage-activated inward I_{Na}. This current has both a fast activating and inactivating transient component (I_{NaT}) and a slow activating, noninactivating persistent component (I_{NaP}). (B) Isolation of VG K$^+$ currents reveals the presence of at least two voltage-activated outward K$^+$ currents: a fast transient (I_{Kfast}) and a slow delayed (I_{Kslow}) current. (C) Isolation of voltage-activated I_{Ca} was achieved by blocking voltage-activated I_K and I_{Na} currents. I_{Ca} was measured using Ba^{++} as the permeant ion. Currents in A–C are from aCC motor neurons in embryos at ~20 h after egg laying. (D) Currents were evoked using voltage steps (10 mV increments; range –60 to +40 mV; 50 msec) applied from a conditioning prepulse of –90 mV (100-msec duration). Traces shown are the average of five trials. (E) For comparison, this panel shows the current carried by either Ca^{2+} or Ba^{++} in an aCC motor neuron from a third-instar larva. The current carried by Ca^{2+} is smaller and inactivates more rapidly. In the presence of Cd^{++}, all Ca^{2+} currents are blocked. (The currents shown in E: Reprinted, with permission, from the American Physiological Society.)

During electrophysiological experiments, Na$^+$ channels may be held in an inactivated state by artificially holding (clamping) the cell membrane potential at 0 mV. However, holding the membrane at this potential for extended periods of time may not be well tolerated. A short inactivation step of several microseconds may be incorporated into a protocol without serious harm to the cell. If it is necessary to block I_{Na} during recordings, tetrodotoxin (TTX) has been proven to be a potent and specific Na$^+$ channel blocker. TTX is effective in the nanomolar range (Saito and Wu 1991).

POTASSIUM CHANNELS

The second class of VG ion channels most biologists will have encountered are VG K$^+$ channels. In addition to being required for the hyperpolarizing phase of the action potential, these channels are critically important for maintaining the resting membrane potential (Hille 2001). It may seem counterintuitive that a channel allowing the flux of positively charged ions could prevent the membrane from becoming more positive. However, K$^+$ channels have a negative reversal or equilibrium potential (~−85 mV): This is the voltage beyond which the direction of flow of ions through the channel effectively reverses. When open, the efflux of K⁺ out of the cell preferentially moves the membrane voltage toward the K⁺ equilibrium potential, although this may never be attained because of the actions of other channels.

The multiplicity of functions served by K⁺ channels is highlighted by the fact that the *Drosophila* genome encodes almost 30 K⁺ channel genes. Four of these make the major contribution to the VG K⁺ current (I_K) in neurons. These genes are *Shaker*, *Shal*, *Shab*, and *Shaw*. *Shaker* and *Shal* lead to rapidly activating and inactivating currents that generate the so-called transient currents, whereas *Shab* and *Shaw* lead to slower activating and noninactivating currents that give rise to the delayed I_K. A typical recording of I_K, showing both the transient and delayed components (termed I_{Kfast} and I_{Kslow}, respectively), is shown in Figure 2B.

In addition to VG K⁺ channels, many neurons also express Ca²⁺-dependent K⁺ channels that are gated by increases in intracellular Ca²⁺ (although some channels are also inherently voltage sensitive) (Hille 2001). These channels, which include *Slowpoke* (the vertebrate BK homolog), often act to terminate repetitive action potential firing through the slow accumulation of intracellular Ca²⁺ (Becker et al. 1995; Lee et al. 2008). The different K⁺ channels present in a neuron can be isolated using pharmacology and, in *Drosophila*, genetics. Widely used pharmacological blockers of K⁺ channels include 4-aminopyridine and tetraethylammonium chloride, both of which are effective at millimolar concentrations (preferentially acting on the transient and delayed components, respectively). *Slowpoke* can be blocked effectively by charybdotoxin or, more simply, through the blockade of Ca²⁺ entry (including removal of external Ca²⁺). Ion block of all I_K channels can also be achieved by use of Cs⁺⁺ ions within the patch pipette, where the Cs⁺⁺ ions act by binding to the cytoplasmic face of the channel proteins.

CALCIUM CHANNELS

Although not primarily involved in the generation of the action potential, VG Ca²⁺ channels nevertheless serve a pivotal role in neuronal function. This is because they often act to couple electrical activity to cellular function. The most apparent of these is the Ca²⁺ dependency of neurotransmitter release (Levitan 2008). Invasion of synaptic terminals by the action potential gates the numerous Ca²⁺ channels present leading to an influx of extracellular Ca²⁺, an initial step that ultimately leads to synaptic vesicle fusion and transmitter release. Influx of Ca²⁺, through VG Ca²⁺ channels, can also lead to increased synthesis of cAMP, activation of calmodulin and Ca²⁺/calmodulin-dependent protein kinases, and/or altered gene regulation (Finkbeiner and Greenberg 1998; Soderling 1999). Indeed, the panoply of roles of intracellular Ca²⁺ are far too vast to be covered here. Suffice to say that Ca²⁺ concentrations can exert a powerful influence on neuronal tissues, from apoptosis to synaptic plasticity.

Several types of Ca²⁺ channels have been identified in mammalian CNS (Catterall 2000b): T-type, a small transient current that opens at negative voltages; L-type, a large persistent current that opens fully at slightly positive membrane potentials; and N-type, which can be further subdivided pharmacologically into N-, P/Q-, and R-type, which are all activated at high voltages, like L-type, but are more transient, reminiscent of T-type channels. The *Drosophila* genome encodes at least four, Ca²⁺-pore-forming subunits (Littleton and Ganetzky 2000), including homologs of N- (*Dmca1A*, *cacophony*) and L-type (*Dmca1D*) mammalian Ca²⁺ channels. Figure 2C shows a typical Ca²⁺ current (I_{Ca}) recorded from *Drosophila* motor neurons using Ba⁺⁺ as the permeant ion. I_{Ca} rapidly inactivates through a mechanism involving the Ca²⁺ ions that traverse the channel (Ca²⁺-dependent inactivation). Although biologically important, for the electrophysiologist this can often prove problematic because it severely reduces the magnitude of I_{Ca}. To overcome this, Ba⁺⁺ is substituted for Ca²⁺ during recording (see Fig. 2E). The advantages of using this substitute ion are that it does not initiate inactivation of the Ca²⁺ channel and, moreover, its extracellular concentration can be raised to as much as 50 mM (or more) without obvious detriment to the cell (unlike Ca²⁺). A high external concentration increases current magnitude by increasing the chemical gradient down which an ion will flow. However, it should be noted that at such high levels, Ba^{++} can inhibit some K^+ channels (e.g., the inward rectifiers [Paradis et al. 2001]).

The various Ca^{2+} currents present in a neuron can be dissected using both pharmacological inhibitors in addition to an array of toxins, most of which originate from cone snails. Examples of such inhibitors include nifedipine (L-type channels), ω-conotoxin (N-type channels), ω-aga iva (P/Q-type channels), and mibefadil (T-type channels). There may be specificity issues with some I_{Ca} blockers; however, if total I_{Ca} block is required, then Cd^{++} can be used.

SYNAPTIC CURRENTS

In addition to isolating and recording specific VG currents in neurons, electrophysiologists often use similar techniques to record synaptic (ligand-gated) currents. This can be done by one of two modalities. The technique of voltage clamp (in which membrane potential is held constant across the neuron) can be used to isolate the synaptic current from any subsequent response that the released neurotransmitter might have, via activation/inactivation of VG currents, in the postsynaptic target cell. Alternatively, current clamp (also known as bridge mode), in which the membrane potential is free to change, can be used to determine how a postsynaptic cell responds to synaptic input (i.e., synaptic and VG currents). The two recording methods are shown in Figure 3. Voltage clamp, a true measure of synaptic current, can also be used to record so-called "minis," which are the small synaptic currents that result from spontaneous release of single vesicles. An example of minis recorded from motor neurons is shown in Figure 3B. It is worth pointing out that determination of quantal content (i.e., the number of vesicles released by a presynaptic action potential) can be problematic in neurons when recording from the soma. This is because the site of synaptic input is both

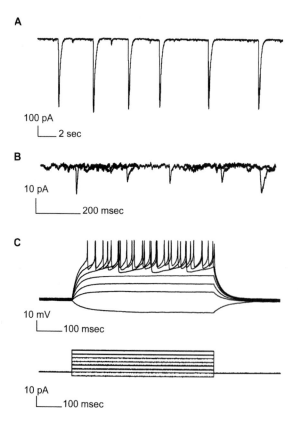

FIGURE 3. Synaptic currents and action potential firing. (*A*) Whole-cell voltage-clamp recordings (V_h = –60 mV) from a wild-type third-instar aCC motor neuron show large inward currents that are the result of the evoked release of presynaptic acetylcholine (Baines et al. 1999; Baines et al. 2001). These currents are relatively long-lived, lasting between 500 and 1000 msec. (*B*) In the presence of 0.1 μM TTX, evoked excitatory currents are abolished leaving only those currents that are elicited by the spontaneous release of vesicles from the presynaptic interneurons (minis). (*C*) In current clamp, injection of depolarizing current is sufficient to fire action potentials in a third-instar aCC motor neuron. Current is injected in 4-pA steps; range –4 to 28 pA; 500 msec.

variable and usually some distance away from the cell soma (from where your recording is being made). This coupled with the dendritic complexity of central neurons means that mini amplitudes recorded are subject to filtering and decrement as the current spreads throughout the neuron. By comparison, mini analysis from muscle is not affected by these issues and, as such, measures of quantal content are more reliable and informative in these cells.

To see how neurons respond to synaptic inputs in a more physiological sense, current clamp recordings are used (Fig. 3C). In these recordings, the membrane potential is free to change. Thus, current clamp recordings will, for example, show whether synaptic inputs are sufficient to fire, or inhibit, action potentials in a postsynaptic neuron.

APPLICATION OF ELECTROPHYSIOLOGY TO CENTRAL NEURONS IN *DROSOPHILA* LARVAE

It is possible to obtain electrophysiological recordings from all stages of *Drosophila*, from early embryos (even before ion channels are expressed at 13 h after egg laying) (Baines and Bate 1998) through to very late wall-climbing third-instar larvae (Worrell and Levine 2008) (see Protocols 1 and 2). Extensive cell lineage studies have been undertaken (particularly for motor neurons), making it possible to identify the same neurons repeatedly in multiple different preparations and in different genetic backgrounds. Reproducibility in the cell recorded, a feat difficult to achieve in mammalian models, greatly reduces variability in results and, moreover, can be used to make very detailed comparisons between different cell types. It is also possible to drive a green fluorescent protein (GFP) marker in many *Drosophila* neurons, but such marking is not routinely required and instead genetics may be devoted to experimental investigation and not to confirming micropipette placement. However, we would urge you to consider having a fluorescent dye in your patch pipette (see Protocol 3) that will enter and label the cell during recording; subsequent examination with fluorescent light will confirm cell identity.

Although *Drosophila* is an established genetic model organism used by ever-increasing numbers of researchers, there still remain issues for the electrophysiologist. Perhaps the most significant of these is that, despite years of work describing the neuronal architecture of both the adult and larvae, the central neuropil remains inaccessible to the electrophysiologist's micropipette. This is a particular concern because the neuropil is where both VG and ligand-gated channels are localized, in addition to synapses. Because the recording pipette is at the soma, which can be, relatively speaking, a very long distance away, this means in practice that control of voltage clamp (i.e., so-called space clamp) and degradation of the signal traveling from neuropil to soma can make it difficult to obtain meaningful current measurements. Thus, the reliability of measurements such as current rise times and voltage dependencies of channel activation can be problematic. Fragility and usable lifetime of the preparation are also problems that must be contended with; mammalian brain slices can be kept viable for hours or even days, but *Drosophila* larval preparations have a much shorter life span of ~30 min. However, although technical difficulties still persist, the advantages of being able to return to the same cell in any genetic background and at all stages of embryonic and larval development considerably outweigh the disadvantages mentioned here.

What follows is a brief description of the basic dissection and recording methods we routinely use to access and record from motor neurons. The reader will appreciate that it is difficult to put into words that which one does with one's hands. Thus, although we have taken every effort to describe our methods accurately, only practice, perseverance, and a little ingenuity will allow you to fully master them.

Dissection of First- and Second-Instar *Drosophila* Larvae: The Flat (or Fillet) Preparation

Exposure of the intact CNS is a prerequisite for electrophysiological recordings. The dissection procedure described here can be applied to both late-stage embryos (stage 16 onward) and larvae. Because of their size, third-instar larvae are more difficult to flatten using this method and, if recording from this stage, the reader might consider using insect pins for the dissection or isolating the CNS using an alternative method (see Protocol 2). The dissection should take <10 min, if all preparation work has been completed (Steps 1–3) in advance. Owing to the short life span of the dissected larva, it is not recommended that the procedure be stopped or the preparation stored for later use.

MATERIALS

CAUTION: See Appendix for proper handling of materials marked with <!>.
See the end of the chapter for recipes for reagents marked with <R>.

Reagents

Drosophila external saline <R>
Mineral oil (or similar)
NaOH (5 M) <!>

Equipment

Borosilicate glass capillaries (outer diameter 1 mm; inner diameter 0.58 mm)
Dissecting microscope (e.g., Leica MZ6 with 25x eye pieces)
Forceps (Dumont #55; Fine Science Tools)
Micropipette puller (e.g., P-97 Flaming/Brown from Sutter Instrument Co.)
Microscope coverslips (22 x 22-mm)
Microscope slides (76 x 26-mm)
Paintbrush (size 00)
Pin vise (aka vice)
Pipette tips (1-mL; blue)
Plastic tubing (flexible; about an arm's length) (Tygon; cat. no. T3601-13)
Power pack able to deliver 15–20 V AC
Sylgard 184 Silicone Elastomer kit (Dow Corning)
Tissue adhesive (Vetbond or GLUture; World Precision Instruments)

> We suspect that reformulation by the manufacturer of the tissue glue (Histoacryl blue; Braun) we previously used now makes it unsuitable for neurophysiological experiments. Histoacryl is an n-butyl derivative that, in high amounts, blocks both cholinergic and voltage-gated K^+ currents. We now recommend using either Vetbond (an n-butyl derivative that is formulated at a lower concentration) or GLUture (a 40:60 mixture of n-butyl and 2-octanyl components).

Tungsten wire (0.1 mm in diameter)

METHOD

1. Prepare Sylgard by combining 10 parts of elastomer base with 1 part of curing agent. Mix together thoroughly. Using the wide end of a pasteur pipette, coat coverslips with an even layer of Sylgard ~1 mm thick. To do this we prefer to use the blunt end of a glass pasteur pipette. We dip the blunt end into the Sylgard and then use the pipette to place a drop of Sylgard on the center of a coverslip (22 x 22-mm). We then spread the drop into a "penny-sized" disc by moving the blunt end of the pipette in a circular motion. Remove bubbles with a sharp pin and cure for 2–3 h at 60°C.

    Unused Sylgard can be kept for some months at −20°C.

2. Prepare glue pipettes by heating and pulling borosilicate glass capillaries with a micropipette puller.

    The settings for the micropipette puller will vary depending on the filament used, the experiment, and laboratory conditions. You are aiming to pull a patch-type electrode (i.e., blunt not sharp) with an opening diameter of ~1 μm. You will need at least 20–50 pipettes and these can be stored indefinitely inside a plastic tray with a lid. Use strips of putty to hold the pipettes in place.

3. Sharpen two tungsten wires. Secure a 2-cm length of tungsten wire (0.1 mm in diameter) in a pin vise. Connect one lead (usually the positive) from a power pack to the shaft of the pin vise (using a crocodile clip) and connect the other lead to a carbon electrode that is in contact with a small volume of 5 M NaOH. Dip the very tip of the tungsten wire into the NaOH to complete the circuit, burning the wire to a point while simultaneously removing any external coating that may be present on the tungsten. To obtain a sharp cutting edge, turn the wire at a 45° angle relative to the surface of the NaOH. View the wire regularly under a dissecting microscope to monitor sharpening.

    This process takes only 10–30 sec if the voltage is sufficient.

    See Troubleshooting.

4. Place 1 drop of mineral oil in the center of a microscope slide and gently place a Sylgard-coated coverslip on top (with the Sylgard surface uppermost).

    This allows the coverslip to be freely rotated (which will not happen if an aqueous droplet is used).

5. Using a small paintbrush, transfer a first-instar larva to the Sylgard-coated coverslip and cover with a drop of *Drosophila* external saline.

6. Half-fill a lid from a 0.5-mL microfuge tube (removed from the remainder of the tube) with tissue adhesive.

    The laboratory environment influences the characteristics of this glue, and it is worthwhile maintaining several samples of different ages (e.g., 1–2 d old and fresh).

7. Attach a blue pipette tip to one end of a length of flexible plastic tubing and a glue pipette to the other end, thereby creating a crude but accurate glue delivery device.

    The plastic tubing should be long enough to give the experimenter free range of movement while holding the blue pipette tip in their mouth and the glass capillary tube in their hand of choice. The glue pipette must be replaced frequently (after each attempt at gluing) because it will block easily.

8. Carefully place the tip of the glue pipette into the tissue adhesive. Gently touch the bottom or side of the container just enough to break the tip of the micropipette, and gently suck the glue into the micropipette shank. Do not overfill the pipette—just a few millimeters of glue are fine.

9. Swiftly, but carefully, move the micropipette tip out of the tissue adhesive and into the saline covering the larva. Maintain a small amount of positive pressure while the pipette tip is out of solution to prevent the tissue adhesive from traveling further up the shank and into the barrel.

The tissue adhesive hardens quickly on contact with saline forming a plug in the tip preventing the tissue adhesive from flowing out of the pipette.

10. Hold the larva in place with one tungsten wire, dorsal side uppermost. Apply the tissue adhesive to the posterior end of the larva to secure it to the Sylgard. To do this, penetrate the Sylgard with the pipette tip until it touches the glass coverslip beneath to remove the plug of hardened glue. This allows the glue once more to flow freely on gentle pressure. As you withdraw the pipette tip from the Sylgard, maintain gentle positive pressure and glue will flow freely. This step is very tricky to master and may take a novice several days or even a few weeks of practice. Sadly, there is no shortcut, other than asking someone else to do all your preparations for you. Once you have secured the posterior end, turn the preparation around and repeat the steps above to secure the anterior. As you get better at this, try to stretch the larvae or embryo a little (be careful) as this will make the following steps much easier.

   See Troubleshooting.

11. To open the larva, gently puncture the cuticle either at the anterior or posterior end (between the trachea) using a sharpened tungsten wire, and gently lift the cuticle up. To cut the cuticle, rub a second tungsten wire against the first, either against the top or bottom surface of the first wire; both strategies work well. This will produce a small cut in the cuticle. Up to a dozen such cuts are required to completely open the cuticle from anterior to posterior.

   Be careful as the cuticle is delicate and prone to tearing. Fragility is a common problem with this preparation (as the observant reader will no doubt have deduced!). While opening the larva, the gut may bulge upward obscuring the interior of the larva and the cuticle. It is acceptable to push the gut to one side, but do not blindly insert the tungsten wire into the interior of the larva because you run the risk of damaging the CNS.

   See Troubleshooting.

12. Glue the cuticle down onto the Sylgard. Using a tungsten wire, gently pull one "corner" of the larval cuticle down to the Sylgard and glue in place. Four glue points are normally needed, one anterior and one posterior at each side (Fig. 4).

   For the best results, create a pile of tissue adhesive near the point you intend to glue, leave it for a fraction of a second allowing it to harden slightly, and then gently move the micropipette tip from

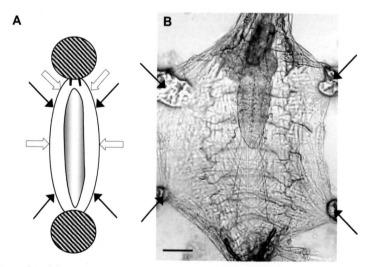

FIGURE 4. A first-instar larval flat preparation. (*A*) The larva is secured at both anterior and posterior ends with glue (cross-hatched area) and the cuticle is opened with sharpened tungsten wires (shaded area). Solid arrows denote optimal glue points for anchoring the cuticle to the Sylgard-coated cover slip. Open arrows show additional points that may be used as necessary to further reveal the CNS. (*B*) A micrograph of an opened larval flat preparation. Solid arrows correspond to *A*. (Reprinted, with permission, from Baines and Pym 2006.)

the pile of glue to the cuticle and pull the cuticle up and down in one smooth half-circle motion to the Sylgard (enough to stretch the cuticle taut and draw it away from the CNS, but not enough to tear).

See Troubleshooting.

13. Remove the gut, body fat, and trachea to allow visualization of the CNS. Cut the intestine with sharpened tungsten wires, as with the cuticle, and remove it with a pair of sharp forceps. Remove any debris (including the two main trachea) by mounting a broken glue pipette into the rubber tubing used in Step 7 and sucking up any free floating debris. Avoid sucking too close to the CNS as it may be drawn in to the pipette and damaged.

See Troubleshooting.

14. Wash the preparation by gently replacing the external saline with fresh saline or other solution appropriate for the experiment (various recipes are given at the end of the chapter) and transfer the preparation to the electrophysiology rig.

See Troubleshooting.

TROUBLESHOOTING

Problem (Step 3): No current appears to be flowing through the tungsten wires.

Solution: The circuit is not complete, likely because of a lack of electrical connection between the tungsten wire and the pin vise. This may be because the wire is not sitting correctly in the vise or there is an accumulation of salt/dirt in the vise. Remove the wire, wash the pin vise in warm ddH$_2$O, and reseat the wire once dry. Retightening the pin vise may also help. Alternatively, swap the leads over.

Problem (Step 10): Glue does not flow out of the pipette tip.

Solution: The glue has traveled up the capillary tube and hardened. Try either to hold your breath while transferring the tip from the glue pot to the saline or gently blowing. Once the tip is in the saline, the pressure from the external saline will cause the glue to rapidly travel up the capillary. Blowing slightly harder will keep the tissue adhesive at the tip allowing it to harden and form a plug. Overcoming this is a matter of practice and dedication.

Problem (Step 11): The larva/cuticle keeps tearing when you are trying to cut it.

Solution: Either resharpen the tungsten wires or be gentler with the preparation; these are very young and fragile animals.

Problem (Step 12): The cuticle is not opened sufficiently or the CNS is obscured.

Solution: The preparation may still be usable; make an extra glue point to try to move the obscuring section of cuticle.

Problem (Step 12–14): A glue point or the cuticle breaks and the preparation closes slightly.

Solution: Make a new glue point to reopen the cuticle. Try not to use the same point as the cuticle will have torn; move slightly anterior or posterior. Any new glue points will be much closer to the CNS and the chance of accidently covering the CNS in tissue adhesive increases.

Problem (Step 14): The preparation is tilted at an angle and is not lying flat against the Sylgard.

Solution: The preparation may still be usable, although the subsequent electrophysiology will be more difficult with a poorly positioned CNS. If the preparation is deemed unusable in its current state, then tissue adhesive may be used to try to maneuver the preparation to a more desirable orientation. It is likely that the preparation will have to be discarded and a new one started, but only take this step if electrophysiology has proved too difficult.

Dissection of Third-Instar Larvae

This dissection procedure is suitable for third-instar larvae. It is somewhat simpler and faster than that described for first-instar larvae in Protocol 1. The dissection should take ~5 min to complete, if all preparation work has been completed (Step 1) in advance. Owing to the short life span of the dissected larva, it is not recommended that the procedure be stopped or the preparation stored for later use.

MATERIALS

See the end of the chapter for recipes for reagents marked with <R>.

Reagents

Drosophila external saline <R>
Mineral oil

Equipment

Dissecting microscope (e.g., Leica MZ6 with 25x eyepiece)
Forceps (Dumont #55; Fine Science Tools) (sharp; the points must meet precisely; at least two pairs)
Microscope coverslips (22 x 22-mm)
Microscope slides (76 x 26-mm)
Paintbrush (size 00)
Plastic dish (shallow; ~2–3 cm in diameter)
Sylgard 184 Silicone Elastomer kit (Dow Corning)
Tissue adhesive (Vetbond or GLUture; World Precision Instruments)
  We suspect that reformulation by the manufacturer of the tissue glue (Histoacryl blue; Braun) we previously used now makes it unsuitable for neurophysiological experiments. Histoacryl is an n-butyl derivative that, in high amounts, blocks both cholinergic and voltage-gated K^+ currents. We now recommend using either Vetbond (an n-butyl derivative that is formulated at a lower concentration) or GLUture (a 40:60 mixture of n-butyl and 2-octanyl components).
Whetstone

METHOD

1. Sharpen forceps as needed using a whetstone and mineral oil for lubrication.

   Forceps should meet precisely, with no overhang from either tip. Sharpening strokes with a shallow angle will produce a long tapering tip and high angles will sharpen only the tips. It is possible to oversharpen one side, such that the other will require grinding down in order for the two tips to meet once more. Slow and steady will produce better forceps than rushing. It is not unusual to spend an hour obtaining an adequate pair of forceps.

2. Fill a shallow plastic dish half full of *Drosophila* external saline and place a third-instar larva in the saline.

3. Using two pairs of sharpened forceps, grip the larva across its dorsal surface about halfway down its length. This placement is important: Too posterior and you will have to remove an

unnecessary excess of tissue taking up valuable time and possibly accidently damaging the CNS; too anterior and the chances of damaging the CNS increase. Tear the larva by moving the forceps apart and slightly upward in a shallow half-circle motion. Discard the posterior part.

See Troubleshooting.

4. To reveal the CNS, use forceps to tear the cuticle of the anterior part between the trachea, starting at the torn end of the larva and working up to the mouth hooks.

   At this stage the preparation should be a flat sheet of cuticle joined to the mouth hooks containing the CNS and a mass of other tissue. It should be possible to visualize the CNS and to confirm it is still intact. It is possible that the posterior portion of the ventral nerve cord has been damaged. Take a moment to ensure that the CNS is intact. If so, proceed to Step 5. If not, return to Step 2 with a fresh larva, washing the shallow plate clean to avoid contamination.

   See Troubleshooting.

5. To remove the cuticle entirely, grip the mouth hooks with one pair of forceps and the cuticle with the other pair, just below the mouth hooks, and while gripping tightly, pull the forceps apart.

   The CNS should not be disturbed by this manipulation, although occasionally an imaginal disc or process will have to be separated manually afterward. Avoid directly pulling on the CNS because, despite the relatively increased toughness of the third-instar CNS compared with the first-instar, the CNS is still fragile.

   See Troubleshooting.

6. To remove the mouth hooks, place a pair of forceps between the mouth hooks and the CNS and pinch firmly. Do not touch the mouth hooks or the CNS with this pair of forceps. This prevents the CNS from being pulled unduly when the mouth hooks are separated. Grip the mouth hooks with a second set of forceps and pull the mouth hooks away from the CNS. Do not pull the CNS away from the mouth hooks.

7. Remove excess tissue or any remaining cuticle still attached to the CNS. Discs may be left on (as they provide a convenient means to grasp and manipulate the CNS) or removed as preferred.

8. Prepare Sylgard by combining 10 parts of elastomer base with 1 part of curing agent. Mix together thoroughly. Using the wide end of a Pasteur pipette, coat coverslips with an even layer of Sylgard ~1 mm thick. Remove bubbles with a sharp pin and cure for 2–3 h at 60°C.

   Unused Sylgard can be kept for some months at −20°C.

9. Place 1 drop of mineral oil in the center of a microscope slide and gently place a Sylgard-coated coverslip on top (with the Sylgard surface uppermost). Add a few drops of *Drosophila* external saline to the coverslip. Transfer the isolated CNS to this saline droplet, placing the CNS in the middle on the Sylgard surface. Fix the CNS to the Sylgard using tissue adhesive. First attach the posterior end of the ventral nerve cord using the posterior nerve roots as anchor points. Next attach the CNS at the anterior end by applying glue to the brain lobes.

   The CNS will actually slightly adhere to the Sylgard without any adhesive, which helps the glueing process. If the CNS needs to be lifted off the Sylgard for repositioning, care should be taken to avoid damage.

   See Troubleshooting.

10. Transfer the preparation to the electrophysiology rig.

TROUBLESHOOTING

Problem (Step 3): The larva will not tear correctly.
Solution: The larva has likely been gripped incorrectly. The forceps should only grip the dorsal surface of the animal, not the full width. First the dorsal surface tears and, as the forceps are pulled further apart, then the ventral surface tears afterward.

Problem (Step 4): After removing the cuticle, the CNS is damaged.

Solution: Damage may have been sustained when tearing the larva in Step 3 or when removing the cuticle. Always tear the dorsal surface, never the ventral, because the CNS lies on the ventral cuticle of the larva. If the CNS is damaged or missing, start the dissection again with a new larva.

Problem (Step 5): The mouth hooks have already come off at an earlier point in the dissection.

Solution: It is still possible to remove the cuticle, but it is more difficult because you will lack an anchor point for the forceps that is close to the CNS to allow maximum removal of the cuticle, yet distinct enough to make damage to the CNS unlikely. It is recommended that the CNS is first clearly visualized and then the cuticle is torn off in strips instead of all at once. This will slowly remove the cuticle and the surrounding tissue still adhering to the CNS, while minimizing the risk of damaging the CNS. If a ready supply of third-instar larvae is at hand, it may be quicker to simply discard the preparation and start over.

Problem (Step 9): The CNS has moved while being glued into position.

Solution: It may be possible to pull the CNS back into position by gluing in the reverse direction, or even pulled out of position the preparation may still be usable. The more gluing that occurs, the greater the chance of damage to the CNS.

Whole-Cell Patch Recording from *Drosophila* Larval Neurons

This protocol describes a procedure for revealing larval motor neurons and applying whole-cell patch recording techniques to these cells. Basic electrophysiological theory will not be covered and it is assumed the reader has some prior experience. The useful lifetime of first-instar larval preparations is ~30 min and that of third-instar CNS preparations is up to 1 h. It is therefore recommended that fresh preparations are used and that no breaks are taken during the procedure, although there may be time to pull and polish a patch pipette.

MATERIALS

See the end of the chapter for recipes for reagents marked with <R>.

Reagents

Drosophila internal patch solutions <R>
Drosophila external saline <R>
Patch pipette dye <R>
Protease type XIV (Sigma; cat. no. P5147-5G)

Equipment

Anotop 10 syringe filters (0.2 μm pore size; 10 mm diameter) (Whatman; cat. no. 6809-1022)
Electrophysiology rig (including an upright compound microscope [e.g., Olympus BX51WI] with
    10x air and 60x long-distance water-immersion lenses and 12.5x eyepieces)
      The setup and maintenance of a rig is beyond the scope of this manual and it is recommended that
      the reader looks to more experienced colleagues for guidance.

Enzyme pipettes (borosilicate glass; 1 mm outer diameter; 0.78 mm inner diameter with filament)
    (Harvard Apparatus; cat. no. 30-0038)
MicroFil 28 AWG needles (World Precision Instruments; cat. no. MF28G67-5)
Microforge
Micropipette puller (e.g., P-97 Flaming/Brown from Sutter Instrument Co.)
Minisart 0.20-μm single-use hydrophilic syringe filters (Sartorius; cat. no. 16534)
Patch pipettes (borosilicate glass; 1 mm outer diameter; 0.58 mm inner diameter with filament)
    (Harvard Apparatus; cat. no. 30-0019)
Software suitable for data capture and analysis in patch-clamp experiments (e.g., pClamp from
    Molecular Devices)
Syringes (1-mL and 10-mL)

METHOD

Preparation of Enzyme and Patch Pipettes

1. Pull enzyme and patch pipettes.

    Enzyme pipettes are patch-type electrodes with an opening of ~5 μm. Patch pipettes have an
    opening, before polishing, of ~1 μm.

2. Fire-polish the patch (but not enzyme) pipettes with the microforge to reduce tip diameter.

   When filled with patch solution, the desired resistance is ~15 MΩ for first-instar larvae and ~10 MΩ for third-instar larvae.

Technique to Reveal Larval Motor Neurons

3. Under low magnification, ensure the CNS is centered in view. Confirm the CNS is also centered under high magnification. Perform a last check on CNS integrity. Check that no damage has been sustained during the dissection.

   Almost any damage to a first-instar larval CNS will render it unusable.

4. Place an enzyme pipette containing a solution of 1% (w/v) protease type XIV in *Drosophila* external saline in the electrode holder on the rig.

   Enzyme pipettes are best filled using a 1-mL syringe with an Anotop 10 0.2-μm filter and a MicroFil 28 AWG needle. Only the tip of the electrode should be filled. The filter is useful to prevent debris from entering the pipette and blocking the tip.

5. Under low magnification, move the pipette gently, but swiftly, down to just above the CNS. Do not move the pipette below the field of view; always move the plane of focus and then move the pipette to match.

   The CNS is very fragile and easily torn by the sharp edge of a pipette.

6. Repeat Step 5 under high magnification until the enzyme pipette is positioned 10–20 μm above the CNS.

7. Select an area along the dorsal midline (between the transverse nerves) for removal of the glial sheath. At first, your disruption of the glia may be overvigorous and will cause unintentional damage, but practice makes perfect. Gently, via a mouth pipette, suck a small section of the outer glial sheath into the tip of the enzyme pipette and hold for ~10 sec to allow the enzyme to work, and then gently exhale. Continue this process until a rupture occurs in the sheath and the neurons are clearly seen.

   If suction is too hard, neurons will be removed from the preparation via the mouth pipette. Similarly, if the operator exhales with too much force, the CNS will be inflated like a balloon, likely disrupting many synaptic connections in the neuropil. First-instar preparations effectively have only a single layer of glial tissue to remove, whereas third-instar preparations have multiple layers that can require multiple applications of enzyme to remove. Neurons should be clearly visible with sharply defined edges. If they appear to be "fuzzy," then there is still glia wrapping to be removed.

Patch Recording from Larval Motor Neurons

8. Fill the tip of a polished patch pipette with patch pipette dye either by back filling or by injecting using a 1-mL syringe with a MicroFil 28 AWG needle. This enables cell identification after patch recording. Fill the rear half of the pipette with *Drosophila* internal patch solution using the MicroFil 28 AWG needle attached to a 10-mL syringe and filtered by a Minisart 0.2-μm filter. Place the filled micropipette into the electrode holder, positioning its tip just above the preparation as for the enzyme pipette (see Steps 5 and 6).

   The tip of the pipette is very narrow and therefore prone to blockage from free-floating debris remaining from the application of protease to the glial sheet. Take care and use extra speed to avoid this from occurring. If any blockage does occur, discard the pipette and repeat this step until an unblocked patch pipette is in position.

9. Before attempting to patch, check the electrode resistance and discard if it lies outside the 10–20-MΩ range (for first- or second-instar larvae) or the 5–15-MΩ range (for third-instar larvae).

10. Select a motor neuron to patch.

   The UAS/GAL4 expression system driving GFP in selected motor neurons will help when starting; for example, RN2-GAL4 drives expression in the aCC and RP2 motor neurons (Fujioka et al. 2003). A full description of how to record from neurons using a patch electrode is beyond the scope of this chapter. We strongly suggest that interested individuals seek help from established electrophysiology groups in their own institutions. Many electrophysiology amplifier manufacturers also provide tutorials for novice users. Patch recording from *Drosophila* neurons is no different to any other neuron, and so guidance can be obtained from any existing electrophysiology group.

11. After a recording, do not immediately change pipettes. Instead, switch to a GFP filter (440–480 nm excitation) and view the recorded Alexa Fluor dye-filled neuron under fluorescent light.

    This will allow visualization of the recorded neuron; its dendritic morphology and muscle target (if a motor neuron) can be clearly observed.

12. Change patch pipettes and select a new neuron to record from.

    First-instar larval preparations will remain viable only for up to 30 min from when the cuticle is first opened, limiting the number of recordings that can be taken. Two good recordings per first-instar preparation is an acceptable aim. However, a third-instar larval preparation remains viable for much longer and three or four good recordings per preparation may be made as long as suitable cells are present.

RECIPES

CAUTION: See Appendix for proper handling of materials marked with <!>.
Recipes for reagents marked with <R> are included in this list.

Drosophila *External Saline (Used to Record Whole-Cell Currents, Action Potentials, and Synaptic Currents)*

Reagent	Final concentration
NaCl	135 mM
KCl <!>	5 mM
MgCl$_2$·6H$_2$O <!>	4 mM
CaCl$_2$·2H$_2$O <!>	2 mM
TES <!>	5 mM
Sucrose	36 mM

If the calcium chloride and sucrose are omitted, then a 10x stock solution can be prepared that will last for months when kept refrigerated. We usually make a 500-mL working solution (50 mL stock + 450 mL ddH$_2$O) to which we add the required amount of calcium and sucrose. The pH (of the 10x stock solution) should be adjusted to 7.15 with 10 M NaOH <!>.

Drosophila *External Calcium Saline (Used to Isolate Ca^{2+} Currents)*

Reagent	Final concentration
NaCl	50 mM
KCl <!>	6 mM
BaCl <!>	50 mM
MgCl$_2$·6H$_2$O <!>	10 mM
Glucose	10 mM
Tetraethylammonium chloride <!>	50 mM
4-aminopyridine <!>	10 mM
HEPES	10 mM
TTX <!>	1 μM

TTX is deadly. Exercise extreme caution. If true Ca^{2+} currents are required (not Ba^{++}), then replace 50 mM Ba^{++} with 2–5 mM Ca^{2+} and increase the concentration of NaCl to compensate. Adjust the pH to 7.1 with 10 M NaOH <!>.

Drosophila *External Potassium Saline (Used to Isolate K⁺ Currents)*

Reagent	Final concentration
Drosophila external saline <R>	1x
TTX <!>	1 µM

TTX is deadly. Exercise extreme caution. If only VG K⁺ channels are to be isolated, add CdCl$_2$ (0.2 mM) to block any Ca^{2+}-dependent K⁺ currents.

Drosophila *External Sodium Saline (Used to Isolate Na⁺ currents)*

Reagent	Final concentration
NaCl	100 mM
KCl <!>	6 mM
MgCl$_2$·6H$_2$O <!>	2 mM
Sucrose	10 mM
Tetraethylammonium chloride <!>	50 mM
4-aminopyridine <!>	10 mM
HEPES	10 mM
CdCl$_2$ <!>	0.2 mM

Adjust the pH to 7.1 with 10 M NaOH <!>.

Drosophila *Internal Cesium Patch Solution (Used to Record Na⁺ and Ca²⁺ Currents)*

Reagent	Final concentration
MgCl$_2$·6H$_2$O <!>	2 mM
EGTA	2 mM
CsCl <!>	5 mM
HEPES	20 mM
CsCH$_3$SO$_3$	140 mM

To make CsCH$_3$SO$_3$, take 2.38 g of CsOH <!> and add 0.91 mL of methanesulfonic acid <!> per 100 mL of patch saline. Adjust the pH to 7.4 with 10 M CsOH <!>.

Drosophila *Internal Potassium Patch Solution (Used to Record K⁺ Currents, Action Potentials, and Synaptic Currents)*

Reagent	Final concentration
MgCl$_2$·6H$_2$O <!>	2 mM
EGTA	2 mM
KCl <!>	5 mM
HEPES	20 mM
KCH$_3$SO$_3$	140 mM

To make KCH$_3$SO$_3$, take 0.78 g of KOH <!> and add 0.91 mL of methanesulfonic acid <!> per 100 mL of patch saline. The internal potassium patch solution will be in contact with the cells' internal environment and therefore great care should be taken to ensure concentrations are as accurate as possible. Adjust the pH to 7.4 with 10 M KOH <!>.

Patch Pipette Dye

Prepare 5-μL aliquots of 0.4% (w/v) Alexa Fluor 488 hydrazide (sodium salt; Invitrogen; cat. no. A10436) by dissolving 1 mg in 250 μL of the required internal patch solution <R>. Store frozen until needed. To make a working solution, add a further 155 μL of required internal patch solution to an aliquot of Alexa Fluor 488.

REFERENCES

Baines RA, Bate M. 1998. Electrophysiological development of central neurons in the *Drosophila* embryo. *J Neurosci* **18:** 4673–4683.

Baines RA, Pym EC. 2006. Determinants of electrical properties in developing neurons. *Semin Cell Dev Biol* **17:** 12–19.

Baines RA, Robinson SG, Fujioka M, Jaynes JB, Bate M. 1999. Postsynaptic expression of tetanus toxin light chain blocks synaptogenesis in *Drosophila*. *Curr Biol* **9:** 1267–1270.

Baines RA, Uhler JP, Thompson A, Sweeney ST, Bate M. 2001. Altered electrical properties in *Drosophila* neurons developing without synaptic transmission. *J Neurosci* **21:** 1523–1531.

Becker MN, Brenner R, Atkinson NS. 1995. Tissue-specific expression of a *Drosophila* calcium-activated potassium channel. *J Neurosci* **15:** 6250–6259.

Catterall WA. 2000a. From ionic currents to molecular mechanisms: The structure and function of voltage-gated sodium channels. *Neuron* **26:** 13–25.

Catterall WA. 2000b. Structure and regulation of voltage-gated Ca^{2+} channels. *Ann Rev Cell Dev Biol* **16:** 521–555.

Finkbeiner S, Greenberg ME. 1998. Ca^{2+} channel-regulated neuronal gene expression. *J Neurobiol* **37:** 171–189.

Fujioka M, Lear BC, Landgraf M, Yusibova GL, Zhou J, Riley KM, Patel NH, Jaynes JB. 2003. Even-skipped, acting as a repressor, regulates axonal projections in *Drosophila*. *Development (Camb)* **130:** 5385–5400.

Hagiwara S, Ohmori H. 1982. Studies of calcium channels in rat clonal pituitary cells with patch electrode voltage clamp. *J Physiol* **331:** 231–252.

Hille B. 2001. *Ion channels of excitable membranes.* Sinauer, Sunderland, MA.

Lampl I, Schwindt P, Crill W. 1998. Reduction of cortical pyramidal neuron excitability by the action of phenytoin on persistent Na^+ current. *J Pharmacol Exp Ther* **284:** 228–237.

Lee J, Ueda A, Wu CF. 2008. Pre- and post-synaptic mechanisms of synaptic strength homeostasis revealed by slow-poke and shaker K^+ channel mutations in *Drosophila*. *Neuroscience* **154:** 1283–1296.

Levitan ES. 2008. Signaling for vesicle mobilization and synaptic plasticity. *Mol Neurobiol* **37:** 39–43.

Li Y, Bennett DJ. 2003. Persistent sodium and calcium currents cause plateau potentials in motoneurons of chronic spinal rats. *J Neurophysiol* **90:** 857–869.

Littleton JT, Ganetzky B. 2000. Ion channels and synaptic organization: Analysis of the *Drosophila* genome. *Neuron* **26:** 35–43.

Nikitin ES, Kiss T, Staras K, O'Shea M, Benjamin PR, Kemenes G. 2006. Persistent sodium current is a target for cAMP-induced neuronal plasticity in a state-setting modulatory interneuron. *J Neurophysiol* **95:** 453–463.

Noda M, Shimizu S, Tanabe T, Takai T, Kayano T, Ikeda T, Takahashi H, Nakayama H, Kanaoka Y, Minamino N, et al. 1984. Primary structure of *Electrophorus electricus* sodium channel deduced from cDNA sequence. *Nature* **312:** 121–127.

Paradis S, Sweeney ST, Davis GW. 2001. Homeostatic control of presynaptic release is triggered by postsynaptic membrane depolarization. *Neuron* **30:** 737–749.

Saito M, Wu CF. 1991. Expression of ion channels and mutational effects in giant *Drosophila* neurons differentiated from cell division-arrested embryonic neuroblasts. *J Neurosci* **11:** 2135–2150.

Soderling TR. 1999. The Ca-calmodulin-dependent protein kinase cascade. *Trends Biochem Sci* **24:** 232–236.

Suzuki DT, Grigliatti T, Williamson R. 1971. Temperature-sensitive mutations in *Drosophila melanogaster*. VII. A mutation (*para^{ts}*) causing reversible adult paralysis. *Proc Natl Acad Sci* **68:** 890–893.

Taglialatela M, Champagne MS, Drewe JA, Brown AM. 1994. Comparison of H5, S6, and H5-S6 exchanges on pore properties of voltage-dependent K^+ channels. *J Biol Chem* **269:** 13867–13873.

Tazerart S, Vinay L, Brocard F. 2008. The persistent sodium current generates pacemaker activities in the central pattern generator for locomotion and regulates the locomotor rhythm. *J Neurosci* **28:** 8577–8589.

Worrell JW, Levine RB. 2008. Characterization of voltage-dependent Ca^{2+} currents in identified *Drosophila* motoneurons in situ. *J Neurophysiol* **100:** 868–878.

18 · In Vivo Whole-Cell Recordings in the *Drosophila* Brain

Mala Murthy[1] and Glenn Turner[2]

[1]*Princeton University, Department of Molecular Biology and Princeton Neuroscience Institute, Princeton, New Jersey 08544;* [2]*Cold Spring Harbor Laboratory, Cold Spring Harbor, New York 11724*

ABSTRACT

Whole-cell patch-clamp recordings provide exceptional access to spiking and synaptic neural activity. This method has recently been applied to neurons in the central nervous system of *Drosophila* and allows researchers the opportunity to study the function of their neurons of interest within the context of native circuits in a genetically tractable model system. In this chapter, we present a preparation method to expose neurons in the fly brain and describe the technique for in vivo whole-cell patch-clamp recordings in this preparation. We also offer technical suggestions and discuss some of the challenges encountered in recording from single neurons in the fly brain.

INTRODUCTION

Drosophila offers many excellent features as a system in which to investigate the function of neuronal circuits. Its brain contains only ~100,000 neurons; subpopulations of cells can be genetically labeled and identified across animals; genes and neurons can be silenced or activated with relative ease; and there exists a wealth of information about various behaviors. We can now add to this list of advantages the application of patch-clamp methods to record the activity of individual neurons in the central nervous system of live, adult flies (Wilson et al. 2004). Monitoring endogenous patterns of neural activity with the high temporal resolution offered by electrophysiology will certainly find numerous and exciting applications for *Drosophila* researchers. From our perspective, examining neuronal function in vivo, in *Drosophila*, is key for studies of sensory processing, for correlating neural activity with behavior, and for examining synaptic properties within the context of native, intact circuits. Below we provide a few examples of such studies that have been performed already in the olfactory system.

Using electrophysiological methods to assay the function of the nervous system is not new for *Drosophila* neurobiologists, as other chapters in this book reveal (studies of synaptic development, transmission, and plasticity in the fly have involved recordings from the neuromuscular junction). However, in vivo whole-cell patch-clamp recordings can be used in the adult fly brain to examine not only properties of the synapse but also sensory responses and circuit dynamics and ultimately to correlate neural activity with behavior. In this recording technique, a glass micropipette forms a tight, high-resistance seal with the neuronal cell membrane. The patch of membrane underneath the pipette opening is then ruptured by suction to provide low-resistance intracellular access to electrical activity within the neuron. Whole-cell recordings provide a high signal-to-noise ratio for recorded events and significant mechanical stability in the face of mechanical movements of the brain in the intact animal. Whole-cell recordings can be obtained from even the smallest of cells in the fly brain, such as the mushroom body Kenyon cells, and both spiking and synaptic events are readily detectable. Example recordings, highlighting such events in two different types of *Drosophila* olfactory neurons, are shown in Figure 1.

Whole-cell recordings are made from the cell bodies of *Drosophila* neurons, which are usually only a few micrometers in diameter. Although patching such small cells is challenging, these recordings reveal the dynamics of neuronal activity with single-spike sensitivity and millisecond temporal resolution (not simply whether or not a neuron responds). Additionally, this technique allows one to examine the rules for synaptic integration that shape how synaptic input turns into spiking output for each cell type. However, whole-cell recordings are limited in their scope, because it is difficult to record from more than one neuron at a time. To access population-level neural activity, functional imaging has been useful, but imaging activity using genetically encoded sensors does not currently offer the sensitivity and temporal resolution of electrophysiology (Pologruto et al. 2004; Reiff et al. 2005; Jayaraman and Laurent 2007). Also, most genetically encoded sensors signal intracellular calcium levels (e.g., GCaMP), not membrane potential, and simultaneous imaging and electrophysiology are needed to establish the relationship between sensor signal and spiking activity. Moreover, because different cell types will possess different complements of calcium channels or different levels of the reporter, this relationship must be established for each neuronal context. Promising studies of this kind using a fast calcium-sensitive dye have shown that imaging techniques can allow neural activity to be followed with high temporal precision (~30 msec) in large populations of neurons (Yaksi and Friedrich 2006).

The value of in vivo recordings in *Drosophila* is nicely illustrated by recent experiments that have examined neuronal function in the olfactory circuit. The olfactory circuit begins with the olfactory sensory neurons (OSNs), which are distributed over the surface of the antennae and maxillary palps. In the adult, there are approximately 1300 OSNs, which fall into approximately 50 different classes

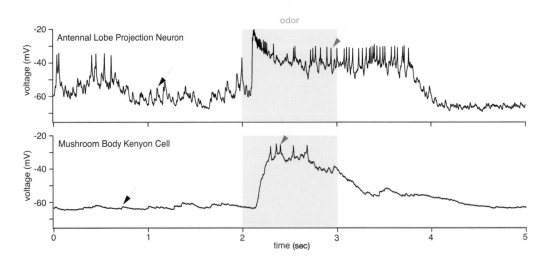

FIGURE 1. Recorded membrane voltage (via whole-cell patch clamp) from an antennal lobe projection neuron (*top*) and a mushroom body Kenyon cell (*bottom*). The odor ethyl butyrate was delivered to each cell for 1 sec (indicated by the light blue bar). Action potentials or spikes are indicated with red arrows and excitatory postsynaptic potentials with black arrows. (Recordings by M.M.)

defined by the identity of the odorant receptor(s) each expresses (Robertson et al. 2003; Vosshall and Stocker 2007; Benton et al. 2009). The in vivo odor responses of these neurons have been extensively characterized with extracellular recordings (another very useful tool for *Drosophila* neurobiologists—methods detailed in Chapter 16) (Kreher et al. 2005; Hallem and Carlson 2006). OSNs send axons to the antennal lobe, where they terminate stereotypically in discrete glomeruli, contacting projection neurons (PNs) and local neurons (LNs). An individual PN extends its dendrites into a single glomerulus and its axon to higher brain centers, called the mushroom body (MB) and the lateral horn (LH) (Tanaka et al. 2004; Komiyama and Luo 2006; Jefferis et al. 2007; Lin et al. 2007). Whole-cell patch-clamp recordings from PNs reveal that each PN responds to a larger number of odors than its presynaptic OSN partner (i.e., PNs are more broadly tuned than OSNs) (Wilson et al. 2004; Bhandawat et al. 2007). Additionally, because each PN averages input from multiple OSNs of a given type, the response of an individual PN is less variable than that of a single OSN (Bhandawat et al. 2007). Together these two operations—broadening of odor tuning and increasing signal to noise—help to ensure that PN responses to different odors are highly distinct. In particular, the broadening of PN tuning curves ensures that, although there is a limited range of response intensities in PNs, spanning 0 to ~250 Hz, responses to different odors are evenly distributed across this range. In contrast, odor responses in OSNs are bimodally skewed to the high and low ends of this range. Consequently, in the OSN population, different odors may evoke rather similar responses. In the PNs, odor responses are consistently distinct, although the magnitude of their difference is intermediate in value. This transformation occurs because weak odor responses in OSNs are amplified in the postsynaptic PNs, but strong OSN inputs are not (Bhandawat et al. 2007). What is the underlying mechanism for this transformation? Recording synaptic responses from PNs while stimulating OSN fibers at different rates showed that the OSN to PN synapse is strong, with a high probability of synaptic vesicle release, but that strong stimuli produce short-term depression at this synapse (Kazama and Wilson 2008).

At the third layer of the circuit, olfactory representations are transformed again. Recordings from MB Kenyon cells (KCs), postsynaptic targets of the PNs, revealed that KCs are highly odor-selective and their responses consist of small numbers of spikes, both in *Drosophila* and in other insects (Perez-Orive et al. 2002; Wang et al. 2004; Szyszka et al. 2005; Turner et al. 2008). This high odor selectivity is theoretically useful for forming accurate olfactory memories (Kanerva 1988; Laurent 2002), an important role for the MB (Heisenberg et al. 1985; Dubnau et al. 2001; McGuire et al. 2001). Recordings from a subset of genetically labeled KCs showed that individual KCs do not consistently possess the same odor response properties in different flies—revealing that the exquisite circuit specification observed in the antennal lobe is likely absent at the third layer of the system (Murthy et al. 2008).

In this chapter, we present methods for performing in vivo patch-clamp recordings in the *Drosophila* brain. The two most challenging elements of these protocols are dissecting the fly to expose the brain (Protocol 1), while keeping the preparation alive and healthy, and achieving stable whole-cell recordings from very small neurons (Protocol 3). For these aspects, we have provided extensive details in the protocols. For other aspects of the methodology (e.g., setting up an electrophysiology rig, de-noising, filtering recorded signals), we refer readers to several excellent books on the topic (Sakmann and Neher 1995; Walz et al. 2002; Molleman 2003). Protocol 2 describes the preparation of pipettes for use in patching.

Dissection of the Head Cuticle and Sheath of Living Flies

The fly is inserted into an opening cut into a piece of tinfoil in a recording platform and carefully fixed in place for dissection. This mounting immobilizes the head of the fly and permits the exposed neurons to be bathed in saline above the surface of the foil, while the sensory structures, including the antennae, the proboscis, and much of the eyes, remain dry below the surface of the foil. The tinfoil is located within a well that allows external saline to be perfused over the fly, and the walls of the well adapted so the ground wire (connected to the amplifier head stage) can be securely positioned within the saline bath. Any platform that serves these functions can be used; our method to construct a recording platform, presented here, makes use of materials at hand in the laboratory. Once the fly is mounted in the platform, the head cuticle and sheath covering the brain are removed. Here we describe how to mount the fly and open up the head cuticle to access the mushroom body Kenyon cells, whose somata are located on the dorsal/posterior surface of the brain. To record from neuronal cell bodies located in other brain regions, the approach to mount the fly must be modified to provide consistent access to the cells of interest.

MATERIALS

See the end of the chapter for recipes for reagents marked with <R>.

Reagents

Collagenase (Sigma, catalog # C0130)
Drosophila flies to be dissected
> The use of younger flies is recommended (see Steps 15 and 17).

External saline <R>
Dental wax
Epoxy (Devcon 5-min epoxy)
Paraffin wax
Super Glue

Equipment

Aluminum foil
Dremel rotary tool
Dissecting microscope with ~150x total magnification
Forceps (Roboz by Dumont, #5 INOX, BIOLOGIE tip)—to be sharpened
Forceps (Roboz by Dumont #3)
Petri dish (100 x 20-mm)
Sharpening stone
Wax melter (Electra Waxer from Almore International or equivalent)
> The probe must be sharpened so that the tip size is compatible for use with *Drosophila*.

Weigh boat

METHOD

Construction of the Recording Platform

1. Create a flat plastic disk by removing the sides from the bottom half of a Petri dish (100 mm diameter) using a Dremel rotary tool.

2. Form an ~15 mm diameter hole in this disk in the location indicated in Figure 2A.

3. Cut an ~25 mm diameter disk from the bottom of a weigh boat, and affix it over the hole in the Petri dish using Super Glue. This creates a well on one side of the Petri dish that will hold the saline bath.

4. Cut a small opening (2 x 3-mm) in the center of the weigh boat piece.

5. Using epoxy, attach a small rectangle of aluminum foil (7 x 7-mm) to the weigh boat piece on the underside of the Petri dish so it covers the opening.

6. Using a 25-gauge needle, form an opening in the foil for the fly, with shape and dimensions as indicated in Figure 2B1. Small adjustments to the shape of the opening may be required

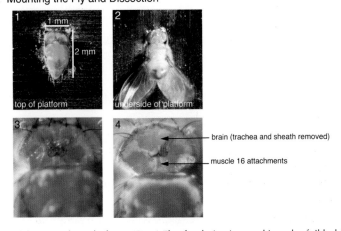

FIGURE 2. (A) Schematic of the recording platform. (B) (1) Fly after being inserted into the foil holder in the recording platform, viewed from the top of platform (dorsal side of fly). (2) View from the underside of platform (ventral side of fly). The fly's legs have been removed in order to take the photo. (3) Dorsal side of the fly after being fixed in place with paraffin wax. The head is completely surrounded by wax and is covered in external saline. (4) The cuticle has been opened to reveal the brain, the fat and air sacs/trachea have been removed, and the brain desheathed. Notice that only a section of cuticle has been removed to reveal the brain region containing the cells to be patched (in this case, mushroom body KCs). The location of muscle 16 attachments is indicated.

depending on the size of the flies used for recording (e.g., males vs. females). The foil will deform and tear after several recordings, and will need to be replaced frequently. Alternatively, a hole for the fly can be milled into a thin sheet of Delrin plastic (used in lieu of aluminum foil), which is less pliable but has a longer operational lifetime (G Maimon, pers. comm.).

7. To the top of the platform, attach paraffin wax to form a circular barrier (~26 mm diameter) outside of the saline well. This wall will constrain the saline covering the fly, preventing it from overflowing the platform. While forming the wax wall, insert a short stretch of tubing to serve as a port for securing the ground wire. Other ports can be inserted into the wall for the perfusion system, which can be an effective configuration.

8. To the edges of both the top and underside of the Petri dish, attach dental wax to raise the platform off the dissecting surface.

Mounting the Fly in the Recording Platform

Flies should be mounted such that the head is stabilized for recordings and as much as possible of the animal remains dry (the brain will of course be bathed in saline).

9. Anesthetize the flies by placing them on ice for ~20–30 sec.

10. Using dull #3 forceps, insert a fly into the hole in the aluminum foil from the underside of the platform, such that the dorsal side of the fly emerges through the topside of the recording platform (Fig. 2B1).

> Aim to keep as much as possible of the ventral side of the fly under the aluminum foil; for example, to observe olfactory responses in KCs, both antennae and maxillary palps must remain dry and fully exposed on the underside of the recording platform (Fig. 2B2). This step can be performed under low magnification.

11. Once the fly is positioned correctly, move to higher magnification. With the dorsal side of the fly facing up (top of the platform), fix the fly in place using molten wax or epoxy, and make sure that the head is well anchored. If using paraffin wax, take care to not damage the fly with the heated probe. The wax or epoxy should completely surround the fly, leaving no gaps between the fly and the aluminum foil (Fig. 2B3).

12. Flip the platform over, and check that wax or epoxy has not leaked through to cover any of the sensory organs on the ventral side of the fly. To prevent the proboscis from extending and creating movement that could disrupt the recording, stabilize the proboscis using a small drop of wax or epoxy to adhere it to the body.

> At this point a healthy fly will be kicking its legs and moving its abdomen.

Dissection: Opening the Head and Removing the Membranous Sheath Covering the Brain

The insect brain is covered in a membranous sheath, consisting mostly of collagen. To patch, this sheath must be delicately removed, without disturbing the neuronal cell bodies at the surface of the brain.

13. Return to the top of the platform and add oxygenated external saline (~1 mL) to the well, covering the dorsal side of the fly.

14. Using sharpened #5 forceps (Fig. 3), carefully remove the cuticle from the surface of the head.

   i. Begin at one area, and gently tear through the cuticle, lifting pieces away as they separate from the head.

   ii. Remove only enough of the cuticle to reveal the region of the brain that contains the cells of interest (e.g., to record from the KCs in the fly's left hemisphere, only a small section of cuticle on the posterior surface of the fly head needs to be removed; see Fig. 2B4).

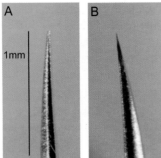

FIGURE 3. Images of a single prong of the forceps used for dissection before (*A*) and after (*B*) sharpening.

iii. From this point on, until the preparation is on the electrophysiology rig, manually exchange the saline with fresh oxygenated external saline approximately every 2–3 min, to maintain the health of the fly brain.

15. Remove any fat cells and any trachea/air sacs obstructing access to the region of the brain to be patched. Because fat accumulates in the brain as flies age, it may be advantageous to record from younger flies.

16. Observe contractions of muscle 16, which connects the aortic funnel to the frontal pulsatile organ (Bate and Martinez-Arias 1993). Because these contractions will cause the entire brain to move, this pair of muscles must be removed to stabilize the preparation. Gently pick at one end of the muscle with sharpened forceps and pull; once removed, the pumping of the brain will cease. Exert care not to damage the esophagus/gut or neck connectives while removing the muscle. Alternatively, muscle 16 can be removed at its site of attachment to the frontal pulsatile organ by tearing a small hole in the cuticle between the two antennae.

    Locating muscle 16 can be difficult—it may be useful to practice with flies expressing green fluorescent protein (GFP) under the control of a muscle-specific GAL4, such as Mhc-GAL4 (Schuster et al. 1996). After sufficient practice, the use of GFP fluorescence to reliably find and remove these muscles will not be necessary.

17. With sharpened forceps, grasp and carefully remove the collagen (perineural) sheath covering the brain (the sheath is easier to remove in young, 1- to 2-d-old, flies), when it is very clear and thin). If necessary, treat the sheath to weaken it by using collagenase at a final concentration of 0.5 mg/mL. After ~1 min, wash the brain with several exchanges of fresh external saline. The amount of time needed to weaken the sheath will vary from fly to fly and must be determined experimentally. Take care not to overdigest the sheath, because collagenase can also damage neurons.

    See Troubleshooting.

18. Wash the brain several times with fresh external saline and move it to the electrophysiology rig (see Protocol 3).

TROUBLESHOOTING

Problem (Step 17): Difficulties are encountered in removing the perineural sheath.

Solution: The most difficult part of the dissection is removing the perineural sheath; however, it must be completely removed from the area of interest for patching attempts to be successful. When the sheath overlies the cells of interest, the neuronal cell membranes will be indistinct and there will be small, highly refractive dots visible that represent proteinaceous deposits in the sheath. When learning the dissection, it can be instructive to remove only the fatty tissue and air sacs and then examine the preparation under the compound microscope to become familiar with the appearance of the sheath.

To remove the sheath properly without damaging the underlying neural tissue, forceps must be extremely sharp, and the two prongs must bite precisely together at the very ends of the tips (Fig. 3). Approach with the forceps at a very shallow angle to the surface of the brain to maximize the area of the forceps that comes in contact with the sheath. Grasp the sheath and then peel it back over the brain area you want to expose. Approaching with the forceps perpendicular to the surface of the brain is more likely to pierce and damage the tissue with the tips of the forceps. Returning to the compound microscope, one should see exposed edges of the sheath that can be drawn into a suction pipette and removed by a combination of suction and pipette movement (see Protocol 3, Step 6). This can be a very effective means of desheathing because you can remove the sheath while ensuring that you do not damage any neurons. Collagenase treatment can be used to weaken the sheath; however, it is important to avoid overdigestion of the tissue, which can lead to neuronal damage.

Protocol 2

Fabrication of Pipettes

Here we offer guidelines for pipette fabrication that work well in our experience, particularly for neurons with somata 2–5 µm in diameter, such as KCs, and 5–10 µm in diameter, such as PNs. It is important to note, however, that pipette shapes must be tailored to match the target cell type. To patch onto and break into very small neurons, it can be helpful to use pipettes that are very blunt with a small opening at the tip (Fig. 4). A blunt taper gives the pipette a low resistance, ensuring both that there is good electrical access to the membrane potential and that, even though the tip opening is small, large pulses of suction can be applied through the pipette to rupture the patch and obtain a whole-cell recording. Blunt pipettes can be fabricated using the pressure polishing technique in which air is forced down the lumen of the pipette while the tip is heated, such that the taper of the pipette actually expands rather than collapses during fire polishing (Goodman and Lockery 2000; Johnson et al. 2008). Although pressure polishing is important for high-quality recordings from neurons <5 µm in diameter, such as KCs, it is not necessary for neurons with larger cell bodies, such as PNs.

MATERIALS

See the end of the chapter for recipes for reagents marked with <R>.

Reagents

Internal saline <R>

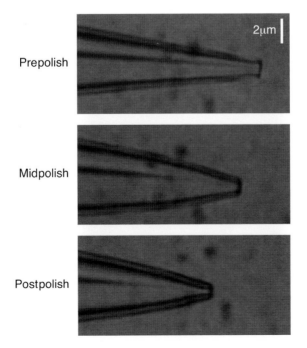

FIGURE 4. Pipette tip morphology during the pressure polishing process. (*Top*) Before any polishing. (*Middle*) After polishing while applying pressure. A blunt taper forms; the tip diameter narrows slightly but is still too large for recording cells <5 µm in diameter. (*Bottom*) After final polishing. After the midpolishing stage, the air pressure is turned off, and polishing continues to narrow the tip opening to a final size of ~0.5 µm.

Equipment

Glass capillaries (WPI thin-wall filament, catalog # TW150F-3)
Micropipette puller
Pipette fire polishing system (inverted microscope with at least 1500x total magnification, polishing microforge, and air pressure regulator)

METHOD

1. Create the pipettes as appropriate for the target cell type:

   For neurons with cell bodies 5–10 μm in diameter, pull pipettes with a tip diameter of ~1 μm, and fire-polish using standard techniques (Molleman 2003).

   For neurons with cell bodies 2–5 μm in diameter, the final tip diameter should be ~0.5 μm, and pipettes should be shaped by pressure polishing. In a microforge equipped with 1500x total magnification, apply heat to the polisher filament while forcing ~35 psi air down the lumen of the pipette.

   > While the tip is polished and the tip opening narrows, the taper of the pipette expands. Often a blunt taper is achieved before the tip opening has narrowed sufficiently. In this case, shut off the air and complete polishing with heat alone to obtain an appropriate tip diameter (Fig. 4).

2. Fill pipettes with ~3-μL internal saline and store in a closed container to prevent accumulation of any dust particles, which could interfere with seal formation.

3. When selecting a pipette for patching, consider that pipette resistances are typically 8–12 MΩ for KCs and 5–8 MΩ for PNs. Note that resistance of the pipette is dictated both by the taper and the size of the tip opening, and a blunt taper can have a very significant impact on the resistance.

Protocol 3

Whole-Cell In Vivo Patch-Clamp Recordings

Neurons are patched following routine recording protocols for whole-cell patch clamp. At the physiology rig, additional cleaning of the brain is performed to allow easy access to the neurons, and the cells can be filled with a diffusible dye during recordings, in order to examine the morphology of the recorded cell post hoc. An electrophysiology rig used for *Drosophila* patch-clamp recordings is shown in Figure 5. The microscope stage has been removed, so that the recording platform instead rests on a ring stand support that is magnetically fixed to the table. Manipulators and stimulus delivery are also in fixed locations, whereas the microscope sits on an *x–y* translation stage.

MATERIALS

See the end of the chapter for recipes for reagents marked with <R>.

Reagents

Drosophila, dissected and immobilized on recording platform (from Protocol 1)
External saline <R>

Equipment

Suppliers are listed for items that may be difficult to find—the brands listed have worked well in our experience but there may be other brands that would suffice.

Air table
Amplifier with head stage feedback resistance of 500 MΩ and capable of true current clamp recordings, such as the Multiclamp700B from Molecular Devices (see Magistretti et al. 1996)

IR-sensitive camera

upright fixed-stage microscope with IR-DIC and fluorescence optics

manipulator

amplifier headstage

ring stand to support preparation

x–y translation stage

FIGURE 5. Whole-cell patch-clamp rig for *Drosophila* central nervous system recordings. The preparation sits on the ring stand, in the light path of a fixed-stage microscope with IR-DIC and fluorescence optics. Micromanipulators are fixed to the air table, while the microscope rests on an *x–y* translation stage so it can be positioned appropriately above the brain area of interest.

Compressed air (for air table)

Data acquisition system

Faraday cage (optional)

Gas tanks (95% O_2–5% CO_2) with regulator

Gel loading pipette tips (1–200 µL)

IR-sensitive camera and video monitor

Micromanipulators

Oscilloscope

Recording electrodes (prepared in Protocol 2)

Ring support (3 in diameter), to support the recording platform, such that it is centered over the microscope condenser

Ring support stand that can be fixed magnetically to the air table

Stimulus delivery system (e.g., odor delivery tube; see Step 4)

Upright fixed-stage compound microscope (Olympus BX51W1 or equivalent) with 40x/0.8 numerical aperture (NA) water immersion objective, Fluor/IR-DIC optics, 0.8 NA dry condenser (stage is not needed)

Vacuum line (for perfusion system)

x–y translation stage for microscope

METHOD

Setting Up the Preparation on the Electrophysiology Rig

1. Place the recording platform with immobilized *Drosophila* on the ring support and center the brain in the field of view by moving the microscope on its translation stage.

2. Secure the ground wire so that it is fully immersed in the bath saline.

3. Start the perfusion system, so that oxygenated external saline continuously flows over the preparation at ~2 mL/min.

   For information on assembling a perfusion system, see Molleman (2003).

4. Position the stimulus delivery system (odor delivery tube) appropriately, if applicable.

5. At higher magnification (40x water-immersion objective) project the image to the video monitor to identify cells of interest and adjust the microscope optics for optimal IR-DIC (infrared differential interference contrast) imaging. If targeting GFP-expressing cells, identify them using fluorescence, and if necessary use neutral density filters to diminish phototoxicity.

6. If necessary, carry out an additional cleaning step to further remove any remaining overlying sheath or cell bodies preventing access to the neurons to be patched. This can best be performed with the use of a large-bore (~10 µm diameter) suction pipette (made from the same glass as the recording electrodes; Protocol 3), and filled with external saline. Gently suction away remaining pieces of sheath, exerting care not to remove nearby cell bodies.

   Refer to the Axon Guide (Molecular Devices) for more information on proper cleaning technique.

Patching

Approach cells in voltage clamp, using a voltage step to monitor resistance at the pipette tip. Apply positive pressure to ensure the tip remains clean before seal formation. For large cells, position the pipette so the positive pressure forms a dimple in the cell membrane. Small cells that are blown away from the pipette must be pursued until the pipette is sufficiently close that releasing pressure causes the cell body to spring back directly onto the tip, forming a tight seal. For information on standard techniques for forming seals and breaking into cells, refer to Marty and Neher (1995) and Walz et al. (2002). For further information on particular challenges associated with recording from small neurons, refer to Lockery and Goodman (1998).

7. Manipulate the recording pipette filled with internal saline (Protocol 2) to the area above the target neuron. Using high-magnification IR-DIC optics and/or fluorescence, position the pipette so that, if driven directly downward, it would make contact with the neuronal soma at about one-quarter of the diameter from the cell body's edge.

8. Approach the cell from directly above, applying sufficient positive pressure that large cells dimple or small cells move away as you approach with the pipette.

    i. Use the manipulator to pursue a cell that moves because of positive pressure.

    ii. Position the pipette sufficiently close so that, when you release pressure and the cell springs back, it immediately contacts the tip of the pipette (i.e., no negative pressure is required for contact). A high-resistance seal (>1 GΩ) should form immediately.

    iii. If necessary, encourage seal formation with gentle but consistent negative pressure, and by setting the command voltage to –60 mV (holding at –60 mV also ensures that when you break into the cell it is voltage-clamped near resting potential).

    iv. If a high-resistance seal does not form, use a new pipette to try again with another cell. The final seal resistance should be in the 10 GΩ range.

9. We find the most effective technique to rupture the patch and obtain a whole-cell recording is to use negative pressure: Try using sharp pulses of suction (which will likely work best), but a gradual ramping of pressure can also be effective. Break-in is indicated by a small increase in the capacitive transient accompanying the voltage step.

    The small size of neurons in the fly can mean that the change in capacitive current is quite small (e.g., KC membrane capacitance values are typically ~2 pF).

    See Troubleshooting.

10. Assess recording quality by measuring the total resistance of electrical access to the cell (R_{access}, the resistance of the pipette plus the resistance of the ruptured patch; also referred to as R_{series}), and by measuring the input resistance of the cell (R_{input}) using standard techniques (Sakmann and Neher 1995; Lockery and Goodman 1998). R_{access} is typically <80 MΩ for PNs and <100 MΩ for KCs, whereas R_{input} is >500 MΩ for PNs and >10 GΩ for KCs.

    For further considerations, also see the Discussion section.

11. If recording in current clamp, it may be necessary to adjust the amount of current injected for the neuron to rest at its appropriate membrane potential. This would occur if R_{seal} is low relative to R_{input}, such that R_{seal} significantly influences the voltage recorded at the tip of the pipette (Sakmann and Neher 1995). If this occurs, use one of two strategies.

    i. Correct the resting membrane potential according to

    $$V_{membrane} = V_{recorded} * [(R_{seal} + R_{input})/R_{seal}]$$

    by adjusting the holding current accordingly, *or*

    ii. Adjust the holding current so that spike rates (spontaneous or stimulus-evoked) in whole-cell mode match those recorded extracellularly in a loose patch configuration (Bhandawat et al. 2007).

12. If using biocytin to dye-fill the neuron, hold the cell for at least 15 min to adequately fill its processes. Remove the brain from the head capsule immediately following the recording, and fix and process according to Wilson et al. (2004).

TROUBLESHOOTING

Problem (Step 9): The pipette seal is lost during attempts to break into the cell.

Solution: If the seal is lost, often the problem is that the neurons are not healthy. The cells should appear clean and smooth with well-defined edges. A dark cell with a rough, uneven appearance and a granular quality to the cytoplasm is not healthy.

If the preparation is not healthy, the main variables to examine are (i) how long the preparation has been kept without oxygen during the dissection phase, (ii) the osmolarity of the saline, (iii) the quality of the dissection itself (whether forceps have damaged neurons, or sometimes opened a hole in the gut releasing bacteria into the neuronal milieu), and (iv) phototoxicity/overheating during fluorescence imaging. To evaluate how long a neuron can be illuminated under fluorescence without damage, patch a GFP[+] neuron using IR-DIC optics alone, and then switch on fluorescent illumination and determine for how long the recording can be maintained.

DISCUSSION

Patching

A schematic of the configuration of elements for whole-cell recording is depicted in Figure 6. The most important parameters for patching efficiently are (1) adjusting the optics to optimally visualize the target neuron, (2) ensuring that the pipette is at the right *z*-depth when attempting to seal, and (3) ensuring that the pipette tip is clear of debris. Adjusting the optics correctly will save hours of frustration. When pipette tip and cell body are at the same *z*-depth, both the tip and the cell membrane will be in sharp focus—the pipette walls around the tip will be crisply visible. It can be extremely valuable to simply practice driving a pipette up and down past a target neuron so it is clear what things look like when the pipette is at the right depth. As one drives the pipette down while exerting positive pressure, a small cell will flutter away from the pipette. At the pipette depth in which the cell is deflected the greatest distance, the pipette and cell are at the same *z*-location. If patching GFP-expressing neurons, imaging with simultaneous fluorescence and IR-DIC optics is particularly effective, because the cell membrane will be clearer under DIC. This requires using a long-pass GFP filter cube to pass both emitted GFP fluorescence and IR.

Another important factor for patching efficiently is clearing the pipette tip of debris. Positioning the pipette quickly can help in this regard, as well as applying positive pressure to blow away debris

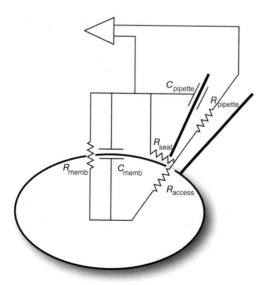

FIGURE 6. Circuit schematic of whole-cell recording configuration showing elements that contribute to recording quality; see text for more information.

in the bath that could attach to the pipette. Positive pressure should be applied to the pipette as it penetrates the meniscus of the saline bath and can also be maintained while positioning the pipette near the cell; however, do not apply so much pressure that particulates from within the pipette are blown down into the tip.

Assessing Recording Quality

R_{seal}, R_{access}, and $R_{membrane}$, schematically depicted in Figure 6, should be monitored to assess recording quality. The effects of each are described below in general terms. Those seeking further detail should consult Sakmann and Neher (1995), Lockery and Goodman (1998), and Walz et al. (2002).

$R_{membrane}$ is an important indicator of neuronal health and can be captured experimentally by measuring R_{input}, which is the sum of $R_{membrane}$, $R_{pipette}$, and R_{access}. R_{input} should be monitored throughout the recording by, for example, injecting a small hyperpolarizing current and recording the resulting change in voltage. R_{input} should remain steady throughout the recording; a drop in R_{input} indicates that neuronal health is failing and $R_{membrane}$ is dropping as the cell membrane is becoming "leakier."

R_{seal} influences the resting potential of the neuron. If a poor seal is formed, significant current can flow through the seal rather than into the pipette, and the neuron will rest at an unnaturally depolarized potential. The voltage at the tip of the pipette is $V_{record} = V_{membrane} * [R_{seal}/(R_{seal} + R_{input})]$. This can be a significant effect for very small neurons, in which R_{input} is extremely high and similar in value to R_{seal}. However, it can be compensated by injecting current to hold the cell at its appropriate resting potential.

R_{access} (or more precisely $R_{access} + R_{pipette}$, a sum that is sometimes referred to as R_{series}) influences the time course of spiking and synaptic events because, together with the capacitance of the pipette, it acts as a low-pass filter with a corner frequency of $\sim 1/(2\pi R_{access} C_{pipette})$. Additionally, R_{access} influences how much the membrane potential actually changes in response to an injected current. Because R_{access} is in series with $R_{membrane}$, when current is injected into the neuron, there will be a voltage drop across R_{access}. This can be compensated with amplifier controls (refer to the Axon Guide [Molecular Devices] for more information). R_{access} should also be monitored periodically during the experiment, to determine if the ruptured patch begins to reseal causing recording quality to drop.

Amplifier Electronics

Amplifier circuitry can have a significant effect on recordings. Many amplifiers designed specifically for voltage clamp do not have circuitry that supports injection of a truly constant level of current in "current-clamp" mode (Magistretti et al. 1996). Instead when rapid voltage changes, such as spikes, occur while recording in current-clamp mode, a small amount of current is injected into the neuron. Although the amplitude of this current is small, R_{input} of many fly neurons is extremely high, so the resulting change in membrane voltage is readily detectable. Because it is difficult to assess the consequences of this effect on the response of a neuron to a sensory stimulus, we recommend using an amplifier that supports true current clamp.

Limitations of the Technique

The frequency of obtaining high-quality recordings depends significantly on the experience of the experimenter and on the target cell type. Initially, one should not be discouraged by success rates <20%. Recording duration for neurons in the 5-μm size range is roughly 1 h, at which point the seal either breaks down or the ruptured patch reseals. Nevertheless, the fly typically appears healthy at this time and continues to move its legs.

Not every neuron in the fly brain is amenable to whole-cell recording—neurons with cell bodies located in the center of the brain can be challenging targets. Although deep neurons can be approached using high pressure to blow overlying neurons out of the way, it can be very hard to keep the pipette tip clean in this situation. Particularly problematic are small cells at deep locations; these will escape the pipette when applying the high pressure necessary to blow away overlying neurons. Low R_{seal} cell-attached (loose patch) recordings may be the only possible approach in these conditions. Whereas cell-attached recordings do not provide access to synaptic events, they offer the advantage of not dialyzing the recorded neuron's cytoplasm with internal saline.

Although in general the whole-cell recording technique provides a low resistance path to the neuronal membrane potential, the resistances involved in patching small neurons in the fly can still be relatively high (both that of the ruptured patch and of some compartments of the neuron itself). These affect the time course of the events recorded as well as the ability to accurately measure membrane potential changes that occur in neuronal compartments far from the somatic recording site. Just as a spiking event can be filtered by a high R_{access} connection of the electrode to the cell, it can also be filtered by the high resistance of a fine neurite that connects the axon to the cell body. We find that synaptic events recorded from KC somata (initiated at dendrites in the calyx) are very rapid, suggesting they are not extensively filtered (Turner et al. 2008). However, spiking events recorded in *Drosophila* have a slower time course than spikes recorded from the mammalian brain and are relatively small in amplitude (~15 mV or less), possibly indicating that spike initiation zones are located some distance from the cell body (Gouwens and Wilson 2009). Further, the high resistance of neurites can affect the experimenter's ability to control membrane potential at sites distal from the recording electrode, giving rise to space clamp problems. These issues are extensively discussed in Armstrong and Gilly (1992).

RECIPES

CAUTION: See Appendix for proper handling of materials marked with <!>.
Recipes for reagents marked with <R> are included in this list.

External Saline

Reagent	Final concentration
NaCl	103 mM
KCl <!>	3 mM
CaCl$_2$ <!>	1.5 mM
MgCl$_2$ <!>	4 mM
NaH$_2$PO$_4$	1 mM
NaHCO$_3$	26 mM
TES <!>	5 mM
Trehalose	10 mM
Glucose	10 mM
Sucrose	9 mM

1. Adjust the osmolarity of the solution to 275–280 mOsm by changing the concentration of sucrose, which is not metabolized by the brain and acts as an inert solute here.

2. Oxygenate by bubbling 95% O$_2$/5% CO$_2$ continuously through the solution. The pH should equilibrate to 7.3.

   This saline composition was derived by attempting to match the amplitude and shape of the electroretinogram (ERG) recorded in intact flies with ERGs from flies whose brains had been dissected according to the protocol described in this chapter. This saline recipe gave the closest match to the ERG recorded in an intact fly; nevertheless, some differences in ERG amplitude and shape were apparent. A variety of different saline recipes in use by other investigators (Wang et al. 2003; Gu and O'Dowd 2006, 2007; Park and Griffith 2006) also work effectively, although neural responses in these different conditions have not been precisely compared.

Internal Saline

Reagent	Final concentration
K-aspartate	125 mM
$CaCl_2$ <!>	0.1 mM
EGTA	1.1 mM
HEPES	10 mM
MgATP	4 mM
Na_3GTP	0.5 mM
Biocytin hydrazide	13 mM
(optional to dye-fill neurons)	

1. Adjust the osmolarity of the solution to 265 mOsm by changing K-aspartate concentration (the lower osmolarity of the internal saline relative to the external facilitates patch rupture). If you do not intend to dye-fill, replace biocytin hydrazide with an equal molar quantity of K-aspartate.

2. Adjust the pH to 7.3 and filter through 0.2-μm membrane to remove any particulates.

   The composition of the internal saline was evaluated by measuring the spike threshold and the reversal potential of γ-aminobutyric acid (GABA)-gated conductance in neurons recorded with different internal saline recipes (Wilson and Laurent 2005). The internal composition below produces a GABA reversal potential more negative than spike threshold, as expected for an inhibitory conductance.

ACKNOWLEDGMENTS

We are extremely grateful to Rachel Wilson, who, with Glenn Turner, codeveloped these recording techniques for *Drosophila*, and to Gilles Laurent for providing an extremely stimulating laboratory environment in which to work. We also thank Rob Campbell, Kyle Honegger, and Eyal Gruntman for comments on this chapter.

REFERENCES

Armstrong CM, Gilly WF. 1992. Access resistance and space clamp problems associated with whole-cell patch clamping. *Methods Enzymol* **207**: 100–122.

Bate M, Martinez-Arias A. 1993. *The development of* Drosophila melanogaster. Cold Spring Harbor Laboratory Press, Cold Spring Harbor, NY.

Benton R, Vannice KS, Gomez-Diaz C, Vosshall LB. 2009. Variant ionotropic glutamate receptors as chemosensory receptors in *Drosophila*. *Cell* **136**: 149–162.

Bhandawat V, Olsen SR, Gouwens NW, Schlief ML, Wilson RI. 2007. Sensory processing in the *Drosophila* antennal lobe increases reliability and separability of ensemble odor representations. *Nat Neurosci* **10**: 1474–1482.

Dubnau J, Grady L, Kitamoto T, Tully T. 2001. Disruption of neurotransmission in *Drosophila* mushroom body blocks retrieval but not acquisition of memory. *Nature* **411**: 476–480.

Goodman MB, Lockery SR. 2000. Pressure polishing: A method for re-shaping patch pipettes during fire polishing. *J Neurosci Methods* **100**: 13–15.

Gouwens NW, Wilson RI. 2009. Signal propagation in *Drosophila* central neurons. *J Neurosci* **29**: 6239–6249.

Gu H, O'Dowd DK. 2006. Cholinergic synaptic transmission in adult *Drosophila* Kenyon cells in situ. *J Neurosci* **26**: 265–272.

Gu H, O'Dowd DK. 2007. Whole cell recordings from brain of adult *Drosophila*. *J Vis Exp* **6**: 248.

Hallem EA, Carlson JR. 2006. Coding of odors by a receptor repertoire. *Cell* **125**: 143–160.

Heisenberg M, Borst A, Wagner S, Byers D. 1985. *Drosophila* mushroom body mutants are deficient in olfactory learning. *J Neurogenet* **2**: 1–30.

Jayaraman V, Laurent G. 2007. Evaluating a genetically encoded optical sensor of neural activity using electrophysiology in intact adult fruit flies. *Front Neural Circuits* **1**: 3.

Jefferis GS, Potter CJ, Chan AM, Marin EC, Rohlfing T, Maurer CR Jr, Luo L. 2007. Comprehensive maps of *Drosophila* higher olfactory centers: Spatially segregated fruit and pheromone representation. *Cell* **128:** 1187–1203.

Johnson BE, Brown AL, Goodman MB. 2008. Pressure-polishing pipettes for improved patch-clamp recording. *J Vis Exp* **20:** 964.

Kanerva P. 1988. *Sparse distributed memory*. MIT Press, Cambridge, MA.

Kazama H, Wilson RI 2008. Homeostatic matching and nonlinear amplification at identified central synapses. *Neuron* **58:** 401–413.

Komiyama T, Luo L. 2006. Development of wiring specificity in the olfactory system. *Curr Opin Neurobiol* **16:** 67–73.

Kreher SA, Kwon JY, Carlson JR. 2005. The molecular basis of odor coding in the *Drosophila* larva. *Neuron* **46:** 445–456.

Laurent G. 2002. Olfactory network dynamics and the coding of multidimensional signals. *Nat Rev Neurosci* **3:** 884–895.

Lin HH, Lai JS, Chin AL, Chen YC, Chiang AS. 2007. A map of olfactory representation in the *Drosophila* mushroom body. *Cell* **128:** 1205–1217.

Lockery SR, Goodman MB. 1998. Tight-seal whole-cell patch clamping of *Caenorhabditis elegans* neurons. *Methods Enzymol* **293:** 201–217.

Magistretti J, Mantegazza M, Guatteo E, Wanke E. 1996. Action potentials recorded with patch-clamp amplifiers: Are they genuine? *Trends Neurosci* **19:** 530–534.

Marty A, Neher E. 1995. Tight-seal whole-cell recording. In *Single-channel recording* (ed. Sakman B, Neher E), pp. 31–51. Plenum Press, New York.

McGuire SE, Le PT, Davis RL. 2001. The role of *Drosophila* mushroom body signaling in olfactory memory. *Science* **293:** 1330–1333.

Molleman A. 2003. *Patch clamping: An introductory guide to patch clamp electrophysiology*. Wiley, New York.

Murthy M, Fiete I, Laurent G. 2008. Testing odor response stereotypy in the *Drosophila* mushroom body. *Neuron* **59:** 1009–1023.

Park D, Griffith LC. 2006. Electrophysiological and anatomical characterization of PDF-positive clock neurons in the intact adult *Drosophila* brain. *J Neurophysiol* **95:** 3955–3960.

Perez-Orive J, Mazor O, Turner GC, Cassenaer S, Wilson RI, Laurent G. 2002. Oscillations and sparsening of odor representations in the mushroom body. *Science* **297:** 359–365.

Pologruto TA, Yasuda R, Svoboda K. 2004. Monitoring neural activity and [Ca^{2+}] with genetically encoded Ca^{2+} indicators. *J Neurosci* **24:** 9572–9579.

Reiff DF, Ihring A, Guerrero G, Isacoff EY, Joesch M, Nakai J, Borst A. 2005. In vivo performance of genetically encoded indicators of neural activity in flies. *J Neurosci* **25:** 4766–4778.

Robertson HM, Warr CG, Carlson JR. 2003. Molecular evolution of the insect chemoreceptor gene superfamily in *Drosophila melanogaster*. *Proc Natl Acad Sci* (Suppl 2) **100:** 14537–14542.

Sakmann B, Neher E, eds. 1995. *Single-channel recording*. Plenum Press, New York.

Schuster CM, Davis GW, Fetter RD, Goodman CS. 1996. Genetic dissection of structural and functional components of synaptic plasticity. I. Fasciclin II controls synaptic stabilization and growth. *Neuron* **17:** 641–654.

Szyszka P, Ditzen M, Galkin A, Galizia CG, Menzel R. 2005. Sparsening and temporal sharpening of olfactory representations in the honeybee mushroom bodies. *J Neurophysiol* **94:** 3303–3313.

Tanaka NK, Awasaki T, Shimada T, Ito K. 2004. Integration of chemosensory pathways in the *Drosophila* second-order olfactory centers. *Curr Biol* **14:** 449–457.

Turner GC, Bazhenov M, Laurent G. 2008. Olfactory representations by *Drosophila* mushroom body neurons. *J Neurophysiol* **99:** 734–746.

Vosshall LB, Stocker RF. 2007. Molecular architecture of smell and taste in *Drosophila*. *Ann Rev Neurosci* **30:** 505–533.

Walz W, Boulton AA, Baker GB. 2002. *Patch-clamp analysis (Neuromethods)*. Humana Press, Clifton, NJ.

Wang JW, Wong AM, Flores J, Vosshall LB, Axel R. 2003. Two-photon calcium imaging reveals an odor-evoked map of activity in the fly brain. *Cell* **112:** 271–282.

Wang Y, Guo HF, Pologruto TA, Hannan F, Hakker I, Svoboda K, Zhong Y. 2004. Stereotyped odor-evoked activity in the mushroom body of *Drosophila* revealed by green fluorescent protein-based Ca^{2+} imaging. *J Neurosci* **24:** 6507–6514.

Wilson RI, Laurent G. 2005. Role of GABAergic inhibition in shaping odor-evoked spatiotemporal patterns in the *Drosophila* antennal lobe. *J Neurosci* **25:** 9069–9079.

Wilson RI, Turner GC, Laurent G. 2004. Transformation of olfactory representations in the *Drosophila* antennal lobe. *Science* **303:** 366–370.

Yaksi E, Friedrich RW. 2006. Reconstruction of firing rate changes across neuronal populations by temporally deconvolved Ca^{2+} imaging. *Nat Methods* **3:** 377–383.

19 | Calcium Imaging

Gregory T. Macleod

Department of Physiology, University of Texas Health Science Center at San Antonio, San Antonio, Texas 78229

ABSTRACT

Calcium imaging uses optical imaging techniques to measure the concentration of free calcium $[Ca^{2+}]$ in live cells. It is a highly informative technique in neurobiology because Ca^{2+} is involved in many neuronal signaling pathways and serves as the trigger for neurotransmitter release. The technique relies on loading Ca^{2+} indicators into cells, measuring the quantity and/or wavelength of the photons emitted by the Ca^{2+} indicator, and interpreting these data in terms of $[Ca^{2+}]$. Here we present several approaches to calcium imaging at the *Drosophila* larval neuromuscular junction (NMJ). We provide three protocols for loading synthetic Ca^{2+} indicators into subcellular compartments: (1) topical application of membrane-permeant Ca^{2+} indicators, (2) forward-filling of dextran conjugates, and (3) direct injection. The calcium-imaging protocols described here for *Drosophila* larvae are readily adaptable to embryo and adult preparations. We also discuss the most powerful experimental designs for examining different aspects of Ca^{2+} signaling in *Drosophila* motor-nerve terminals. A fourth, simple protocol is provided for collecting and analyzing a set of imaging data from a loaded preparation. Finally, we consider options for building a Ca^{2+}-imaging rig according to your budget.

INTRODUCTION

Here we review the basic principles of Ca^{2+} signaling with reference to nerve terminals, and we follow this with an introduction to fluorescent Ca^{2+} indicators and a short history of their application at the *Drosophila* larval NMJ.

Ca^{2+} Signaling

Ca^{2+} is ubiquitous as a secondary messenger in the nervous system. It is most well known for its role in triggering neurotransmitter release from nerve terminals. Yet, it is also critical in most forms of synaptic plasticity and a multitude of other cellular pathways. This highlights an enigma: How can the same molecule be an important signal in so many different pathways within the same space (e.g.,

315

the presynaptic terminal)? The answer is that Ca^{2+} signals are coded. Each pathway is specified, not just within the amplitude of the Ca^{2+} change, but also according to the location and time course of the Ca^{2+} change. Signaling often occurs in spatially restricted microdomains in which Ca^{2+} can reach very high concentrations. If the Ca^{2+} microdomain overlaps with the distribution of a Ca^{2+}-binding molecule, such that significant Ca^{2+} binding occurs, pathways downstream from this molecule will be activated.

An example of such specified signaling is seen in the colocalization of voltage-gated Ca^{2+} channels (VGCCs) in nerve terminals with the Ca^{2+} sensor for fast-triggered neurotransmitter release. The distribution of the VGCCs defines the location of the Ca^{2+} microdomains. The time course of Ca^{2+} entry is a function of the duration of the action potential (AP) and VGCCs' gating kinetics, whereas cytosolic Ca^{2+} buffering proteins confine the spatial spread of the microdomain. The Ca^{2+} change in this microdomain is so restricted in space and time that the Ca^{2+} sensor will be activated only if it is located within tens of nanometers of VGCCs. Beyond this radius, the Ca^{2+} concentration ($[Ca^{2+}]$) will be insufficient to trigger a low-affinity Ca^{2+} sensor.

A second key concept in Ca^{2+} signaling is that a coding system based on differences in the space, time, and amplitude of $[Ca^{2+}]$ changes can only work if such signals can be reliably discriminated from random changes in $[Ca^{2+}]$. Neurons use up a lot of energy regulating their $[Ca^{2+}]$ levels to provide a stable Ca^{2+} environment for effective Ca^{2+} signaling. Although there are numerous examples of amplification in Ca^{2+} signals, such as that seen in Ca^{2+}-induced Ca^{2+} release, the tendency is for dampening of $[Ca^{2+}]$ changes. Within compartments (e.g., the cytosol, mitochondrial matrix, and endoplasmic reticulum [ER] lumen), Ca^{2+}-binding proteins are crucial for dampening $[Ca^{2+}]$ changes. Between compartments (e.g., between the compartments listed above and between the cytosol and extracellular compartment), membrane proteins such as Ca^{2+}/Na^+ exchangers and the Ca^{2+}-ATPase pumps work to maintain a stable $[Ca^{2+}]$ environment. Most of these elements are illustrated in *Drosophila* motor-nerve terminals in Figure 1.

Ca^{2+} Indicators

Ca^{2+} indicators are Ca^{2+}-binding molecules that "indicate" when they bind to Ca^{2+} through a change in either the intensity of their fluorescence or the wavelength of their fluorescence. Some Ca^{2+} indicators report Ca^{2+} binding through bioluminescence rather than fluorescence but only fluorescent Ca^{2+} indicators will be discussed here. Ca^{2+} and Ca^{2+} indicators bind reversibly, and the equilibrium dissociation constant (K_D) is a useful measure of the indicator's strength of binding to Ca^{2+}. The lower the K_D value, the greater the affinity of the Ca^{2+} indicator for Ca^{2+}. For example, the salt forms of Oregon green BAPTA-1 (OGB-1) and fura-2 have low K_D values (~0.2 µM) indicating high affinities for Ca^{2+}, whereas dextran conjugates of rhod-2 and fluo-4 have low affinities (K_D ~3 µM). Generally, Ca^{2+} indicators yield the best quantitative data when the Ca^{2+} change to be measured approaches the K_D value but always remains below it.

The Ca^{2+} indicator should be chosen with consideration given to the expected amplitude of the change in $[Ca^{2+}]$, the speed of the change, and the location of the change. The Ca^{2+} indicator also must be chosen with a view to the equipment available. As Ca^{2+} indicators are not endogenous they are referred to as exogenous Ca^{2+} buffers and will inevitably perturb Ca^{2+} handling within the loaded compartment, particularly when present at a high concentration. A particularly helpful discussion of the impact of these exogenous buffers is given by Tank et al. (1995).

Synthetic Ca^{2+} indicators are small organic molecules, whereas genetically encoded Ca^{2+} indicators (GECIs) are proteins transcribed and translated within the cells being imaged. Synthetic Ca^{2+} indicators are preferred as most have a high dynamic range, a close to linear response to Ca^{2+} near their K_D, and predictable Ca^{2+} binding kinetics. However, synthetic Ca^{2+} indicators can be difficult to load with specificity. GECIs, on the other hand, can be reliably expressed in specific cell types. In *Drosophila*, the Gal4/UAS system (Brand and Perrimon 1993) is commonly used to express GECIs in subsets of neurons. Sequences added to GECIs can target them to, and retain them within, subcellular compartments such as mitochondria (Chouhan et al. 2010) and the ER (Snapp et al. 2004).

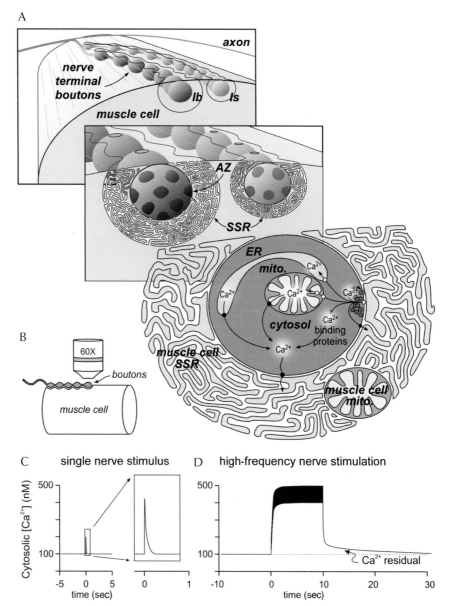

FIGURE 1. The flow of Ca²⁺ through a *Drosophila* motor nerve terminal. (*A*) An exploding illustration of *Drosophila* larval neuromuscular junctions (NMJs). Type-Ib (type "one" b) and type-Is (type "one" s) terminal boutons are shown on body-wall muscle 13, along with important elements of the Ca²⁺-regulation machinery in a type-Ib bouton. For clarity, only a single active zone (AZ) is represented as the site of Ca²⁺ entry. The arrival of an action potential (AP) opens voltage-gated Ca²⁺ channels at the AZ allowing in Ca²⁺ to trigger the exocytosis of neurotransmitters from synaptic vesicles. The arrows show the flow of Ca²⁺ after the AP. Ca²⁺, which is not immediately removed from the cytosol across the plasma membrane, is bound by cytosolic proteins or taken up into the ER or mitochondria. The sarcoendoplasmic reticulum Ca²⁺-ATPase (SERCA) pumps Ca²⁺ into the lumen of the ER, while the uniporter "gates" Ca²⁺ entry into the mitochondria. Ca²⁺ is subsequently released from these organelles, and from the Ca²⁺-binding proteins, and is removed across the plasma membrane. Ryanodine receptors (RyRs) and inositol 1,4,5-triphosphate receptors (IP₃Rs) release Ca²⁺ from the ER, whereas Ca²⁺ release from the mitochondrion is mediated by a Na⁺/Ca²⁺ exchanger. At the plasma membrane, Ca²⁺ is removed by Ca²⁺-ATPases, Na⁺/Ca²⁺ exchangers, and Na⁺/Ca²⁺-K⁺ exchangers. A muscle mitochondrion is shown encroaching on the terminal in the muscle subsynaptic reticulum (SSR). The cytosol, mitochondrial matrix, and ER lumen may be separately loaded with Ca²⁺ indicators. (*B–D*) Ca²⁺ imaging. Illustrative plots showing the typical time course and amplitude of changes in the concentration of cytosolic free Ca²⁺ ([Ca²⁺]) in terminal boutons, in response to a single AP (*C*), and a train of APs (*D*), e.g., 40 Hz.

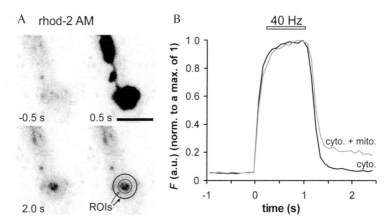

FIGURE 2. Ca^{2+}-imaging data from a terminal loaded by topical application of rhod-2 AM (Protocol 1). (A) Four images showing fluorescence (inverted grayscale look-up table) from a rhod-2 AM loaded terminal (type-Ib boutons) before stimulation (–0.5 sec), during a 1-sec nerve stimulation train (0.5 sec), and after the stimulus train (2.0 sec). Note a centrally located mitochondrion in the most distal bouton, apparent once stimulation ceases (2.0 sec). The fourth image, like the third, was captured at 2 sec and is used to illustrate the two regions of interest (ROIs) used for fluorescence quantification. The central ROI contains the putative mitochondrion, along with cytosol. The outer ROI has the shape of a "donut" and fluorescence from the cytosol between the two rings is quantified. Scale bar, 5 μm. (B) A plot of fluorescence (F) in the two ROIs. F is defined as the average pixel intensity value within an ROI minus the average pixel intensity value of the background outside the ROIs. F traces were normalized to a maximum value of 1. As rhod-2 is effectively invisible at resting levels of cytosolic free Ca^{2+}, and mitochondrial matrix free Ca^{2+}, it is not helpful to express the fluorescence response in terms of conventional "$\Delta F/F$" ($[F_{stim} - F_{rest}]/F_{rest}$) as the denominator ($F_{rest}$) is close to 0 (greatly amplifying any variance). Details of loading are identical to Protocol 1. The nerve was stimulated at 40 Hz for 1 sec, 78 min after the fourth rinse. Images were collected at 11 frames/sec on a laser-scanning confocal microscope (Olympus FluoView 300), using an Olympus 60x water-immersion objective (0.9 NA).

A Short History of Cytosolic Ca^{2+} Imaging at the *Drosophila* Larval NMJ

The first report of cytosolic Ca^{2+} imaging at the *Drosophila* NMJ described the topical application of the acetoxymethyl ester (AM) membrane-permeant form of fluo-3 to the filleted larval preparation (Karunanithi et al. 1997). This study was soon followed by other studies that used the same technique utilizing the AM form of other Ca^{2+} indicators: calcium crimson (Umbach et al. 1998), fura-2 and fluo-4 (Dawson-Scully et al. 2000), and rhod-2 (Kuromi and Kidokoro, 2002). Although it is apparent that the cytosol of nerve terminals gets loaded, it has become clear that some of these membrane-permeant Ca^{2+} indicators also accumulate within the muscle subsynaptic reticulum (SSR) (calcium crimson), and presynaptic and postsynaptic mitochondria (rhod-2 [Fig. 2], and possibly fluo-4). In 2002, two studies introduced techniques that achieve highly specific loading of the cytosol of *Drosophila* nerve terminals. Rieff et al. (2002) showed the use of a genetically encoded Ca^{2+} indicator (GECI; yellow cameleon-2), whereas Macleod et al (2002) showed forward filling of dextran-conjugated Ca^{2+} indicators (OGB-1 and fura-2). Reiff and colleagues have since described the performance of approximately a dozen more GECIs in *Drosophila* motor-nerve terminals (Reiff et al. 2005; Mank et al. 2006; Hendel et al. 2008). Most recently, Hendel et al. (2008) have directly injected the salt form of Ca^{2+} indicators into nerve terminals (OGB-1 and magnesium green).

EXPERIMENTAL DESIGN

Any Ca^{2+}-imaging experiment must be designed with the limitations of Ca^{2+} imaging in mind. We know that Ca^{2+} signals are coded in terms of amplitude, location, and time course; however, direct quantification of any of these parameters is no easy feat. Consider for a moment the challenge of trying to quantify the Ca^{2+} signals relevant to triggering neurotransmitter release. First, the [Ca^{2+}] change close to the VGCCs is from ~100 nM to ~100 μM, three orders of magnitude. Even Ca^{2+} indicators best suited for this range will not report effectively over the entire range. Second, the diameter of the microdomain relevant to triggering the Ca^{2+} sensor is likely to be <100 nm, yet the limit of resolution of standard wide-field and confocal light microscopes is ~200 nm. Lastly, the duration of Ca^{2+} entry during an AP is ~1 msec; therefore, to capture Ca^{2+} entry, imaging data must be collected at a rate of more than 1000 samples per second. Although this rate has been achieved (Sugimori et al. 1994), it is far from commonplace. Despite the limitations of Ca^{2+} imaging, experiments may still be designed to measure parameters of Ca^{2+} change, or correlates thereof, relevant to the question of interest.

Under each heading below we consider the best choice of Ca^{2+} indicator, loading protocol, and equipment for commonly encountered questions.

Amplitude

Perhaps the most commonly asked question is whether differences in the number of quanta released at a mutant *Drosophila* larval NMJ, relative to wild type, are due to differences in Ca^{2+} entry. Although an electrophysiological technique has been described for measuring presynaptic Ca^{2+} current in type-III terminal boutons (Morales et al. 1999), no such technique has been developed for glutamatergic type-Ib or type-Is terminals responsible for most ionotropic receptor mediated neurotransmission at *Drosophila* NMJs. The best Ca^{2+}-imaging approach is to measure the amplitude of Ca^{2+}-indicator responses to individual APs (Macleod et al. 2004, 2006; Bao et al. 2005; Lnenicka et al. 2006; He et al. 2009). The change in the volume-averaged cytosolic Ca^{2+} concentration $[Ca^{2+}]$ should be proportional to Ca^{2+} entry. However, many assumptions are made here, such as similarities in Ca^{2+} extrusion rates, endogenous Ca^{2+} buffering, and bouton size.

Some synthetic Ca^{2+} indicators are well suited to make quantitative estimates of single AP changes in $[Ca^{2+}]$ (e.g., rhod-dextran, fluo-4 dextran, OGB-1, fura-2). Preference is given to those Ca^{2+} indicators with a K_D well above the expected peak of single AP $[Ca^{2+}]$ in which the relationship between the change in $[Ca^{2+}]$ and the change in fluorescence is close to linear. With any analysis of single AP changes in fluorescence, care has to be taken to keep the concentration of the synthetic Ca^{2+} indicator low. Although increases in the concentration will increase the signal-to-noise ratio, high-affinity Ca^{2+} indicators such as the salt forms of OGB-1 and fura-2 will reduce the amplitude and slow the kinetics of single AP changes in fluorescence (Helmchen et al. 1996). Dextran conjugation of both of these Ca^{2+} indicators reduces their affinity for Ca^{2+} to a more suitable range, and provides the added bonus of preventing compartmentalization. The low-affinity versions of rhod-dextran ($K_D \sim 3.0$ μM) and fluo-4 dextran ($K_D \sim 2.6$ μM) are probably the best-suited Ca^{2+} indicators for this purpose. Both are effectively invisible at resting $[Ca^{2+}]$ but coloading with a Ca^{2+}-insensitive dextran conjugate such as AF647 overcomes this problem and serves other functions: It allows a direct comparison of responses between terminal types and a calibration of fluorescence responses to $[Ca^{2+}]$. Quantitative estimates of single AP changes in fluorescence require data acquisition speeds of >200 sample/sec (Hz), because $[Ca^{2+}]$ is predicted to peak in <10 msec in small boutons (Atluri and Regehr 1996). These speeds are commonly attainable on high-speed electron-multiplying charge-coupled device (EMCCD) cameras, laser-scanning (see example in Fig. 3), multiphoton, and spinning-disk confocal microscopes.

GECIs cannot be used to make quantitative estimates of single AP changes in $[Ca^{2+}]$, because although incremental improvements are being made all the time, they still have either insufficient sensitivity, response time, or linearity in their response (Hendel et al. 2008). GCaMP, however, may well be an exception (Tian et al. 2009). The sensitivity of different forms of green fluorescent protein (GFP) to changes in pH, halides, and superoxide also make most GECIs based on GFP unsuitable for quantitative studies.

A less direct approach to assessing Ca^{2+} entry is to measure the amplitude of the fluorescence response to a train of APs. This experimental design relies on the same assumptions as those above, and makes the further assumption that the change in Ca^{2+} current from one AP to the next shows the same trend between terminals/larvae being compared. Quantitative estimates here are not so

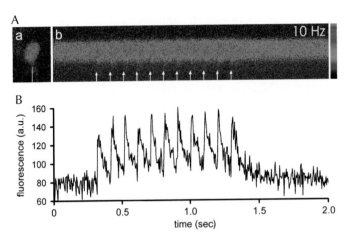

FIGURE 3. Changes in Ca^{2+}-indicator fluorescence in response to action potentials. (*Aa*) A scanned image of a type-Ib bouton filled with OGB-1 on muscle 6. The bouton is ~3 µm in diameter. (*Ab*) A serial line scan through the center of the bouton, as marked in *Aa*, 4 msec per vertical line that composes the image. *Left to right*, the image is composed over a period of 2 sec. Each arrow "tick" indicates the time at which a pulse was delivered to the nerve (2 V, 0.3 msec). The look-up table used for *F* intensity is shown on the *right*. (*B*) A plot of *F* for each line in *Ab* (a.u.). Images were collected on the same system as described for Fig. 5. (Modified, with permission, from Macleod et al. 2002.)

sensitive to high loading levels (Tank et al. 1995) and require considerably less acquisition speed (perhaps as low as 2 Hz).

As the quantification of Ca^{2+} changes relies on fluorescence quantification, movement must be minimized, even with ratiometric Ca^{2+} indicators such as fura. Adding glutamate to the bath is probably the least invasive means of preventing movement. The continual presence of glutamate maintains glutamate receptors in a desensitized state. Augustin and colleagues (2007) concluded that glutamate is usually present in larval hemolymph at ~2 mM, and that more than one-half of the receptors are constitutively desensitized. Indeed, with 2 mM glutamate in the bathing solution, we are unable to discern miniature excitatory junction potential in the muscle membrane potential, and evoked excitatory junction potential amplitudes are reduced to ~10% of their usual size (AK Chouhan and GT Macleod, unpubl.). At 7 mM glutamate, even high-frequency stimulation will not cause muscle contraction (Macleod et al. 2004).

Ratiometric GECIs are well suited for testing for differences in resting levels of Ca^{2+} between neurons. This technique does not require mechanically disruptive loading approaches, but it does require prior expression of the GECI (protocol not provided here). The ratio signal should not be sensitive to differences in expression levels between terminals, because although GECI expression should increase the Ca^{2+}-buffering capacity, resting levels of cytosolic Ca^{2+} ultimately reflect an equilibrium state that is influenced more by pumps and exchangers than by buffering capacity per se. Further, the ability to report this steady state level should not be compromised by the poor kinetics of GECIs. The ratiometric GECIs yellow cameleon 3.60 and D3cpv (Nagai et al. 2004; Palmer et al. 2006; Hendel et al. 2008) and TN-XXL (Mank et al. 2008) should be particularly useful for this application.

Location

Questions about changes in Ca^{2+} handling in the mitochondria and ER can be experimentally addressed through specific targeting of Ca^{2+} indicators.

Mitochondria are commonly loaded using the AM forms of cationic Ca^{2+} indicators. Rhod-2 AM (Guo et al. 2005), rhod-FF AM, rhod-5N AM (Chouhan et al. 2010), and mag-fluo-4 AM (GT Macleod and KE Zinsmaier, unpubl.) are all very effective at loading mitochondria, but a significant incubation period is required after the application of the dye to clear it from the cytosolic compartment before the signals can be interpreted. Within the mitochondrial matrix, the [Ca^{2+}] level will range between ~100 nM (similar to the cytosol at rest) and several micromolars after nerve activity (David et al. 2003). However, the amplitude of responses seen with rhod-5N (Chouhan et al. 2010), with an estimated K_D of 320 μM in vitro, suggests that the [Ca^{2+}] in the mitochondrial matrix of *Drosophila* motor-nerve terminals may substantially exceed [Ca^{2+}] levels seen in lizard motor-nerve terminals. Rhod-FF AM (K_D ~ 19 μM) is a good choice for imaging mitochondrial [Ca^{2+}] changes at the *Drosophila* NMJ because it is very responsive to nerve stimulation and does not appear to saturate. The use of 7 mM glutamate as described above not only prevents movement, but also prevents changes in mitochondria in the postsynaptic muscle. Mitochondria close to and within the sarcoplasmic reticulum load strongly with AM forms of cationic Ca^{2+} indicators and always respond to nerve stimulation if Ca^{2+} flow through the postsynaptic receptor/channels (as shown by Guerrero et al. 2005) is not minimized (GT Macleod and KE Zinsmaier, unpubl.).

Using the targeting strategy described by Filippin and colleagues (2005), the GECIs ratiometric pericam (RP), yellow cameleon-2, and camgaroo-2 (CG2) have been targeted to the mitochondrial matrix with high specificity (Chouhan et al. 2010). All three GECIs are responsive to nerve stimulation but those containing cpYFP (circularly permuted yellow fluorescent protein) such as RP and CG2 are very sensitive to changes in superoxide levels (Wang et al. 2008) and possibly changes in pH within the mitochondria.

Within the ER lumen [Ca^{2+}] changes are likely to be much larger than in mitochondria, although there are no estimates from motor-nerve terminals. ER [Ca^{2+}] has been reported as high as 500 μM (Pinton and Rizzuto, 2006). ER targeting of GECIs is likely to be the most effective way to target Ca^{2+} indicators to this compartment. The best-suited GECIs are discussed by Palmer and colleagues (Palmer et al. 2004; Palmer and Tsien 2006). An ER GECI targeting strategy that has shown some promise for *Drosophila* motor-neuron terminals is that described by Snapp et al. (2004) (GT Macleod and KE Zinsmaier, unpubl.).

A common ambition is to resolve differences in [Ca^{2+}] changes within the same compartment according to amplitude, space, and time. For example, what is the amplitude of the change in Ca^{2+} just inside the plasma membrane relative to the "volume-averaged" Ca^{2+} change across an entire bouton? Resolution of these sorts of questions will require targeting of fast, low-affinity Ca^{2+} indicators to the microdomain of interest. Indiscriminate loading in the compartment is a poor experimental design, as it puts the burden of resolution on the ability to discriminate fast localized changes from a slower global change. Considering that Ca^{2+}, entering through VGCCs, may equilibrate across a bouton ~1.5 μm in diameter in <10 msec (Atluri and Regehr 1996), an exceptionally fast and sensitive detection system would be required. Although GECIs can be targeted to many domains (Palmer and Tsien 2006), it is not clear that any of the currently available GECIs are sufficiently fast to adequately resolve such events.

The value of your Ca^{2+}-imaging data is highly dependent on being able to identify the compartment in which your Ca^{2+} indicator is located. Topical application of a Ca^{2+} indicator risks indiscriminate loading that can produce confusing Ca^{2+} signals from multiple cellular and subcellular compartments. Cells with the highest surface-to-volume ratio (e.g., neuronal processes) generally show the most rapid accumulation of membrane permeant Ca^{2+} indicators. Because of the high surface-to-volume ratio of the muscle SSR, you should expect strong loading in the muscle in the region of the NMJ. Forward-filling, direct injection, and genetic expression generally guarantee positive identification of the cells loaded, but the identity of the subcellular compartment still requires confirmation. Ca^{2+} indicators with a net positive charge (cations), when either directly injected into the cytosol or topically applied, will usually find their way into the negatively charged mitochondria. Dextran-conjugated Ca^{2+} indicators with a neutral charge are favored for measuring Ca^{2+} changes in the cytosol as dextran conjugation prevents the Ca^{2+} indicator being taken up into other compartments. The time course of Ca^{2+} changes has been characterized for most cell types and subcellular compartments, and this information, along with a pharmacological approach, should be used to test your assumption about which compartment(s) have been loaded.

Time Course

Questions regarding the function of components of the Ca^{2+}-handling machinery (shown in Fig. 1) may be answered with reference to data that allow quantification of rates of change in $[Ca^{2+}]$ in the cytosol, mitochondrion, or ER. Beyond ensuring that your Ca^{2+} indicator is in the correct compartment, the key factors are to select a Ca^{2+} indicator that will not perturb Ca^{2+}-handling kinetics, and to sample changes in fluorescence at rates adequate to quantify the kinetics.

Rhod-dextran and fluo-4 dextran are suitable choices for quantification of the rate of change in cytosolic Ca^{2+} and may be forward-filled. They should be cofilled with an easily visible dextran-conjugated Ca^{2+}-insensitive dye with complementary spectra (e.g., AF488, AF568, or AF647) to allow focusing and calibration. Magnesium green (Hendel et al. 2008) (hexapotassium salt, $K_D \sim 6\ \mu M$) and the low-affinity forms of fura-2 (e.g., fura-FF pentapotassium salt, $K_D \sim 5.5\ \mu M$) are good choices for direct injection. Fura-FF as a ratiometric Ca^{2+} indicator yields very useful data but requires excitation at low wavelengths (340–380 nm), which excludes its use on most laser-scanning confocal microscopes.

The decay rate of cytosolic $[Ca^{2+}]$ decay after a single action potential has a Tau of ~60 msec (Macleod et al. 2002) which dictates a minimum sampling rate of ~100 Hz, and preferably >200 Hz (Fig. 3), to adequately describe the time course. High-speed cameras, laser-scanning, multiphoton, and spinning-disk confocal microscopes are best suited to this task. A much faster sampling rate (several thousand hertz) would be required to describe the rate of increase in Ca^{2+}-indicator fluorescence in response to a single AP, but these data would be difficult to interpret as they would reveal more about the rate of Ca^{2+} diffusion and equilibration across a bouton than the rate of Ca^{2+} entry. Dedicated photodiodes and photomultiplier are best suited for such acquisition speeds but much information is lost in the spatial dimension.

The rate of decay after a train of APs is slower than the rate of decay after a single AP (Macleod et al. 2002; Lnenicka et al. 2006), and it may reflect several process with different time courses. Data revealing rates of Ca^{2+} accumulation from one AP to another during an AP train have the potential to tell us much about not only the Ca^{2+}-buffering processes, but also regulation of VGCCs (Xu et al. 2007). Such data could inform on the degree of calcium-dependent inactivation (CDI) of presynaptic Ca^{2+} channels in *Drosophila*. Although data have been published showing the amplitude of consecutive impulses in a train (Macleod et al. 2002, 2006 [Suppl. Fig. 1]; Bao et al. 2005), these data are of limited value on the topic of CDI because a high-affinity Ca^{2+} indicator with a rapid rate of association (OGB-1) was used, which may extinguish CDI (Lee et al. 2000).

The following protocols describe three Ca^{2+}-indicator-loading techniques. A common question in *Drosophila* physiology is the following: "Which bath solution should I be using? Standard solution, HL3, HL6, or HL3.1?" (Jan and Jan 1976; Stewart et al. 1994; Macleod et al. 2002; Feng et al. 2004). The answer is that it does not really matter for Ca^{2+} imaging, as long as you are consistent in its use. It does matter, however, if you need the preparation to survive for >90 min, as you would require for the forward-filling technique or for the complete cytosolic clearance of topically applied Ca^{2+} indicators. For these applications you should use either Schneider's or HL6 for long incubation periods. The final Ca^{2+} imaging after the incubation can be performed in HL3, HL3.1, or HL6. Schneider's is not appropriate for Ca^{2+} imaging because it is autofluorescent (because of the presence of 10 mM yeastolate) and contains high levels of Ca^{2+} (5.4 mM).

Topical Application of Ca^{2+} Indicators

This is the simplest technique to execute and yields data quickly (see example of the imaging data in Fig. 2). The drawback is that these data are the most difficult to interpret, primarily because one cannot be certain which cellular and subcellular compartment(s) are loaded (e.g., muscle, nerve, or glia; cytosol, mitochondrion, or ER).

The following articles describe the use of this technique in *Drosophila* larvae: Karunanithi et al. (1997); Umbach et al. (1998); Dawson-Scully et al. (2000); Bronk et al. (2001); Kuromi and Kidokoro (2002); Kuromi et al. (2004); Sanyal et al. (2005).

MATERIALS

CAUTION: See Appendix for proper handling of materials marked with <!>.

Reagents

CaCl$_2$ <!> in H$_2$O (1.0 M)
DMSO (dimethyl sulfoxide) <!> (100% anhydrous) with 20% pluronic acid
Drosophila larva, wandering third-instar stage
Hemolymph Like No. 6 (HL6) (Macleod et al. 2002), freshly made or freshly defrosted
> Our laboratory makes up 2 L of HL6 every 6 months. Immediately after it is made, it is pH adjusted and filtered, the entire "batch" is then dispensed in 10- and 20-mL aliquots into plastic Falcon tubes and frozen at –80°F.

L-glutamic acid monosodium salt hydrate (L-GA) in H$_2$O (0.7 M)
Rhod-2 AM (50 µg)
Schneider's insect medium containing Ca^{2+} and L-glutamine (Sigma-Aldrich; Cat. No. 50146)

Equipment

Dental wax (Modern Materials, Cat. No. 50094491)
Dissection microscope and cold light source
Dissection tools (Vannas microscissors and two pairs of fine forceps)
Entomology pins (FST, Cat. No. 26002-10)
Epoxy resin (5-min set time)
Glass slide (frosted at one end)
Plumber's gasket (Product No. 634; Master Plumber, Brantford, ON)
Sylgard 184 silicone elastomer kit

METHOD

Making the Preparation Bath

This is required for all three loading techniques; see Figure 3.

1. Clean a glass slide and a plumber's gasket.
2. Mix the epoxy resin.

3. Using a continuous fine bead of epoxy, glue the plumber's gasket flat against the glass slide in the center of the slide (shim side away from glass).

4. Once the glue has set, lay the slide on a flat surface and fill the gasket with water to test whether the gasket is bound to the slide with a watertight seal.

5. Dry slide and gasket.

6. Mix Sylgard elastomer and use it to fill the gasket, such that it forms a strongly convex surface with a depth in the center of ~1.5 mm.

7. "Pop" all bubbles remaining in the elastomer 1 h after filling.

8. Leave elastomer to cure for 48 h.

9. Break the epoxy bead with a single-sided razor blade and remove it from the glass without disturbing the bed of Sylgard in the center.

10. Populate the Sylgard bed with seven pins cut to 6 mm in length and bent to a right angle.

11. Form two dental wax tablets, as shown in Figure 4, to support a glass pipette.

Preparing the Loading Mixture

12. Prepare a rhod-2 AM loading "stock" solution by adding 22 µL of DMSO to 50 µg of rhod-2 AM (2 mM final concentration).

13. Vortex this solution vigorously for 5 sec, and then spin for 2 sec at 1000 rpm; repeat three times. Store this solution in a dark location at room temperature (RT).

14. Prepare HL6 by adding 1 M $CaCl_2$ at 1:500 (2 mM final concentration), and adding 0.7 M L-GA at 1:100 (7 mM final concentration). Chill prepared HL6 on ice.

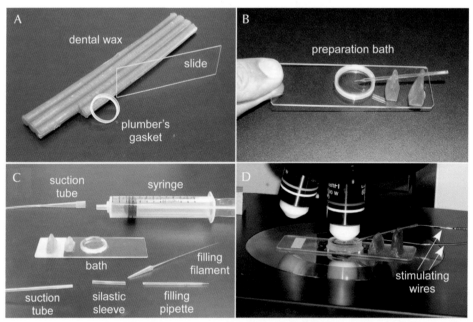

FIGURE 4. Photographs of the preparation bath and peripherals. (A) Raw materials for making the preparation bath (not showing the Sylgard elastomer or the epoxy resin). (B) Preparation bath showing the dental wax tablets that form the ramp to support the filling pipette. (C) Peripherals required for drawing a nerve into a glass filling pipette and applying Ca^{2+} indicator to the end of the nerve. Fabrication of most items is described in Protocol 2. (D) Bath in position under a microscope objective, with one stimulating wire inserted in the filling pipette and another wire in contact with the bath.

Dissecting and Loading the Preparation

15. Fillet dissect a wandering third-instar *Drosophila* larva in chilled Schneider's (Jan and Jan 1976; Rossano and Macleod 2007). Do not cut the segment nerves. Do not damage muscle cells 6, 7, 13, or 12 in segments 2–8.

16. Rinse the preparation in chilled prepared HL6 solution.

17. Prepare the "final" rhod-2 AM loading solution by adding 3 μL of rhod-2 AM loading stock solution to 1.2 mL of prepared HL6 (5 μM final concentration, 0.25% DMSO, 0.05% pluronic acid). Store this solution in a dark location at RT.

18. Replace HL6 covering the preparation with the final rhod-2 AM loading solution.

19. Incubate the preparation in this solution in a 4°C fridge in a darkened box for only 10 min.

20. After 10 min of incubation, rinse the preparation with one bath volume change of prepared HL6.

21. Repeat the rinse after 5, 10, and 15 min. Allow the preparation to sit in a darkened location at RT between rinses.

22. After the fourth rinse, sever the segment nerves and draw a nerve to the fourth segment into a glass pipette full of HL6 to allow nerve stimulation; follow the instructions for drawing a nerve into a glass pipette for stimulation in Protocol 3, Steps 4–9.

    This step may be performed any time, from immediately after the fourth rinse to 5 h afterward (see note to Step 23).

    See Troubleshooting.

23. The preparation will be ready for Ca^{2+} imaging 20 min after severing the nerve.

    Immediately after rhod-2 AM loading, the rhod-2 signal will come predominantly from the cytosol with some contribution from the mitochondria (Fig. 2). With time incubated at RT in rhod-2 AM–free HL6, the signal will come predominantly from the mitochondrion with little if any contribution from the cytosol.

TROUBLESHOOTING

Problem (Step 22): Nerve terminal cytosol and/or mitochondria do not load.
Solution: Ensure that DMSO is anhydrous and contains pluronic acid.

Problem (Step 22): The nerve is unresponsive to nerve stimulation. There is muscle movement.
Solution: Solutions for these problems are given in the General Troubleshooting section following Protocol 4.

Forward-Filling of Dextran-Conjugated Ca²⁺ Indicators

This technique is particularly well suited for imaging changes in cytosolic Ca^{2+} as dextran conjugation prevents compartmentalization of the Ca^{2+} indicator. The major drawback is that the nerves must be severed at the start of the loading process, several hours before nerve terminals are ready to examine.

The following articles describe the use of this technique in *Drosophila* larvae: Macleod et al. (2002, 2003, 2004, 2006); Bao et al. (2005); Bronk et al. (2005); Gou et al. (2005); Verstreken et al. (2005); Lnenicka et al. (2006); Rossano and Macleod (2007); Shakiryanova et al. (2007); Klose et al. (2008, 2009); Giagtzoglou et al. (2009); He et al. (2009); Yao et al. (2009); Chouhan et al. (2010).

MATERIALS

CAUTION: See Appendix for proper handling of materials marked with <!>.

Reagents

AF647-dextran (10,000 MW) (5 mg)
$CaCl_2$ <!> in H_2O (1.0 M)
Drosophila larva, wandering third-instar stage
Hemolymph Like No. 6 solution (HL6), freshly made or freshly defrosted (Macleod et al. 2002)
L-glutamic acid monosodium salt hydrate (L-GA) in H_2O (0.7 M)
Rhod-dextran (10,000 MW) (5 mg)
Schneider's insect medium containing Ca^{2+} and L-glutamine (Sigma-Aldrich; Cat. No. 50146)

Equipment

Bunsen burner
Capillary glass without internal filament (Sutter Instrument; Cat. No. B150-86-10)
Dissection microscope and cold light source
Dissection tools (Vannas microscissor and two pairs of fine forceps)
Glass filling pipette (fabrication described in protocol)
Micropipette puller (e.g., Flaming/Brown P-97; Sutter Instruments Co.)
Microforge (e.g., Narishige, MF-900)
Plastic filling filament (fabrication described in Steps 1–4)
Plastic syringe (10 mL)
Polyethylene tubing—1.5 mm outer diameter (30 cm)
Silastic tubing (sleeve), soft thin-walled with 1.5 mm inner diameter (4 cm)
Sylgard bottom bath with pins and dental wax ramp (Fig. 4; see Protocol 1, Steps 1–11 for fabrication).

METHOD

Making the Plastic Filling Filament

See Figure 4C.

1. Using a struck match, heat a 200 µL yellow pipette tip halfway along its length until it becomes transparent and bends under its own weight. Immediately stop heating, grab the other end of the yellow tip between your fingertips, and pull the molten portion out into a fine filament.

2. Once the molten portion cools, cut the filament 8 cm from the original tip of the yellow tip.

3. Against a bench top, secure the end of the filament with the downward pressure of your thumbnail and pull on the yellow tip as if to stretch it, drawing out a fine 1 cm length adjacent to your thumbnail.

4. Using a new razor blade, trim off the portion of the filament crushed by your fingernail. The filament must be ~80 µm in outer diameter and open to pass solutions.

Making the Glass Filling Pipette

See Figure 4, B–D.

5. Pull a glass capillary tube on a micropipette puller to form a tapered tip as for intracellular muscle cell electrophysiology. To start, the capillary should be 1.5 mm in outside diameter, 0.86 or 1.10 mm inner diameter, 10 cm in length, and *without* an internal filament.

6. Using a grain of abrasive on the edge of a piece of sandpaper, score the side of the taper at a point in which the internal diameter is slightly greater than 80 µm, then push the tip to break it off, leaving a clean break perpendicular to the long axis of the capillary.

7. Using a microforge fire-polish the break such that the internal diameter shrinks from >80 µm down to ~12–13 µm.

8. Score and break the nontapered end of the capillary, so that the total length of the capillary is now 7 cm, and lightly heat-polish the broken end over a Bunsen burner.

Making the "Suction" Tube

See Figure 4C.

9. Trim the blunt end of another yellow tip and fit it to the end of a 10-mL plastic syringe.

10. Fit the end of the stiff polyethylene tubing over the free pointed end of the yellow tip.

11. Fit the end of the soft silastic tubing (sleeve) over the free end of the polyethylene tubing.

Preparing the Loading Mixture

12. Prepare a 5 mM rhod-dextran loading solution by adding 100 µL of ddH$_2$O to 5 mg of rhod-dextran. Vortex and spin down several times to mix thoroughly.

13. Prepare a 5 mM AF647-dextran-loading solution by adding 100 µL of ddH$_2$O to 5 mg of AF647-dextran. Mix thoroughly.

14. Add 5.3 µL of the AF647-dextran-loading solution to 100 µL of the rhod-dextran loading solution to create a mixture for loading (final ratio 1:20).

15. Dispense the mixture in 5 µL aliquots into microcentrifuge tubes and store in the dark at 4°C (do not freeze).

Dissecting and Loading the Preparation

16. Fillet dissect a wandering third-instar *Drosophila* larva in chilled Schneider's as in Protocol 1. Do not cut the segment nerves. Do not damage muscle cells 6, 7, 13, or 12 in segments 2–8.

17. Ensure that your filling pipette is not obstructed and ensure that when it rests on the dental wax ramp it can be secured with the tip near the center of the preparation.

18. Draw the loading mixture into the filament (~2 cm).

19. Sever the segment nerve(s) you wish to fill (segment 4) close to the ventral ganglion and cut all others midway along their length.

20. Place the filling pipette on the dental wax ramp with its tip close to the cut end of the nerve you wish to draw up.

21. Loosely fit the polyurethane end of the suction tube over the blunt end of the filling pipette.

22. Draw a single nerve into the filling pipette along with some Schneider's (Fig. 5A–C). Do not pinch the nerve in a loop while attempting to draw it into the pipette. If it is not a snug fit that completely stops further entry of Schneider's to the pipette, then draw up the nerves to both hemisegments.

23. Remove the suction tube from the blunt end of the filling pipette, and then fit it over the blunt end of the filling filament.

24. Touch the side of the filling filament to the meniscus of Schneider's in the bath (to discharge any static electricity), and then immediately insert it into the end of the filling pipette.

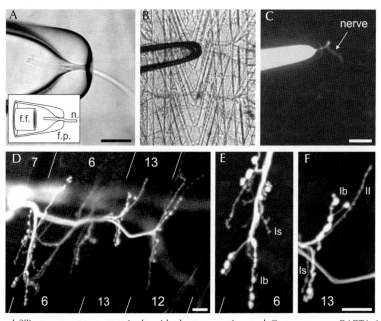

FIGURE 5. Forward-filling motor-nerve terminals with dextran-conjugated Oregon green BAPTA-1 (Protocol 2). (*A*) The severed end of a hemisegment nerve fits snugly in the filling pipette. Scale bar, 50 μm. (*Inset*) Nerve (n.) that has been drawn inside the tip of a filling pipette (f.p.), and showing the tip of the plastic filling filament (f.f.). (*B*) A filling pipette in place over the midline of a filleted larval preparation. (*C*) The same field of view as in *B*, observed using epifluorescence optics. The fluorescent Ca²⁺ indicator is visible in the lumen of the pipette and in the nerve. Scale bar, 100 μm (*B* and *C*). (*D*) A segment nerve (3) at body-wall muscles 7, 6, 13, and 12, forward-filled with 1 mM Oregon green BAPTA-1 (OGB-1) over a 40-min period and viewed on a confocal microscope after 2 h. Ca²⁺-indicator fluorescence is clearly visible in boutons of the nerve terminals. Scale bar, 20 μm. (*E*) Bouton types Ib and Is on muscles 7 and 6. (*F*) Bouton types Ib, Is, and II on muscle 13. Scale bar, 20 μm (*E* and *F*). The laser-scanning confocal microscope used in *D–F* consisted of a BioRad-600 scan head mounted on a Nikon Optiphot-2 microscope with a Nikon 40x water-immersion objective (0.55 NA). (Reprinted, with permission, from Macleod et al. 2002.)

25. Displace the Schneider's adjacent to the nerve with loading mixture. The amount ejected should be no more than one-half the volume of the Schneider's already present in the pipette tip (i.e., final concentration <2.5 mM).

    The loading mixture must be applied to the end of the nerve within 5 min of severing the nerve as axons seal rapidly to exclude large molecules (Eddleman et al. 2000). The 5-min time limit is a critical requirement for the success of this technique.

26. Incubate the preparation at RT in a darkened location for 40 min.

27. After 40 min, discharge any static electricity from the filling filament and insert it into the filling pipette to extract the loading mixture.

28. Use the filling filament to completely fill the filling pipette with Schneider's. Ensure there are no bubbles. The pipette will be used to stimulate the nerve.

    See Troubleshooting.

29. Allow rhod-dextran and AF647-dextran to equilibrate in the nerve for at least 1 h, but no more than 4 h, before commencing Ca^{2+} imaging (Fig. 5D–F).

30. Rinse the preparation with Schneider's every 30 min while it is equilibrating.

31. Twenty minutes before imaging replace Schneider's with prepared HL6 containing 2 mM $[Ca^{2+}]$ and 7 mM L-GA.

TROUBLESHOOTING

Problem (Step 28): The nerve terminal cytosol does not load.
Solution: The pipette was too loose on the nerve and the Ca^{2+} indicator was pushed away from the end of the nerve by Schneider's entering the pipette.

Problem (Step 28): The nerve terminal cytosol does not load but the nerve glial sheath fills.
Solution: The nerve was too tightly constricted by the pipette, or the nerve was drawn up too vigorously and axons were broken.

Problem (Step 28): The nerve is unresponsive to nerve stimulation. There is muscle movement.
Solution: Solutions for these problems are given in the General Troubleshooting section following Protocol 4.

Protocol 3

Direct Injection of Ca^{2+} Indicators

This technique allows rapid loading of most Ca^{2+} indicators and does not require that the nerves be severed. The drawback is that it is arguably the most difficult technique to master and requires additional electrophysiological equipment. Also, Ca^{2+} indicators that are easily injected are usually susceptible to compartmentalization.

The use of this technique is described in Hendel et al. (2008).

MATERIALS

CAUTION: See Appendix for proper handling of materials marked with <!>.

Reagents

CaCl$_2$ <!> in H$_2$O (1.0 M)
Drosophila larva, wandering third-instar stage
Hemolymph Like No. 6 solution (HL6) (Macleod et al. 2002), freshly made, or, freshly defrosted
KCl <!> (100 mM)
KCl:potassium acetate solution [1:1 mixture of 3 M KCl: 3 M potassium acetate (v/v)]
L-glutamic acid monosodium salt hydrate (L-GA) in H$_2$O (0.7 M)
Oregon green 488 BAPTA-1 hexapotassium salt (OGB-1) (500 µg)
Schneider's insect medium containing Ca^{2+} and L-glutamine (Sigma-Aldrich; Cat. No. 50146)

Equipment

Amplifier
Compound microscope with epifluorescence
Dissection microscope and cold light source
Dissection tools (Vannas microscissor and two pairs of fine forceps)
Glass capillary (1.5 mm in outside diameter, ~0.86 mm inner diameter, 10 cm in length), must contain an internal filament (Sutter Instrument; Cat. No. BF150-86-10)
Glass pipette for nerve stimulation (fabrication of a glass filling pipette described in Protocol 2)
Micromanipulator
Micropipette puller (e.g., Flaming/Brown P-97; Sutter Instruments Co.)
Oscilloscope, or data acquisition system with real-time display of micropipette potential
Plastic filling filament (fabrication described in Protocol 2)
Suction tube and syringe (fabrication described in Protocol 2)
Sylgard bottom bath with pins and dental wax ramp (Fig. 4; see Protocol 1, Steps 1–11, for fabrication)

METHOD

Preparing the Ca^{2+} Indicator for Injection

1. Prepare a 5 mM OGB-1 solution for injection by adding 90 µL of 100 mM KCl to 500 µg of OGB-1. Vortex and spin down several times to mix thoroughly.

2. Dispense the 5 mM OGB-1 solution in 10 µL aliquots into microcentrifuge tubes and store in the dark at 4°C.

Dissecting the Preparation and Placing It on the Microscope

3. Fillet dissect a wandering third-instar *Drosophila* larva in chilled Schneider's as in Protocol 1. Do not cut the segment nerves. Do not damage muscle cells 6, 7, 13, or 12 in segments 2–8.

4. Sever the segment nerve(s) you wish to fill (segment 4) close to the ventral ganglion and cut all others midway along their length.

5. Place the glass pipette on the dental wax ramp with its tip close to the cut end of the nerve you wish to draw up.

6. Loosely fit the polyurethane end of the suction tube over the blunt end of the glass pipette.

7. Draw a single nerve into the glass pipette along with some Schneider's.

8. Remove the suction tube from the blunt end of the filling pipette then fit it over the blunt end of the plastic filling filament.

9. Use the filling filament to completely fill the glass pipette with Schneider's. Ensure there are no bubbles. The pipette will be used to stimulate the nerve.

10. Place the bath on the stage of an epifluorescence compound microscope such that the sharp micropipette (below) will be able to approach perpendicular to the larva's long axis.

Preparing a Sharp Glass Micropipette for Ca^{2+}-Indicator Injection

See Figure 6A.

11. Pull a glass capillary tube on a micropipette puller to form a tapered tip (as for intracellular muscle cell electrophysiology). To start (before pulling), the capillary should be 1.5 mm in outside diameter, ~0.86 mm inner diameter, 10 cm in length, and must contain an internal filament.

12. Examine the tip of the sharp glass micropipette under the 40x air objective of a compound microscope (Fig. 6) to ensure that the final 20 μm length has a concave taper, ending in a very fine tip (~0.1 μm). These tips should have a resistance of >70 MΩ if they were to be filled with 3 M KCl.

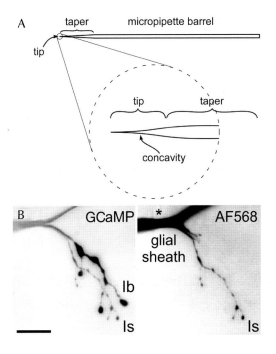

FIGURE 6. Injection of motor-nerve axons (Protocol 3). (*A*) An illustration indicating a critical feature of micropipette tips for injecting fine neuronal process such as axons. The micropipette must not only have a very sharp tip with a high resistance, but the micropipette walls should appear to be slightly concave as they converge into a tip over the final 20 μm. (*B*) *Left:* GCaMP (Wang et al. 2003) expressed in the type-Ib and -Is nerve terminals on muscle 13. *Right:* AF568 hydrazide fluorescence in the type-Is terminal 5 min after injection. Injection was for 2 min and impalement with the micropipette was in the *top left corner* of the image (asterisk, micropipette removed for clarity of imaging). Scale bar, 20 μm.

The fine tip with its "concave" taper is a critical requirement for the success of this technique (Fig. 6).

13. Draw 5 mM Ca^{2+}-indicator injection solution into the plastic filling filament (~2 cm).

14. Insert the tip of the filament 5 mm inside the blunt end of the micropipette and eject the Ca^{2+} indicator forcefully. Allow 5 min for Ca^{2+} indicator to fill the tip.

15. Fill the barrel of the micropipette with a 1:1 mixture of 3 M KCl and 3 M potassium acetate.

Inject an Axon

16. Fit the micropipette to the holder of an amplifier head stage, which in turn is fitted to a micromanipulator.

    For more information on the use of electrophysiological equipment, refer to Chapter 12.

17. Under the 40x water-dipping objective of a compound microscope, advance the micropipette tip to the location where a branch of the SNb/d nerve (Hoang and Chiba 2001) courses over the dorsal surface of muscle 13. The micropipette tip must approach from the lateral margin of the preparation (i.e., not the midline).

18. Advance the tip of the micropipette until it touches in the center of the nerve then advance the tip a further 5 μm until it "catches" on the nerve, pushing in the direction of its long axis.

19. Tap the end micromanipulator with a 10-cm length of stiff polyethylene tubing (1.5 mm inner diameter) until the micropipette tip pierces one of the five axons in the nerve. The preferred method is to advance the tip in a series of short (2 μm) steps at maximum velocity using a piezostepper (Macleod et al. 2001). Successful impalement registers as a sudden depolarization to −40 mV or more negative.

    See Troubleshooting.

20. Monitor the progress in axon filling with periodic inspections of the axons using epifluorescence microscopy (Fig. 6B).

    See Troubleshooting.

21. Hold the impalements for as short a time as is required to fill the axon (i.e., <5 min). Current of 0.1–0.5 nA may be injected for several seconds and repeated to increase filling rate.

22. Withdraw the micropipette tip and allow the preparation to equilibrate for at least 30 min before imaging. At least 20 min before imaging, transfer the preparation to prepared HL6 containing 2 mM $[Ca^{2+}]$ and 7 mM L-GA.

TROUBLESHOOTING

Problem (Step 19): Axon is impaled, but the nerve terminal cytosol does not load. This suggests a blockage in micropipette tip.
Solution: Test for Ca^{2+}-indicator expulsion by current injection in the bath before attempting to impale the axon. Also, establish correct polarity for Ca^{2+}-indicator expulsion at this time.

Problem (Step 20): The axon and boutons fill with Ca^{2+} indicator, but the boutons do not respond to nerve stimulation. This occurs in 33% of our injections. The axon may have been "killed" by mechanical damage or too much current injection, or an AP can no longer propagate past the site of injection.
Solution: Improve the quality of the micropipette tip. Make gentler impalements. Pass less current.

Problem (Step 20): The nerve is unresponsive to nerve stimulation. There is muscle movement.
Solution: Solutions for these problems are given in the General Troubleshooting section following Protocol 4.

Protocol 4

Imaging and Analysis of Nonratiometric Ca^{2+} Indicators

The *Drosophila* larval NMJ preparation, loaded with Ca^{2+} indicator, is set up for imaging of the muscle fiber during stimulation of its innervating nerve cell. The final steps of the protocol describe the sequence of calculations involved in image analysis.

The change in the intensity of the Ca^{2+} indicator must be quantified to obtain an estimate of the change in the concentration of free Ca^{2+} ($\Delta[Ca^{2+}]$). The change in intensity is conventionally represented as the expression "$\Delta F/F$." Simply put, this is the change in fluorescence intensity relative to the resting fluorescence intensity. If the K_D of the Ca^{2+} indicator is in excess of the maximum value of $[Ca^{2+}]$ during the response, then $\Delta F/F$ is considered to be linearly related to $\Delta[Ca^{2+}]$. In practice, $\Delta F/F$ is calculated for each image using a simple algorithm ($[F_{stim} - F_{rest}]/F_{rest}$), where F_{stim} is the intensity of the Ca^{2+} indicator in each image, and F_{rest} is the intensity before nerve stimulation.

Some image acquisition packages are integrated with data analysis software; however, it is recommended that all propriety image analysis software be validated. ImageJ software, available in the public domain (http://rsbweb.nih.gov/ij/), can be used to open almost any image file and has a straightforward interface that allows systematic and transparent image analysis.

MATERIALS

Reagents

Drosophila wandering third-instar larva dissected preparation, loaded with Ca^{2+} indicators as described in Protocol 1, 2, or 3

Equipment

Ca^{2+}-imaging rig including an upright compound microscope
> For details of the imaging equipment, see the section entitled Imaging Equipment Options following this protocol.

Isolated-pulse stimulator (e.g., A-M Systems; Model 2100)
Preparation bath and peripherals, as described in Protocol 1, Steps 1–11, and Figure 4

METHOD

Setting Up the Preparation for Nerve Stimulation

1. Place the bath containing your loaded *Drosophila* preparation, with the pipette in place for nerve stimulation (Fig. 4, B and D), on the stage of an upright microscope (Fig. 4D).

2. Ground the two output wires from the isolated-pulse stimulator.

   i. Place the first of these two wires in the solution in the bath.

   ii. Insert the second wire in the lumen of the pipette (Fig. 4D) after it is touched to the first wire (to dissipate any charge).

   > Insertion of the negative pole wire in the pipette will reduce the risk of electrotonic depolarization of the terminal.

Finding the Loaded Nerve Terminals

3. Using transmitted light only, find the muscles fibers innervated by the nerve in the pipette. Focus on the top surface of the muscle fibers halfway along their length.

4. Use epi-illumination only to bring the loaded terminals into sharp focus and center them in the field of view.

Minimizing Illumination Power

5. Reduce illumination power until the fluorescence intensity of the Ca^{2+} indicator is still above the intensity of the muscle surface, but not excessively so (see note below). The rate of bleach over the time course of your experiment should be ≤5%. Tune your system to maximize the collection efficiency of photons from the Ca^{2+} indicator: for example, optimize the dichroic mirror and filter sets and the detection settings for your camera or photomultiplier. Any automatic gain control for the camera must be switched off for Ca^{2+} imaging.

   > An important principle differentiates epifluorescence imaging of live preparations from immunofluorescence microscopy. As intense illumination destroys both Ca^{2+} indicators and endogenous molecules and produces destructive by-products that further disturb cellular biology, live imaging should be performed with the least excitation power required (e.g., generally ≤2% laser power on confocal systems and ≤12.5% on wide-field systems). As the identity of the compartment loaded is usually established during the loading procedure (at least in Protocols 2 and 3), it is not necessary to maintain illumination levels that allow resolution of fine spatial features.

6. Crop the image to include the terminals of interest and a portion of the muscle surface where no Ca^{2+}-indicator fluorescence can be detected (the latter will serve as your background).

   > Cropping reduces the size of image files and often increases acquisition speed.

Collecting the Data

7. Perform a trial run (Fig. 7A) (i.e., collect a series of images during epifluorescence illumination).

   > This trial run will be used to estimate the "bleach trend" of the Ca^{2+} indicator.

8. Perform a stimulation run (Fig. 7B), during which the nerve is stimulated while a series of images is collected. Start the stimulation impulses at a consistent interval after beginning image acquisition; this allows data from several image series to be averaged, if necessary. Allow several seconds to expire before stimulation to provide for a good estimation of the resting level of fluorescence emission.

   > Depending on the seal of the pipette on the nerve, an impulse of as little as 0.7 V for 0.3 msec will initiate an AP in the nerve. We recommend an impulse amplitude 20%–50% above the threshold of AP initiation, with 0.3 msec duration.

Image Analysis

Refer to Figure 7 for an illustration of the steps described below. An excerpt from an Excel spreadsheet, shown in Figure 8, is included to help you organize your data. The annotations will assist you with the required calculations. All columns referred to in the text below can be found in Figure 8.

9. Determine the value of S (the signal in Fig. 7B) in each image of the trial run, and then do the same for the stimulation run. The signal (fluorescence intensity of the loaded terminal boutons) is determined by drawing a region of interest (ROI) around the boutons and calculating the average pixel intensity within the ROI (column B and column J). Ensure that other loaded structures, below or adjacent, do not contribute fluorescence to the ROI. For each image, a time

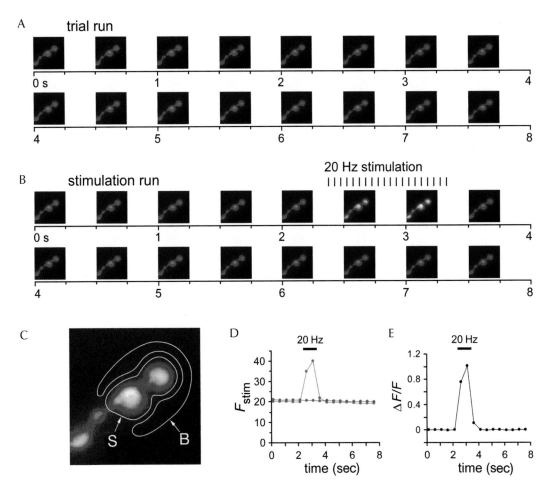

FIGURE 7. Ca^{2+}-imaging data and analysis, illustrated. (*A*) A series of images of a fluorescent nerve terminal, loaded with a Ca^{2+} indicator, collected over a period of 8 sec. Images are collected at a rate of 2/sec and the space between images represents an acquisition delay. (*B*) A series of images as in *A*, but with a stimulus train commencing 2.375 sec after image acquisition commences. Each "tick" represents a stimulus pulse to the nerve of 0.3 msec duration. (*C*) A close-up view of the seventh image in *B* with two regions of interest (ROIs) superimposed. The first ROI (S) is drawn around the boutons to quantify the average pixel intensity from the Ca^{2+} indicator. The second ROI (B) defines a region of the image in which the fluorescence appears to be representative of the background of the boutons. (*D*) A plot of the Ca^{2+}-indicator fluorescence in the nerve terminals (F_{stim}) during the trial run (green; column D in Fig. 8), and during a stimulation run (red; column L in Fig. 8). (*E*) A plot of $\Delta F/F$ during the stimulation run (column P in Fig. 8). $\Delta F/F$ has been corrected for the effect of bleaching.

of image acquisition should accompany the measurement of average pixel intensity (column A and column I).

> A common mistake is to fit the ROI too snugly around the boutons. This will not increase the signal-to-noise ratio, but rather, it will make your determinations of the signal intensity susceptible to any movements of the nerve terminal.

10. Determine the value of *B* (the background in Fig. 7C) in each image of the trial run, and then do the same for the stimulation run. The background (fluorescence intensity of a portion of the muscle surface where no Ca^{2+}-indicator fluorescence can be detected) is usually uneven in the field of view, so the ROI should be selected close to the terminal boutons (column C and column K).

> As the average pixel intensity is being measured, the area of the ROI in Step 9 (*S*) does not have to be equal to the area of the ROI in Step 10 (*B*).

columns

		A	B	C	D	E	F	G	H	I	J	K	L	M	N	O	P
			trial run								stimulation run						
		time	S	B	F_{stim}	exp.fit	y	F_{rest}	corr.	time	S	B	F_{stim}	F_{stim}'	F_{rest}	ΔF	ΔF/F
rows	1	0.13	26.0	5.0	21.0		21.0		1.01	0.13	25.0	4.8	20.2	20.0		0.02	0.00
	2	0.63	25.9	5.0	20.9	$y=Ae^{-bt}$	20.9		1.01	0.63	24.9	4.8	20.1	20.0		-0.01	0.00
	3	1.13	25.8	4.9	20.9		20.8		1.00	1.13	24.8	4.7	20.1	20.0		0.06	0.00
	4	1.63	25.6	4.9	20.7	A	20.8	20.76	1.00	1.63	24.6	4.7	19.9	19.9	19.97	-0.07	0.00
	5	2.13	25.6	4.9	20.7	21	20.7		1.00	2.13	24.6	4.7	19.9	20.0		0.00	0.00
	6	2.63	25.5	4.9	20.6	b	20.6		0.99	2.63	39.5	4.7	34.8	35.0		15.08	0.76
	7	3.13	25.5	4.9	20.6	0.007	20.5		0.99	3.13	44.5	4.7	39.8	40.2		20.25	1.01
	8	3.63	25.2	4.8	20.4		20.5		0.99	3.63	26.4	4.6	21.8	22.1		2.14	0.11
	9	4.13	25.3	4.8	20.5		20.4		0.98	4.13	24.3	4.6	19.7	20.0		0.08	0.00
	10	4.63	25.1	4.8	20.3		20.3		0.98	4.63	24.1	4.6	19.5	19.9		-0.05	0.00
	11	5.13	25.1	4.8	20.3		20.3		0.98	5.13	24.1	4.6	19.5	20.0		0.02	0.00
	12	5.63	24.9	4.8	20.1		20.2		0.97	5.63	24.1	4.6	19.5	20.1		0.09	0.00
	13	6.13	25.0	4.8	20.2		20.1		0.97	6.13	23.9	4.6	19.3	19.9		-0.05	0.00
	14	6.63	24.8	4.8	20.0		20.0		0.97	6.63	23.8	4.6	19.2	19.9		-0.08	0.00
	15	7.13	24.8	4.8	20.0		20.0		0.96	7.13	23.8	4.6	19.2	20.0		-0.01	0.00
	16	7.63	24.7	4.8	19.9		19.9		0.96	7.63	23.8	4.6	19.2	20.0		0.06	0.00

column	formula		column	formula		column	formula	
D	=B1-C1		H	=F1/G$4		N	=AVERAGE(M3:M5)	
F	=E$5*POWER(2.7183,-E$7*I1)		L	=J1-K1		O	=M1-N$4	$(F_{stim} - F_{rest})$
G	=AVERAGE(F3:F5)		M	=L1/H1		P	=O1/N$4	$(\Delta F / F_{rest})$

FIGURE 8. Excerpt from an Excel spreadsheet displaying the data and analysis from the sequence of images shown in Fig. 7. The columns are designated by letters and rows are numbered. Annotations at the *bottom* show the formulas as they appear in the first cell of each column.

11. Calculate values of F_{stim} for the trial run and then for the stimulation run. F_{stim} is the fluorescence intensity of the boutons with the background intensity subtracted (columns: D = B – C, and, L = J – K). F_{stim} for both runs is plotted in Figure 7D.

12. Estimate the bleach trend by fitting an exponential decay to the F_{stim} values of the trial run. The equation for the curve fit will be used to remove the influence of bleaching from the stimulation run. Both SigmaPlot and Excel spreadsheets provide least-squares-fitting algorithms. The equation for an exponential decay takes the form $y = Ae^{-bt}$ and the value of A (the initial value) and b (the decay constant) are obtained from the curve fit (column E).

13. Calculate values on the bleach curve for the time of each image of the stimulation run. Times in the stimulation run are used (column I) rather than those in the trial run, because it is the stimulation run F_{stim} values that need correction rather than the trial run values, and the timing between the trial and stimulation runs can be different. A second reason a curve fit equation is used for removing the influence of bleaching, rather than the F_{stim} values of the trial run, is that the later values can compound noise in the final estimates of ΔF/F.

14. Calculate F_{rest} for the curve fit values; that is, average several curve fit values in column F immediately before the time of stimulation in the stimulation run.

15. Calculate the correction factor for each F_{stim} value; that is, divide curve fit values by F_{rest} (column F/column G).

16. Apply the correction factor to F_{stim} values of the stimulation run; that is, divide each F_{stim} value by its correction factor (column L/column H).

17. Calculate F_{rest} for F_{stim} values of the stimulation run using a similar procedure to that in Step 14.

18. Calculate values of ΔF (the change in fluorescence intensity). As $\Delta F = F_{stim} - F_{rest}$, subtract F_{rest} from each value of F_{stim} (column M – column N).

19. Calculate values of $\Delta F/F$ (the change in fluorescence intensity relative to the resting fluorescence intensity). As $\Delta F/F = (F_{stim} - F_{rest})/F_{rest}$, divide each value of ΔF by F_{rest} (column O/column N). These values are plotted in Figure 7E. A $\Delta F/F$ value of 1 indicates a 100% change in the fluorescence of the Ca^{2+} indicator.

> Although not discussed here, nonratiometric Ca^{2+} indicators can be calibrated against known values of $[Ca^{2+}]$ (see Maravall et al. 2000; Hendel et al. 2008).

GENERAL TROUBLESHOOTING

We highlight here a number of general threats to the success of your Ca^{2+}-imaging experiments with suggested solutions.

Photodestruction

A common mistake is to overilluminate the preparation.
Symptom: There are inconsistent/anomalous responses and a rapidly deteriorating signal-to-noise ratio.
Solution: In future experiments, unless you are capturing data, do not illuminate the preparation. Focusing is best performed with transmitted light and only intermittently with epifluorescence.

Electrocution

Static electricity is an insidious threat. Once the nerve is isolated within the filling pipette it is very susceptible to electrocution. A discharge of only tens of volts down the lumen of the pipette is probably sufficient to render the nerve unresponsive to more controlled stimulus pulses.

Symptom: The nerve is unresponsive to nerve stimulation. Muscles show continual, slow movements.
Solution: In future, the output wires from the isolated-pulse stimulator should be earthed, and the wire to be inserted into the filling pipette must always be touched to the other wire in the bath before it is inserted into the pipette.

Fixatives

Any residue of fixatives on your dissection instruments, pins, bath, or microscope objective will slowly permeabilize and fix your preparation.

Symptom: There is erratic, ongoing movement of the muscles and fine vacuolation of the muscles.
Solution: Before dissecting the next larva, discard any baths that have contained fixatives, no matter how well rinsed subsequently. Do not use dissection instruments (especially scissors) that have been used to cut or manipulate fixed tissues. Do not use dipping objectives that are used to examine tissue that has been fixed.

Anoxia

Once your primary focus is on gathering imaging data you may forget to replace/exchange or even circulate the bath solution. This leads either to anoxia or perhaps even to an unstirred K^+ layer effect on the preparation.

Symptom: There is movement of the muscle. There is muscle contraction on nerve stimulation, even when L-GA is present at 7 mM. A slow increase in resting $[Ca^{2+}]$ is seen with cytosolic fura-dextran.

Solution: When not using a superfusion bath, exchange the bathing solution at least every 30 min, even when not imaging. When imaging, exchange or circulate the solution regularly. Even raising a dipping objective out of the bathing solution and replacing it back in the bath can sometimes reverse some of the effects described above. The most consistent responses are seen under conditions in which the preparation is allowed to rest for several minutes between runs, and the bath solution is exchanged or circulated regularly.

DISCUSSION: IMAGING EQUIPMENT OPTIONS

Many options are available for putting together a Ca^{2+}-imaging rig and not all of them are expensive. We describe three imaging rigs, each utilizing a different technology, starting with the most inexpensive. At the heart of each imaging rig is the compound microscope, but what really differentiates rigs is the source of light for fluorophore excitation and the emission detector.

The manufacturers, suppliers, and brands mentioned below are recommended on the basis of the author's direct experience in Ca^{2+} imaging in several laboratories and are given as a guide only. These recommendations are not the result of a comprehensive survey and are not intended to exclude any other manufacturers, suppliers, or brands.

Option 1

The simplest Ca^{2+}-imaging rig uses an upright compound microscope with an Hg bulb epi-illuminator and a charge-coupled device (CCD) camera.

Fluorescence microscopes can be quite expensive, but many reputable brand fluorescence microscopes built in the last 25 years will still deliver high-quality images when fitted with the right optics (see note below). A standard 100-W Hg bulb, with epi-illuminator housing and power source, is sufficient to provide most wavelengths for fluorescence imaging. By way of contrast, the specifications of the objective lens, filters, and dichroic mirror are more critical. Select a water-dipping objective with a high magnification ($\geq$60X) and a high numerical aperture ($\geq$0.8 NA). Excitation filters, dichroic mirrors, and emission filters can be purchased as sets optimized for specified Ca^{2+} indicators (e.g., Semrock, Inc., www.semrock.com; Chroma Technology Corp., www.chroma.com). In many cases, conventional monochromatic CCD cameras will be sufficient to capture emission fluorescence from the Ca^{2+} indicator. The rate at which data can be acquired will depend on the light sensitivity of the camera. A high sensitivity allows a fast rate of data acquisition. The camera should have a range of $\geq$10 bits and a linear response over the appropriate range of fluorescence intensities (e.g., Scion Corporation, www.scioncorp.com; Photometrics, www.photomet.com). Other components of the rig include a framegrabber card and imaging software, which are usually specified, if not sold, by the camera supplier.

The imaging software should be able to signal to, or accept a signal from, the nerve stimulator. If single trains only are required for your experiment, a stand-alone isolated-pulse generator should suffice (e.g., A-M Systems, Inc., www.a-msystems.com). If a series of trains is required, and the software cannot be programmed to coordinate the stimulator, one solution is to use a programmable pulse generator (e.g., Master-8/A.M.P.I., www.ampi.co.il) to control an isolated constant voltage stimulator (e.g., Digitimer, www.digitimer.com).

The major microscope manufacturers have now adopted infinity corrected optics as the standard, replacing the earlier standard in which the objective lens formed an image at a fixed distance (160 mm) along the microscope tube. Older objectives designed for 160-mm tube length microscopes (stamped 160/-) are not compatible with newer microscopes designed to accommodate infinity corrected objectives. Similarly, objectives with infinity corrected optics (stamped ∞) are not compatible with older 160-mm microscopes (e.g., Nikon Optiphot series or Olympus BH2 series).

Online tools are available to test filter selections for each fluorophore: http://www.bdbiosciences.com/spectra/, http://www.invitrogen.com/site/us/en/home/support/Research-Tools/Fluorescence-SpectraViewer.html.

Option 2

A more powerful wide-field Ca^{2+}-imaging rig would use a more flexible illumination system along with a faster and more sensitive camera.

In Option 1 excitation and emission light must be manually shuttered, a process which is slow, highly variable, and introduces significant vibration. Manual shuttering makes ratiometric Ca^{2+} imaging almost impossible. On the excitation side, a number of systems provide solutions. The most inexpensive solution is to interpose a motorized filter wheel between the light source and the filter cubes (e.g., Sutter Instrument, www.sutter.com; Newport, www.newport.com). Systems that provide superior performance, but cost commensurately more, are wavelength-switching devices (e.g., Sutter Instrument) or monochromators (e.g., TILL Photonics, www.till-photonics.com). All three of these options allow selection of the excitation light wavelength, switching between wavelengths, and shuttering. Common Ca^{2+} indicators requiring wavelength switching on the excitation side are fura-2 and ratiometric pericam. On the emission side, a motorized filter wheel interposed between the filter cubes and the camera provides an easy solution. A device that is very effective for collecting two emission wavelengths simultaneously (e.g., FRET imaging) is the Optosplit-II (Cairn Research, www.cairn-research.co.uk). Ca^{2+} indicators requiring wavelength switching on the emission side are yellow cameleons, TN-XXL, and other FRET-based Ca^{2+} indicators.

The solution to the shortcomings of the camera in Option 1 is to invest in a superior camera. EMCCD cameras with their high quantum efficiency and fast readout rates (e.g., Andor Technology, www.andor.com; Hamamatsu Photonics, www.hamamatsu.com) are gradually replacing conventional CCD and intensified CCD cameras.

The illumination systems described above, along with the cameras, require computer support that also allows them to be coordinated to collect data. The manufacturers/suppliers will recommend not only the computer specifications to support these peripherals but also software packages to coordinate them.

Option 3

Perhaps the most powerful systems for Ca^{2+} imaging are the confocal laser-scanning microscope (CLSM) and the multiphoton fluorescence microscope.

These systems are generally sold as fully integrated packages and few choices, except for an inverted stage microscope, will significantly limit your Ca^{2+}-imaging options for the *Drosophila* larval NMJ. Similar to fluorescence microscopes, CLSMs can be well over a decade old yet generate high-quality Ca^{2+}-imaging data (e.g., BioRad 600; Leica TCS 4D). Your choice of lasers to populate a laser launch will dictate which fluorophores you can image on a CLSM. The 488-nm laser is probably the most versatile, but a laser line ~440 nm will be required for imaging with FRET Ca^{2+} indicators containing CFP (cyan fluorescent protein), and a laser line ~543 nm will be required for imaging most of the rhod derivatives. Perhaps the most significant limitation of a CLSM is that it is rarely supplied with a UV laser capable of Ca^{2+} imaging with fura-2. One of the most powerful tools on a CLSM is the ability to repeatedly make scans through a compartment loaded with a Ca^{2+} indicator to collect data with high temporal resolution (e.g., Fig. 3).

Not mentioned above is the use of photodiodes or photomultiplier tubes (e.g., Hamamatsu Photonics; TILL Photonics), in conjunction with wide-field microscopy systems (Options 1 and 2) to obtain very high temporal resolution data (Habets and Borst 2006; Lin 2008). An analogy can be made to a very fast CCD or EMCCD camera with a 1 x 1-pixel array, in which spatial information is lost but the speed of data acquisition may increase to thousands of "frames" (intensity values) per second.

ACKNOWLEDGMENTS

G.T.M. is supported by NIH RO1 NS061914. Thanks to Amit K. Chouhan for comments on this manuscript.

REFERENCES

Atluri PP, Regehr WG. 1996. Determinants of the time course of facilitation at the granule cell to Purkinje cell synapse. *J Neurosci* **16:** 5661–5671.

Augustin H, Grosjean Y, Chen K, Sheng Q, Featherstone DE. 2007. Nonvesicular release of glutamate by glial xCT transporters suppresses glutamate receptor clustering in vivo. *J Neurosci* **27:** 111–123.

Bao H, Daniels RW, MacLeod GT, Charlton MP, Atwood HL, Zhang B. 2005. AP180 maintains the distribution of synaptic and vesicle proteins in the nerve terminal and indirectly regulates the efficacy of Ca^{2+}-triggered exocytosis. *J Neurophysiol* **94:** 1888–1903.

Brand AH, Perrimon N. 1993. Targeted gene expression as a means of altering cell fates and generating dominant phenotypes. *Development (Cam)* **118:** 401–415.

Bronk P, Wenniger JJ, Dawson-Scully K, Guo X, Hong S, Atwood HL, Zinsmaier KE. 2001. *Drosophila* Hsc70-4 is critical for neurotransmitter exocytosis in vivo. *Neuron* **30:** 475–488.

Chouhan AK, Zhang J, Zinsmaier KE, Macleod GT. 2010. Presynaptic mitochondria in functionally different motor neurons exhibit similar affinities for Ca^{2+} but exert little influence as Ca^{2+} buffers at nerve firing rates in situ. *J Neurosci* **30:** 1869–1881.

David G, Barrett EF. 2003. Mitochondrial Ca^{2+} uptake prevents desynchronization of quantal release and minimizes depletion during repetitive stimulation of mouse motor nerve terminals. *J Physiol* **548:** 425–438.

Dawson-Scully K, Bronk P, Atwood HL, Zinsmaier KE. 2000. Cysteine-string protein increases the calcium sensitivity of neurotransmitter exocytosis in *Drosophila*. *J Neurosci* **20:** 6039–6047.

Eddleman CS, Bittner GD, Fishman HM. 2000. Barrier permeability at cut axonal ends progressively decreases until an ionic seal is formed. *Biophys J* **79:** 1883–1890.

Feng Y, Ueda A, Wu CF. 2004. A modified minimal hemolymph-like solution, HL3.1, for physiological recordings at the neuromuscular junctions of normal and mutant *Drosophila* larvae. *J Neurogenet* **18:** 377–402.

Filippin L, Abad MC, Gastaldello S, Magalhaes PJ, Sandona D, Pozzan T. 2005. Improved strategies for the delivery of GFP-based Ca^{2+} sensors into the mitochondrial matrix. *Cell Calcium* **37:** 129–136.

Guerrero G, Reiff DF, Agarwal G, Ball RW, Borst A, Goodman CS, Isacoff EY. 2005. Heterogeneity in synaptic transmission along a *Drosophila* larval motor axon. *Nat Neurosci* **8:** 1188–1196.

Guo X, Macleod GT, Wellington A, Hu F, Panchumarthi S, Schoenfield M, Marin L, Charlton MP, Atwood HL, Zinsmaier KE. 2005. The GTPase dMiro is required for axonal transport of mitochondria to *Drosophila* synapses. *Neuron* **47:** 379–393.

Helmchen F, Imoto K, Sakmann B. 1996. Ca^{2+} buffering and action potential-evoked Ca^{2+} signaling in dendrites of pyramidal neurons. *Biophys J* **70:** 1069–1081.

Hendel T, Mank M, Schnell B, Griesbeck O, Borst A, Reiff DF. 2008. Fluorescence changes of genetic calcium indicators and OGB-1 correlated with neural activity and calcium in vivo and in vitro. *J Neurosci* **28:** 7399–7411.

Hoang B, Chiba A. 2001. Single-cell analysis of *Drosophila* larval neuromuscular synapses. *Dev Biol* **229:** 55–70.

Jan LY, Jan YN. 1976. Properties of the larval neuromuscular junction in *Drosophila melanogaster*. *J Physiol* **262:** 189–214.

Karunanithi S, Georgiou J, Charlton MP, Atwood HL. 1997. Imaging of calcium in *Drosophila* larval motor nerve terminals. *J Neurophysiol* **78:** 3465–3467.

Kuromi H, Kidokoro Y. 2002. Selective replenishment of two vesicle pools depends on the source of Ca^{2+} at the *Drosophila* synapse. *Neuron* **35:** 333–343.

Kuromi H, Honda A, Kidokoro Y. 2004. Ca^{2+} influx through distinct routes controls exocytosis and endocytosis at *Drosophila* presynaptic terminals. *Neuron* **41:** 101–111.

Lee A, Scheuer T, Catterall WA. 2000. Ca^{2+}/calmodulin-dependent facilitation and inactivation of P/Q-type Ca^{2+} channels. *J Neurosci* **20:** 6830–6838.

Lnenicka GA, Grizzaffi J, Lee B, Rumpal N. 2006. Ca^{2+} dynamics along identified synaptic terminals in *Drosophila* larvae. *J Neurosci* **26:** 12283–12293.

Macleod GT, Dickens PA, Bennett MR. 2001. Formation and function of synapses with respect to Schwann cells at the end of motor nerve terminal branches on mature amphibian (*Bufo marinus*) muscle. *J Neurosci* **21:** 2380–2392.

Macleod GT, Hegstrom-Wojtowicz M, Charlton MP, Atwood HL. 2002. Fast calcium signals in *Drosophila* motor neuron terminals. *J Neurophysiol* **88:** 2659–2663.

Macleod GT, Suster ML, Charlton MP, Atwood HL. 2003. Single neuron activity in the *Drosophila* larval CNS detected with calcium indicators. *J Neurosci Methods* **127:** 167–178.

Macleod GT, Marin L, Charlton MP, Atwood HL. 2004. Synaptic vesicles: Test for a role in presynaptic calcium regulation. *J Neurosci* **24:** 2496–2505.

Macleod GT, Chen L, Karunanithi S, Peloquin JB, Atwood HL, McRory JE, Zamponi GW, Charlton MP. 2006. The *Drosophila* cac[ts2] mutation reduces presynaptic Ca^{2+} entry and defines an important element in $Ca_V2.1$ channel inactivation. *Eur J Neurosci* **23**: 3230–3244.

Mank M, Reiff DF, Heim N, Friedrich MW, Borst A, Griesbeck O. 2006. A FRET-based calcium biosensor with fast signal kinetics and high fluorescence change. *Biophys J* **90**: 1790–1796.

Mank M, Santos AF, Direnberger S, Mrsic-Flogel TD, Hofer SB, Stein V, Hendel T, Reiff DF, Levelt C, Borst A, et al. 2008. A genetically encoded calcium indicator for chronic in vivo two-photon imaging. *Nat Methods* **5**: 805–811.

Morales M, Ferrus A, Martinez-Padron M. 1999. Presynaptic calcium-channel currents in normal and *csp* mutant *Drosophila* peptidergic terminals. *Eur J Neurosci* **11**: 1818–1826.

Nagai T, Yamada S, Tominaga T, Ichikawa M, Miyawaki A. 2004. Expanded dynamic range of fluorescent indicators for Ca^{2+} by circularly permuted yellow fluorescent proteins. *Proc Natl Acad Sci* **101**: 10554–10559.

Palmer AE, Tsien RY. 2006. Measuring calcium signaling using genetically targetable fluorescent indicators. *Nat Protocols* **1**: 1057–1065.

Palmer AE, Jin C, Reed JC, Tsien RY. 2004. Bcl-2-mediated alterations in endoplasmic reticulum Ca^{2+} analyzed with an improved genetically encoded fluorescent sensor. *Proc Natl Acad Sci* **101**: 17404–17409.

Palmer AE, Giacomello M, Kortemme T, Hires SA, Lev-Ram V, Baker D, Tsien RY. 2006. Ca^{2+} indicators based on computationally redesigned calmodulin-peptide pairs. *Chem Biol* **13**: 521–530.

Pinton P, Rizzuto R. 2006. Bcl-2 and Ca^{2+} homeostasis in the endoplasmic reticulum. *Cell Death Differ* **13**: 1409–1418.

Reiff DF, Thiel PR, Schuste CM. 2002. Differential regulation of active zone density during long-term strengthening of *Drosophila* neuromuscular junctions. *J Neurosci* **22**: 9399–9409.

Reiff DF, Ihring A, Guerrero G, Isacoff EY, Joesch M, Nakai J, Borst A. 2005. In vivo performance of genetically encoded indicators of neural activity in flies. *J Neurosci* **25**: 4766–4778.

Rossano AJ, Macleod GT. 2007. Loading *Drosophila* nerve terminals with calcium indicators. *J Vis Exp* **6**: 250.

Sanyal S, Consoulas C, Kuromi H, Basole A, Mukai L, Kidokoro Y, Krishnan KS, Ramaswami M. 2005. Analysis of conditional paralytic mutants in *Drosophila* sarco-endoplasmic reticulum calcium ATPase reveals novel mechanisms for regulating membrane excitability. *Genetics* **169**: 737–750.

Snapp EL, Iida T, Frescas D, Lippincott-Schwartz J, Lilly MA. 2004. The fusome mediates intercellular endoplasmic reticulum connectivity in *Drosophila* ovarian cysts. *Mol Biol Cell* **15**: 4512–4521.

Stewart BA, Atwood HL, Renger JJ, Wang J, Wu CF. 1994. Improved stability of *Drosophila* larval neuromuscular preparations in haemolymph-like physiological solutions. *J Comp Physiol A* **175**: 179–191.

Sugimori M, Lang EJ, Silver RB, Llinas R. 1994. High-resolution measurement of the time course of calcium-concentration microdomains at squid presynaptic terminals. *Biol Bull* **187**: 300–303.

Tank DW, Regehr WG, Delaney KR. 1995. A quantitative analysis of presynaptic calcium dynamics that contribute to short-term enhancement. *J Neurosci* **15**: 7940–7952.

Tian L, Hires A, Mao T, Huber D, Chiappe ME, Chalasani SH, Petreanu L, Akerboom J, McKinney SA, Schreiter ER, et al. 2009. Imaging neural activity in worms, flies and mice with improved GCaMP calcium indicators. *Nat Methods* **6**: 871–872.

Umbach JA, Saitoe M, Kidokoro Y, Gundersen CB. 1998. Attenuated influx of calcium ions at nerve endings of *csp* and *shibire* mutant *Drosophila*. *J Neurosci* **18**: 3233–3240.

Verstreken P, Ly CV, Venken KJ, Koh TW, Zhou Y, Bellen HJ. 2005. Synaptic mitochondria are critical for mobilization of reserve pool vesicles at *Drosophila* neuromuscular junctions. *Neuron* **47**: 365–378.

Wang JW, Wong AM, Flores J, Vosshall LB, Axel R. 2003. Two-photon calcium imaging reveals an odor-evoked map of activity in the fly brain. *Cell* **112**: 271–282.

Wang W, Fang H, Groom L, Cheng A, Zhang W, Liu J, Wang X, Li K, Han P, Zheng M, et al. 2008. Superoxide flashes in single mitochondria. *Cell* **134**: 279–290.

Xu J, He L, Wu LG. 2007. Role of Ca^{2+} channels in short-term synaptic plasticity. *Curr Opin Neurobiol* **17**: 352–359.

20 Imaging Neuropeptide Release and Signaling in the *Drosophila* Neuromuscular Junction with Green Fluorescent Protein

Edwin S. Levitan and Dinara Shakiryanova

Department of Pharmacology & Chemical Biology, University of Pittsburgh, Pittsburgh, Pennsylvania 15261

ABSTRACT

Electrophysiological studies of synaptic function cannot directly reveal the internal workings of the nerve terminal and do not robustly report release of neuropeptides and neurotrophins. These limitations can now be overcome with the presynaptic expression of green fluorescent protein (GFP) indicators of vesicle motion, release, and signaling. Here we describe how to image single wavelength and ratiometric FRET (fluorescence resonance energy transfer)-based GFP indicators with fluorescence microscopy in living synaptic boutons of the *Drosophila* neuromuscular junction.

INTRODUCTION

GFP imaging has revolutionized the study of synaptic function because release, vesicle dynamics, and signal transduction can be monitored directly with optics. The Introduction to Chapter 7 in this volume provides an excellent description of the features and ultrastructure of the *Drosophila* neuromuscular junction (NMJ), and Chapter 17 describes electrophysiological recordings at the NMJ. Here we describe how to optimize fluorescence imaging of the NMJ preparation. We first discuss basic optical principles and the equipment for epifluorescence microscopy, and then we present a detailed protocol for adjusting the imaging setup with a description of experimental parameters.

EPIFLUORESCENCE MICROSCOPY: BASIC OPTICAL PRINCIPLES AND EQUIPMENT

The Wide-Field Epifluorescence Microscope

The first issue confronting the investigator interested in GFP imaging in *Drosophila* NMJ synaptic boutons is the choice of microscope. Surprisingly to many novices, a standard simple upright wide-field epifluorescence microscope is best. Confocal microscopes collect only a fraction of the emitted fluorescence through a pinhole to exclude out-of-focus light. However, because synaptic boutons are small and the postsynaptic muscle need not be labeled (because of the availability of specific presynaptic GAL4 drivers), there is no significant out-of-focus fluorescence emission with presynaptic

GFP expression in motor neurons. Therefore, the benefit of slightly better resolution with confocal microscopy is outweighed by the dramatic loss of signal, which has a number of drawbacks. First, detection of faint signal may not be possible. Second, the bright laser beams typically used with scanning confocal microscopes can cause photobleaching and photodamage. Third, because the signal-to-noise ratio for any measurement is proportional to the square root of the signal, collecting less light with a confocal microscope is associated with noisier measurements. Finally, confocal microscopes are very expensive and thus not as accessible to the typical fly laboratory. This latter difference is even greater for multiphoton microscopes. Therefore, wide-field epifluorescence microscopy is optimal for quantitative assays of presynaptic function. An inverted microscope is best for observing neurons cultured on coverslips, but an upright microscope is required for the filleted larval NMJ preparation, so that nerve terminals can be viewed directly.

The High-Numerical-Aperture Objective

Once the microscope is selected, the most important component of the optical setup is the microscope objective. Here one parameter—numerical aperture (NA)—is critical. NA determines the ability of the objective lens to concentrate illumination and to collect emitted fluorescence. In fact, each of these parameters is proportional to NA^2, resulting in dependence of the signal on NA^4. Furthermore, resolution also depends on NA: The limit of resolution is inversely proportional to NA in the plane of focus (i.e., the x–y plane) and is inversely proportional to NA^2 axially (i.e., in the z-axis). In contrast, magnification (M) does not affect resolution or the efficiency of collecting light from a small source (e.g., a GFP-labeled vesicle). Therefore, the highest NA possible should be used. Because the filleted larval NMJ is viewed directly through saline, a "dipping" objective, rather than a coverslip-corrected objective, should be used. For example, an Olympus 1.1 NA 60x direct water-immersion objective with a working distance of 1.5 mm is well suited for imaging the response to electrical nerve stimulation of the NMJ. Because objectives are matched for each manufacturer's microscope, fly researchers with a microscope should use an objective with the highest NA available for their instrument.

The importance of NA is illustrated in Figure 1, which shows images of the same motor neuron type Ib synaptic bouton preparation acquired with 1.1 and with 0.9 NA 60x objectives. Note that the 1.1 NA objective yields a brighter image (compare the left and center panels) so that dim features are more likely to be detected and the signal-to-noise ratio for intensity measurements is increased. Furthermore, even after contrast enhancement (shown in the right panel), the 0.9 NA image is blurrier. Hence, the 1.1 NA objective is better for quantifying loss of fluorescence associated with release, as well as vesicle redistribution in the bouton (Fig. 2).

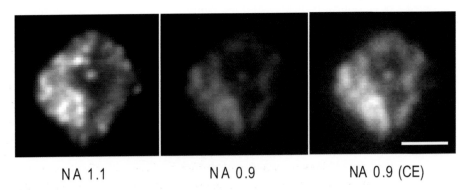

NA 1.1   NA 0.9   NA 0.9 (CE)

FIGURE 1. The importance of objective numerical aperture (NA). Images of a motor neuron type Ib synaptic bouton from an elav-GAL4 UAS-preproANF-EMD third-instar larva obtained with 1.1 NA and 0.9 NA 60x water-immersion objectives. Scale bar, 2 µm. Note that the image with the higher NA (*left*) is brighter than with the lower NA (*center*). Even after contrast enhancement (CE), the lower NA image is blurrier (*right*) so that puncta from neuropeptide vesicles are harder to resolve.

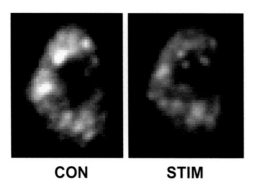

CON   **STIM**

FIGURE 2. The effect of stimulation on the GFP-tagged neuropeptide signal from a single bouton. 70 Hz stimulation for 30 sec (STIM) of a type Ib bouton from an elav-*GAL4 UAS-preproANF-EMD* third-instar larva causes a decrease in fluorescence compared with the control (CON) that is indicative of neuropeptide release. Furthermore, vesicles redistribute to enter the central dark region.

Digital Camera

The next important component is the camera. For the quantitative studies considered here, cooled CCD (charged couple device) cameras are best; however, a range of specifications dramatically affects camera price. More cooling is desirable to minimize instrument noise, but with short exposures (e.g., 100 msec), such noise has limited impact. Electron multiplying or cascade cameras offer greater sensitivity, but these produce complex noise characteristics that can confound quantitative intensity measurements. In general, when fluorescence can be seen through a microscope eyepiece, this camera technology is not advantageous. Camera speed is also a critical factor. Very fast frame rates require more expensive technology and are accompanied by less signal per frame. However, extremely fast image acquisition is not required for most release and signaling experiments with the fly NMJ. Therefore, a camera equipped with a 1000 x 1000 array of ~7-μm-wide pixels is well suited for detecting vesicle mobility, release, and signaling. We have had success with cameras that are cooled by >20°C from ambient temperature. Sensitivity is the final parameter. When only a few images are required, moderate quantum efficiency (e.g., 40%, when 100% is the theoretical maximum) is acceptable: Exposures must be longer, but photobleaching is minimal with only a few exposures. However, high sensitivity becomes important for time-lapse studies requiring many images. Back-thinned (back-illuminated) cameras are capable of >90% quantum efficiency but are very expensive and often feature a lower number of larger pixels. Thus, a good compromise is a camera that has an interline chip with >60% quantum efficiency (e.g., Hamamatsu Orca AG). It is important to test candidate cameras with the preparation of interest because imaging quality can vary, even when the cameras feature the same light-detecting chip, owing to other internal gain settings and electronics. Furthermore, camera technology is upgraded at a rapid pace, unlike the case for optics in which substantive improvements occur only rarely.

Other Microscope Components

Another required component is an electronically controlled shutter that allows illumination to be controlled during time-lapse experiments by software that also communicates with the camera. Also, for ratiometric FRET experiments, it is necessary to synchronize the camera and shutter with equipment that varies emission fluorescence collection (Fig. 3). We have used a Sutter Instruments filter wheel to vary emission filters. The use of this wheel, however, introduces a speed limitation because time must be allocated for the wheel to change positions. If very fast acquisition rates are required, an alternative is to use a device based on a beam splitter that projects duplicate images through different emission filters onto the camera (Optical Insights). With this latter approach, both emission wavelengths are acquired simultaneously without the need for computer control, but magnification is reduced, resulting in larger effective pixel width (see below). Finally, there is the choice of light source. We use xenon lamps because their emission spectra are fairly uniform, and because the bulbs have long lives and contain an inert gas. In contrast, less expensive mercury lamps produce a spectrum with large peaks, have shorter lives, and are problematic for disposal and cleanup, if breakage occurs.

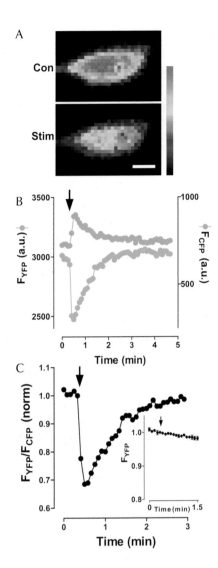

FIGURE 3. FRET shows activity-dependent cyclic nucleotide response in fly boutons. (*A*) Pseudocolor YFP/CFP FRET ratio images of muscle 12 type III bouton expressing Epac1-cAMPs before (Con) and 5 sec after 70-Hz stimulation for 5 sec (Stim). Scale bar, 1 μm. (*B*) Rapid decrease in YFP and increase in CFP fluorescence from a single type III bouton on 5 sec of activity (arrow). (*C*) Normalized YFP/CFP ratio time course shows a loss in FRET for the experiment in *B*. (*Inset*) No response is seen with direct excitation of YFP (*n* = 4). (Adapted, with permission, from Shakiryanova and Levitan 2008, © National Academy of Sciences.)

GFP Indicators in Flies

Finally, we consider the use of transgenic flies; a variety of these are available that express desired indicators. For tracking neuropeptides, the *UAS-preproANF-EMD* line (#7001, Bloomington Stock Center) (Rao et al. 2001) is used; EMD refers to the emerald variant of GFP, ANF represents the rat atrial natriuretic factor (also known as atrial natriuretic peptide [ANP]). Because ANF does not exist in the *Drosophila* genome, it is unlikely to activate postsynaptic receptors. Potentially, ANF-EMD (also called AnfGFP) could compete with native peptides to affect their packaging and release, but such a side effect has not been evident in experiments with a variety of GAL4 drivers (e.g., *elav-GAL4, 386-GAL4, CCAP-GAL4*). Instead, the reporter accurately reports native peptide release (Haifetz and Wolfner 2004; Husain and Ewer 2004). AnfGFP expressed in *Drosophila* motor neurons has been imaged in two ways. First, because fluorescent intensity reports content, a drop in fluorescence is indicative of release. Second, the motion of neuropeptide-containing vesicles can be directly imaged; such imaging studies have led to discoveries concerning vesicle mobilization, capture, and motor-mediated transport (Shakiryanova et al. 2005, 2006, 2007; Pack-Chung et al. 2007; Barkus et al. 2008; Wong et al. 2009).

Small synaptic vesicles (SSVs) are labeled with GFP-tagged synaptotagmin. However, because synaptotagmin is an intrinsic membrane protein, this construct cannot be used for optical detection of SSV exocytosis. SynaptopHluorin has been used instead for this purpose in the larval NMJ, but the sensitivity of this indicator is diminished by the relatively neutral pH of secretory vesicles in *Drosophila* (Sturman et al. 2006). Thus, non-GFP indicators, such as the styryl dye FM1-43, are most suitable for studying SSV exocytosis in the fly.

Transgenic flies have been generated for numerous second messenger indicators. Some of these are variants of GFP that require measuring a single emission wavelength; however, signals from such indicators are difficult to calibrate. An alternative is to use ratiometric indicators based on FRET between cyan and yellow variants of GFP (i.e., CFP and YFP). FRET-based indicators work by reporting conformational changes that alter the distance or angle between the donor (CFP) and the acceptor (YFP). An increase in FRET is apparent with an increase in the YFP/CFP emission ratio when CFP is excited. Specifically, 440-nm light is used for excitation and 535- and 480-nm light is collected to monitor acceptor (YFP) and donor (CFP) emission signals, respectively. With ideal FRET, changes in the YFP/CFP emission ratio reflect opposite changes in the YFP and CFP emissions (Fig. 3). It is important to perform a control to verify that a ratio change actually reflects FRET, rather than some other effect (e.g., the pH dependence of some YFPs). First, the donor YFP can be directly excited at 510 nm (Fig. 3C, inset). Because CFP does not absorb this light, FRET cannot occur and therefore the response should not occur. Second, under conditions that are thought to produce FRET, YFP can be photobleached. If the donor emission signal (480 nm for CFP) becomes brighter, this result implies that energy transfer had been occurring before to photobleaching the acceptor.

We have performed signaling studies using cameleon, a calcium sensor that reports calcium increases with an increase in FRET (Shakiryanova et al. 2007). We have also used an optical indicator based on the cyclic AMP (cAMP) binding domain of Epac1 (an exchange protein that is directly activated by cAMP); this Epac1-cAMP sensor (Epac1-cAMPs) reports cyclic nucleotides with a decrease in FRET (Shakiryanova and Levitan 2008). Interestingly, Epac1-cAMPs has a 10-fold higher affinity for cAMP than for cGMP, but the cGMP signal predominates in the larval motor neuron boutons (Shakiryanova and Levitan 2008). Thus, indicators originally validated in mammalian cells should be used with consideration of the *Drosophila* milieu.

Imaging Neuropeptide Release in the *Drosophila* NMJ: Adjusting the Imaging Setup

We describe here the steps for setting up the imaging equipment for epifluorescence microscopy followed by special considerations for preparing the larval NMJ for peptide release studies.

MATERIALS

See the end of the chapter for recipes for reagents marked with <R>.

Reagents

Drosophila larvae and reagents for preparation of NMJ as described in Chapter 7, Protocol 1 (see Step 6 below)

> The larvae are obtained from transgenic flies expressing the desired indicator.

HL3 supplemented with glutamate <R> (for use as bathing medium, see Step 6)
Zero-calcium HL3 <R> (for filleting the larvae, see Step 6)

Equipment

Equipment for the preparation of larval NMJ, as described in Chapter 7, Protocol 1
Equipment for electrical stimulation of the larval preparation (stimulation electrode)
Upright epifluorescence microscope equipped with a high-NA water-immersion objective (Olympus), an electronic shutter (Uniblitz or Sutter Instruments), a cooled CCD camera (Hamamatsu), an emission filter wheel (Sutter Instruments), and a computer with software for controlling image acquisition (SimplePCI, Hamamatsu). Note that other vendors produce comparable components.

METHOD

1. Align and focus epifluorescence illumination. Quantitative comparisons between different regions in the field of view require even illumination.

    i. To set up even illumination, remove the microscope objective and view the epifluorescence excitation illumination on a piece of paper on the microscope stage.

    ii. Adjust focus knob and positioning screws on the epifluorescence lamphouse to focus and center the image of the light source (e.g., the arc from the bulb in the lamphouse). Many lamphouses have a parabolic mirror to produce a second inverted image of the light source. In these cases, focus the reflected image to be adjacent to the original image.

    iii. Replace the objective and verify that illumination is even across the field of view.

    See Troubleshooting.

2. Set optical magnification and camera binning.

    i. Choose a microscope objective with the highest-NA objective available.

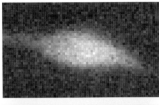

No Binning

2X2 Binning

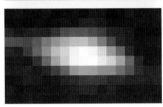

4X4 Binning

FIGURE 4. Binning improves signal-to-noise ratio when spatial resolution is not important. Images of a type III bouton expressing AnfGFP with three different camera binning settings. Binning lowers spatial resolution, which is not important with a homogeneous GFP signal. However, the difference between the bouton and the background is greater and the signal-to-noise ratio is improved.

ii. When spatial resolution is not important, use binning to improve the quantification of intensity in dim samples.

Pixel binning pools the signals of multiple pixels into a single larger effective pixel to improve the signal-to-noise ratio. Because boutons must be identifiable, there is an upper limit to useful binning. Figure 4 compares contrast-enhanced images of a type III peptidergic bouton acquired with no (i.e., 1 × 1) binning, 2 × 2 binning, and 4 × 4 binning. Note that the increase in pixel size is associated with a greater difference from background. Therefore, in this example in which too many vesicles are present to resolve individuals, binning is advantageous for release measurements. This is typically true also with mobile indicators that diffuse more quickly than the frame rate and so give a homogeneous signal throughout the bouton.

iii. For measurements that require spatial resolution (e.g., vesicle motion), adjust the magnification and binning, according to the guideline that effective pixel width should equal $M\lambda/4(NA)$, in which M is the magnification in the plane of the camera, λ is the wavelength of the emitted light, and NA is the numerical aperture of the objective. Note that effective pixel width can be increased with binning. To adjust M, include another lens in the light path between the objective and the camera (e.g., with a magnification changer in the microscope).

The importance of this guideline is revealed in two examples described in Box 1.

3. Optimize illumination.

BOX 1. EXAMPLES

Example 1
Consider that a 100x, 1.0 NA objective being used to view emission at 500 nm (0.5 µm) with a camera having a chip containing 6.25-µm-wide pixels. The relationship indicates that 12.5-µm-wide pixels are needed; however, the camera's pixels are one-half the optimal size. This situation would produce oversampling, giving rise to more noise without any gain in optical information. Hence, 2 × 2 binning should be used to double effective pixel width.

Example 2
Consider the same situation, but in which a 25x objective is being used. In this case, the camera's pixels are twice the required size. Hence, a 2x magnification changer should be used to give the correct sampling for optimal spatial resolution.

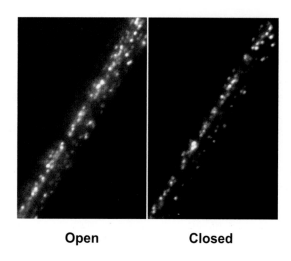

Open   **Closed**

FIGURE 5. Closing the field diaphragm improves visualization of AnfGFP-labeled peptidergic vesicles in the motor nerve. (*Left*) The field diaphragm is open; (*right*) the diaphragm is closed.

i. Adjust the epifluorescence field diaphragm to be as small as possible, while maintaining even illumination over the region of interest.

As illustrated in images of vesicles undergoing axonal transport in the motor nerve (Fig. 5), this adjustment produces better image quality.

ii. One limitation, however, is that the NMJ can produce small twitches even when transmission is inhibited. Therefore, set the field diaphragm size so that such small movements do not alter the fluorescence in the region of interest.

iii. Many epifluorescence microscopes have aperture diaphragms in the epifluorescence light path as well. Use the aperture diaphragm in concert with neutral density filters for adjusting illumination intensity.

These diaphragms lower light intensity, which is desirable for reducing photobleaching.

4. Set up an illumination protocol for time-lapse experiments.

i. Use controlled and timed illumination to manage standard irreversible photobleaching, as well as reversible changes in fluorescence associated with the equilibrium between dark and fluorescent states of fluorescent proteins.

ii. Set the exposure time to take into account possible motion in the preparation.

For example, because transiting vesicles move at ~1 µm/sec (Shakiryanova et al. 2006), exposures should be short (<0.1 sec) to yield sharp images.

iii. Adjust the exposures so that there are no saturated pixels, to ensure linearity of intensity measurements.

For release and signaling indicators, a series of 3–5 frames with 80–200-msec exposures acquired every 3 sec is appropriate.

5. Adjust the objective.

i. If the objective has an adjustable collar, use it to correct for the immersion medium (e.g., water vs. air or oil) or the presence of a coverslip in the light path.

ii. Alternatively, collars can control an aperture diaphragm. For fluorescence imaging of synaptic boutons, always open the objective aperture diaphragm fully to maximize NA.

iii. Correction collars can correct for subtle variations in refractive index of a tissue. In these cases, set the collar empirically to give the best image.

Note that, because the nerve terminal of the NMJ is fairly superficial (i.e., located near the surface of the muscle), this correction is not usually required.

6. Prepare the larval NMJ.

The preparation of larval NMJ is described in Chapter 7, Protocol 1, and should be generally followed with the omission of fixation, taking into account the following adjustments for use of the preparation in neuropeptide release studies.

i. Fillet the larva in zero-calcium HL3, then add normal HL3 supplemented with 10 mM glutamate and stretch the filleted preparation taut before the experiment.

> Because it is beneficial to inhibit muscle contraction during stimulation, HL3 supplemented with glutamate is used as the bathing medium to desensitize postsynaptic receptors.

ii. The motor nerves cannot be cut; therefore, to prevent unstimulated motor neuron activity, cut the ventral ganglion to sever inputs from a central pattern generator, then withdraw or suck a loop of motor nerve snugly into a stimulation electrode.

iii. Because neuropeptide release requires bouts of activity, stimulate the preparation.

> We typically stimulate at 70 Hz for 1–30 sec. In fact, release can be evoked with 2 min of continuous electrical stimulation. However, for longer periods of neuropeptide release, bursting protocols are effective.

See Troubleshooting.

7. Image the preparation.

i. Focus on one bouton and acquire a series of images in which the focus is continually adjusted. Repeat these bouts of acquisition at fixed intervals before and after stimulation.

ii. Photobleach a region of interest to track vesicles within the chosen bouton (see Discussion section below for special considerations in photobleaching and FRET and FRAP analysis).

iii. Select the images in which the chosen bouton is in focus for analysis (see the Discussion section for the use of ImageJ in data analysis).

> Even with the precautions described in Step 6, some movement of the NMJ can be expected during the experiment. Changes in focus can occur because the nerve terminal is three dimensional. Therefore, the investigator should focus on a single bouton.

See Troubleshoooting.

TROUBLESHOOTING

Problem (Step 1): There is difficulty in distinguishing the original and reflected images of the light source.

Solution: Turn adjustment screws on the lamphouse. If the screw is for the lamp, then both images will move. If the screw is for the parabolic mirror, only one image will move. With the identity of the adjustment screws assigned, first center and focus the lamp image. Then adjust the parabolic mirror.

Problem (Step 6): There is no optical response to electrical stimulation.

Solution: The suction electrode may not be sealed onto the loop of the motor nerve, an electrical connection may be broken, or the preparation may not be viable. First, try the stimulation without glutamate in the HL3 bath. If muscle contraction is elicited, then the electrical circuit is intact. If no response is seen, then check the electrical connections. If there is still no response, check for viability by depolarizing the preparation with bath K^+ (Levitan et al. 2007). For more efficient stimulation, try a smaller suction electrode to form a tighter seal.

Problem (Step 7): Reversible photobleaching occurs.

Solution: True photobleaching is an irreversible reaction. However, many GFP variants including emerald GFP are reversibly driven into a dark state by prolonged or repetitive excitation. In such cases, cessation of excitation results in reacquisition of fluorescence. There are two situations in which this "reversible photobleaching" must be taken into account. First, during time-lapse experiments it is critical that the period between images is not varied so that the effect of reversible photobleaching is at a steady state. If the time base must be altered, then controls to correct the effect of reversible photobleaching must be performed. Second, fluorescence recovery after photobleaching (FRAP) data must be calibrated for the small recovery that happens over seconds that is due to reversible photobleaching rather than movement of GFP-tagged proteins or organelles.

Problem (Step 7): Camera vibration produces blurry images.

Solution: Camera vibration can originate from the fan in Peltier-cooled cameras. When possible, turn off the fan with software. Otherwise, change the camera mounting. Finally, if the signal of interest is bright, switch to a camera with moderate cooling that does not use a fan.

DISCUSSION

The protocol describes the setup for GFP-based imaging in *Drosophila* motor neuron boutons of vesicle dynamics, signaling, and peptide release. These assays are powerful because other methods such as electrophysiology can only indirectly reveal these events. Furthermore, the guidelines for optimization of optics and data acquisition apply to any fluorescence microscopy measurement. Therefore, this protocol will be applicable to new fluorescent protein indicators expressed in a variety of fly neurons.

Photobleaching

Photobleaching has been an important tool for imaging the *Drosophila* NMJ. First, photobleaching a bouton lowers background fluorescence so that arrival and movement of new unbleached vesicles can be detected. In fact, this approach was used to detect activity-dependent capture of transiting vesicles. Second, FRAP can also be used to measure changes in vesicle mobility in *Drosophila* boutons (Shakiryanova et al. 2005, 2006; Levitan et al. 2007). Finally, a standard control for FRET is to test whether photobleaching the acceptor increases the donor's signal.

The easiest way to photobleach a region of interest is to use a laser scanning confocal microscope. For GFP, maximal 488-nm laser power is used for repetitive scans of a region of interest until fluorescence is reduced by >70%. Unbleached vesicles can then be detected after they enter the photobleached region. Furthermore, further photobleaching can be used to limit the number of vesicles in the region of interest so that an individual vesicle's trajectory can be followed in detail.

Following photobleaching, the confocal microscope can acquire data when GFP labeling is robust. However, dim GFP signals may be more readily detected by wide-field epifluorescence microscopy because of the efficient collection of light and the low noise and longer exposures possible with a cooled CCD camera. In such cases, a CCD camera can be added to the confocal microscope so that laser illumination is used only for photobleaching, whereas epifluorescence is used for data collection.

Analysis of Intensity Changes

Once data acquisition is complete, imaging data must be analyzed. ImageJ is a free cross-platform software package that runs on Java and features many tools for displaying and quantifying images (see http://rsbweb.nih.gov/ij/). In addition, an ever-growing list of plug-ins is available for powerful data analysis (e.g., correlation analysis) (Levitan et al. 2007). The most basic measurement is to determine the average brightness in a region of interest (e.g., in a bouton). To ensure that such quantification is valid, there must be no saturated pixels in the image. Then background fluorescence must be subtracted. For this purpose, fluorescence in a nearby nonsynaptic region can be quantified. Brightness measurements can then be normalized to control values to quantify changes indicative of release or FRET.

ImageJ can also be used for generating figures. To accentuate changes without altering the numerical data embedded in the images (i.e., the values of individual pixels), images can be contrast enhanced or presented with a pseudocolor scale. In the latter case, false colors are assigned corresponding numerical values: Blue is usually a low intensity, whereas red is a high intensity. Finally, the image calculator function of ImageJ can be used to subtract a control image from an experimental response image to reveal the difference.

RECITES

CAUTION: See Appendix for proper handling of materials marked with <!>.
Recipes for reagents marked with <R> are included in this list.

HL3 Supplemented with Glutamate

Component	Amount per liter	Final concentration
NaCl	4.09 g	70 mM
MgCl$_2$·6H$_2$O <!>	4.07 g	20 mM
NaHCO$_3$	0.84 g	10 mM
Trehalose	1.89 g	5 mM
Sucrose	39.36 g	115 mM
HEPES	1.19 g	5 mM
KCl <!> (1 M)	5 mL	5 mM
CaCl$_2$ <!> (1 M)	1.5 mL	1.5 mM
Glutamate (1 M)	10 mL	10 mM

1. Add all of the dry components to ~750 mL ddH$_2$O. Stir to dissolve.

2. Add 5 mL of 1 M KCl solution, 1.5 mL of 1 M CaCl$_2$ solution, and 10 mL of 1 M glutamate solution.

3. Adjust the pH to 7.2 with NaOH <!> or HCl <!>.
   Do not use KOH because it will alter [K$^+$] in the saline.

4. Adjust the final volume to 1000 mL with H$_2$O.

Zero-Calcium HL3

Component	Amount per liter	Final concentration
NaCl	4.09 g	70 mM
MgCl$_2$·6H$_2$O <!>	4.07 g	20 mM
NaHCO$_3$	0.84 g	10 mM
Trehalose	1.89 g	5 mM
Sucrose	39.36 g	115 mM
HEPES	1.19 g	5 mM
KCl <!> (1 M)	5 mL	5 mM
Na$_3$EGTA (100 mM, pH 9.2)	5 mL	0.5 mM

1. Add all of the dry components to ~750 mL ddH$_2$O. Stir to dissolve.

2. Add 5 mL of 1 M KCl solution and 5 mL of 100 mM Na$_3$EGTA solution (pH 9.2).

3. Adjust the pH to 7.2 with NaOH <!> or HCl <!>.
   Do not use KOH because it will alter [K$^+$] in the saline.

4. Adjust the final volume to 1000 mL with H$_2$O.

ACKNOWLEDGMENTS

Research for this chapter is supported by a grant from the National Institutes of Health (R01 NS32385).

REFERENCES

Barkus RV, Klyachko O, Horiuchi D, Dickson BJ, Saxton WM. 2008. Identification of an axonal kinesin-3 motor for fast anterograde vesicle transport that facilitates retrograde transport of neuropeptides. *Mol Biol Cell* **19:** 274–283.

Heifetz Y, Wolfner MF. 2004. Mating, seminal fluid components, and sperm cause changes in vesicle release in the *Drosophila* female reproductive tract. *Proc Natl Acad Sci* **101:** 6261–6266.

Husain QM, Ewer J. 2004. Use of targetable GFP-tagged neuropeptide for visualizing neuropeptide release following execution of a behavior. *J Neurobiol* **59:** 181–191.

Levitan ES, Lanni F, Shakiryanova D. 2007. In vivo imaging of vesicle motion and release at the *Drosophila* neuromuscular junction. *Nat Protoc* **2:** 1117–1125.

Pack-Chung E, Kurshan PT, Dickman DK, Schwarz TL. 2007. A *Drosophila* kinesin required for synaptic bouton formation and synaptic vesicle transport. *Nat Neurosci* **10:** 980–989.

Rao S, Lang C, Levitan ES, Deitcher DL. 2001. Visualization of neuropeptide expression, transport, and exocytosis in *Drosophila melanogaster. J Neurobiol* **49:** 159–172.

Shakiryanova D, Levitan ES. 2008. Prolonged presynaptic posttetanic cyclic GMP signaling in *Drosophila* motoneurons. *Proc Natl Acad Sci* **105:** 13610–13613.

Shakiryanova D, Tully A, Hewes RS, Deitcher DL, Levitan ES. 2005. Activity-dependent liberation of synaptic neuropeptide vesicles. *Nat Neurosci* **8:** 173–178.

Shakiryanova D, Tully A, Levitan ES. 2006. Activity-dependent synaptic capture of transiting peptidergic vesicles. *Nat Neurosci* **9:** 896–900.

Shakiryanova D, Klose MK, Zhou Y, Gu T, Deitcher DL, Atwood HL, Hewes RS, Levitan ES. 2007. Presynaptic ryanodine receptor-activated calmodulin kinase II increases vesicle mobility and potentiates neuropeptide release. *J Neurosci* **27:** 7799–7806.

Sturman DA, Shakiryanova D, Hewes RS, Deitcher DL, Levitan ES. 2006. Nearly neutral secretory vesicles in *Drosophila* nerve terminals. *Biophys J* **90:** L45–47.

Wong MY, Shakiryanova D, Levitan ES. 2009. Presynaptic ryanodine receptor-CamKII signaling is required for activity-dependent capture of transiting vesicles. *J Mol Neurosci* **37:** 146–150.

21

In Vivo Imaging of *Drosophila* Larval Neuromuscular Junctions to Study Synapse Assembly

Till F.M. Andlauer[1,2] and Stephan J. Sigrist[2]

[1]Bio-Imaging Center at the Rudolf Virchow Center/DFG Research Center for Experimental Biomedicine, University of Würzburg, 97080 Würzburg, Germany; [2]Institute for Biology/Genetics, Free University Berlin, 14195 Berlin, Germany

ABSTRACT

In the past decade, a significant number of proteins involved in the developmental assembly and maturation of synapses have been identified. However, detailed knowledge of the molecular processes underlying developmental synapse assembly is still sparse. Here, we present a protocol that makes extended in vivo imaging of selected proteins in live *Drosophila* larvae feasible at a single-synapse resolution. The intact larvae are anesthetized and noninvasively imaged with confocal microscopy. This method allows for both protein trafficking and protein turnover kinetics to be studied at various points in time during the development of an animal. These data contribute to our understanding of synaptic assembly under in vivo conditions.

INTRODUCTION

Chemical synapses are specialized for rapid directional signaling and are among the most elaborate signaling machines of cells. Recent work suggests that in vivo synaptogenesis is a long and intricate process involving multiple interrelated steps with reciprocal induction, as well as independent assembly of pre- and postsynaptic structures. Although data acquired in neuron cultures show that synapses can assemble quickly (Garner et al. 2006), other studies analyzing mammalian tissue samples suggest that this process takes ~1 d (Nägerl et al. 2007). Thus, it is conceivable that regulation of synapse formation differs between in vitro and in vivo models. At present, a lack of knowledge concerning the detailed spatiotemporal sequence of in vivo synaptic assembly remains a barrier to a comprehensive understanding of the development of synaptic circuits. This issue could be overcome by extended molecular intravital imaging of individual synaptic proteins at identified sets of synapses.

The *Drosophila* larval neuromuscular junction (NMJ) is a leading model system to study synapses and their molecular foundation. It is especially well suited for the analysis of synaptic

355

assembly for several reasons: Larval muscles are built and arranged in a stereotypical, repetitive manner, as are the motor neurons innervating them (Crossley 1978; Landgraf and Thor 2006). Therefore, the same individual NMJs can be easily identified among different individual larvae as well as within one particular larva at several points in time. It is thus possible to follow a defined NMJ and its population of synapses over time. The basic geometry of an individual NMJ is established during embryonic stages. With time, however, NMJs grow and expand continuously, constantly gaining additional boutons and individual synapses. Thus, at *Drosophila* NMJs, assembly of newly forming synapses can be investigated throughout all larval stages over relatively short periods of time.

The study of *Drosophila* larval NMJs is attractive due to the availability of many methodological approaches (Prokop and Meinertzhagen 2006). The ultrastructure of the synapses is thoroughly described and they are easily accessible for electrophysiological methods, which have been applied in numerous studies. In addition, the GAL4-UAS system (Brand and Perrimon 1993) allows for the restriction of expression of transgenic constructs to either the presynaptic (motor neuron) or postsynaptic (muscle) cell, allowing for the functional definition of the place of action of synaptic proteins by rescue experiments. Moreover, *Drosophila* NMJs show substantial, genetically evoked, and experience-mediated structural and functional plasticity (Zhong et al. 1992; Sigrist et al. 2003) and have significant similarities to glutamatergic excitatory synapses of the vertebrate central nervous system (Featherstone and Broadie 2000), and many of their key molecules are phylogenetically highly conserved (Keshishian et al. 1996).

An NMJ appears like beads on a string with each bead representing a bouton, which can be as large as 5 µm in diameter (Broadie and Richmond 2002) (Fig. 1A–C). Different types of motor neuron boutons have distinct physiological as well as morphological properties and are categorized into three classes. Type-I boutons form glutamatergic synapses; they show a stereotypic morphology and can be further subdivided into type-Ib and -Is boutons. The superficial muscles 26 (VA1) and 27 (VA2) are well suited for intravital imaging (Fig. 1A). They have only type-Ib boutons, which are larger boutons that mainly contain clear SVs (Landgraf et al. 2003; Prokop 2006).

Throughout larval development, the number of synaptic boutons at NMJs and individual synapses decorating these boutons increases significantly on a timescale of hours (Schuster et al. 1996; Collins and DiAntonio 2007). Moreover, similar to central synapses of mammals, synapses at *Drosophila* NMJs are not static structures but, rather, undergo activity-dependent and experience-dependent changes regarding the number of individual boutons and the overall number of synapses per NMJ (Zhong et al. 1992; Sigrist et al. 2003; Schmid et al. 2008).

Initial in vivo experiments of developing NMJs revealed that individual synapses (consisting of the presynaptic active zone [AZ] and the respective adjunct postsynaptic density [PSD]) assemble on a timescale of several hours, with pre- and postsynaptic proteins joining into the assembly of synapses in a temporally defined order (Rasse et al. 2005; Schmid et al. 2008; Fouquet et al. 2009). Clearly, it is intrinsically problematic to use exclusively static methods like immunohistochemistry when trying to reveal dynamic aspects of synapse assembly. This is particularly true because the assembly processes are not synchronized between individual synapses. Instead, the NMJ synapse population is a mosaic in terms of the age and maturation status of individual synapses.

Therefore, the temporal sequence of synaptic assembly and, moreover, turnover rates of proteins and axonal transport phenomena are questions that have to be addressed with approaches that make it feasible to track individually identified situations (e.g., distinct identified synapses or cargo containing vesicles within axons) over extended periods in vivo. Because the growth of individual synapses takes place over hours (Rasse et al. 2005; Schmid et al. 2008) and days, invasive methods that result in an immediate halt in developmental processes (e.g., opening the cuticle of a larva to image with water-immersion lenses) are also problematic. With noninvasive in vivo imaging on the other hand, larvae can be anesthetized for a limited amount of time and subsequently woken and fully recovered within minutes. Consequentially, they can be imaged repeatedly during their life span as they develop (Rasse et al. 2005; Schmid et al. 2008; Fouquet et al. 2009) (Fig. 1E-G). The development of individual synapses as well as the history of particular proteins can readily be tracked and reconstructed on a molecular level.

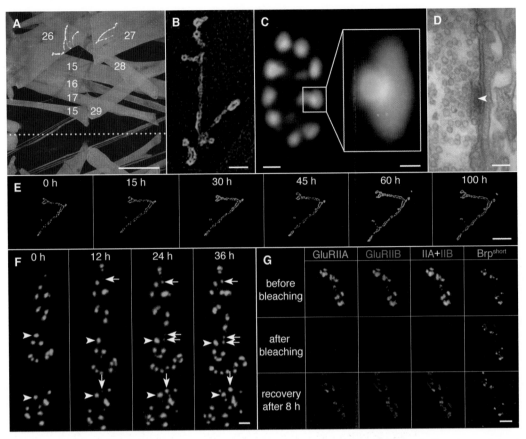

FIGURE 1. Intravital imaging of neuromuscular junctions (NMJs). (*A*) Selected body wall muscle fibers in an abdominal segment as seen from the exterior (outside-out view). The dotted line indicates the ventral midline. The image was modified to highlight NMJs at muscles 26 (VA1) and 27 (VA2). Construct used for the visualization of muscle fibers and NMJs: *uas-dlgS97-gfp*, expressed using *c57-gal4*. Scale bar, 100 μm. (*B*) Morphological structure of a *Drosophila* larval NMJ at muscle 27 (VA2), outlined by the expression of *uas-dlgS97-gfp*, using *c57-gal4* as a driver. Scale bar, 10 μm. (*C*) Immunohistochemical stainings of a bouton of a larval NMJ and an individual synapse in lateral view (*right* box). Green, monoclonal antibody Bruchpilot[Nc82] (Kittel et al. 2006); magenta, antibody against the glutamate receptor subunit DGluRIID (Qin et al. 2005). Scale bar (bouton), 1 μm; scale bar (synapse), 100 nm. (*D*) Ultrastructure of an active zone. The arrowhead points at the T-bar, synaptic vesicles cluster next to it. Scale bar, 100 nm. (*E,F*) In vivo imaging of synapse formation at an identified NMJ at muscle 27, tracked using the GFP-labeled glutamate receptor subunit DGluRIIA. (Adapted, with permission, from Rasse et al. 2005.) (*E*) The development of an identified NMJ at muscle 27 tracked over 100 h at 16°C. Scale bar, 10 μm. (*F*) Higher magnification of the image shown in *E*: New synapses form de novo (arrows). Mature synapses (arrowheads) remain stable. Scale bar, 2 μm. (*G*) In vivo fluorescent bleach experiment (FRAP) of DGluRIIA-mRFP (green), DGluRIIB-GFP (magenta) and Brp[short]-Cerulean (cyan). (*Top row*) Images taken directly before bleaching. (*Middle row*) Images taken directly after bleaching of DGluRIIA and DGluRIIB. (*Bottom row*) Images taken 8 h after bleaching. Recovery of both DGluRIIA and DGluRIIB can be observed. Scale bar, 2 μm.

Key Players in Synaptic Assembly and Development

Each synapse consists of a presynaptic active zone (the site of transmitter release), a postsynaptic density, and a synaptic cleft separating them. The AZ is defined as the presynaptic region in which synaptic vesicles (SVs) fuse with the presynaptic membrane and release their cargo into the synaptic cleft. It can also be regarded as the area where pre- and postsynaptic membranes lie in planar opposition. The AZ can be further subdivided into the protein-rich cytomatrix at the active zone (CAZ), formed by several specialized proteins that play a role as a scaffold for other proteins (Zhai and Bellen 2004; Schoch and Gundelfinger 2006), and the associated machinery for the SV exocycle and endocycle (Siksou et al. 2007). In addition, a cytoplasmic pool of SVs is associated with the AZ.

TABLE 1. Selected constructs suitable for in vivo imaging

Protein	Fluorophore	Localization	Reference(s)
Atrial natriuretic peptide (ANP)	GFP/Emerald	Neuropeptidergic vesicles	Rao et al. 2001
Bruchpilot	CFP, GFP, mStrawberry	Active zone	Wagh et al. 2006; Schmid et al. 2008; Fouquet et al. 2009
Calcium channel Cacophony (Cac1)	GFP	Active zone	Kawasaki et al. 2004; Rasse et al. 2005; Fouquet et al. 2009
Cytochrome c oxidase	GFP	Mitochondria	Pilling et al. 2006
Discs large (DlgS97)	GFP	Subsynaptic reticulum and muscle fibers	Bachmann et al. 2004; Rasse et al. 2005
DLiprin-α	GFP	Active zone	Fouquet et al. 2009
Fasciclin II	GFP	Periactive zone	Zito et al. 1999; Rasse et al. 2005
Glutamate receptor subunit DGluRIIA	GFP, mRFP	Postsynaptic density	Rasse et al. 2005; Schmid et al. 2008; Fouquet et al. 2009
Glutamate receptor subunit DGluRIIB	GFP, mRFP	Postsynaptic density	Schmid et al. 2008
Glutamate receptor subunit DGluRIIE	GFP	Postsynaptic density	Qin et al. 2005
PAK-kinase	GFP	Postsynaptic density	Rasse et al. 2005
Potassium channel construct CD8-Shaker	GFP	Subsynaptic reticulum	Zito et al. 1999; Rasse et al. 2005
Synaptobrevin	GFP	Synaptic vesicles	Zhang et al. 2002
Synaptotagmin	GFP	Synaptic vesicles	Zhang et al. 2002

For a table of additional constructs, see Füger et al. (2007).

In the following paragraphs, several key proteins in synaptic development and function will be introduced and, if available, constructs suitable for in vivo imaging will be mentioned (these constructs are also listed in Table 1).

The fusion of SVs with the presynaptic membrane takes place upon calcium influx through voltage-gated calcium channels. A marker suitable for in vivo imaging of calcium channels is Cac1-GFP (Kawasaki et al. 2004). A very prominent structure of *Drosophila* AZs is an electron-dense projection called T-bar (Fig. 1D). The protein Bruchpilot, a member of the Cast/ERC/ELKS family (Ohtsuka et al. 2002), is required for T-bar formation (Kittel et al. 2006). The monoclonal antibody Nc82, which labels Bruchpilot, serves as a canonical marker for AZs in *Drosophila* (Kittel et al. 2006; Wagh et al. 2006). Fluorescent fusion proteins of Bruchpilot are also available, which allow for the identification of AZs in vivo (Wagh et al. 2006; Schmid et al. 2008; Fouquet et al. 2009) (Fig. 1G). Transgenically expressed full-length Bruchpilot localizes to AZs but also aggregates at other sites. We found a labeled fragment of the protein (Brp[short], corresponding to amino acids 473–1226 of the 1740-amino-acid Bruchpilot protein) that is especially well suited to identify AZs in vivo because it does not aggregate inappropriately. This fragment colocalizes with endogenous Bruchpilot and requires endogenous full-length Bruchpilot for its localization to AZs (Schmid et al. 2008; Fouquet et al. 2009).

An additional protein associated with the AZ is DLiprin-α, which interacts with several other AZ proteins (Kaufmann et al. 2002) and is also relevant for the transport of synaptic cargo (Miller et al. 2005). Recently, it was shown using in vivo imaging that DLiprin-α is incorporated into nascent synapses several hours before the arrival of Bruchpilot (Fouquet et al. 2009).

The presynaptic protein Syd-1 and the RIM family of proteins are known to play important roles in synapse development and function in other organisms (Owald and Sigrist 2009). To date, their role in *Drosophila* remains unclear. Bassoon and Piccolo, important components of the CAZ in mammals (Schoch and Gundelfinger 2006), have no homologs in *Drosophila*. Accordingly, an unknown combination of other proteins must take over their scaffolding role in the fly, with Bruchpilot being a primary candidate.

After the arrival of DLiprin-α at AZs, but before the incorporation of presynaptic Bruchpilot, ionotropic glutamate receptors accumulate at the postsynaptic site (Fouquet et al. 2009). Transgenic lines derived from genomic clones carrying fusions of fluorescent proteins with the receptor subunits DGluRIIA (dominant at immature synapses), DGluRIIB (incorporated during the maturation of synapses), and DGluRIIE (essential subunit of all DGluR receptors) are available. These fusion proteins have been shown to be functional at genetic, immunochemical, and electrophysiological levels (Qin et al. 2005; Rasse et al. 2005; Schmid et al. 2008) (Fig. 1E–G).

In Vivo Imaging of Larval NMJs

Drosophila larval NMJs have a simple architecture, which makes it possible to re-identify single synapses during successive imaging time points. This situation provides the opportunity to monitor subtle changes over time (e.g., the number of synapses, or the amount and distribution of a certain protein). It also means that individual synapses with a size of merely a few hundred nanometers can be re-identified during subsequent imaging sessions and, thus, protein trafficking and diffusion dynamics can be reconstructed. But to achieve this level of resolution, even the slightest movements of the larva have to be prevented throughout the imaging session.

The technique we describe here allows for the examination of live, intact larvae with a confocal microscope. Typically larvae are imaged at several points in time during larval development, with the larvae waking up and roaming freely in between imaging time points. In this way, the normal, physiological development of identified synapses over extended periods of time can be studied in an intact animal.

To ensure high optical accessibility of the NMJs during the imaging procedure, the larvae are embedded in halocarbon oil, which has a refractive index very similar to their waxy cuticle. Thus, little reflection occurs on the interface between oil and cuticle, with the cuticle of the larvae becoming virtually transparent. The most superficial NMJs (at muscle fibers 26 [VA1] and 27 [VA2]; Fig. 1A) are the easiest to investigate as they lie directly below the cuticle, and no further light is lost by dispersion while passing through other tissues. However, deeper lying structures such as axon bundles can also be easily investigated.

The proteins of interest are labeled with a fluorescent marker tag (optimally monomeric fluorescent proteins to avoid aggregation [Shaner et al. 2005]). These proteins can then be expressed either under the control of their endogenous promoter at physiological expression levels (Rasse et al. 2005; Venken et al. 2008) or as overexpression constructs by using the flexibility of the GAL4-UAS system (Brand and Perrimon 1993). Furthermore, *Drosophila* allows for the evaluation of the functionality of the tagged proteins by rescue assays.

Although photobleaching can be problematic for weakly expressed proteins, the described imaging method is sensitive enough to typically image the same synapses several times or to observe the trafficking of proteins in time-lapse recordings.

The assay presented here is not restricted to the analysis of AZ assembly but can be applied to every field of interest in developing larvae that allows for reasonable optical access. For example, axonal transport is observed using markers for SVs (synaptobrevin-GFP and synaptotagmin-GFP [Zhang et al. 2002]), for neuropeptidergic vesicles (ANF-GFP) (Rao et al. 2001), or for mitochondria (Mito-GFP) (Pilling et al. 2006).

For a beginner who tries to orient in the "confusing jungle" of larval muscles and NMJs, starting with the observation of the expression of the protein Discs large (Dlg) with a postsynaptic driver (e.g., C57-GAL4 or MHC-GAL4) is recommended. Dlg is a major scaffolding component at larval NMJs and is mainly associated with the postsynaptic, complexly folded muscle membrane ("subsynaptic reticulum"). Two isoforms exist, DlgA, important for synapse development in embryonic stages, and DlgS97, the predominant form in adult flies (Mendoza-Topaz et al. 2008). Both proteins regulate AZ structure by influencing AZ length and spacing (Guan et al. 1996; Mendoza-Topaz et al. 2008). GFP-tagged constructs of Dlg have been created (Bachmann et al. 2004) and applied for in vivo imaging (Rasse et al. 2005). Dlg overexpression in muscles outlines NMJs brightly and also stains muscle fibers at a more basal level (Fig. 1A,B) (Bachmann et al. 2004).

Table 1 lists a selection of constructs suitable for in vivo imaging. For additional constructs, see Füger et al. (2007). Also helpful is the GFP protein trap database (FlyTrap) derived from exon trap screens (http://flytrap.med.yale.edu).

This assay is also amenable to study protein turnover and diffusion dynamics by FRAP (fluorescence recovery after photobleaching), preferably with a second, unbleached channel as a control (Fig. 1G), and by photoconversion (e.g., by using a protein tagged with the fluorescent marker protein Eos [Wiedenmann et al. 2004]).

For protocols concerning FRAP, please consult Chapter 20 or previously published articles (Füger et al. 2007; Schmid and Sigrist 2008; Schmid et al. 2008).

Protocol 1

Building an Imaging Chamber for In Vivo Imaging

This protocol describes the assembly of an imaging chamber (Fig. 2) as needed for in vivo imaging of *Drosophila* larvae (see Protocol 2). It also covers the construction of an appropriate anesthetization device.

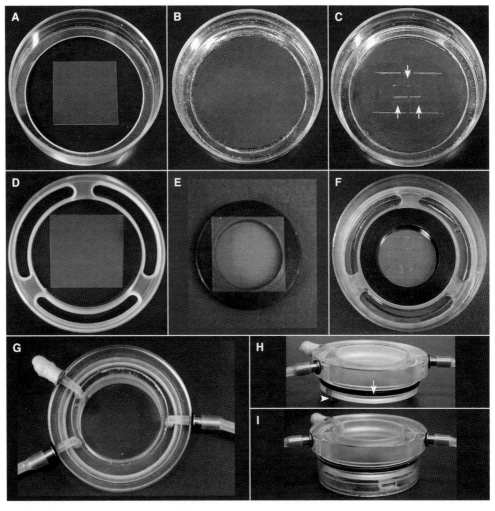

FIGURE 2. Components of the imaging chamber. (*A*) A 60-mm Petri dish with its center cut out. In the center a 24 x 24-mm coverslip indicates the size. (*B*) The Petri dish shown in *A* with a round 50-mm coverslip circle glued to its undersurface. (*C*) The Petri dish shown in *B* with adhesive tape inside. A slit and lanes have been cut into the tape (see Fig. 4C,D). Arrows indicate additional layers of tape serving as a spacer. (*D*) The plastic centering ring. A 24 x 24-mm coverslip indicates its size. (*E*) The iron ring. A 24 x 24-mm coverslip indicates its size. (*F*) The half-assembled chamber: A 24 x 24-mm coverslip has been placed on top of the adhesive tape; the plastic centering ring and the iron ring lie above it. (*G*) Top view of the lid with its three outlets. Silicone tubes are connected to two of the outlets; the third one has been sealed with Parafilm. (*H*) Side view of the lid. The arrow points at the O-ring that ensures a tight seal. The arrowhead points at the ending of an outlet tube on the undersurface. (*I*) The assembled imaging chamber. The lid has been placed on top of the lower part of the chamber (shown in *F*).

361

MATERIALS

Reagents

Two-component glue (containing epoxide resin and polyamine hardener)

Equipment

Adhesive tape (one-sided transparent)

Airtight lid with three outlets/tube connections (Fig. 2G,H):
- Upper plastic part: diameter 60.3 mm, height 8.4 mm
- Lower plastic part: diameter 52.1 mm, height 9 mm
- O-ring seal: ID 46 mm, thickness 3 mm

  The thickness of the O-ring can be adapted, depending on the exact diameter of the Petri dish.
- Metal outlet: diameter of the outer tube (connected to the silicone tubes) 4.5 mm, thread M4 drilled into the plastic, diameter of the inner tube (pointing downward) 4 mm, height of the inner tube sticking out from the undersurface of the lower plastic part 3.8 mm, angles between the outlet tubes 16°, 172°, 172°

Anesthetization device (vaporizer), custom-built, for in vivo imaging (Fig. 3)

  Anesthetization systems are also commercially available from, for example, Dräger Medical or GE Healthcare (a low airflow rate, <50 mL min^{-1}, is necessary).

Coverslip circles, round, diameter 50 mm, thickness 0.12 mm (Menzel-Gläser, Germany)

Iron ring (as a weight, fitting into the centering ring; Fig. 2E), ID 22.4 mm, OD 35.2 mm, height 4 mm, weight 18 g

Petri dish, diameter 60 mm (Sarstedt)

Plastic centering ring (fitting into the Petri dish and around the iron ring; Fig. 2D), ID 35.5 mm, OD 50.8 mm, height 4.5 mm

  This ring should contain holes on its inside into which the lower parts of the metal outlets of the lid fit loosely (see Fig. 2D).

  If the plastic ring fits around the iron ring too tightly, pressure has to be applied when placing the iron ring into the chamber, which might hurt the larva. If it fits too loosely, the iron ring might move when lifting the imaging chamber and, in turn, the position of the larva might be altered.

Scalpel

Soft silicone tubes, ID 2 mm, thickness 1 mm, hardness 60° Shore

METHOD

Building the Imaging Chamber

1. Cut out the center of the base plate of a Petri dish, leaving a circular border (a segment of the outer edge of the dish remains, about 3–5 mm in size; see Fig. 2A).

2. Glue the round glass coverslip to the outside of the remaining border of the excised Petri dish (see Fig. 2B). The seal must be airtight; therefore avoid air bubbles.

3. Stick one layer of adhesive tape to the center of the round coverslip (to the side facing inward).

4. Use the scalpel to cut a rectangular slit into the center of the tape (about 8 x 3-mm). Cut three additional lanes leading to the center to assure the access of air (see Figs. 2C and 4C).

5. Put three small pieces of tape on the stretch of tape already present. Use one or two layers, depending on the thickness of the tape. This will create a spacer between the lower and upper coverslip to avoid squashing of the larva (see Fig. 2C).

6. Ensure that the custom-built lid fits exactly onto the modified Petri dish, as the seal must be airtight (see Fig. 2I). A rubber O-ring should assure the formation of this seal (see Fig. 2H). In

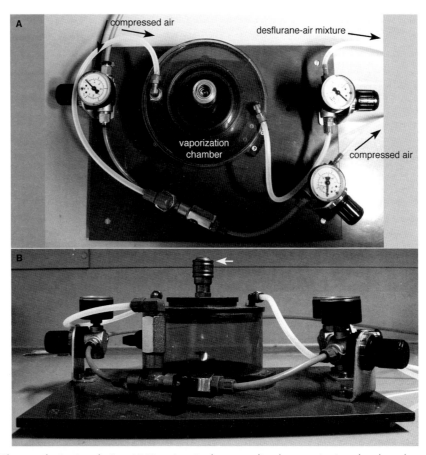

FIGURE 3. The anesthetization device. (*A*) Top view. In the center lies the vaporization chamber; the connection to the outlet adapter valve of the desflurane bottle is located in the middle of the chamber (see Fig. 4B). Compressed air enters through the tube at the *upper left*. It can either be routed to the vaporization chamber (where it enters at the *left side*) or to the imaging chamber (tube at the *lower right*). The desflurane–air mixture leaves the anesthetization chamber on the right side and is routed to the imaging chamber via the tube on the upper right side of the image. (*B*) Side view. The arrow points at the connection to the outlet adapter valve of the desflurane bottle.

addition, the lid needs three outlets to which silicone tubes can be connected (see Fig. 2G). For a mechanical drawing of the lid, see Füger et al. (2007).

> The exact diameter of Petri dishes varies. If the seal is not tight enough with a certain Petri dish, dental wax can be applied under the O-ring to increase its diameter.

Anesthetization Device

7. Assemble the anesthetization device, connecting tubing to the appropriate valves as shown in Figure 3. If you plan to build your own anesthetization device, please consider the following recommendations:

   i. Use transparent PVC (polyvinyl chloride) for the central vaporization chamber. Materials that are corroded by desflurane (e.g., acrylic glass) cannot be used.

   ii. Ensure that the bottom and the top of the chamber are concave, so that the desflurane is collected in the middle of the chamber. This arrangement also helps with mixing desflurane with air.

   iii. Check that the custom-built anesthetization device is absolutely airtight to prevent exposure to desflurane (see Protocol 2).

   > Acute exposure to desflurane may cause eye and skin irritation. Overexposure by inhalation can lead to headaches, dizziness, drowsiness, unconsciousness, or death.

In Vivo Imaging of *Drosophila* Larvae

This protocol describes the repetitive, noninvasive imaging of a live, intact, anesthetized *Drosophila* larva over extended periods of time. The method has proven highly useful for the study of synaptic assembly and the trafficking of proteins. Proteins of interest must be tagged with a fluorescent label and have to be expressed in transgenic fly strains.

MATERIALS

CAUTION: See Appendix for proper handling of materials marked with <!>.

Reagents

Desflurane <!> (Suprane; Baxter, USA)

> Isoflurane <!> can also be used; isoflurane is less volatile, which makes it more difficult to reach an appropriate level of anesthetization.

> The anesthetic desflurane interferes with all internal movement within the larva, including the heartbeat, which is a prerequisite for imaging structures as small as individual synapses. It does not appear to have any negative long-time effects on survival and performance of the larvae; thus it is suited for repetitive imaging on a single-synapse level.

Drosophila larvae, transgenic strain appropriate for analysis

> See Table 1 for examples of constructs expressed in transgenic strains.

Halocarbon oil (Voltalef 10S [Atofina, France and Arkema, USA] or Halocarbon Oil 400 [Halocarbon, USA])

> Halocarbon oil has a refractive index very similar to the waxy cuticle of the larva and solves oxygen well, thereby assuring good imaging conditions as well as the survival of the larva.

Equipment

Anesthetization device (vaporizer, see Protocol 1 and Fig. 3)
Beaker (small, e.g., 100 mL)
Compressed air or a bicycle pump
Coverslip, square, 24 x 24-mm or 22 x 22-mm
Dissectoscope
Imaging chamber (custom-built, see Protocol 1 and Fig. 2)
Inverted confocal microscope (e.g., Leica SP5 [Leica Microsystems] or Zeiss LSM 710 [Carl Zeiss MicroImaging])

> The imaging chamber, as described in this protocol, can only be used with an inverted microscope. Although a similar chamber can be built for use with an upright microscope, the use of an inverted microscope is recommended. In an upright setup the imaging chamber itself has to be inverted and the larva has to be placed directly below the top of the inverted chamber. In such a setup the larva has to be pressed against the coverslip circle (the Petri dish) from below, which makes the mounting procedure more difficult.

Oil-immersion objective with a high numerical aperture (NA > 1.3; e.g., 63x oil, NA 1.4)
Outlet adapter valve fitting on the desflurane bottle, custom-built (see Fig. 4B)
Paintbrushes (small: size 0, 1, or 2)

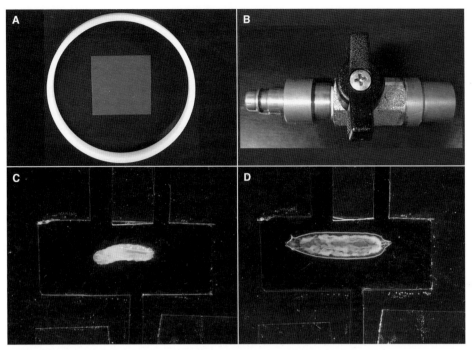

FIGURE 4. Additional components for in vivo imaging and mounting the larva in the imaging chamber. (*A*) Plastic ring for holding the imaging chamber in place on the microscope. A 24 x 24-mm coverslip indicates its size. (*B*) Outlet adapter valve for the connection of a desflurane bottle to the vaporization chamber. The right side of the adapter valve gets connected to the bottle, the left side to the vaporization chamber. (*C,D*) Mounting of a larva. The larva is placed in the modified Petri dish within the slit cut into adhesive tape (see Fig. 2C). (*D*) A coverslip, the plastic centering ring, and the iron ring have been placed on top of the larva; it appears slightly squeezed (same situation as in Fig. 2F).

Parafilm
Petri dishes, diameter 60 mm
Plastic ring holding the imaging chamber in place on the microscope or, alternatively, plasticine
    (see Fig. 4A), ID 55.2 mm, OD 59 mm, height 6.9 mm (measurements for a Leica SP5)

METHOD

Mounting the Larva in the Imaging Chamber

1. Put the lower part of the imaging chamber (i.e., the modified Petri dish from Protocol 1, shown in Fig. 2C) under a dissectoscope.

2. Place a small drop of halocarbon oil within the slit cut into the adhesive tape on the modified Petri dish; preferably use a paintbrush to apply the oil.

3. Place a clean, dry larva into the slit, with its dorsal side facing upward (see Fig. 4C); an early third-instar larva is suited best. It does not have to lie perfectly straight yet. Be careful that the posterior endings of the tracheae are not covered in oil. Use a paintbrush to move the larva into position.

   Always place the larva in the same orientation (e.g., mouthparts to the left) to ease the identification of NMJs.

4. Spread a very small amount of halocarbon oil onto a square coverslip; place it with its oily side pointing downward on top of the larva.

5. Carefully move the coverslip around to stretch out the larva and to adjust its position. The larva's dorsal side must face upward (the dorsal side can easily be recognized by the tracheae). You should

give the animal a little tilt upward or downward (up to 30°) for optimal optical access to one larval hemisphere.

6. Put the plastic centering ring (Fig. 2D) into the chamber and place the iron ring (Fig. 2E) into its middle (on top of the coverslip) (Fig. 2F). The larva should be *slightly* squeezed between the oily coverslips (see Fig. 4D). Adjust the larva's position if necessary.

See Troubleshooting.

7. Place the custom-built lid on top of the modified Petri dish, making sure that the chamber is closed tightly.

Anesthetization of the Larva

8. Place the imaging chamber onto the inverted confocal microscope. To assure that it cannot move, fix it with a custom-built plastic ring (Fig. 4A) or with plasticine in its position.

See Troubleshooting.

9. Check the position of the larva under the microscope. Make sure that your region of interest (ROI) is optically accessible. If it is not, repeat the mounting procedure (Steps 1–7).

See Troubleshooting.

10. If possible, place the anesthetization device into a fume hood to avoid exposure to desflurane.

11. Connect two plastic tubes of the imaging chamber to the anesthetization device (Fig. 3A). One tube is connected to the vaporization chamber containing the anesthetic. The second one will be connected to the valve with access to fresh air. You will only need this second connection if you plan to induce fresh air into the chamber to wake up the larva at a later point. Make sure to seal this second tube if you do not plan to use it (see Fig. 2G and Step 30).

12. Place the third tube into a beaker filled with water.

13. Close all valves of the anesthetization device.

14. Install the outlet adapter valve (Fig. 4B) on top of the desflurane bottle.

15. Connect the outlet adapter to the vaporization chamber of the anesthetization device and fill desflurane into the vaporization chamber until the ground of the chamber is covered.

16. Apply pressure to the vaporization chamber, using either compressed air or a bicycle pump. Close the according valve afterward to maintain the pressure.

17. Open the valve controlling anesthetic flow to the imaging chamber. Bubbles will appear in the beaker containing water unless the imaging chamber is leaky or the vaporization chamber is not under pressure.

18. Apply a number of short pulses of the desflurane–air mixture by opening and closing the valve repeatedly (try 50, the amount of desflurane mixture needed varies with the size of the larva).

19. Close all valves.

20. Remove the third tube from the water; otherwise water can ascend to the inside of the imaging chamber.

Imaging the Larva

21. Allow 2-3 min for the anesthetic to fully take its effect.

22. Check under the microscope that all internal movements within the larva have stopped; otherwise apply more pulses of desflurane.

See Troubleshooting.

23. Use the microscope to move to your ROI within the larva.

24. For imaging NMJs, the abdominal muscles 26 (VA1) and 27 (VA2) are suited best (Fig. 1A), because they lie directly below the cuticle and are, thus, optically most accessible (Crossley 1978; Bate 1993). Any abdominal segment between A2 and A6 can be used.

    To find these muscles, first identify the ventral midline (when having trouble finding the ventral midline, a good indication is the region in which the denticle belts of the larval cuticle separating the segments have the largest diameter); the pointy edge of muscle 27 will then be recognizable in a more lateral position (see Fig. 1A).

25. Switch to confocal mode. Try to find the NMJs of interest quickly to avoid bleaching.

26. Scan a *z*-stack of the NMJ; saturate the signal but avoid oversaturation. Use a pseudocolored lookup table with range indicators when adjusting laser intensities.

    i. Scan as quickly as the desired image quality allows for (e.g., use bidirectional scanning and an image size of 512 × 512-pixels).

    See Troubleshooting.

    ii. Choose a voxel size for about twofold oversampling (when the efficient resolution is 200 nm, use a voxel size of 100 nm, according to the Nyquist–Shannon theorem); we usually use a pinhole of 1 Airy unit (AU).

    iii. When performing FRAP, select a slower scan speed for the ROI that will be bleached, to increase bleaching efficiency.

27. If you plan to wake up the larva and re-image it at a later point in time, limit the imaging time. That is, do not exceed an overall anesthetization time of 30 min, and ideally stay below 10 min.

    See Troubleshooting.

"Recovering" Larvae from an Imaging Experiment

28. After an imaging session is completed, detach the imaging chamber from the anesthetization device as well as from the confocal microscope.

29. Carefully open the chamber, remove the iron and plastic centering ring.

    Do not remove the coverslip on top of the larva before the animal has woken up.

30. Wait until the larva has woken up from the anesthetization. It is also possible to connect the second tube of the imaging chamber to the air valve of the anesthetization device (see Step 11). In this case, you can wake the larva by opening this valve and allowing fresh air to flow through the chamber.

    See Troubleshooting.

31. Place the larva into a Petri dish containing a small amount of mashed fly cultivation medium. Seal the Petri dish with Parafilm. Do not press the larva into the food or it might suffocate.

32. Clean the imaging chamber.

33. Before a second round of imaging, check the vitality of the larva. Exclude larvae that have not grown adequately since the previous round of anesthetization (for a third-instar larva at 25°C, expect a length increase of at least 10% after 12 h, and of 20% after 24 h).

TROUBLESHOOTING

Problem (Steps 6, 8, 26, 27, 30): The larva becomes injured or dies.
Solution:

• Too little halocarbon oil has been applied onto the modified Petri dish or onto the coverslip—apply more.

• If the adhesive tape spacer is too low and therefore the larva gets squeezed too strongly, apply a second layer of tape.

- When doing FRAP, do not use excessively high laser power or wavelengths <450 nm to bleach the signal, otherwise the larva might take permanent damage.
- Make sure that the larva wakes up before you remove it from the imaging chamber.
- Decrease the anesthetization time.

Problem (Steps 9, 22): The larva can still crawl inside the fully assembled imaging chamber.
Solution:
- The adhesive tape spacer is too high—remove a layer of tape.
- Too much halocarbon oil has been applied or the oil is not viscous enough.

Problem (Step 22): The larva cannot be anesthetized.
Solution:
- Make sure that the imaging chamber is sealed airtight.
- Too much halocarbon oil might inhibit proper anesthetization (do not cover the posterior endings of the tracheae in oil).
- Make sure that you have established sufficiently high air pressure within the vaporization chamber to force the desflurane–air mixture out when opening the valve.

Problem (Step 26): The image quality is poor.
Solutions:
- The larva is not flat in the imaging chamber—use a lower spacer.
- Too little halocarbon oil has been applied onto the modified Petri dish or onto the coverslip—apply more.
- The larva is not slightly tilted to one side; thus optical access to NMJs 26 and 27 is not optimal.
- The larva is not properly anesthetized, and residual movements occur.
- The NMJs imaged lie too deeply within the larva.
- If your signal is too weak, try increasing the pinhole to 1.5 or 2 AU.

DISCUSSION

With this method, it is possible to image NMJs or other structures of interest repetitively during the lifetime of a larva while waking up the larva between imaging sessions. Dynamic events, such as the movement of vesicles in an axon, can also be readily observed in an intact, anesthetized animal; FRAP and photoconversion experiments can easily be conducted to directly measure protein trafficking.

Multiple proteins, tagged with a fluorophore, can be imaged simultaneously (e.g., Cerulean, eGFP, and mStrawberry). When imaging pre- and postsynaptic signals at the same time, we recommend using a construct that is fused to a promoter rather than using the GAL4-UAS system for the expression of both constructs. This is important because, if pre- (e.g., OK6 or D42) and postsynaptic (e.g., MHC or C57) GAL4 driver lines were used simultaneously, each protein expressed by the UAS construct will accumulate both pre- and postsynaptically.

The structures to be imaged should not lie too deeply within the larva, to avoid dispersion of the fluorescent light and spherical aberrations. Thus, the NMJs 4 and 6/7, which are often used in immunohistochemistry and electrophysiology, are not suitable for imaging with this method.

Acute effects of desflurane on synaptic transmission in *Drosophila* larvae are likely, because the highly related compound isoflurane leads to presynaptic hyperpolarization and, thereby, to a reduction in excitability at *Drosophila* larval NMJs (Sandstrom 2008). Effects on the dynamics of synaptic proteins cannot fully be excluded. It should be emphasized that we always wake up the larvae in between imaging sessions. For long-term imaging, larvae are demounted and free locomotion is allowed. Later, the same animal is subjected to another round of intravital imaging and anesthesia. Therefore, in vivo imaging, as described in this protocol, constitutes an unchallenged approach to image proteins repetitively over time within an intact animal.

Protocol 3

Quantitative Analysis of NMJ Morphology

Image analysis and quantification are best performed in 3D (e.g., with the software Imaris by Bitplane), but they can also be performed using a simpler method for 2D analysis using the free software ImageJ as presented in this protocol. Our intention is to propose various possibilities for how an analysis may be performed with ImageJ, rather than providing an inflexible protocol, in which the steps must be followed without modifications.

Although execution of most of the tools will be described via the ImageJ menu, most are also readily accessible through icons in ImageJ MBF toolsets. These toolsets can also very easily be adapted for higher efficiency.

MATERIALS

Equipment

Image files (8-bit) (from Protocol 2)
ImageJ software (http://rsb.info.nih.gov/ij).
> Use of ImageJ MBF is highly recommended; this version of ImageJ includes a very useful collection of plug-ins and toolsets for microscopy (http://www.macbiophotonics.ca/imagej/).
> This protocol is based on ImageJ version 1.42q.

METHOD

Adjusting the Settings and Background Correction

1. Make sure to acquire your data according to the Nyquist–Shannon theorem (i.e., with twofold oversampling). Saturate the signal but avoid oversaturation. Use a pseudocolored lookup table with range indicators when adjusting laser intensities.

2. Open your images, preferably from the original file generated by the microscope software (e.g., .lif, .lsm, or .oif) and not from exported TIFFs (this preserves scale settings). ImageJ MBF supports opening these file types. Check the box *Split channels*.

3. Set measurements (*Analyze/Set Measurements…*). Check *Area, Min & Max Gray Value, Mean Gray Value*, and *Limit to Threshold*. Choose *Redirect To: None, Decimal Places: 2*.

4. Check whether the scale settings are correct (*Analyze/Set Scale…*). These settings will be preserved even during image resize operations and in TIFF files saved by ImageJ.

5. If you have acquired *z*-stacks, generate maximum intensity projections of all channels (*Image/Stacks/Z Project…, Projection Type: Max Intensity*).

6. Change the lookup table (LUT) to *Fire* (*Image/Lookup Tables/Fire*).

7. Optionally: Remove irrelevant (nonsynaptic) structures from the image (e.g., tracheae) by using the *Freehand selections* (standard toolbar) and *Fill* tools (*Edit/Fill* or *f*). For this you have to set the *Color Picker* tool to 0,0,0 (black) first (*Image/Color/Color Picker…*).

8. Remove background: Mark part of the image background using *Freehand selections* and copy it to the *ROI manager* by pressing *t*. Repeat this twice to obtain three representative regions. If you

369

did not remove tracheae, make sure not to include them in your selections. Measure the ROIs in all three channels (*Analyze/Measure* or *m*). Subtract the *Mean Gray Value* (*Process/Math/Subtract...*) of one of the ROIs (i.e., of the image background) from the whole image (i.e., de-select any selection in your image first).

9. Shift the intensity distribution back to its original scale (i.e., the range prior to background correction): Multiply the whole image by the factor 255/(255 – *n*) (*Process/Math/Multiply...*), where *n* is the value used for background correction in the previous step. If you compare several genotypes, you must use the same subtraction and multiplication factors for all genotypes. Repeat Steps 8 and 9 for all channels; use the same ROIs.

10. Save the images.

Creating a Mask of the Synaptic Area

Perform Steps 11–18 for each of the channels.

11. Apply a threshold to remove irrelevant lower intensity pixels: Set the lower threshold to a fixed gray value *n* that cuts off unwanted structures but preserves your actual signal (*Image/Adjust/Threshold...*, *Lower Threshold Level: n, Upper Threshold Level: 255*, press *apply*). It is best to zoom in while choosing the lower threshold. The result will be a black and white image.

12. To facilitate segmentation, generate minimum overlays of the black and white masks generated in the previous step (Step 11) with the images saved in Step 10 (*Process/Image Calculator*), *Image1*: Mask from Step 11, *Image2*: Saved image from Step 10. *Operation: Min*, check *Create New Window*. In the image produced in this step, the white areas of the mask created in Step 11 will again show the signal from your original image.

13. Change the LUT to *Fire* (*Image/Lookup Tables/Fire*).

14. Manually segment the remaining fused synapses in the overlay images. Draw lines using *Freehand line selections* with a line width of 1 pixel (*Image/Adjust/Line Width...*). The *Color Picker* tool has to be set to 0,0,0 (black) (*Image/Color/Color Picker...*).

    Freehand line selections can be accessed by right-clicking on the *Straight line selections* tool. The selections are filled with color using the *Fill* tool (*Edit/Fill* or *f*). You can also perform this step semiautomatically via the command *Process/Binary/Find Maxima..., Output type: Segmented Particles*, and then refine the result manually.

15. Save the segmented masks.

16. Remove high-frequency noise from the masks by applying a gaussian blur filter (*Process/Filters/Gaussian Blur...*). Note that here you do not apply this filter to your original data but only to the mask.

    Do not choose a *Sigma (Radius)* that merges neighboring signals (AZs or PSDs). Zoom in to observe this.

17. Apply another threshold to create black and white masks again, set the lower threshold to the same level *n* as in Step 11, the upper threshold to 255 (*Image/Adjust/Threshold...*, *Lower Threshold Level: n, Upper Threshold Level: 255*, press *apply*). Now you have created masks that can be used for analysis of single AZs or PSDs.

18. Repeat Step 12: Create overlays of the masks you created during the last steps with the images you saved in Step 10.

Analysis

19. For the next step, *Analyze Particles*, to work correctly, set the lower threshold of each segmented image to a gray value of 1, set the upper threshold to 255, and do not press apply. (*Image/Adjust/Threshold..., Lower Threshold Level: 1, Upper Threshold Level: 255*, do not press *apply!*)

20. Analyze the dimension and intensity of all remaining individual structures (AZs or PSDs) (*Analyze/Analyze Particles…*). Choose a reasonable (minimum) *Size* (e.g., 4 pixels [2 * 2]). *Show Outlines*; check *Display Results, Clear Results, Exclude on Edges.*

21. Save the outline drawing and compare the outlines of the areas used for the analysis with your original image and, if necessary, remove additional structures and segment further (repeat Steps 14–20 on the basis of the mask saved in Step 15).

22. Copy the results to a spreadsheet or statistics application and continue the analysis with these applications. If the scale settings were correct in Step 4, these settings have been preserved throughout the segmentation and normalization steps. Thus, the results (in micrometers) should need no further conversion. To be sure, double-check the scale settings (*Analyze/Set Scale…*) of the images you performed the particle analysis on.

23. Measure the total synaptic area and synaptic density in a similar manner. Instead of analyzing particles, measure the total area highlighted in the mask from Step 17 (*Analyze/ Measure*). In case of immunostainings, pre- and postsynaptic area can be related to a generic neuronal marker such as Hrp to assess synaptic density. The normalization to an Hrp signal will cancel out variations in animal size to some degree. Optimally, the values should be normalized to the size of the respective muscle or to the segment length to minimize errors due to variations in animal size. Because this is often difficult to achieve under in vivo imaging conditions, using NMJ sizes for normalization is another option.

ACKNOWLEDGMENTS

This work was supported by grants from the Deutsche Forschungsgemeinschaft to SJS (Exc 257, SI849/2-1 and 2-2, TP A16/SFB 551, TP B23/SFB581). We would like to thank Wernher Fouquet, Omid Khorramshahi, and Carolin Wichmann for contributions to Figure 1, as well as Richard W. Cho, Phillip A. Vanlandingham, and members of the Sigrist laboratory for critical comments on the manuscript. In addition, we would like to thank Gunther Tietsch (MSZ, University of Würzburg) for technical assistance regarding the construction of the imaging chamber and the anesthetization device.

REFERENCES

Bachmann A, Timmer M, Sierralta J, Pietrini G, Gundelfinger ED, Knust E, Thomas U. 2004. Cell type–specific recruitment of *Drosophila* Lin-7 to distinct MAGUK-based protein complexes defines novel roles for Sdt and Dlg-S97. *J Cell Sci* **117:** 1899–1909.

Bate M. 1993. The mesoderm and its derivatives. in *The development of* Drosophila melanogaster (ed. M Bate, A Martínez Arias), pp. 1013–1090. Cold Spring Harbor Laboratory Press, Cold Spring Harbor, NY.

Brand AH, Perrimon N. 1993. Targeted gene expression as a means of altering cell fates and generating dominant phenotypes. *Development* **118:** 401–415.

Broadie KS, Richmond JE. 2002. Establishing and sculpting the synapse in *Drosophila* and *C. elegans. Curr Opin Neurobiol* **12:** 491–498.

Crossley CA. 1978. The morphology and development of the *Drosophila* muscular system. in *The genetics and biology of* Drosophila (ed. M Ashburner, T Wright), pp. 499–560. Academic Press, New York.

Featherstone DE, Broadie K. 2000. Surprises from *Drosophila:* Genetic mechanisms of synaptic development and plasticity. *Brain Res Bull* **53:** 501–511.

Fouquet W, Owald D, Wichmann C, Mertel S, Depner H, Dyba M, Hallermann S, Kittel RJ, Eimer S, Sigrist SJ. 2009. Maturation of active zone assembly by *Drosophila* Bruchpilot. *J Cell Biol* **186:** 129145.

Füger P, Behrends LB, Mertel S, Sigrist SJ, Rasse TM. 2007. Live imaging of synapse development and measuring protein dynamics using two-color fluorescence recovery after photo-bleaching at *Drosophila* synapses. *Nat Protocols* **2:** 3285–3298.

Garner CC, Waites CL, Ziv NE. 2006. Synapse development: Still looking for the forest, still lost in the trees. *Cell Tissue*

Res **326:** 249–262.

Guan B, Hartmann B, Kho YH, Gorczyca M, Budnik V. 1996. The *Drosophila* tumor suppressor gene, *dlg*, is involved in structural plasticity at a glutamatergic synapse. *Curr Biol* **6:** 695–706.

Kaufmann N, DeProto J, Ranjan R, Wan H, Van Vactor D. 2002. *Drosophila* Liprin-α and the receptor phosphatase Dlar control synapse morphogenesis. *Neuron* **34:** 27–38.

Kawasaki F, Zou B, Xu X, Ordway RW. 2004. Active zone localization of presynaptic calcium channels encoded by the *cacophony* locus of *Drosophila*. *J Neurosci* **24:** 282–285.

Keshishian H, Broadie K, Chiba A, Bate M. 1996. The *Drosophila* neuromuscular junction: A model system for studying synaptic development and function. *Annu Rev Neurosci* **19:** 545–575.

Kittel RJ, Wichmann C, Rasse TM, Fouquet W, Schmidt M, Schmid A, Wagh DA, Pawlu C, Kellner RR, Willig KI, et al. 2006. Bruchpilot promotes active zone assembly, Ca^{2+} channel clustering, and vesicle release. *Science* **312:** 1051–1054.

Landgraf M, Thor S. 2006. Development and structure of motoneurons. *Int Rev Neurobiol* **75:** 33–53.

Landgraf M, Sánchez-Soriano N, Technau GM, Urban J, Prokop A. 2003. Charting the *Drosophila* neuropile: A strategy for the standardised characterisation of genetically amenable neurites. *Dev Biol* **260:** 207–225.

Mendoza-Topaz C, Urra F, Barría R, Albornoz V, Ugalde D, Thomas U, Gundelfinger ED, Delgado R, Kukuljan M, Sanxaridis PD, et al. 2008. DLGS97/SAP97 is developmentally upregulated and is required for complex adult behaviors and synapse morphology and function. *J Neurosci* **28:** 304–314.

Miller KE, DeProto J, Kaufmann N, Patel BN, Duckworth A, Van Vactor D. 2005. Direct observation demonstrates that Liprin-α is required for trafficking of synaptic vesicles. *Curr Biol* **15:** 684–689.

Nägerl UV, Köstinge, G, Anderson JC, Martin KA, Bonhoeffer T. 2007. Protracted synaptogenesis after activity-dependent spinogenesis in hippocampal neurons. *J Neurosci* **27:** 8149–8156.

Ohtsuka T, Takao-Rikitsu E, Inoue E, Inoue M, Takeuchi M, Matsubara K, Deguchi-Tawarada M, Satoh K, Morimoto K, Nakanishi H, et al. 2002. CAST: A novel protein of the cytomatrix at the active zone of synapses that forms a ternary complex with RIM1 and Munc13-1. *J Cell Biol* **158:** 577–590.

Owald D, Sigrist SJ. 2009. Assembling the presynaptic active zone. *Curr Opin Neurobiol* **19:** 311–318.

Pilling AD, Horiuchi D, Lively CM, Saxton WM. 2006. Kinesin-1 and dynein are the primary motors for fast transport of mitochondria in *Drosophila* motor axons. *Mol Biol Cell* **17:** 2057–2068.

Prokop A. 2006. Organization of the efferent system and structure of neuromuscular junctions in *Drosophila*. *Int Rev Neurobiol* **75:** 71–90.

Prokop A, Meinertzhagen IA. 2006. Development and structure of synaptic contacts in *Drosophila*. *Semin Cell Dev Biol* **17:** 20–30.

Qin G, Schwarz T, Kittel RJ, Schmid A, Rasse TM, Kappei D, Ponimaskin E, Heckmann M, Sigrist SJ. 2005. Four different subunits are essential for expressing the synaptic glutamate receptor at neuromuscular junctions of *Drosophila*. *J Neurosci* **25:** 3209–3218.

Rao S, Lang C, Levitan ES, Deitcher DL. 2001. Visualization of neuropeptide expression, transport, and exocytosis in *Drosophila melanogaster*. *J Neurobiol* **49:** 159–172.

Rasse TM, Fouquet W, Schmid A, Kittel RJ, Mertel S, Sigrist CB, Schmidt M, Guzman A, Merino C, Qin G, et al. 2005. Glutamate receptor dynamics organizing synapse formation in vivo. *Nat Neurosci* **8:** 898–905.

Sandstrom DJ. 2008. Isoflurane reduces excitability of *Drosophila* larval motoneurons by activating a hyperpolarizing leak conductance. *Anesthesiology* **108:** 434–446.

Schmid A, Sigrist SJ. 2008. Analysis of neuromuscular junctions: Histology and in vivo imaging. *Methods Mol Biol* **420:** 239–251.

Schmid A, Hallermann S, Kittel RJ, Khorramshahi O, Frölich AM, Quentin C, Rasse TM, Mertel S, Heckmann M, Sigrist SJ. 2008. Activity-dependent site-specific changes of glutamate receptor composition in vivo. *Nat Neurosci* **11:** 659–666.

Schoch S, Gundelfinger ED. 2006. Molecular organization of the presynaptic active zone. *Cell Tissue Res* **326:** 379–391.

Schuster CM, Davis GW, Fetter RD, Goodman CS. 1996. Genetic dissection of structural and functional components of synaptic plasticity. I. Fasciclin II controls synaptic stabilization and growth. *Neuron* **17:** 641–654.

Shaner NC, Steinbach PA, Tsien RY. 2005. A guide to choosing fluorescent proteins. *Nat Methods* **2:** 905–909.

Sigrist SJ, Reiff DF, Thiel PR, Steinert J., Schuster CM. 2003. Experience-dependent strengthening of *Drosophila* neuromuscular junctions. *J Neurosci* **23:** 6546–6556.

Siksou L, Rostaing P, Lechaire JP, Boudier T, Ohtsuka T, Fejtová A, Kao HT, Greengard P, Gundelfinger ED, Triller A, et al. 2007. Three-dimensional architecture of presynaptic terminal cytomatrix. *J Neurosci* **27:** 6868–6877.

Venken KJ, Kasprowicz J, Kuenen S, Yan J, Hassan BA, Verstreken P. 2008. Recombineering-mediated tagging of *Drosophila* genomic constructs for in vivo localization and acute protein inactivation. *Nucleic Acids Res* **36:** e114.

Wagh DA, Rasse TM, Asan E, Hofbauer A, Schwenkert I, Dürrbeck H, Buchner S, Dabauvalle MC, Schmidt M, Qin G, et al. 2006. Bruchpilot, a protein with homology to ELKS/CAST, is required for structural integrity and function of synaptic active zones in *Drosophila*. *Neuron* **49:** 833–844.

Wiedenmann J, Ivanchenko S, Oswald F, Schmitt F, Röcker C, Salih A, Spindler KD, Nienhaus GU. 2004. EosFP, a fluorescent marker protein with UV-inducible green-to-red fluorescence conversion. *Proc Natl Acad Sci* **101:** 15905–15910.

Zhai RG, Bellen HJ. 2004. The architecture of the active zone in the presynaptic nerve terminal. *Physiology (Bethesda)* **19:** 262–270.

Zhang YQ, Rodesch CK, Broadie K. 2002. Living synaptic vesicle marker: Synaptotagmin-GFP. *Genesis* **34:** 142–145.

Zhong Y, Budnik V, Wu CF. 1992. Synaptic plasticity in *Drosophila* memory and hyperexcitable mutants: Role of cAMP cascade. *J Neurosci* **12:** 644–651.

Zito K, Parnas D, Fetter RD, Isacoff EY, Goodman CS. 1999. Watching a synapse grow: Noninvasive confocal imaging of synaptic growth in *Drosophila*. *Neuron* **22:** 719–729.

22 | Experimental Methods for Examining Synaptic Plasticity

Douglas P. Olsen and Haig Keshishian

Molecular, Cellular, and Developmental Biology Department, Yale University, New Haven, Connecticut 06511

ABSTRACT

The *Drosophila* neuromuscular junction (NMJ) ranks as one of the preeminent model systems for studying synaptic development, function, and plasticity. In this chapter, we review the experimental genetic methods that include the use of mutated or reengineered ion channels to manipulate the synaptic connections made by motor neurons onto larval body-wall muscles. We also provide a consideration of environmental and rearing conditions that phenocopy some of the genetic manipulations.

INTRODUCTION

The *Drosophila* larval NMJ has been extensively studied as a model system for examining all aspects of synaptic development, function, and plasticity. Several excellent and up-to-date reviews of this system have been published over the last few years (Budnik and Ruiz-Canada 2006; Collins and Diantonio 2007). In this chapter, we provide an overview of the genetic methods most often used for manipulating the NMJ for the purpose of studying synaptic development and plasticity. There is a multitude of mutations that yield NMJ phenotypes. Here we focus on the better understood genetic approaches: those that directly affect neurophysiological functions to alter levels of membrane excitability; constructs that alter either presynaptic transmitter release or postsynaptic receptor function; and manipulations of second messengers and signal transduction cascades, cell adhesion molecules (CAMs), retrograde growth factors, and transcription factors.

In addition, we consider environmental and rearing conditions that phenocopy the genetic approaches that affect synaptic growth and function. The chapter is organized around seven tables that summarize the major experimental tools, the genes that are involved, comments about the method, source material, and relevant references.

METHODS FOR MANIPULATING SYNAPTIC CONNECTIONS

Membrane Excitability: Table 1

We begin with a review of the genetic tools that are commonly used to manipulate membrane excitability, summarized in Table 1. Two approaches have been adopted. The earliest methods made use of mutations that affect either the expression or function of ion channels. The more recent second approach is to use the directed expression of modified ion channels in neurons or muscles. These channels have often been reengineered to introduce useful properties, such as altered activation voltages or GFP tagging.

Ion Channel Mutations

In general, these approaches involve mutations that have one of three primary effects: They alter the main conductance subunit of the channel itself, such as *paralytic (para)* or *Shaker (Sh)*; they affect other genes required for channel function, such as *temperature-induced paralytic E (tipE)*; or they affect the expression of the ion channel subunits, such as *maleless (mle^{nap-ts1})*. An important property

TABLE 1. Methods to alter membrane excitability

Effect	Tool (gene symbol)	Comment	Commonly used alleles (allele type: stock #)[a]	Reference(s)[b]
Increased excitability	*Shaker (Sh)*	K_V1.2 K$^+$ channel	Sh^{14} (null: 3563); Sh^7 (anti); Sh^{21} (hypo); Sh^{16}; Sh^5 (111)	Budnik et al. 1990
	ether-a-gogo (eag)	K$^+$ channel subunit	eag^{sc29} (null: 1442); eag^1 (hypo: 3561); eag^{4PM}	Budnik et al. 1990
	Hyperkinetic (Hk)	Channel subunit; oxidoreductase	Hk^1 (3562); Hk^2 (55)	Budnik et al. 1990
	comatose; Ca-P60A (comt; Ca-P60A)	Temperature-dependent seizure	$Comt^6$; $CaP60A^{Kum170}$	Hoeffer et al. 2003
	Dp *para*+	Duplication of chromosomal region containing *para* locus	Dp(1;4)r$^+$l (5273)	Budnik et al. 1990
	UAS-SDN	Shaker dominant negative		Mosca et al. 2005
	UAS-eag-DN	eag dominant negative	(8178)	Broughton et al. 2004
	UAS-TrpA1	Warm-activated trp channel	(26263, 26264)	Hamada et al. 2008
	UAS-TRPM8	Cold-activated trp channel		Peabody et al. 2009
	UAS-NaChBac	Bacterial Na channel		Nitabach et al. 2006
	UAS-P2X$_2$	Light activation of channel agonist		Lima and Miesenbock 2005
	UAS-Channelrhodopsin-2 (UAS-ChR2)	Light-gated cation channel		Schroll et al. 2006; Ataman et al. 2008
Decreased excitability	*maleless (mle)*	Regulates Na$^+$ channel expression	$mle^{nap-ts1}$ (GOF)	Budnik et al. 1990; Jarecki and Keshishian 1995
	paralytic (para)	Na$^+$ channel	$para^{lk2}$ (null); $para^{ts1}$ (hypo: 1572)	Budnik et al. 1990; Jarecki and Keshishian 1995
	Temperature-induced paralytic E (tipE)	Channel subunit	$tipE^1$	Jarecki and Keshishian 1995
	UAS-Kir2.1	Inward rectifier K$^+$ channel		Baines et al. 2001; Paradis et al. 2001
	UAS-EKO	Voltage-gated reengineered Sh channel		White et al. 2001
	UAS-dORK	Constitutively open K$^+$ channel		Nitabach et al. 2002

[a]The allele nomenclature and annotation of allele type listed in this and subsequent tables is derived from FlyBase. Please refer to FlyBase (http://flybase.org/) for synonymous alleles. The stock number listed is from the Bloomington *Drosophila* Stock Center (http://flystocks.bio.indiana.edu/). Abbreviations of allele type: null, amorph; hypo, hypomorph; anti, antimorph; GOF, gain of function.

[b]In general, the references cited in this and subsequent tables refer to examples of the use of the genetic tools at the neuromuscular junction (NMJ). In cases in which a tool has not been used at the NMJ, the reference cited refers to the original characterization/development of the genetic tool.

of many of the classical ion channel mutations is the availability of temperature-sensitive (ts) conditional alleles. In most cases the ts mutations have a modest phenotype at the permissive temperature; however, at the restrictive temperature a full-fledged phenotype is obtained. The conditional quality of the ts mutations is of value for studies of neural plasticity and function. For example, paralytic hypoactivity mutations that would otherwise be lethal at the embryonic stage are used to acutely suppress activity at later times of development for phenocritical analyses.

Elevated membrane excitability is obtained using loss of function of ion channels that carry outward currents that repolarize the membrane voltage, such as the K^+ channel mutations *Sh* and *ether-a-gogo (eag)*, which are often used in combination (*eag Sh*). Hyperexcitability can also be obtained by increasing the dosage of Na^+ channels, for example, by a duplication of the *para* gene. The most powerful bouts of acute hyperexcitability involve mutant combinations of ts "seizure" mutations, such as the *comatose Kum* double mutations *com^{ts}; CaP60A^{kumts}*. At the restrictive temperature these animals experience brief bouts of very high-frequency action potential firing, followed by paralysis (Hoeffer et al. 2003). In general, presynaptic membrane hyperexcitability leads to the expansion of the synapse, including an increase in the number of synaptic boutons and nerve terminal branches, as well as increased neurotransmitter release (Budnik et al. 1990; Schuster et al. 1996b).

Reduced membrane excitability is often achieved by loss of function of channels carrying inward currents, most commonly the Na^+ channel *para*, or through loss of function of genes required for Na^+ channel function or expression, such as *tipE* or *mle^{nap-ts1}*. In contrast to the growth effects observed with hyperactivity, reduced neural activity impairs the proper refinement of synaptic connections at the NMJ (Jarecki and Keshishian 1995; White et al. 2001). Although hypoactive mutants have relatively normal NMJ growth, their individual synaptic boutons are significantly smaller (Lnenicka et al. 2003).

Expression of Ion Channels

The directed expression of modified ion channels to manipulate neural excitability represents the most promising direction for controlling neural activity, in both a constitutive and a conditional manner. The tools are expressed as upstream activating sequence (UAS) effectors, controlled by either conventional Gal4 drivers or by inducible methods, such as the steroid-activated GeneSwitch Gal4 drivers (Osterwalder et al. 2001; Roman et al. 2001). Three channels have been widely used to suppress membrane excitability: UAS-EKO, UAS-Kir2.1, and UAS-dORK. Each is a K^+ channel that shunts the membrane voltage toward the equilibrium potential of potassium (E_K), a voltage that is usually more negative than the resting voltage. Whereas the dORK channel introduces a constitutively open K^+ shunt, both the Kir2.1 and EKO channels have voltage dependence (Baines et al. 2001; Paradis et al. 2001; White et al. 2001; Nitabach et al. 2002). Kir2.1 is a modified GFP-tagged inward rectifier channel, and it is most effective near the resting membrane voltage (Baines et al. 2001; Paradis et al. 2001). In contrast, EKO is an extensively modified variant of the fast activating Shaker K^+ channel, in which the fast inactivation function has been disabled. Although EKO has a constitutive K^+ conductance at the resting voltage, it rapidly activates with depolarization, intercepting the membrane voltage to drive it back toward E_K when the cell initiates an action potential. The construct also has a GFP tag, to help monitor its localization. As EKO is based on a native K^+ channel (Shaker), it also localizes to membrane sites in which the Dlg adapter protein is found, such as the subsynaptic reticulum of the NMJ (White et al. 2001).

Several effective tools have also been introduced for acutely elevating membrane excitability. For example, both the eag and Sh proteins can be suppressed by dominant negative constructs (Broughton et al. 2004; Mosca et al. 2005). SDN (Shaker Dominant Negative) is a truncated Shaker subunit that abolishes Sh function when expressed in neurons or muscles. It functions as a dominant negative construct, presumably by disrupting the proper assembly of the Sh channel during its biogenesis (Mosca et al. 2005). However, SDN will have an effect only in cells that express the Shaker channel. Moreover, the degree of increased excitation is limited to that achieved by a *Sh* mutant. As an alternative, a more general tool for elevating membrane excitability is NaChBac, a bacterially derived Na$^+$ channel that depolarizes any cell in which it is expressed (Nitabach et al. 2006).

Several recently introduced approaches allow the investigator to control neural excitability in a conditional fashion using either light or temperature. These involve the expression of either the light-activated protein Channelrhodopsin-2 (ChR2) or the temperature-activated channels TrpA1 and TRPM8 (Schroll et al. 2006; Ataman et al. 2008; Hamada et al. 2008; Peabody et al. 2009). These tools allow the investigator to elevate membrane excitability transiently in targeted neurons or in muscles, upon illumination with the appropriate wavelength or by a temperature shift. A caveat of ChR2 is that *Drosophila* requires that retinal be provided to assemble a functional light-gated channel (Schroll et al. 2006; Ataman et al. 2008). An alternative and effective method for light activation involves P2X$_2$, a purine receptor that can be activated by using photolytically cleaved ligands. However, this approach requires that the caged ligand be provided to the animal before light activation (Lima and Miesenbock 2005).

Finally, the researcher has the option of using the directed expression of double-stranded UAS RNA interference (RNAi) constructs to knock down specific ion channels (Dietzl et al. 2007). When combined with inducible Gal4 expression systems, this knockdown can be performed in a conditional manner in the cells of interest, bypassing early lethality that might result from constitutive channel knockdown. The increasing availability of these lines for a wide range of *Drosophila* ion channels provides a powerful alternative to the several methods described above. The use of transgenic RNAi lines is also a powerful tool for loss of function studies for any of the signaling cascades described later in the chapter.

Overall, a wide range of options is available to the researcher interested in manipulating neural excitability, but care must be exercised in selecting the appropriate tool for the task at hand. If the goal is to examine the effects of long-term changes in membrane excitability, any of the classic ion channel mutations are a good choice. However, these will usually affect neurons and/or muscles throughout the organism, complicating cell-specific analyses. Also, to have a direct effect, the gene being mutated must be expressed in the neuron or muscle of interest. For example, the para Na$^+$ channel is not expressed in muscles, whereas the Shaker K$^+$ channel is not expressed in all neurons nor at all developmental stages. A second consideration is to control for genetic background effects, because many synaptic phenotypes have limited penetrance or result in modest changes in NMJ size or function. One solution to this problem is to use genetic suppression as a control. Thus the triple mutant *eag Sh mle$^{nap-ts1}$* line is often used as a control for *eag Sh*, in which the reduced Na$^+$ channel expression due to *mle$^{nap-ts1}$* suppresses the membrane hyperexcitability phenotype caused by *eag Sh* (Budnik et al. 1990).

The conditionally activated channels such as UAS-ChR2, UAS-TrpA1, or UAS-TRPM8 have tremendous power, and they are expected to become the methods of choice when rapid and acute excitation of specific neurons is desired. It is likely that corresponding methods will be introduced for acute suppression of activity, also based on optical methods, such as those that involve the use of halorhodopsins. An exciting prospect for the investigator will be animals in which distinct wavelengths of light could be used to acutely elevate and suppress membrane excitability in specific neurons.

Synaptic Transmission: Table 2

The next class of genetic tools target chemical synapses, with the goal of either elevating or decreasing synaptic transmission (Table 2). As we saw with methods used to manipulate membrane excitability (Table 1), these tools make use of both classic mutations as well as a variety of transgenic constructs.

Functional plasticity at the NMJ typically involves changes in either quantal content (the amount of neurotransmitter quanta released per action potential) or quantal size (the postsynaptic response due to the release of a single transmitter quantum). There are numerous manipulations that affect either parameter, as will be discussed in later sections. However, here we focus on genetic tools that affect either the exocytosis/endocytosis machinery of the synaptic terminal or neurotransmitter receptors in muscle in a direct, specific, and well-controlled fashion.

TABLE 2. Methods to alter synaptic transmission

Effect	Tool (gene symbol)	Comment	Commonly used alleles (allele type: stock #)[a]	Reference
Decreased presynaptic release	UAS-shi[ts1]	Temperature-sensitive Dynamin mutation		Kitamoto 2001
	UAS-TNT	Tetanus toxin light chain		Sweeney et al. 1995
Decreased postsynaptic sensitivity	*DGluRIIA*	Mutations in postsynaptic glutamate receptor subunit	*DGluRIIA[AD9]*; *DGluRIIA[SP16]*	Petersen et al. 1997
	UAS-DGluRIIA M614R	Channel dead glutamate receptor subunit		Diantonio et al. 1999
	UAS-DGluRIIB	Wild-type glutamate receptor subunit		Diantonio et al. 1999
Increased postsynaptic sensitivity	UAS-DGluRIIA subunit	Wild-type glutamate receptor		Petersen et al. 1997

Tools Targeting Presynaptic Function

The first widely used genetic tool for controlling synaptic transmission was the ts allele of the dynamin gene *Shibire* (*Shi[ts]*). At the restrictive temperature, *Shi[ts]* animals rapidly collapse and remain paralyzed until cooled to the permissive temperature (Grigliatti et al. 1973). For acute blockade of chemical synaptic transmission throughout the nervous system, this is a powerful tool, but it has been supplanted by the introduction of a UAS Shi[ts] effector, permitting dominant, cell-specific, and conditional silencing of synaptic release on shifting to the restrictive temperature (Waddell et al. 2000; Kitamoto 2001). For example, the rapid activation and conditional nature of the construct has allowed researchers to examine with impressive detail when cell-specific synaptic activity is required during the establishment of long-term memories (Waddell et al. 2000). However, because dynamin function is required for membrane turnover and neuronal growth, suppressing function for prolonged periods during development is problematic, especially when the structural plasticity of synapses is being studied.

A direct approach for silencing synaptic transmission is to disrupt the SNARE (soluble NSF attachment protein receptor) complex involved in vesicle exocytosis. The method of choice uses tetanus toxin light chain (TetTxLC), expressed in a cell-specific manner under the control the UAS-TNT effector. TetTxLC is a protease that degrades the vSNARE synaptobrevin, resulting in a complete blockade of evoked synaptic transmission (Sweeney et al. 1995). The silencing is essentially irreversible; therefore, to restrict expression to specific postembryonic stages of development requires the use of inducible Gal4 expression systems, such as GeneSwitch or the Gal80-based TARGET system. Even so, the researcher should be aware that all existing inducible systems have a low level of leakiness. Given the very potent nature of TetTxLC, there may be an effect on evoked transmission even in the absence of induction. Expression of inactive forms of the tetanus toxin light chain can be used as a control.

Tools Targeting Postsynaptic Function

The suppression and elevation of the postsynaptic response at the NMJ is most directly achieved by altering glutamate receptor function. *Drosophila* body-wall muscles express two tetrameric DGluRII receptor complexes, which differ by whether they include a DGluRIIA or DGluRIIB subunit. Quantal size increases in proportion to the ratio of DGluRIIA to DGluRIIB. Thus, an effective method to suppress synaptic drive at the NMJ is to use either the *DGluRIIA* mutation or alternatively to drive expression of the dominant negative transgene UAS DGluRIIA-DN. Quantal size can also be decreased by driving expression of DGluRIIB, once again reducing the ratio of DGluRIIA/DGluRIIB. It was through the use of these lines that the first evidence for a retrograde homeostatic regulation of presynaptic neurotransmitter release at the NMJ was discovered (Petersen et al. 1997; Diantonio et al. 1999). Quantal size is effectively increased by driving expression of DGluRIIA.

Signal Transduction and Second Messengers: Table 3

Cyclic AMP

The cyclic AMP (cAMP) cascade is a potent regulator of both functional and structural synaptic plasticity at the NMJ. The genetic tools most widely used to manipulate the levels of intracellular cAMP at the synapse are the adenylyl cyclase *rutabaga* (*rut*) and the cAMP phosphodiesterase *dunce* (*dnc*). Levels of cAMP can be increased by using *dnc* mutations or expression of UAS-rut, whereas low levels of cAMP can be achieved by using *rut* mutations or UAS-dnc (Table 3). Loss-of-function mutations of either rut or dnc disrupt short-term plasticity, as measured by paired pulse facilitation (PPF) or posttetanic potentiation (PTP) (Zhong and Wu 1991). However, the increased levels of presynaptic cAMP in *dnc* mutant- or UAS-rut-expressing animals enhance evoked transmitter release and lead to an expansion of the synaptic arbor (Zhong and Wu 1991, 2004; Zhong et al. 1992). Furthermore, the activity- and environment-dependent expansion of the synaptic arbor is tightly regulated by presynaptic cAMP levels (Zhong and Wu 2004).

Ca^{2+}

Calcium signaling is required for the proper function of evoked transmitter release and is an important player in different forms of synaptic plasticity. A variety of alleles of the presynaptic $Ca_V2.1$ Ca^{2+}

TABLE 3. Methods to manipulate signal transduction and second messenger systems

Effect	Tool (gene symbol)	Comment	Commonly used alleles (allele type: stock #)[a]	Reference(s)
Increased cAMP	*dunce* (*dnc*)	Phosphodiesterase mutation	dnc^{M14} (null: 4714); dnc^{M11} (null); dnc^{1} (hypo: 6020)	Zhong and Wu 1991; Zhong et al. 1992
	UAS-rut	Used for overexpression of wild-type adenylyl cyclase	(9405; 9406)	Zhong and Wu 2004
Decreased cAMP	*rutabaga* (*rut*)	Adenylyl cyclase mutation	rut^{1} (9404)	Zhong and Wu 1991; Zhong et al. 1992
	UAS-dnc	Used for overexpression of wild-type phosphodiesterase		Cheung et al. 1999; Sanyal et al. 2002
Decreased Ca^{2+}	*cacophony* (*cac*)	Presynaptic $Ca_V2.1$ Ca^{2+}l channel	cac^{L13}; cac^{HC129}; cac^{S}; cac^{NT27}	Kawasaki et al. 2000; Rieckhof et al. 2003; Frank et al. 2006
	UAS-PV (parvalbumin)	Intracellular Ca^{2+} buffer		Harrisingh et al. 2007
Increased CaMKII activity	UAS-CaMKII^{T287D}	Constitutively active CaMKII		Jin et al. 1998; Koh et al. 1999
Decreased CaMKII activity	UAS-Ala UAS-CaMKIINtide	CaMKII inhibitory peptide CaMKII inhibitory peptide		Jin et al. 1998; Koh et al. 1999 Haghighi et al. 2003
Increased PKA activity	UAS-PKAact	Constitutively active PKA catalytic subunit		Davis et al. 1998
Decreased PKA activity	UAS-PKAinh	PKA regulatory subunit with mutation in cAMP binding site		Davis et al. 1998
Increased PAR protein activity	UAS-PKM UAS-par-1	Persistently active form of aPKC		Ruiz-Canada et al. 2004 Zhang et al. 2007
Decreased PAR protein activity	*aPKC*	Atypical protein kinase C	$aPKC^{k06403}$ (10622) ; $aPKC^{Exc55}$; $aPKC^{EX48}$	Ruiz-Canada et al. 2004
	UAS-DN-PKM	Dominant negative kinase dead aPKC		Ruiz-Canada et al. 2004
	par-6	Component of PAR-aPKC complex	$par-6^{D226}$	Ruiz-Canada et al. 2004
	bazooka (*baz, par-3*)	Component of PAR-aPKC complex	baz^{4} (3295)	Ruiz-Canada et al. 2004
	par-1	Serine threonine kinase	$par-1^{k06323}$ (10615) ; $par-1^{W3}$; $par-1^{D-16}$	Zhang et al. 2007

channel *cacophony* (*cac*) have been used to impair calcium influx at the NMJ (Table 3). Null mutations of *cac* are lethal but heteroallelic combinations and a variety of hypomorphic alleles can circumvent this problem. Reduced *cac* function causes a decrease in the size of the synaptic arbor. This effect is likely not a direct result of reduced synaptic transmission, because other mutants that inhibit transmitter release do not have synaptic growth phenotypes (Rieckhof et al. 2003). Normal *cac* function is also required for the homeostatic up-regulation of transmitter release that is induced when postsynaptic GluR function is impaired (Frank et al. 2006). Another tool that could be used to reduce the levels of intracellular Ca^{2+} at the synapse is the transgenic expression of the intracellular Ca^{2+} buffer protein parvalbumin (UAS-PV) (Harrisingh et al. 2007) (Table 3). This tool, to our knowledge, has not yet been used at the NMJ.

CaMKII

The role of CaMKII at the NMJ has been investigated using transgenic constructs that inhibit its activity (UAS-Ala and UAS-CaMKIINtide) or produce constitutive activation of the enzyme (UAS-CaMKII^{T287D}) (Table 3). A role for CaMKII in regulating the synaptic localization of the PDZ protein Discs large (Dlg) and bouton morphology was shown by altering CaMKII activity both pre- and postsynaptically (Koh et al. 1999). Postsynaptic CaMKII has been shown to negatively regulate the retrograde signaling that controls the homeostasis of transmitter release (Haghighi et al. 2003).

Protein Kinase A

The role of protein kinase A (PKA) in postsynaptic muscles has been investigated using transgenic tools that increase (UAS-PKAact) or inhibit (UAS-PKAinh) PKA activity (Table 3). These studies have shown that postsynaptic PKA regulates quantal size in a GluRIIA-dependent fashion. Increased PKA activity decreases quantal size, whereas inhibition of PKA leads to an increase in quantal size. These effects are eliminated in the absence of the GluRIIA subunit (Davis et al. 1998).

Proline- and Acidic Amino Acid–Rich Proteins

Experiments have shown roles for the baz/par-6/aPKC complex as well as the par-1 serine-threonine kinase at the NMJ. The tools available to study the proline- and acidic amino acid–rich (PAR) proteins include loss of function mutations (*par-6, baz, aPKC, par-1*) and transgenic lines that express dominant negative (UAS-DN-PKM), persistently active (UAS-PKM), or wild-type forms of these proteins (UAS-par-1) (Table 3). Because null mutations in the baz/par-6/aPKC complex are embryonic lethal, hypomorphic alleles of *aPKC* and heterozygous mutation of *baz* and *par-6* have been used (Ruiz-Canada et al. 2004). Both loss- and gain-of-function manipulations of the baz/par-6/aPKC complex impair synaptic growth and disrupt postsynaptic GluR distribution (Ruiz-Canada et al. 2004). The regulation of the cytoskeleton by the baz/par-6/aPKC complex is important for the normal growth of the synapse but the specific role of baz/par-6/aPKC during activity-dependent plasticity is not well understood. Loss of function and overexpression studies have shown that *par-1* negatively regulates synapse formation and synaptic strength via its ability to phosphorylate Dlg (Zhang et al. 2007).

CAMs and Growth Factors: Tables 4 and 5

CAMs

The growth and plasticity of the larval NMJ is directly influenced by both CAMs expressed at the synaptic contacts and by at least two transsynaptic growth factor signaling systems.

An important insight into the regulation of NMJ growth emerged from the discovery that the immunoglobulin cell adhesion molecule Fasciclin 2 (Fas2) is abundant at the NMJ, and that its expression scales inversely with membrane excitability. Thus, in hyperactive mutant backgrounds, in which the size of the NMJ increases substantially, the level of synaptic Fas2 decreases significantly (Schuster

TABLE 4. Methods to alter cell adhesion

Effect	Tool (gene symbol)	Comment	Commonly used alleles (allele type: stock #)[a]	Reference(s)
Increased Fas2 levels	UAS-Fas2			Schuster et al. 1996a,b; Davis et al. 1997
Decreased Fas2 levels	*Fasciclin 2 (Fas2)*	Series of mutant Fas2 alleles that express different levels of Fas2	*Fas2^{EB112}* (null) > *Fas2^{e76}* (hypo) > *Fas2^{e86}* (hypo) > *Fas2^{e93}* (precise P-element excision)	Grenningloh et al. 1991; Schuster et al. 1996a,b; Davis et al. 1997

et al. 1996b). Alleles of Fas2 that reduce expression to the levels found in hyperactive mutants also have significantly expanded NMJs (Schuster et al. 1996b). This observation shows that the reduction of synaptic Fas2 observed with elevated neuromuscular activity is sufficient to account for most of the resulting NMJ overgrowth. A well-characterized allelic series of Fas2 mutants is available, with phenotypes ranging from mild hypomorphs to complete nulls (Table 4) (Grenningloh et al. 1991).

Fas2 levels can also be elevated by expression of the UAS-Fas2 effector in either motor neurons or muscles. Although overexpression of Fas2 in muscle generally suppresses NMJ growth, it should be noted that a balanced overexpression of Fas2 on both sides of the synapse results, surprisingly, in the expansion of the NMJ (Ashley et al. 2005).

Growth Factors

The transsynaptic signaling systems that affect the size of the *Drosophila* NMJ include the TGF-β growth factor Glass bottom boat (Gbb), a member of the BMP family, and the Wnt growth factor Wingless (Wg). Both growth factors have been shown to be essential for normal NMJ development (Aberle et al. 2002; Marques et al. 2002; Packard et al. 2002; McCabe et al. 2003). In addition to mutant alleles of the Gbb ligand and its type II receptor Wishful Thinking (Wit), the TGF-β signaling cascade can be manipulated using a variety of transgenic lines (Table 5). It should be noted that *wit* and *gbb* mutations are generally not homozygosed, but are used in heteroallelic combinations, such as *gbb^1/gbb^2* and *wit^{A12}/wit^{B11}*.

For gain of function using directed cell-specific expression the most common approach has been to drive the constitutively active BMP receptor in motor neurons (the BMP type I co-receptor UAS-Act-Tkv). It is also possible to overexpress the ligand in muscle, using the UAS-Gbb effector. For loss of function studies in specific cells there are dominant-negative transgenes that suppress downstream signaling components, such as the transcription factor Mad (UAS-Mad^1), as well as expressible forms of inhibitory proteins that suppress TGF-β signaling (UAS-Dad).

Wg signaling can be suppressed using both classic mutations targeting Wg or its receptor frizzled, as well as a DN form of the frizzled receptor. To bypass early lethality caused by Wg loss of function, hypomorphic or ts alleles are used. Retrograde signaling from synapse to cell body can be disrupted using dominant negative transgenes that target the p150 dynactin component of the dynein-dynactin complex (UAS-GluedDN or UAS-ΔGl) (McCabe et al. 2003; Mathew et al. 2005). This approach has been extensively used to show the need for retrograde growth factor transport to the cell body, but must be applied with caution, as it undoubtedly disrupts multiple retrograde signals, as well as potentially affecting orthograde transport as well.

Transcription Factors: Table 6

Activator Protein 1 (AP-1: Fos/Jun)

The transcription factor AP1, a heterodimer of Fos and Jun, plays a key role in synaptic plasticity at the NMJ. The genetic tools that have been used to manipulate AP-1 function at the NMJ include transgenic lines that allow for expression of wild-type (UAS-Fos, UAS-Jun) or dominant negative (UAS-fbz, UAS-jbz) forms of these transcription factors (Table 6). Overexpression of both Fos and

TABLE 5. Methods to alter growth factor signaling

Effect	Tool (gene symbol)	Comment	Commonly used alleles (allele type: stock #)	Reference(s)
Increased BMP signaling	*Daughters against dpp* (*Dad*)	Loss of function of inhibitory Smad	Dad^{271-68} ; Dad^{j1E4}	Sweeney and Davis 2002; McCabe et al. 2004
	UAS-Gbb	Used for overexpression of BMP ligand		McCabe et al. 2004
	UAS-Act-Tkv	Constitutively active type I BMP receptor		McCabe et al. 2004; Collins et al. 2006; Wang et al. 2007; O'Connor-Giles et al. 2008
Decreased BMP signaling	*glass bottom boat* (*gbb*)	Secreted BMP ligand	gbb^1 (null); gbb^2 (null); gbb^3 (hypo); gbb^4 (hypo)	McCabe et al. 2003
	wishful thinking (*wit*)	Type II BMP receptor	wit^{A12} (5173); wit^{B11} (5174); wit^{HA1} (hypo); wit^{HA2} ; wit^{HA3} ; wit^{HA4} ; wit^{HA5} ; $wit^{S126215}$	Aberle et al. 2002; Marques et al. 2002
	UAS-Dad	Used for overexpression of an inhibitory Smad		McCabe et al. 2004; Goold and Davis 2007
	UAS-Mad[1]	Mutant Smad that lacks DNA-binding activity		Takaesu et al. 2005
Decreased Wg signaling	*wingless* (*wg*)	Secreted growth factor	wg^{1-12} (null: 7000); wg^1 (hypo: 2978)	Packard et al. 2002; Ataman et al. 2008
	frizzled 2 (*fz2*)	wg receptor	$fz2^{C1}$	Mathew et al. 2005
	UAS-DFz2DN	Dominant negative frizzled		Packard et al. 2002; Mathew et al. 2005
Disrupted retrograde transport	UAS-GluedDN/UAS-ΔGl	DN disruption of dynein/dynactin complex		McCabe et al. 2003; Mathew et al. 2005

Jun (referred to as UAS-AP1) in neurons increases both the size of the synapse and the strength of transmission; however, expression of either Fos or Jun alone does not produce these effects. Conversely, the expression of either dominant negative construct in neurons leads to a smaller synapse with reduced synaptic strength (Sanyal et al. 2002). Directed expression of AP-1 or dominant negative molecules is an effective means of influencing synaptic plasticity, but little is known about the in vivo mechanisms that regulate the activity of these important transcription factors at the NMJ.

cAMP Response Element Binding Protein

Transgenic cAMP response element binding (CREB) protein activator (Creb2a) and CREB blocker (Creb2b) lines have been used to alter CREB activity at the NMJ (Table 6). These transgenic tools are available in either heat shock–driven forms or as UAS lines. Activation or inhibition of CREB

TABLE 6. Methods to alter transcription factor signaling

Effect	Tool (gene symbol)	Comment	Commonly used alleles (allele type: stock #)	Reference(s)
Increased AP-1 activity	UAS-AP1	UAS-fos and UAS-jun together		Sanyal et al. 2002
Decreased AP-1 activity	UAS-Fbz	Dominant inhibitory form of Fos	(7214 ; 7215)	Sanyal et al. 2002
	UAS-Jbz	Dominant inhibitory form of Jun	(7217 ; 7218)	Sanyal et al. 2002
Increased CREB activity	Creb2a	Transgenic CREB activator; heat shock or UAS driven		Davis et al. 1996
Decreased CREB activity	Creb2b	Transgenic CREB repressor; heat shock or UAS driven		Davis et al. 1996; Sanyal et al. 2002

using the heat shock–driven lines in an otherwise wild-type background has little effect on the structure or function of the synapse (Davis et al. 1996). Inhibition of CREB in a *dnc* mutant- or AP-1-overexpressing background prevents the increase in synaptic strength but does block the growth of the synapse (Davis et al. 1996; Sanyal et al. 2002). Conversely, CREB activation can increase synaptic strength when the synapse is structurally expanded via reduction in Fas2 levels (Davis et al. 1996).

Environmental Manipulations: Table 7

Elevated Rearing Temperature

Chronic rearing of wild-type larvae at 29°C or 30°C leads to increased larval locomotion, growth of the synapse, and increased synaptic strength (Sigrist et al. 2003; Zhong and Wu 2004). The temperature-dependent growth and potentiation of the synapse phenocopies the effects of activity-dependent growth in many ways, with reduced levels of Fas2 and increases in postsynaptic GluRs (Schuster et al. 1996b; Sigrist et al. 2000, 2003). The temperature-dependent expansion and strengthening of the synapse is blocked when neural activity is silenced with mutations that affect Na^+ channels or when cAMP levels are reduced (Sigrist et al. 2003; Zhong and Wu 2004). Rearing temperature also interacts with hyperexcitability as shown by the enhanced growth of *Sh* synapses at 25°C when compared with *Sh* animals at 18°C or wild-type animals raised at 25°C (Zhong and Wu 2004). Although increased rearing temperature is a potent inducer of plasticity phenomena at the synapse, elevated temperature is likely to have a broad range of effects on larvae and the molecular mechanisms that underlie temperature-dependent events at the synapse are not well understood. Thus, elevated rearing temperature is often best used as a complementary approach, along with genetic approaches, to manipulate synaptic plasticity at the NMJ.

Acute Increases in Locomotor Activity

The increased locomotor activity of animals raised at 30°C is likely a key component of the temperature-dependent growth and strengthening of the synapse. When larvae are transferred from food slurry to moist, food-free agar plates, they show greatly increased locomotor activity with significant individual variability (Sigrist et al. 2003; Steinert et al. 2006). Larvae that maintain a high level of locomotor activity under these conditions for up to 2 h have enhanced synaptic transmission. This experience-dependent potentiation of the synapse has distinct phases. No changes in synaptic transmission are observed until after a period of ~40 min of fast crawling, at which point increased quantal sizes are observed. After ~90 min of fast crawling, quantal sizes have returned to baseline whereas evoked responses remain elevated, suggesting increased transmitter release (Steinert et al. 2006). Larvae raised on agar plates for 12–18 h have only slightly larger synapses than their vial-raised counterparts (Sigrist et al. 2003). Thus, this method can effectively potentiate synaptic transmission without the potentially pleiotropic effects of elevated temperature. This method is, however, more labor intensive, requiring monitoring the locomotor activity of individual larvae for up to 2 h, and it does not produce detectable synaptic growth.

TABLE 7. Environmental conditions that affect synaptic plasticity

Environmental condition	Effect	References
Elevated rearing temperature	Increased neural excitability and increased locomotor activity	Sigrist et al. 2003; Zhong and Wu 2004; Schuster 2006; Steinert et al. 2006
Transfer of larvae to moist food-free agar plate	Acute increases in locomotor activity	Sigrist et al. 2003; Schuster 2006; Steinert et al. 2006

CONCLUSIONS AND PROSPECTS

At present the researcher has an impressive array of well-characterized genetic methods to effectively control the function and plasticity of the NMJ. In many cases the manipulations can be performed both acutely or over longer developmental periods. However, experimental control of the NMJ is only part of the problem facing the investigator: To complement these genetic tools we need transgenes that report either membrane electrical excitability, levels of synaptic transmission, or the activity of molecules in the downstream signaling cascades. The optical clarity of the larval NMJ suggests that the new tools will likely involve transgenic fluorescent reporters. We therefore look forward to the development of effective, expressible constructs for reporting synaptic function at the single-cell level to complement the impressive genetic tools described in this chapter.

Manipulation of Membrane Excitability by Expression of Modified Shaker Constructs

We describe here the use of the two-electrode voltage clamp (TEVC) to examine potassium currents mediated by voltage-gated ion channels, and we show several genetic and pharmacological methods that are used to study the currents. *Drosophila* larval muscle fibers possess three major K$^+$ currents: a fast voltage-activating and inactivating I_A current, mediated by the Shaker channel; a slow, voltage-activated I_K or "delayed rectifier" current that is responsible for the repolarization of the action potential and mediated by the Shab channel; and finally a Ca^{2+}-activated K$^+$ current or I_{CF}, mediated by the Slowpoke channel.

In addition to differences in the way the channels are activated, the Shaker and Shab channels also differ with respect to inactivation. The Shaker channel rapidly inactivates, whereas the Shab channel remains substantially open during prolonged depolarization (it shows a much slower "C" type inactivation). In addition, the Shaker channel is characterized by its sensitivity to the drug 4-aminopyridine (4-AP).

Two useful transgenic tools for altering membrane excitability have been developed by making specific modifications of the Shaker channel. The SDN construct is a truncated and GFP-tagged Shaker protein that suppresses the I_A current when it is expressed in muscles or neurons (Mosca et al. 2005). The SDN-expressing cell develops a *Sh* phenotype, and as a result becomes electrically hyperexcitable. The EKO (Electrical KnockOut) protein includes most of the native Shaker protein sequence; however, its amino-terminal inactivation domain (known as the "inactivation ball") is replaced with GFP (White et al. 2005). When EKO is expressed in neurons or muscles, it introduces a large, rapidly activating outward current that does not inactivate rapidly. The construct is therefore used to suppress membrane excitability in the targeted cell to study the effect of expressing these transgenes in muscle.

TEVC is a powerful tool for monitoring transmembrane currents (Box 1). Briefly, both a voltage-monitoring and a current-passing microelectrode are inserted into the muscle fiber, with a third reference electrode in the bath. The instrument rapidly shifts the membrane voltage to any "command" value the investigator chooses (in μsec), and then reads out the resulting transmembrane membrane current. Figure 1 shows examples of TEVC recordings from larval muscles in wild-type *Drosophila* and in transgenic strains expressing mutant Shaker proteins.

BOX 1. VOLTAGE CLAMP

A voltage clamp is an electronic instrument that allows the researcher to measure the currents that flow across the membrane of a cell as a function of voltage, and it is indispensable for studying voltage-gated ion channels. The transmembrane current that flows when a cell changes its membrane voltage includes two components. The first is the current that flows through open ion channels, and it is defined by the product of the ion channel conductances and the driving forces. For a K$^+$ current this is $I_K = g_K(V_m - E_K)$. This is the current we usually want to measure when we study ion channel function. However, whenever the membrane voltage changes there is in addition a capacitive current I_C that is due to the flow of charges on and off of the cell membrane surfaces, $I_C = C_m \, dV/dt$, where C_m is the membrane capacitance (in Farads) and dV/dt is the rate of change of the membrane voltage. We need a way to tease these two currents apart. The solution is to very rapidly force the membrane voltage to a specific experimental value and then *hold it steady* so that $dV/dt = 0$, and thus there is no further capacitive current. The remaining transmembrane current observed when the voltage is held steady is that flowing through ion channels.

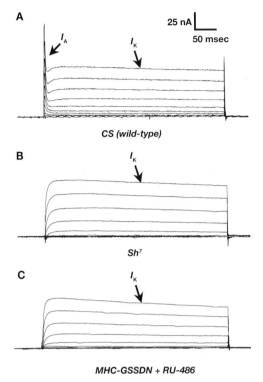

FIGURE 1. Examples of two-electrode voltage-clamp (TEVC) recordings from *Drosophila* larval muscles. (*A*) The wild-type currents shown include a rapidly inactivating *Shaker* current (I_A) as well as persistent delayed rectifier current (I_K). (*B*) In the Sh^7 mutant the I_A current is absent without affecting I_K. (*C*) The Shaker Dominant Negative (SDN) transgene abolishes the I_A current, phenocopying the Sh^7 mutation (genotype of larva shown: *yw/UAS-SDN/+; MHC-GS/+*). Outward currents were recorded from larval muscles in a saline with 0 mM Ca^{2+} using a TEVC. Depolarizing steps of 10 mV were used from a holding potential of –80 mV to +30 mV. (Adapted from Mosca and Keshishian 2005, © National Academy of Sciences, U.S.A.)

MATERIALS

See the end of the chapter for recipes for reagents marked with <R>.

Reagents

4-aminopyridine (4-AP), 400 mM stock. Add to the saline bath to achieve a concentration of 4 mM 4-AP (see Step 4)

Drosophila melanogaster mutant stocks (see Table 1)

The UAS-SDN and UAS-EKO effectors are crossed to either the MHC-Gal4 or the MHC-GS GAL4 to generate animals expressing either transgene in muscle (Osterwalder et al. 2001). Use either the UAS lines or wild-type animals for controls. Sh^7 mutants may also be used to observe a phenotype similar to the effects of SDN.

Reagents for preparation and dissection of body wall muscles of the NMJ as described in Chapter 12, Protocol 2

Voltage-clamp (physiological) saline <R>

The recordings are performed in a modified *Drosophila* saline that has 0 Ca^{2+}, 10 mM Mg^{++}, and 0.5 mM EGTA to suppress all inward Ca^{2+} currents. As a result the muscles only show outward voltage-gated K^+ currents upon depolarization, as well as voltage-independent leakage currents. Synaptic transmission is also suppressed in this saline.

Equipment

Electrodes

The voltage and current electrodes should each be sharp glass micropipettes, filled with 3 M KCl <!>, and have resistances of 20 MΩ or less. The same electrodes used for recording synaptic activity at the larval NMJ may be used here. Avoid high electrode resistances, especially for the current electrode, because this will affect your ability to pass large amounts of current. The reference electrode may be either a silver chloride pellet or a broken tip micropipette filled with saline. To avoid cross talk, the current and voltage electrodes should be positioned 180° apart on either side of the microscope objective.

Electronic hardware

The voltage-clamp instrument applies large voltages to the current electrode, as high as 180 V with some models. This high level is needed to pass the large amount of current required to clamp the very large *Drosophila* longitudinal muscles.

Important hardware considerations:

1. Use the voltage clamp with the reference electrode as a virtual ground, as opposed to connecting the reference electrode to the circuit ground.

2. The clamp usually has high-frequency filter cutoffs. For most applications described below, a relatively "slow clamp" can be used, with a cutoff at ≤3 kHz.

3. Many clamps allow for correction for series resistance. Because these recordings are being performed on an exposed muscle in the bath, set this off or to its minimal value.

Warning: Many TEVCs will provide a powerful and potentially dangerous electric shock if the current output is touched while the feedback gain control is set to a high value. Grounding the current output should also be avoided as this may result in instrument damage. Check your wiring carefully!

Equipment for preparation and dissection of body wall muscles of the NMJ as described in Chapter 12, Protocol 2

Microscope and optics

A microscope is used to visualize the preparation and electrodes. These studies are best performed with differential interference contrast (DIC) optics, using water-immersion objectives. The immersion objective should have a nonconducting barrel. Modern objectives have ceramic tips or barrels, whereas older objectives use plated metal barrels that will short out your preparation. Electrically isolate the immersion objectives by using a nonconducting mount on the nosepiece (this usually has to be custom fabricated by your machine shop out of Teflon or PVC—trying to isolate the objective by wrapping the threads with plumber's Teflon tape *does not* work well).

Software

A TEVC can be controlled manually; however, the most convenient approach is to use computer control. Programs such as the Axon Instruments pClamp package (Molecular Devices) provide an excellent environment for setting up routine voltage-clamp protocols. A holding potential of –80 mV is used. Command steps of 10-mV increments are used to activate the voltage-gated K^+ currents. Command steps of as long as 500 msec are needed to show some of the more slowly activating currents.

METHOD

Preparing the Body Wall Fillet

1. Select a mutant to be studied, and dissect a body wall fillet preparation, following the Steps 1–3 of Chapter 12, Protocol 2. Dissection should be done in voltage-clamp saline.

2. For experiments where the I_A current is to be blocked, add 4-AP to a final concentration of 4 mM in voltage-clamp saline. After washing, transfer the fillet preparation into the saline bath. Larvae will survive for up to 2–3 h in this saline if the saline solution is regularly exchanged and aerated.

   The I_A current is most robust in a newly dissected preparation. It will degrade over time following dissection. Thus, a preparation that is a couple hours old may only reveal the delayed rectifier current.

Performing a Voltage-Clamp Experiment

3. Begin by impaling the muscle with both the current and voltage electrodes, with the amplifier set to a voltage monitoring or "current clamp" mode for the current electrode. Place the two

electrodes in the muscle within ~50 μm of each other. Impalements may be performed by either tapping or by ringing the electrodes, but be careful not to dislodge one electrode upon impaling the muscle with the other.

> Ideally, both electrodes should report the same membrane potential, but reject any impalements in which the two differ by >5 mV. The resting potential should be in excess of –50 mV for healthy muscles.

> On most instruments the current electrode can be switched to a voltage-monitoring (sometimes called a "current-clamp") mode. The impalement of the muscle with the current electrode is performed in this voltage-sensing mode.

4. Once a good, stable impalement is achieved (as determined by a resting potential of at least –50 mV with both electrodes), switch the instrument to its voltage-clamp mode; that is, switch the current electrode from voltage-monitoring to current-output mode.

> In this configuration, the voltage electrode provides a monitor of the membrane potential, whereas the current electrode now functions to pass current to the clamp and hold the cell to the desired voltage.

5. You are now ready to voltage-clamp the cell: Carefully increase the feedback gain. As you *slowly* adjust the gain upward, the membrane potential will clamp toward the holding potential of –80 mV. Assuming that the impalements are good and the current is passing well, you should be able to get to within 1 mV of the desired holding potential.

> If you increase the gain to too high a value, you will establish a positive feedback between the electrodes, resulting in potentially damaging oscillations of the membrane voltage. With practice you should be able to increase gain to the point at which you have good compliance with the holding potential but are safely below the point at which feedback oscillations take over.

6. Always monitor the voltage step waveform on the computer display or the oscilloscope. In a good clamp, the command voltage should begin with a square step. If you observe the voltage sagging, oscillating, or taking a few milliseconds to reach the command voltage, it is likely that there are electrode or impalement problems and/or that the feedback gain is too low. Switch back to the current-clamp mode and make sure both electrodes have good impalements.

7. In the final step of the sequence, turn the experiment over to the computer. Use a protocol that shifts the membrane potential in 10-mV steps from the holding value of –80 mV to +30 mV, in which each step lasts 500 msec. Allow a 1-sec pause between each step.

8. Repeat this protocol (Steps 1–7) several times to obtain averaged traces.

DISCUSSION

In wild-type muscle, the I_A current that passes through Shaker channels activates first, at command voltage steps of +30 to +40 mV depolarized relative to the holding potential. At more positive voltages the delayed rectifier current activates. This current is distinguished from I_A by its late activation and relative lack of inactivation. Plot the peak currents for each against the command voltage to generate a standard current-voltage or I–V plot of the data.

With SDN expression in muscle or in *Sh* mutants, the I_A currents are reduced or eliminated, with no effect on the delayed rectifier current. Make an I–V plot of the remaining current and compare it with the wild-type control I–V plots.

With EKO expression the current traces are dominated by a fast activating current that does *not* inactivate. Bear in mind that the delayed rectifier current is also activated. You can distinguish the two currents by a method termed "current subtraction." Both the I_A and the EKO currents can be abolished by bath application of 4-AP. Apply the 400-mM stock solution to the bath to a final dilution of 4 mM and run the voltage-clamp protocol during the bath application. It should be possible to observe the gradual elimination of the EKO currents. From an experiment of this sort it is possible to reconstruct the EKO current by subtracting the post-4-AP currents from the initial ones. The same procedure can be performed with a wild-type muscle preparation, leading to the eventual elimination of the Shaker current. The use of current subtraction is a standard method for dissecting apart multiple transmembrane currents.

RECIPE

CAUTION: See Appendix for proper handling of materials marked with <!>.
Recipes for reagents marked with <R> are included in this list.

Voltage-Clamp Saline

Component	Final concentration
NaCl	150 mM
KCl <!>	5 mM
MgCl$_2$ <!>	10 mM
NaHCO$_3$	4 mM
EGTA	0.5 mM
TES (2-[(2-hydroxy-1, 1-bis (hydroxymethyl)ethyl)amino] ethanesulfonic acid) <!>	5 mM
Trehalose	5 mM
Sucrose	50 mM

Adjust to pH 7.2 with 1 N NaOH <!> solution.

REFERENCES

Aberle H, Haghighi AP, Fetter RD, McCabe BD, Magalhaes TR, Goodman CS. 2002. *wishful thinking* encodes a BMP type II receptor that regulates synaptic growth in *Drosophila*. *Neuron* **33:** 545–558.

Ashley J, Packard M, Ataman B, Budnik V. 2005. Fasciclin II signals new synapse formation through amyloid precursor protein and the scaffolding protein dX11/Mint. *J Neurosci* **25:** 5943–5955.

Ataman B, Ashley J, Gorczyca M, Ramachandran P, Fouquet W, Sigrist SJ, Budnik V. 2008. Rapid activity-dependent modifications in synaptic structure and function require bidirectional Wnt signaling. *Neuron* **57:** 705–718.

Baines RA, Uhler JP, Thompson A, Sweeney ST, Bate M. 2001. Altered electrical properties in *Drosophila* neurons developing without synaptic transmission. *J Neurosci* **21:** 1523–1531.

Broughton SJ, Kitamoto T, Greenspan RJ. 2004. Excitatory and inhibitory switches for courtship in the brain of *Drosophila melanogaster*. *Curr Biol* **14:** 538–547.

Budnik V, Zhong Y, Wu CF. 1990. Morphological plasticity of motor axons in *Drosophila* mutants with altered excitability. *J Neurosci* **10:** 3754–3768.

Budnik V, Ruiz-Canada C. 2006. The fly neuromuscular junction: Structure and function. In *International review of neurobiology,* 2nd ed. (ed. Bradley R, et al.), Vol. 75, pp. 1–406. Academic Press, New York.

Cheung US, Shayan AJ, Boulianne GL, Atwood HL. 1999. *Drosophila* larval neuromuscular junction's responses to reduction of cAMP in the nervous system. *J Neurobiol* **40:** 1–13.

Collins CA, Diantonio A. 2007. Synaptic development: Insights from *Drosophila*. *Curr Opin Neurobiol* **17:** 35–42.

Collins CA, Wairkar YP, Johnson SL, DiAntonio A. 2006. Highwire restrains synaptic growth by attenuating a MAP kinase signal. *Neuron* **51:** 57–69.

Davis GW, Schuster CM, Goodman CS. 1996. Genetic dissection of structural and functional components of synaptic plasticity. III. CREB is necessary for presynaptic functional plasticity. *Neuron* **17:** 669–679.

Davis GW, Schuster CM, Goodman CS. 1997. Genetic analysis of the mechanisms controlling target selection: Target-derived Fasciclin II regulates the pattern of synapse formation. *Neuron* **19:** 561–573.

Davis GW, Diantonio A, Petersen SA, Goodman CS. 1998. Postsynaptic PKA controls quantal size and reveals a retrograde signal that regulates presynaptic transmitter release in *Drosophila*. *Neuron* **20:** 305–315.

Diantonio A, Petersen SA, Heckmann M, Goodman CS. 1999. Glutamate receptor expression regulates quantal size and quantal content at the *Drosophila* neuromuscular junction. *J Neurosci* **19:** 3023–3032.

Dietzl G, Chen D, Schnorrer F, Su KC, Barinova Y, Fellner M, Gasser B, Kinsey K, Oppel S, Scheiblauer S, et al. 2007. A genome-wide transgenic RNAi library for conditional gene inactivation in *Drosophila*. *Nature* **448:** 151–156.

Frank CA, Kennedy MJ, Goold CP, Marek KW, Davis GW. 2006. Mechanisms underlying the rapid induction and sustained expression of synaptic homeostasis. *Neuron* **52:** 663–677.

Goold CP, Davis GW. 2007. The BMP ligand Gbb gates the expression of synaptic homeostasis independent of synaptic growth control. *Neuron* **56:** 109–123.

Grenningloh G, Rehm EJ, Goodman CS. 1991. Genetic analysis of growth cone guidance in *Drosophila*: Fasciclin II functions as a neuronal recognition molecule. *Cell* **67:** 45–57.

Grigliatti TA, Hall L, Rosenbluth R, Suzuki DT. 1973. Temperature-sensitive mutations in *Drosophila melanogaster*. XIV. A selection of immobile adults. *Mol Gen Genet* **120:** 107–114.

Haghighi AP, McCabe BD, Fetter RD, Palmer JE, Hom S, Goodman CS. 2003. Retrograde control of synaptic transmission by postsynaptic CaMKII at the *Drosophila* neuromuscular junction. *Neuron* **39:** 255–267.

Hamada FN, Rosenzweig M, Kang K, Pulver SR, Ghezzi A, Jegla TJ, Garrity PA. 2008. An internal thermal sensor controlling temperature preference in *Drosophila*. *Nature* **454:** 217–220.

Harrisingh MC, Wu Y, Lnenicka GA, Nitabach MN. 2007. Intracellular Ca^{2+} regulates free-running circadian clock oscillation in vivo. *J Neurosci* **27:** 12489–12499.

Hoeffer CA, Sanyal S, Ramaswami M. 2003. Acute induction of conserved synaptic signaling pathways in *Drosophila melanogaster*. *J Neurosci* **23:** 6362–6372.

Jarecki J, Keshishian H. 1995. Role of neural activity during synaptogenesis in *Drosophila*. *J Neurosci* **15:** 8177–8190.

Jin P, Griffith LC, Murphey RK. 1998. Presynaptic calcium/calmodulin-dependent protein kinase II regulates habituation of a simple reflex in adult *Drosophila*. *J Neurosci* **18:** 8955–8964.

Kawasaki F, Felling R, Ordway RW. 2000. A temperature-sensitive paralytic mutant defines a primary synaptic calcium channel in *Drosophila*. *J Neurosci* **20:** 4885–4889.

Kitamoto T. 2001. Conditional modification of behavior in *Drosophila* by targeted expression of a temperature-sensitive *shibire* allele in defined neurons. *J Neurobiol* **47:** 81–92.

Koh YH, Popova E, Thomas U, Griffith LC, Budnik V. 1999. Regulation of DLG localization at synapses by CaMKII-dependent phosphorylation. *Cell* **98:** 353–363.

Lima SQ, Miesenbock G. 2005. Remote control of behavior through genetically targeted photostimulation of neurons. *Cell* **121:** 141–152.

Lnenicka GA, Spencer GM, Keshishian H. 2003. Effect of reduced impulse activity on the development of identified motor terminals in *Drosophila* larvae. *J Neurobiol* **54:** 337–345.

Marques G, Bao H, Haerry TE, Shimell MJ, Duchek P, Zhang B, O'Connor MB. 2002. The *Drosophila* BMP type II receptor Wishful Thinking regulates neuromuscular synapse morphology and function. *Neuron* **33:** 529–543.

Mathew D, Ataman B, Chen J, Zhang Y, Cumberledge S, Budnik V. 2005. Wingless signaling at synapses is through cleavage and nuclear import of receptor DFrizzled2. *Science* **310:** 1344–1347.

McCabe, BD, Marques G, Haghighi AP, Fetter RD, Crotty ML, Haerry TE, Goodman CS, O'Connor MB. 2003. The BMP homolog Gbb provides a retrograde signal that regulates synaptic growth at the *Drosophila* neuromuscular junction. *Neuron* **39:** 241–254.

McCabe BD, Hom S, Aberle H, Fetter RD, Marques G, Haerry TE, Wan H, O'Connor MB, Goodman CS, Haghighi AP. 2004. Highwire regulates presynaptic BMP signaling essential for synaptic growth. *Neuron* **41:** 891–905.

Mosca TJ, Carrillo RA, White BH, Keshishian H. 2005. Dissection of synaptic excitability phenotypes by using a dominant-negative Shaker K$^+$ channel subunit. *Proc Natl Acad Sci* **102:** 3477–3482.

Nitabach MN, Blau J, Holmes TC. 2002. Electrical silencing of *Drosophila* pacemaker neurons stops the free-running circadian clock. *Cell* **109:** 485–495.

Nitabach MN, Wu Y, Sheeba V, Lemon WC, Strumbos J, Zelensky PK, White BH, Holmes TC. 2006. Electrical hyperexcitation of lateral ventral pacemaker neurons desynchronizes downstream circadian oscillators in the fly circadian circuit and induces multiple behavioral periods. *J Neurosci* **26:** 479–489.

O'Connor-Giles KM, Ho LL, Ganetzky B. 2008. Nervous wreck interacts with thickveins and the endocytic machinery to attenuate retrograde BMP signaling during synaptic growth. *Neuron* **58:** 507–518.

Osterwalder T, Yoon KS, White BH, Keshishian H. 2001. A conditional tissue-specific transgene expression system using inducible GAL4. *Proc Natl Acad Sci* **98:** 12596–12601.

Packard M, Koo ES, Gorczyca M, Sharpe J, Cumberledge S, Budnik V. 2002. The *Drosophila* Wnt, wingless, provides an essential signal for pre- and postsynaptic differentiation. *Cell* **111:** 319–330.

Paradis S, Sweeney ST, Davis GW. 2001. Homeostatic control of presynaptic release is triggered by postsynaptic membrane depolarization. *Neuron* **30:** 737–749.

Peabody NC, Pohl JB, Diao F, Vreede AP, Sandstrom DJ, Wang H, Zelensky PK, White BH. 2009. Characterization of the decision network for wing expansion in *Drosophila* using targeted expression of the TRPM8 channel. *J*

Neurosci **29**: 3343–3353.

Petersen SA, Fetter RD, Noordermeer JN, Goodman CS, Diantonio A. 1997. Genetic analysis of glutamate receptors in *Drosophila* reveals a retrograde signal regulating presynaptic transmitter release. *Neuron* **19**: 1237–1248.

Rieckhof GE, Yoshihara M, Guan Z, Littleton JT. 2003. Presynaptic N-type calcium channels regulate synaptic growth. *J Biol Chem* **278**: 41099–41108.

Roman G, Endo K, Zong L, Davis RL. 2001. P{Switch}, a system for spatial and temporal control of gene expression in *Drosophila melanogaster*. *Proc Natl Acad Sci* **98**: 12602–12607.

Ruiz-Canada C, Ashley J, Moeckel-Cole S, Drier E, Yin J, Budnik V. 2004. New synaptic bouton formation is disrupted by misregulation of microtubule stability in aPKC mutants. *Neuron* **42**: 567–580.

Sanyal S, Sandstrom DJ, Hoeffer CA, Ramaswami M. 2002. AP-1 functions upstream of CREB to control synaptic plasticity in *Drosophila*. *Nature* **416**: 870–874.

Schroll C, Riemensperger T, Bucher D, Ehmer J, Voller T, Erbguth K, Gerber B, Hendel T, Nagel G, Buchner E, et al. 2006. Light-Induced activation of distinct modulatory neurons triggers appetitive or aversive learning in *Drosophila* larvae. *Curr Biol* **16**: 1741–1747.

Schuster C. 2006. Glutamatergic synapses of *Drosophila* neuromuscular junctions: A high-resolution model for the analysis of experience-dependent potentiation. *Cell Tissue Res* **326**: 287–299.

Schuster CM, Davis GW, Fetter RD, Goodman CS. 1996a. Genetic dissection of structural and functional components of synaptic plasticity. I. Fasciclin II controls synaptic stabilization and growth. *Neuron* **17**: 641–654.

Schuster CM, Davis GW, Fetter RD, Goodman CS. 1996b. Genetic dissection of structural and functional components of synaptic plasticity. II. Fasciclin II controls presynaptic structural plasticity. *Neuron* **17**: 655–667.

Sigrist SJ, Thiel PR, Reiff DF, Lachance PE, Lasko P, Schuster CM. 2000. Postsynaptic translation affects the efficacy and morphology of neuromuscular junctions. *Nature* **405**: 1062–1065.

Sigrist SJ, Reiff DF, Thiel PR, Steinert JR, Schuster CM. 2003. Experience-dependent strengthening of *Drosophila* neuromuscular junctions. *J Neurosci* **23**: 6546–6556.

Steinert JR, Kuromi H, Hellwig A, Knirr M, Wyatt AW, Kidokoro Y, Schuster CM. 2006. Experience-dependent formation and recruitment of large vesicles from reserve pool. *Neuron* **50**: 723–733.

Sweeney ST. Davis GW. 2002. Unrestricted synaptic growth in spinster-a late endosomal protein implicated in TGF-β-mediated synaptic growth regulation. *Neuron* **36**: 403–416.

Sweeney ST, Broadie K, Keane J, Niemann H, O'Kane CJ. 1995. Targeted expression of tetanus toxin light chain in *Drosophila* specifically eliminates synaptic transmission and causes behavioral defects. *Neuron* **14**: 341–351.

Takaesu NT, Herbig E, Zhitomersky D, O'Connor MB, Newfeld SJ. 2005. DNA-binding domain mutations in SMAD genes yield dominant-negative proteins or a neomorphic protein that can activate WG target genes in *Drosophila*. *Development* **132**: 4883–4894.

Waddell S, Armstrong JD, Kitamoto T, Kaiser K, Quinn WG. 2000. The *amnesiac* gene product is expressed in two neurons in the *Drosophila* brain that are critical for memory. *Cell* **103**: 805–813.

Wang X, Shaw WR, Tsang HT, Reid E, O'Kane CJ. 2007. *Drosophila* spichthyin inhibits BMP signaling and regulates synaptic growth and axonal microtubules. *Nat Neurosci* **10**: 177–185.

White BH, Osterwalder TP, Yoon KS, Joiner WJ, Whim MD, Kaczmarek LK, Keshishian H. 2001. Targeted attenuation of electrical activity in *Drosophila* using a genetically modified K+ channel. *Neuron* **31**: 699–711.

Zhang Y, Guo H, Kwan H, Wang JW, Kosek J, Lu B. 2007. PAR-1 kinase phosphorylates Dlg and regulates its postsynaptic targeting at the *Drosophila* neuromuscular junction. *Neuron* **53**: 201–215.

Zhong Y, Wu CF. 1991. Altered synaptic plasticity in *Drosophila* memory mutants with a defective cyclic AMP cascade. *Science* **251**: 198–201.

Zhong Y, Wu CF. 2004. Neuronal activity and adenylyl cyclase in environment-dependent plasticity of axonal outgrowth in *Drosophila*. *J Neurosci* **24**: 1439–1445.

Zhong Y, Budnik V, Wu CF. 1992. Synaptic plasticity in *Drosophila* memory and hyperexcitable mutants: Role of cAMP cascade. *J Neurosci* **12**: 644–651.

23 Acute Inactivation of Proteins Using FlAsH-FALI

Ron L.P. Habets and Patrik Verstreken

Laboratory of Neuronal Communication, Katholieke Universiteit Leuven, Center for Human Genetics, 3000 Leuven, Belgium; VIB, Department of Molecular and Developmental Genetics, 3000 Leuven, Belgium

ABSTRACT

Fluorescein-assisted light inactivation (FALI) is a powerful method to study acute loss of protein function, even if the corresponding mutations lead to early lethality. Here we describe FALI mediated by the membrane-permeable FlAsH (4′,5′-bis(1,3,2-dithioarsolan-2-yl)fluorescein) compound that binds with high specificity to the genetically encoded tetracysteine tag. FlAsH-FALI allows the inactivatation of protein function in vivo with exquisite spatial (<40 Å) and temporal (<30 sec) resolution and enables the analysis of kinetically distinct processes such as synaptic vesicle exocytosis and endocytosis. We describe methods to express tetracysteine-tagged proteins under endogenous control and describe how FlAsH-FALI can be performed at the neuromuscular junction of third-instar larvae.

INTRODUCTION

Loss-of-function studies in *Drosophila* are invaluable for analyzing protein function in vivo. However, numerous genes are essential and null mutants of these genes cause lethality, precluding analyses of their phenotype at later stages. Therefore, several technologies have been developed to circumvent early lethality or compensatory mechanisms that may arise during development, allowing analyses of gene function well beyond the lethal phase of the null mutant. Temperature-sensitive (ts) alleles of essential genes, producing proteins that are functional at low temperature but become inactive at high temperature, have been widely used (Suzuki et al. 1971; Poodry et al. 1973). However, it is usually not straightforward to design or isolate such alleles, and ts alleles may not always operate in a clean "on/off" manner (Rao et al. 2001). Other technologies used to analyze protein function beyond the lethal phase of the corresponding gene mutations are mitotic recombination induced by X rays or by using FLP (flippase)-FRT (flippase recognition target) (Chapter 8; Golic and Lindquist 1989; Xu and Rubin 1993; Stowers and Schwarz 1999; Newsome et al. 2000), recombination-mediated cassette exchange (RMCE; Choi et al. 2009), and RNA interference (RNAi)-mediated gene knockdown (Boutros et al. 2004; Dietzl et al. 2007). These techniques, however, suffer from long inactivation times as lingering wild-type protein must turn over before loss-of-function phenotypes will become visible. Furthermore, RMCE has thus far been performed only in a limited number of *Drosophila* tissues (Choi et al. 2009); mitotic recombination is not efficient in all cell types (e.g., in motor neurons, but see Vogler and Urban 2008); and RNAi appears in general less efficient in neurons, compared with other tissues, and may sometimes be subject to poor specificity (Ivanov et al. 2004; Neuser et al. 2008). Hence, although several methodologies to inactivate protein function in vivo are in use, each suffers from specific disadvantages.

To circumvent the long periods of time typically required to inactivate protein function, researchers have used acute pharmacological inhibition. Clearly this approach may be hampered by low specificity to the protein studied: All targets of a given pharmacological agent at a particular concentration usually are not known, and most agents have been tested to inhibit vertebrate proteins and not their *Drosophila* counterparts (Nichols 2006). Finally, chemical inhibitors for most proteins are not available. Other approaches used to acutely inactivate protein function are chromophore-assisted light inactivation (CALI) and fluorescein-assisted light inactivation (FALI). Here, non-function-blocking antibodies conjugated to malachite green (CALI) or fluorescein (FALI) are injected in the cell and target a protein of interest (Jay 1988; Surrey et al. 1998; Castelo and Jay 1999; Jay and Sakurai 1999; Liu et al. 1999; Beck et al. 2002; Horstkotte et al. 2005; Guo et al. 2006). Excitation of the photosensitizers (malachite green or fluorescein) results in production of local reactive oxygen species (ROS) that inactivate the targeted protein (Jay and Sakurai 1999; Yan et al. 2006; Hoffman-Kim et al. 2007). Although the half maximum destructive distance of FALI has been estimated to range up to 40 Å (Surrey et al. 1998; Beck et al. 2002), well below the size of an average protein, the possibility that other closely localized proteins may suffer damage remains; therefore, including the proper controls in these experiments is critical (e.g., Castelo and Jay 1999; Liu et al. 1999; Guo et al. 2006). These techniques clearly provide unprecedented spatial and temporal resolution; however, as tagged antibodies must be injected in the cell, CALI and FALI are not universally applicable and, to a large extent, depend on the specificity of the antibodies used.

A related but alternative approach came with the development of the "tag" and its membrane-permeable ligand 4',5'-bis(1,3,2-dithioarsolan-2-yl)fluorescein (FlAsH) by the Tsien group (Griffin et al. 1998, 2000; Adams et al. 2002; Gaietta et al. 2002; Tour et al. 2003; Martin et al. 2005; Madani et al. 2009). The use of these reagents has provided a widely applicable and highly efficient method to acutely inactivate proteins using FlAsH-mediated FALI in genetic model organisms including *Drosophila* (Marek and Davis 2002). With this technology a protein of interest is transgenically labeled with a short tag that binds the fluorescein derivative compound FlAsH with high specificity. Excitation of FlAsH leads to ROS production and inactivation of the tagged protein. The simplest method for expressing a tagged protein in *Drosophila* is to use UAS (upstream activation sequence)/GAL4-mediated overexpression of a modified complementary DNA (Fig. 1A) (Brand and Perrimon 1993; Marek and Davis 2002; Poskanzer et al. 2003, 2006; Heerssen et al. 2008). The use of "recombineering," however, allows the creation of tagged genes in a genomic context and the expression of proteins at endogenous levels (Fig. 1B) (Venken et al. 2006, 2008). Hence, FlAsH-FALI combines the spatial and temporal resolution of CALI and FALI with the ability to target any protein of interest using genetics, thereby circumventing the use of antibodies to target a protein of interest.

FlAsH-FALI has been used in several different organisms and tissues to study protein function following acute inactivation (Marek and Davis 2002; Poskanzer et al. 2003, 2006; Tour et al. 2003; Yan et al. 2006; Heerssen et al. 2008; Kasprowicz et al. 2008; Venken et al. 2008). A number of synaptic proteins have been targeted in *Drosophila melanogaster* using the technique. Acute inactivation of synaptotagmin facilitated further understanding of the role of this protein in synaptic vesicle exocytosis (Marek and Davis 2002), and allowed the Davis group to analyze the in vivo role of the protein in synaptic vesicle endocytosis at a live synapse (Poskanzer et al. 2003, 2006; Poskanzer and Davis 2004). Highlighting the strength of the technology in these studies, synaptotagmin could be inactivated in as little as 30 sec using wide field illumination.

More recently, FlAsH-FALI of two endocytic proteins further underscored the usefulness of the technology. Both clathrin light chain (clc) and clathrin heavy chain (chc) were inactivated using FlAsH-FALI, allowing for the first time the analysis of the role of clathrin during synaptic vesicle for-

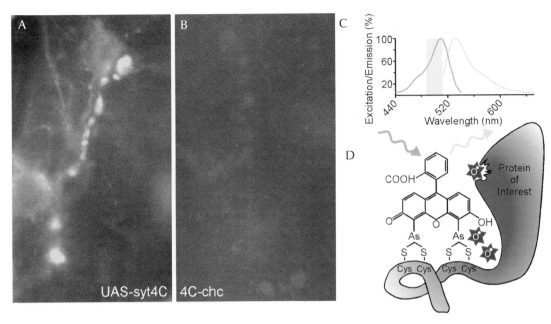

FIGURE 1. FlAsH-FALI of tetracysteine-tagged proteins under UAS and endogenous control. (*A*) Live image of a FlAsH-stained *Drosophila* third-instar NMJ, expressing tetracysteine (4C)-tagged synaptotagmin under control of UAS/C155 Gal4. (*B*) Live image of FlAsH-stained NMJ, expressing tetracysteine-tagged clathrin heavy chain under endogenous control. Images *A* and *B* were taken using the same microscope settings. Note that endogenous expression leads to much dimmer staining, but that staining can still be clearly seen. (*C*) Excitation (green) and emission (yellow) spectra for FlAsH. The width of the YFP excitation filter is shown as a light green bar. (*D*) Schematic representation of a protein fused to the tag containing the four cysteine (Cys) residues bound to the FlAsH molecule. Excitation and emission are depicted as green and yellow arrows, respectively. Reactive oxygen molecules are shown as purple stars.

mation (Heerssen et al. 2008; Kasprowicz et al. 2008; Venken et al. 2008). A block in membrane uptake had been observed in studies of many other endocytic mutants (Zhang et al. 1998; Simpson et al. 1999; Fergestad and Broadie 2001; Verstreken et al. 2002, 2003). In contrast, FlAsH-FALI of both clc and chc showed massive membrane invaginations, indicating that clathrin does not function merely as a scaffold during vesicle formation, but that the role of the clathrin lattice is to prevent uncontrolled membrane uptake during intense neuronal activity. The work also provided further evidence of the specificity of the FlAsH-FALI technology: Inactivation of both clc and chc resulted in very similar phenotypes, whereas the inactivation of synaptotagmin, shown to interact in the clathrin network (Zhang et al. 1994), reveals different phenotypes.

FlAsH-FALI is therefore a versatile method to acutely and specifically inactivate protein function in seconds, precluding developmental defects (Marek and Davis 2002). The small size of the tetracysteine tag, combined with recently developed genome manipulation methods, allows us to create transgenic animals that express FALI-ready proteins in a genomic context (Kasprowicz et al. 2008; Venken et al. 2008) and, likely in the future, from their endogenous locus (Choi et al. 2009), minimally impacting on the normal cellular and protein function.

Construction and Expression of Tetracysteine-Tagged Proteins

In this section we provide a protocol to create a tagged construct that can be used in FlAsH-FALI. Tagged proteins can be expressed using the UAS/GAL4 system, leading to overexpression of the protein. In one study, overexpression of tetracysteine-tagged clc (4C-clc) was shown to largely replace the wild-type clc, and the levels of 4C-clc at the synapse appeared similar to the endogenous levels, indicating functional replacement. Nonetheless, it is probably advisable not to overexpress proteins, to avoid artifacts, but also because inactivation of the lower overall amount of tetracysteine-tagged protein will most likely occur more quickly. We will not describe how to create tetracysteine-tagged proteins for overexpression as the tetracysteine sequence can be simply added to a polymerase chain reaction (PCR) primer and the resulting product cloned in a suitable UAS transformation vector. Instead we provide here an overview of how to create genomic constructs with a tetracysteine tag. The protocol describes the tagging of genomic constructs that are cloned in a conditional amplifiable bacterial artificial chromosome (BAC), such as P(acman), using recombineering (Venken et al. 2006). For a protocol on retrieving large stretches of DNA using recombineering, see Venken et al (2006).

Before you start:

- The methodology enables the construction of amino- or carboxy-terminally tagged proteins. Before deciding on where to insert the tag, consider the proximity of functional domains within the protein of interest. The small size of the tetracysteine tag may not interfere with protein or domain function, but it has been postulated that only domains near the tetracysteine tag (within 40 Å) will be efficiently inactivated (Beck et al. 2002), possibly leaving more distal domains intact. However, this phenomenon has not yet been observed with the proteins inactivated in *Drosophila*, and thus the likely best approach is to create one construct tagging the amino-terminal end and another tagging the carboxy-terminal end.

- Although we believe it is better to express tetracysteine-tagged proteins in a null mutant background, FlAsH-FALI may lead to a dominant loss of protein function (Marek and Davis 2002; Heerssen et al. 2008; Kasprowicz et al. 2008). Nonetheless, expression of the tetracysteine construct in a null mutant background allows you to optimally test the functionally of the tagged protein in rescue experiments, and on induction of FlAsH-FALI, will likely lead to more extensive loss of protein function.

MATERIALS

CAUTION: See Appendix for proper handling of materials marked with <!>.
See the end of the chapter for reagents marked <R>.

Reagents

CopyControl (Epicentre Biotechnologies)
    CopyControl is used to induce plasmid copy number in EPI300 cells (see Step 3).
DpnI restriction endonuclease
Escherichia coli strains: DY380 (Lee et al. 2001); EPI300 (Epicentre Biotechnologies); EL350 (Lee et al. 2001)
Enzymes for creating transgenic flies: Δ 2,3 transposase or φ-C-31 recombinase

Ethanol (70%)

Glycerol, sterile (10%)

Isopropanol <!>

L(+)-arabinose (Sigma-Aldrich A3256)

> Addition of L(+)-arabinose induces Cre expression in EL350 cells to promote excision of the neomycin cassette (see Step 4).

Luria Broth (LB) <R>

LB <R> supplemented with 10 μg/mL tetracycline and 100 μg/mL ampicillin

LB <R> supplemented with 15 μg/mL kanamycin and 100 μg/mL ampicillin

LB plates <R> supplemented with 15 μg/mL kanamycin and 100 μg/mL ampicillin

LB plates <R> supplemented with 100 μg/mL ampicillin

Minipreps or midipreps for preparation of plasmid DNA, including P1, P2, N3 buffers (QIAGEN)

PCR clean-up kit (QIAGEN) for purifying PCR products

Primers for amino-terminal fusions

> i. The left homology arm (LA) for recombineering consists of the 50 bp upstream of the start codon of the gene of interest (ATG may or may not be included); the right homology arm (RA) consists of the 50 bp downstream from the start codon. To the 5′ end of "ATGGATTACAAGGATGACGAC" (sequence homologous to the FLAG-4C end of N-FLAG-4C) add the 50 bp LA sequence (resulting primer name: *your_gene*-N-4C-F). To the 5′ end of "ACTAGTGGATCCCCTCGAGGGAC" (sequence homologous to the PL452-Neo end of N-FLAG-4C), add the 50 bp RA sequence (in reverse complement) (resulting primer name: *your_gene*-N-term-PL452-R). Both primers will be ~70 bp long. PAGE/HPLC purification are not required.
>
> ii. Screening/sequencing primers; one forward primer (*your_gene*-recocheck-F) located ~50 bp upstream of the LA sequence, and a reverse primer (*your_gene*-recocheck-R) located ~50 bp downstream from the RA sequence. Make sure that both these primers can anneal in the rescue construct present in the P(acman) vector that will be modified.
>
> iii. Cassette screening/sequencing primers:
>
> PL452-5′-Seq-R: TAAAGCGCATGCTCCAGACTG
>
> PL452-3′-Seq-F: GGTGGGCTCTATGGCTTCTGA

Primers for carboxy-terminal fusions (see Fig. 2)

> i. The LA for recombineering consists of the 50 bp before the stop codon of the gene of interest (note that the stop codon should *not* be included in the right arm); the RA consists of the 50 bp downstream from the stop codon (stop codon included). To the 5′ end of "GCAGCCCAATTCCGAT-CATATTC" (sequence homologous to the PL452-Neo end of C-FLAG-4C), add the 50 bp LA sequence (primer name: *your_gene*-C-term-PL452-F). To the 5′ end of "TTAGGGCTCCAT-GCAGCAG" (sequence homologous to the PL452-Neo end of C-FLAG-4C), add the 50 bp RA sequence (in reverse complement) (primer name: *your_gene*-C-4C-R). Both primers will be ~70 bp long. PAGE/HPLC purification are not required.
>
> ii. Screening/sequencing primers; one forward primer (*your_gene*-recocheck-F) located ~50 bp upstream of the LA sequence, and a reverse primer (*your_gene*-recocheck-R) located ~50 bp downstream from the RA sequence. Make sure that both these primers can anneal in the rescue construct present in the P(acman) vector that will be modified.
>
> iii. Cassette screening/sequencing primers:
>
> PL452-5′-Seq-R: TAAAGCGCATGCTCCAGACTG
>
> PL452-3′-Seq-F: GGTGGGCTCTATGGCTTCTGA

Reagents for amplification by PCR (polymerase, deoxyribonucleotide triphosphates [dNTPs], buffer)

Reagents for DNA sequencing of minipreps

Tris-EDTA (TE) <R>

Vectors: N-FLAG-4C or C-FLAG-4C template vector DNA, available from Addgene

> (www.addgene.com); *Your_gene* carried in P(acman) and transformed into *E. coli* DY380 cells

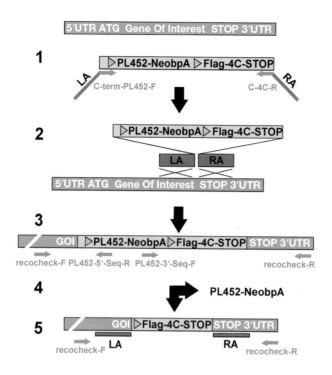

FIGURE 2. Recombineering of a carboxy-terminal Flag-tetracystein (4C) tag. The figure shows the overview of steps for creating Flag-4C-tagged proteins under control of their endogenous regulatory sequences. The numbers on the *left* correspond with the numbered steps in Protocol 1. The gene of interest (GOI) in an appropriate BAC is shown in purple; the template vector DNA for carboxy-terminal tagging is shown in yellow. LoxP sites are indicated in blue, primers in green, and the sequence homologous to the GOI in red. LA, Left arm; RA, right arm.

Equipment

Centrifuge
Incubator (preset to 30°C)
Thermal cycler for amplification (DNA engine thermal cycler, Bio-Rad, Hercules, CA)
Waterbath, shaking

METHOD

See Figure 2 for an illustration of the method; note that the numbers on the left in the figure match the numbered steps of the protocol.

1. Set up a reaction to amplify the tetracysteine cassette from the template vector.

   For amino-terminal fusions: Use N-FLAG-4C template vector DNA and primers *your_gene*-N-4C-F and *your_gene*-N-term-PL452-R.

   For carboxy-terminal fusions: Use C-FLAG-4C template vector DNA and primers *your_gene*-C-term-PL452-F and *your_gene*-C-4C-R.

   i. Amplify the template DNA using the following settings:

      One denaturation cycle: 94°C for 10 min
      35 cycles: 94°C for 30 sec; 44°C–66°C for 30 sec; 72°C for 2 sec
      One postamplification cycle: 72°C for 10 min.

   ii. Purify the PCR product by gel electrophoresis and digest contaminating (methylated) plasmid DNA with DpnI.

   iii. Purify the resulting PCR fragment and use 1–2 μL for transformation (see Step 2v).

2. Modify the gene of interest (*your_gene*) with the PCR product from Step 1 in DY380 cells.

   i. Innoculate a 5-mL culture of LB (10 μg/mL tetracycline and 100 μg/mL ampicillin) with a single colony of DY380 carrying P(acman) encoding *your_gene*. Grow the culture overnight at 30°C; dilute 300 μL of the overnight culture into each of two 15-mL tubes of LB (10 μg/mL tetracycline and 100 μg/mL ampicillin) and grow for 3 h at 30°C.

ii. Place one of the two 15-mL tubes in a 42°C shaking water bath for 15 min; maintain the other tube for 15 min at 30°C.

iii. Prepare electrocompetent cells by washing the cells of both samples twice in ice-cold sterile water (centrifuge cells, decant supernatant, and resuspend cells). Add 1 mL of 10% glycerol to each of the samples, collect the cells by centrifugation at 4°C and resuspend the pellets in 120 μL 10% glycerol (total volume). Divide each sample into two aliquots.

> You will need only one of each of the two aliquots; the other aliquot can be flash frozen and stored.

iv. Electrotransform 1–2 μL of the PCR product from Step 1 into an uninduced control sample and 1–2 μL of the PCR product into an induced sample; plate on LB plates (15 μg/mL kanamycin and 100 μg/mL ampicillin) and grow overnight at 30°C.

> More colonies should grow on the plate with induced sample.

v. Carry out colony PCR screening on colonies from the plate with the induced sample, using *your_gene*-recocheck-F/PL452-5′-Seq-R and PL452-3′-Seq-F/*your_gene*-recocheck-R.

3. Isolate the modified plasmid.

i. Innoculate a positive colony (identified in Step 2) into a 5-mL culture of LB (15 μg/mL kanamycin and 100 μg/mL ampicillin) and grow the culture overnight at 30°C.

ii. Collect the cells by centrifugation, lyse the cells, and prepare plasmid DNA using the P1, P2, and N3 reagents from QIAGEN. Centrifuge quickly once or twice to isolate supernatant without contaminants. Add an equal volume isopropanol to precipitate the DNA, centrifuge, and wash the pellet with 70% ethanol. Resuspend the DNA pellet in 50 μL TE.

iii. Transform 1 μL of DNA into electrocompetent EPI300 cells and plate on LB plates with 15 μg/mL kanamycin and 100 μg/mL ampicillin. Grow at 37°C and screen the colonies by PCR with *your_gene*-recocheck-F/PL452-5′-Seq-R and PL452-3′-Seq-F/*your_gene*-recocheck-R.

iv. Select a single colony and inoculate it into 1 mL LB (15 μg/mL kanamycin and 100 μg/mL ampicillin) and grow overnight at 37°C. Add 9 mL LB (15 μg/mL kanamycin and 100 μg/mL ampicillin) and 10 μL CopyControl to the culture and grow for 5 h longer at 37°C.

v. Isolate the modified plasmid using a miniprep or midiprep kit (or, for large constructs, special "large construct kits") and sequence the plasmid with *your_gene*-recocheck-F, PL452-5′-Seq-R, PL452-3′-Seq-F and *your_gene*-recocheck-R.

4. Remove the neomycin cassette using Cre recombinase.

i. Innoculate a single colony of EL350 into 5 mL LB, and grow overnight at 30°C.

ii. Dilute 300 μL of the overnight culture into 15 mL LB; grow for 2 h at 30°C, add 150 μL 10% L(+)-arabinose, and grow for 1 h at 30°C.

iii. Prepare electrocompetent cells as described in Step 2iv. You will need to use only one aliquot; the other can be snap frozen and stored.

iv. Transform 1 μL of a 1:100 dilution of the modified P(acman) plasmid (from Step 3v) into electrocompetent EL350 cells, and plate on LB plates (100 μg/mL ampicillin).

v. Verify removal of the neomycin cassette by screening with colony PCR using *your_gene*-recocheck-F and *your_gene*-recocheck-R.

5. Isolate the modified tetracysteine-tagged plasmid without neomycin marker as described in Step 3 and create transgenic flies by using either δ 2,3 transposase or φ-C-31 recombinase with the appropriate host strain (for details, see Venken et al. 2008).

Protocol 2

FlAsH-FALI at the Third-Instar Neuromuscular Junction

This protocol describes the different steps needed for efficient inactivation of a protein using FlAsH-FALI at the neuromuscular junction (NMJ) of third-instar larvae. Note that FlAsH-FALI in other tissues is also theoretically possible with minor adaptations to the protocol described here. We explain controls for positional effects, for unspecific FlAsH binding to endogenous proteins, and for phototoxicity. Following FlAsH-FALI, protein function can be studied using a number of secondary assays, many of which are described in this volume, including electrophysiology (Chapter 12), immunohistochemistry (Chapter 10), and electron microscopy or FM1-43 labeling of synaptic vesicle pools (Verstreken et al. 2008).

Before you start: The genetic control larvae should be grown under the same conditions as the experimental groups (same temperature and food). Compare with the phenotype of experimental flies that have not been treated with FlAsH and have not been illuminated to the wild-type negative control.

Consider the following additional specific controls for FlAsH-mediated FALI.

1. Control for unspecific FlAsH binding to nontagged proteins. Treat an appropriate wild-type negative control with FlAsH and illuminate the preparation (details see below).

2. Control for phototoxic effects. Omit FlAsH and illuminate the preparation containing your tetracysteine-tagged protein.

3. Control for specificity of phenotypes. Use FlAsH-FALI to inactivate an unrelated tetracysteine-tagged protein and test the specificity of the recorded phenotype to your protein of interest.

MATERIAL

CAUTION: See Appendix for proper handling of materials marked with <!>.
See the end of the chapter for reagents marked <R>.

Reagents

BAL buffer (2,3-Dimercapto-1-propanol, 25 mM, 100x stock, included in T34561kit, Invitrogen)
   Store the BAL buffer in small aliquots in –20°C
Drosophila third-instar larvae
FlAsH-EDT$_2$ ($C_{24}H_{18}O_5S_4As_2$; Fig. 1D)
HL-3, freshly prepared (from powder) the day of the experiment <R>
In-cell Tetracysteine Tag Detection Kit for live-cell imaging (product number T34561, Invitrogen), 2 mM, 2000x stock in DMSO (dimethyl sulfoxide) <!>
   Store the FlAsH solution in small aliquots at –20°C in the dark.

Equipment

Culture dish, 35–55-mm-diameter (product number 351008, VWR International), filled halfway with Sylgard (product number SYLG184, World Precision Instruments [WPI], Sarasota, FL)
Fluorescent microscope (Eclipse F1, Nikon, Japan) equipped with 40x 0.8-NA (numerical aperture) water-immersion objective (Nikon); Intensilight C-HGFI excitation source (Nikon); YFP

filter set (excitation filter [500/24 nm], dichroic mirror [520 nm], and emission filter [542/27 nm], Semrock, Rochester, NY)

Insect pins (product number 26002-10, Fine Science Tools, Foster City, CA)
   Cut most of the top part of the pins and use the sharp bottom part.

Shaker (Unimax 1010, product number 444-1310, VWR International)

Stereo microscope with 8X–50X magnification (SMZ660, Nikon, Japan)

Tweezer, Dumont #3 (product number 14095, WPI)

Tweezer, Dumont #5 (product number 500338, WPI)

Vannas scissors, 0.025 × 0.015-mm (product number 500086, WPI)

METHOD

1. Dissect third-instar larvae in HL-3 as described in Verstreken et al. (2008) and summarized here.

   i. Place a larva in a Sylgard-covered culture dish and add ~2–3 mL HL-3. Place the culture dish under the stereo microscope.

   ii. Turn a third-instar larvae dorsal side up using the #3 tweezers. Stretch the larva by placing one pin anteriorly and one posteriorly, taking care that the mouth hooks are extended when placing the anterior pin.

   iii. Using a third pin, rub gently on the dorsal skin in the middle of the larvae until a small hole forms.

   iv. Using the dissection scissors, insert one of the blades in the hole and make a longitudinal cut across the dorsal side of the larva in both directions (posterior and anterior). Take care not to damage the tissue on the ventral side of the larva.

   v. Remove the fat tissue and intestinal tract of the larva.

   vi. Using four additional pins, spread the larva open.

   vii. Using the #5 tweezers, remove the two main trachea.
      Do NOT cut the motor nerves.

2. Wash the dissected larvae with HL-3 to remove debris left after removing the digestive system and fat tissue.

3. Dilute 0.5 μL FlAsH-EDT$_2$ (from a 2000X [2 mM] stock) into 999.5 μL HL3.

4. Incubate for 10 min on the shaker at 75 rotations per minute. Keep the preparation in the dark by covering it with a small box or aluminum foil.

5. Wash the larvae with HL-3 (wash quickly to remove most of the FlAsH).

6. Dilute 10 μL BAL buffer from a 100X (25 mM) stock into 990 μL HL-3. Wash the larvae with the diluted BAL buffer for 5 min on the shaker.

7. Wash the larvae 3X with HL-3.

8. Cut the motor nerves (if necessary for the experiment).

   i. It is not advisable to cut the motor nerves that emerge from the ventral nerve cord during or before FlAsH-FALI labeling. We noticed that the preparation stays healthier when the nerves remain intact during this period.

   ii. If needed (e.g., when performing FM1-43 labeling experiments or electrophysiological measurements), motor nerves can at this point be carefully severed by cutting them with the dissection scissors, taking care to cut all the nerves while not touching the muscles that lie below.

9. Transfer the culture dish containing the dissected larvae to the microscope, locate the desired muscle in the desired segment, and focus on the muscle.

   We regularly use muscle 6 and 7 or muscle 4 in segments A3 and A4.

10. Illuminate NMJs using epifluorescent light (YFP filter emission; Fig. 1C) for ≤ 10 min.

    For all proteins studied using FlAsH-FALI, 10 min is more than sufficient to inactivate most of the tagged protein; it is, however, advisable to perform a time series to identify the ideal inactivation period.

    See Troubleshooting.

TROUBLESHOOTING

Problem (Step 10): No phenotype is observed.
Solutions:

1. As with any other mutant, it is crucial to determine the appropriate assay. No observed phenotype does not necessarily mean that your protein is not inactivated (Heerssen et al. 2008; Kasprowicz et al. 2008).

2. Even if the phenotype of the null or hypomorphic mutant is known, acute disruption of the protein could yield a different phenotype because developmental compensatory mechanisms did not occur.

3. Determine if the fusion protein is expressed and localized correctly. FlAsH bound to a tetracysteine tag is visible in unfixed larvae and can be imaged using YFP excitation/emission filters (fluorescence will be visible before inactivation; Fig. 1A,B) (Marek and Davis 2002; Heerssen et al. 2008; Kasprowicz et al. 2008).

4. If the tagging method described here was followed and the FlAsH fluorescence is too dim, it is also possible to fix the preparation and use anti-FLAG antibodies to label and localize the tagged protein in situ or to determine expression using western blotting (Kasprowicz et al. 2008).

DISCUSSION

In this chapter we describe a protocol to tag proteins with a FlAsH binding site as well as a method that allows acute inactivation of the tagged protein using FlAsH-FALI. The method is efficient, specific, and should in principle be applicable to every protein in the *Drosophila* proteome. We describe inactivation of proteins at the third-instar neuromuscular junction; however, it is relatively easy to adapt the protocol for experiments in different tissues, such as photoreceptors (Venken et al. 2008). Given that the correct controls for specificity are performed, we believe that FlAsH-FALI complements and significantly extends the existing toolkit to study *Drosophila* proteins and is ideally suited to analyze the acute loss of protein function.

Nonspecific binding of FlAsH to cysteine-rich proteins has been reported in cell culture (Stroffekova et al. 2001; Tour et al. 2003; Berens et al. 2005; Hearps et al. 2007) and was shown to be concentration dependent, resulting in up to 21% nonspecific protein inactivation (Tour et al. 2003). In *Drosophila*, however, such nonspecific protein inactivation has not been reported. Neurotransmitter release measured using electrophysiology in wild-type *Drosophila* treated with FlAsH and illuminated were identical to wild-type animals not treated with FlAsH (Marek and Davis 2002; Heerssen et al. 2008; Kasprowicz et al. 2008). In addition, synaptic vesicle cycling visualized using FM1-43 and synaptic ultrastructure determined by electron microscopy of FlAsH-treated wild-type NMJs that were illuminated did not reveal overt differences from nontreated controls (Heerssen et al. 2008; Kasprowicz et al. 2008). These data suggest that, based on the analyses performed, in *Drosophila* FlAsH does not cause significant toxicity, in contrast to data in cell culture

(Langhorst et al. 2006). Furthermore, illumination of FlAsH-treated *Drosophila* synapses that do not express a tetracysteine-tagged protein does not cause obvious synaptic defects.

FlAsH-FALI is mediated by creation of ROS (Fig. 1D), and although the inactivation radius of FALI was estimated to be on the order of 4 nm (Beck et al. 2002), it is, in general, difficult to ascertain local, protein-specific inactivation. Indeed, most proteins are active in complexes or bind lipids that may also be oxidized by membrane lipids. Although the actual inactivation radius of FlAsH-FALI has not been determined in vivo, where ROS quenchers may limit the inactivating effect, in *Drosophila* available evidence does not indicate off-target effects of FlAsH-FALI, even if the proteins are present in complexes or bind lipids. Indeed, FlAsH-FALI of synaptotagmin, a membrane-bound protein involved in exocytosis and endocytosis of synaptic vesicles, results in almost identical phenotypes to those observed in *synaptotagmin* null mutants (Littleton et al. 1993; Marek and Davis 2002). Furthermore, inactivation of both clc and chc independently show very similar phenotypes. Although synaptotagmin is found in a multiprotein complex with clathrin (Zhang et al. 1994), FlAsH-FALI of both proteins show very different phenotypes in multiple assays (Marek and Davis 2002; Heerssen et al. 2008; Kasprowicz et al. 2008). Together, these data indicate local and specific inactivation of tetracysteine-tagged proteins in *Drosophila*.

We have described here the specific use of FlAsH in combination with the tetracysteine tag to inactivate protein function. However, a number of additional developments allow for further analyses of tetracysteine-tagged proteins. The tetracysteine tag binds both the green/yellow fluorescent FlAsH as well as the red fluorescent ReAsH, allowing for pulse-chase experiments and fluorescence resonance energy transfer (Gaietta et al. 2002, 2006). Furthermore, ReAsH-mediated photoconversion of diaminobenzidine into electron-dense precipitates enables one to analyze tagged protein localization at ultrastructural resolution using electron microscopy (Gaietta et al. 2006). Finally, new superior fluorophores that bind tetracysteine tags, as well as recent improvements of genetically encoded tags that bind these fluorophores, will undoubtedly further expand on the applications of this technology and should allow one to inactivate different proteins at different time points using specific wavelength light (Marks et al. 2004; Cao et al. 2006; Marks and Nolan 2006; Spagnuolo et al. 2006; Hauser and Tsien 2007; Tour et al. 2007; Genin et al. 2008; Lin et al. 2008; Taguchi et al. 2009).

RECIPES

CAUTION: See Appendix for proper handling of materials marked with <!>.
Recipes for reagents marked with <R> are included in this list.

HL-3

Reagent	Final concentration (mM)
NaCl	110
KCl <!>	5
NaHCO$_3$	10
HEPES	5
Sucrose	30
Trehalose	5
MgCl$_2$ <!>	10

Adjust the pH to 7.2 and the final volume to 1 L. Prepare fresh (from powder) the day of the experiment.

LB Medium

Reagent	Amount per liter
Tryptone	10 g
NaCl	10 g
Yeast extract	5 g
H_2O	950 mL

Combine the reagents and shake until the solutes have dissolved. Adjust the pH to 7.0 with 5 N NaOH <!> (~0.2 mL). Adjust the final volume of the solution to 1 L with H_2O. Distribute into bottles and sterilize by autoclaving for 20 min at 15 psi (1.05 kg/cm^2) on liquid cycle.

LB Agar Plates

Prepare LB (see previous recipe). Just before autoclaving, add 15 g/L Bacto Agar and sterilize by autoclaving for 20 min at 15 psi (1.05 kg/cm^2) on liquid cycle. Swirl the medium gently to distribute the melted agarose. Allow the medium to cool to ~50°C–60°C before adding thermolabile substances such as antibiotics. Swirl again to mix and pour into Petri dishes (~35 mL/100-mm plate).

TE

Reagent	Quantity (for 100 mL)	Final concentration
EDTA (0.5 M, pH 8.0)	0.2 mL	1 mM
Tris-Cl (1 M, pH 8.0)	1 mL	10 mM
H_2O	to 100 mL	

ACKNOWLEDGMENTS

We thank Koen Venken (Baylor College of Medicine, Houston, TX) and Jarek Kasprowicz for their help and members of the Verstreken Laboratory for critical comments. Work in the Verstreken Laboratory is supported by VIB, the Research Fund KULeuven, Fonds voor Wetenschappelijk Onderzoek Vlaanderen (FWO #G.0747.09), Instituut voor Wetenschap en Technologie (IWT), and a Marie Curie Excellence grant (#MEXT-CT-2006-042267).

REFERENCES

Adams SR, Campbell RE, Gross LA, Martin BR, Walkup GK, Yao Y, Llopis J, Tsien RY. 2002. New biarsenical ligands and tetracysteine motifs for protein labeling in vitro and in vivo: Synthesis and biological applications. *J Am Chem Soc* **124:** 6063–6076.

Beck S, Sakurai T, Eustace BK, Beste G, Schier R, Rudert F, Jay DG. 2002. Fluorophore-assisted light inactivation: A high-throughput tool for direct target validation of proteins. *Proteomics* **2:** 247–255.

Berens W, Van Den Bossche K, Yoon TJ, Westbroek W, Valencia JC, Out CJ, Naeyaert JM, Hearing VJ, Lambert J. 2005. Different approaches for assaying melanosome transfer. *Pigment Cell Res* **18:** 370–381.

Boutros M, Kiger AA, Armknecht S, Kerr K, Hild M, Koch B, Haas SA, Paro R, Perrimon N. 2004. Genome-wide RNAi analysis of growth and viability in *Drosophila* cells. *Science* **303:** 832–835.

Brand AH, Perrimon N. 1993. Targeted gene expression as a means of altering cell fates and generating dominant phenotypes. *Development* **118:** 401–415.

Cao H, Chen B, Squier TC, Mayer MU. 2006. CrAsH: A biarsenical multi-use affinity probe with low non-specific fluorescence. *Chem Commun (Camb)* **2006:** 2601–2603.

Castelo L, Jay DG. 1999. Radixin is involved in lamellipodial stability during nerve growth cone motility. *Mol Biol Cell* **10**: 1511–1520.

Choi CM, Vilain S, Langen M, van Kelst S, De Geest N, Yan J, Verstreken P, Hassan BA. 2009. Conditional mutagenesis in *Drosophila*. *Science* **324**: 54.

Dietzl G, Chen D, Schnorrer F, Su KC, Barinova Y, Fellner M, Gasser B, Kinsey K, Oppel S, Scheiblauer S, et al. 2007. A genome-wide transgenic RNAi library for conditional gene inactivation in *Drosophila*. *Nature* **448**: 151–156.

Fergestad T, Broadie K. 2001. Interaction of stoned and synaptotagmin in synaptic vesicle endocytosis. *J Neurosci* **21**: 1218–1227.

Gaietta G, Deerinck TJ, Adams SR, Bouwer J, Tour O, Laird DW, Sosinsky GE, Tsien RY, Ellisman MH. 2002. Multicolor and electron microscopic imaging of connexin trafficking. *Science* **296**: 503–507.

Gaietta GM, Giepmans BNG, Deerinck TJ, Smith WB, Ngan L, Llopis J, Adams SR, Tsien RY, Ellisman MH. 2006. Golgi twins in late mitosis revealed by genetically encoded tags for live cell imaging and correlated electron microscopy. *Proc Natl Acad Sci* **103**: 17777–17782.

Genin E, Carion O, Mahler B, Dubertret B, Arhel N, Charneau P, Doris E, Mioskowski C. 2008. CrAsH—Quantum dot nanohybrids for smart targeting of proteins. *J Am Chem Soc* **130**: 8596–8597.

Golic KG, Lindquist S. 1989. The FLP recombinase of yeast catalyzes site-specific recombination in the *Drosophila* genome. *Cell* **59**: 499–509.

Griffin BA, Adams SR, Tsien RY. 1998. Specific covalent labeling of recombinant protein molecules inside live cells. *Science* **281**: 269–272.

Griffin BA, Adams SR, Jones J, Tsien RY. 2000. Fluorescent labeling of recombinant proteins in living cells with FlAsH. *Methods Enzymol* **327**: 565–578.

Guo J, Chen H, Puhl HL III, Ikeda SR. 2006. Fluorophore-assisted light inactivation produces both targeted and collateral effects on N-type calcium channel modulation in rat sympathetic neurons. *J Physiol* **576**: 477–492.

Hauser CT, Tsien RY. 2007. A hexahistidine-Zn^{2+}-dye label reveals STIM1 surface exposure. *Proc Natl Acad Sci* **104**: 3693–3697.

Hearps AC, Pryor MJ, Kuusisto HV, Rawlinson SM, Piller SC, Jans DA. 2007. The biarsenical dye Lumio exhibits a reduced ability to specifically detect tetracysteine-containing proteins within live cells. *J Fluoresc* **17**: 593–597.

Heerssen H, Fetter RD, Davis GW. 2008. Clathrin dependence of synaptic-vesicle formation at the *Drosophila* neuromuscular junction. *Curr Biol* **18**: 401–409.

Hoffman-Kim D, Diefenbach TJ, Eustace BK, Jay DG. 2007. Chromophore-assisted laser inactivation. *Methods Cell Biol* **82**: 335–354.

Horstkotte E, Schröder T, Niewöhner J, Thiel E, Jay DG, Henning SW. 2005. Toward understanding the mechanism of chromophore-assisted laser inactivation—Evidence for the primary photochemical steps. *Photochem Photobiol* **81**: 358–366.

Ivanov AI, Rovescalli AC, Pozzi P, Yoo S, Mozer B, Li HP, Yu SH, Higashida H, Guo V, Spencer M, Nirenberg M. 2004. Genes required for *Drosophila* nervous system development identified by RNA interference. *Proc Natl Acad Sci* **101**: 16216–16221.

Jay DG. 1988. Selective destruction of protein function by chromophore-assisted laser inactivation. *Proc Natl Acad Sci* **85**: 5454–5458.

Jay DG, Sakurai T. 1999. Chromophore-assisted laser inactivation (CALI) to elucidate cellular mechanisms of cancer. *Biochim Biophys Acta* **1424**: M39–M48.

Kasprowicz J, Kuenen S, Miskiewicz K, Habets RLP, Smitz L, Verstreken P. 2008. Inactivation of clathrin heavy chain inhibits synaptic recycling but allows bulk membrane uptake. *J Cell Biol* **182**: 1007–1016.

Langhorst MF, Genisyuerek S, Stuermer CA. 2006. Accumulation of FlAsH/Lumio Green in active mitochondria can be reversed by β-mercaptoethanol for specific staining of tetracysteine-tagged proteins. *Histochem Cell Biol* **125**: 743–747.

Lee EC, Yu D, Martinez de Velasco J, Tessarollo L, Swing DA, Court DL, Jenkins NA, Copeland NG. 2001. A highly efficient *Escherichia coli*-based chromosome engineering system adapted for recombinogenic targeting and subcloning of BAC DNA. *Genomics* **73**: 56–65.

Lin MZ, Glenn JS, Tsien RY. 2008. A drug-controllable tag for visualizing newly synthesized proteins in cells and whole animals. *Proc Natl Acad Sci* **105**: 7744–7749.

Littleton JT, Stern M, Schulze K, Perin M, Bellen HJ. 1993. Mutational analysis of *Drosophila* synaptotagmin demonstrates its essential role in Ca^{2+}-activated neurotransmitter release. *Cell* **74**: 1125–1134.

Liu CWA, Lee G, Jay DG. 1999. Tau is required for neurite outgrowth and growth cone motility of chick sensory neurons. *Cell Motil Cytoskelet* **43**: 232–242.

Madani F, Lind J, Damberg P, Adams SR, Tsien RY, Graslund AO. 2009. Hairpin structure of a biarsenical-tetracysteine motif determined by NMR spectroscopy. *J Am Chem Soc* **131**: 4613–4615.

Marek KW, Davis GW. 2002. Transgenically encoded protein photoinactivation (FlAsH-FALI): Acute inactivation of synaptotagmin I. *Neuron* **36**: 805–813.

Marks KM, Nolan GP. 2006. Chemical labeling strategies for cell biology. *Nat Methods* **3**: 591–596.

Marks KM, Braun PD, Nolan GP. 2004. A general approach for chemical labeling and rapid, spatially controlled protein inactivation. *Proc Natl Acad Sci* **101:** 9982–9987.

Martin BR, Giepmans BNG, Adams SR, Tsien RY. 2005. Mammalian cell-based optimization of the biarsenical-binding tetracysteine motif for improved fluorescence and affinity. *Nat Biotechnol* **23:** 1308–1314.

Neuser K, Triphan T, Mronz M, Poeck B, Strauss R. 2008. Analysis of a spatial orientation memory in *Drosophila*. *Nature* **453:** 1244–1247.

Newsome TP, Asling B, Dickson BJ. 2000. Analysis of *Drosophila* photoreceptor axon guidance in eye-specific mosaics. *Development* **127:** 851–860.

Nichols CD. 2006. *Drosophila melanogaster* neurobiology, neuropharmacology, and how the fly can inform central nervous system drug discovery. *Pharmacol Ther* **112:** 677–700.

Poodry CA, Hall L, Suzuki DT. 1973. Developmental properties of Shibire: A pleiotropic mutation affecting larval and adult locomotion and development. *Dev Biol* **32:** 373–386.

Poskanzer KE. Davis GW. 2004. Mobilization and fusion of a non-recycling pool of synaptic vesicles under conditions of endocytic blockade. *Neuropharmacology* **47:** 714–723.

Poskanzer KE, Marek KW, Sweeney ST, Davis GW. 2003. Synaptotagmin I is necessary for compensatory synaptic vesicle endocytosis in vivo. *Nature* **426:** 559–563.

Poskanzer KE, Fetter RD, Davis GW. 2006. Discrete residues in the C_2B domain of synaptotagmin I independently specify endocytic rate and synaptic vesicle size. *Neuron* **50:** 49–62.

Rao SS, Stewart BA, Rivlin PK, Vilinsky I, Watson BO, Lang C, Boulianne G, Salpeter MM, Deitcher DL. 2001. Two distinct effects on neurotransmission in a temperature-sensitive SNAP-25 mutant. *EMBO J* **20:** 6761–6771.

Simpson F, Hussain NK, Qualmann B, Kelly RB, Kay BK, McPherson PS, Schmid SL. 1999. SH3-domain-containing proteins function at distinct steps in clathrin-coated vesicle formation. *Nat Cell Biol* **1:** 119–124.

Spagnuolo CC, Vermeij RJ, Jares-Erijman EA. 2006. Improved photostable FRET-competent biarsenical-tetracysteine probes based on fluorinated fluoresceins. *J Am Chem Soc* **128:** 12040–12041.

Stowers RS, Schwarz TL. 1999. A genetic method for generating *Drosophila* eyes composed exclusively of mitotic clones of a single genotype. *Genetics* **152:** 1631–1639.

Stroffekova K, Proenza C, Beam KG. 2001. The protein-labeling reagent FLASH-EDT$_2$ binds not only to CCXXCC motifs but also non-specifically to endogenous cysteine-rich proteins. *Pflugers Arch* **442:** 859–866.

Surrey T, Elowitz MB, Wolf PE, Yang F, Nédélec F, Shokat K, Leibler S. 1998. Chromophore-assisted light inactivation and self-organization of microtubules and motors. *Proc Natl Acad Sci* **95:** 4293–4298.

Suzuki DT, Grigliatti T, Williamson R. 1971. Temperature-sensitive mutations in *Drosophila melanogaster*. VII. A mutation (para[ts]) causing reversible adult paralysis. *Proc Natl Acad Sci* **68:** 890–893.

Taguchi Y, Shi ZD, Ruddy B, Dorward DW, Greene L, Baron GS. 2009. Specific biarsenical labeling of cell surface proteins allows fluorescent- and biotin-tagging of amyloid precursor protein and prion proteins. *Mol Biol Cell* **20:** 233–244.

Tour O, Meijer RM, Zacharias DA, Adams SR, Tsien RY. 2003. Genetically targeted chromophore-assisted light inactivation. *Nat Biotechnol* **21:** 1505–1508.

Tour O, Adams SR, Kerr RA, Meijer RM, Sejnowski TJ, Tsien RW, Tsien RY. 2007. Calcium Green FlAsH as a genetically targeted small-molecule calcium indicator. *Nat Chem Biol* **3:** 423–431.

Venken KJ, He Y, Hoskins RA, Bellen HJ. 2006. P[acman]: A BAC transgenic platform for targeted insertion of large DNA fragments in *D. melanogaster*. *Science* **314:** 1747–1751.

Venken KJT, Kasprowicz J, Kuenen S, Yan J, Hassan BA, Verstreken P. 2008. Recombineering-mediated tagging of *Drosophila* genomic constructs for in vivo localization and acute protein inactivation. *Nucleic Acids Res* **36:** e114. doi:10.1093/ nar/gkn486

Verstreken P, Kjaerulff O, Lloyd TE, Atkinson R, Zhou Y, Meinertzhagen IA, Bellen HJ. 2002. Endophilin mutations block clathrin-mediated endocytosis but not neurotransmitter release. *Cell* **109:** 101–112.

Verstreken P, Koh TW, Schulze KL, Zhai RG, Hiesinger PR, Zhou Y, Mehta SQ, Cao Y, Roos J, Bellen HJ. 2003. Synaptojanin is recruited by endophilin to promote synaptic vesicle uncoating. *Neuron* **40:** 733–748.

Verstreken P, Ohyama T, Bellen HJ. 2008. FM 1-43 labeling of synaptic vesicle pools at the *Drosophila* neuromuscular junction. *Methods Mol Biol* **440:** 349–369.

Vogler G, Urban J. 2008. The transcription factor Zfh1 is involved in the regulation of neuropeptide expression and growth of larval neuromuscular junctions in *Drosophila melanogaster*. *Dev Biol* **319:** 78–85.

Xu T, Rubin GM. 1993. Analysis of genetic mosaics in developing and adult *Drosophila* tissues. *Development* **117:** 1223–1237.

Yan P, Xiong Y, Chen B, Negash S, Squier TC, Mayer MU. 2006. Fluorophore-assisted light inactivation of calmodulin involves singlet-oxygen mediated cross-linking and methionine oxidation. *Biochemistry* **45:** 4736–4748.

Zhang JZ, Davletov BA, Südhof TC, Anderson RG. 1994. Synaptotagmin I is a high affinity receptor for clathrin AP-2: Implications for membrane recycling. *Cell* **78:** 751–760.

Zhang B, Koh YH, Beckstead RB, Budnik V, Ganetzky B, Bellen HJ. 1998. Synaptic vesicle size and number are regulated by a clathrin adaptor protein required for endocytosis. *Neuron* **21:** 1465–1475.

24 Studying Behavior in *Drosophila*
An Introduction to Section 3

Scott Waddell

Department of Neurobiology, University of Massachusetts Medical School, Worcester, Massachusetts 01605

Animal behavior is investigated from a variety of vantage points by researchers representing several disciplines including psychology, physiology, zoology, anatomy, ecology, and genetics. Despite this apparent segregation of the investigators, the same grand questions motivate inquiry: What is the cause and purpose of behavior? How does the brain/mind control behavior? What are the underlying physiological mechanisms that provide the control?

Insects have long proven to be popular subjects for the study of both innate and adaptive behavior; however, the basis for this interest now extends well beyond the age-old belief that they are simply automata. Insects have elaborate sensory systems, are affected by internal state, and exhibit a wide array of stereotyped and nonstereotyped behaviors. They display several phenomena that, from our human standpoint, we would describe as thirst, hunger, lust, aggression, attention, sleep, and learning and memory. The fact that these same terms have been central to decades of debate by psychologists to explain the purpose and driving force of animal behavior in general encourages one to believe that studying insects can indeed inform us of the driving forces of behavior.

Although this book focuses on the fruit fly *Drosophila melanogaster*, it is important to recognize that fruit flies are not the best "model" insect for the study of all behaviors. Seminal work has been performed using ants, bees, locusts, butterflies, crickets, blowflies, etc. Thus far, social structure, sophisticated foraging strategies, migration and navigation skills, and elaborate methods of communication are the realm of some of these bigger bugs. Furthermore, there were, and often are, practical reasons to study larger insects. For example, the larger the brain and the individual neurons, the easier it is to implant multiple electrodes for electrophysiological studies.

Nevertheless, many investigators have taken Seymour Benzer's lead and turned to studying behavior in *Drosophila*. There is no doubt that genetics provides a unique precision with which to peer inside the workings of the brain and potentially brings our understanding of the underlying mechanisms to the resolution of individual molecules. Furthermore, recent technological advances have allowed investigators to utilize both electro- and optophysiological analyses in live intact animals, allowing them to monitor activity and specific cell-signaling processes within defined individual neurons in the brains of behaving, although tethered, animals.

Despite the very great excitement that comes with technological advance, it remains important to recognize the relative strengths and weaknesses of studying behavior in a particular way. Although the assays described in the forthcoming chapters are used routinely in laboratories around the world, this indicates by no means that they are the best and only way to study behavior in flies. The reader should view all the protocols with a cynical eye and question whether the analysis is optimal, well controlled, and adequate to draw the desired conclusions. Also, animals do very unique things within their natural environment that they may not exhibit in the laboratory. Conversely, the laboratory lacks the dimension of field studies but experiments can often be designed that take advantage of the artificial environment and allow one to ask very specific questions in a well-controlled manner. Animals can of course influence each other and therefore may behave very differently when alone than when in groups. In this section of the book the reader will encounter examples of behavioral assays monitoring single, pairs of, and populations of flies or larvae. It is worth questioning, however, whether the conclusions drawn are ethologically relevant, and ultimately the suitability of a particular means of study depends on the nature of the question the investigator wishes to ask.

Most behaviors are a series of temporally organized maneuvers and these sequences are classically described using an ethogram as a sort of catalog. It is always worthwhile to describe the behavior of interest in this way initially, and then to make the nontrivial decision of which feature(s) of the behavior to quantify. In some cases the reader will notice that behavioral "end points" are quantified. For example, when studying courtship one might display the percentage of flies copulating in a given time, but this score does not inform how the flies got to that point or whether they went through the usual steps of the courtship ritual. Other assays have developed an "index" as a measure of performance (e.g., courtship index or a learning [or performance] index). Scoring behavior with an index is useful because it allows one to perform a statistical analysis on the numbers but it unavoidably omits the richness of the behavior behind the analysis. Of equal importance to developing or having a quantifiable behavioral assay, one must devise appropriate control assays to measure task-relevant accessory behaviors (e.g., locomotion, olfactory acuity, vision). Only by including these controls can one conclude a certain mechanism underlies the observed mutant defect.

For many investigators, the primary purpose of designing a new quantifiable *Drosophila* behavior assay is to provide a mutant screen. The ability to quantify the effects of genetic lesions is critical unless one is looking for an overt gain-of-function behavior. For example, initial behavioral screens for circadian rhythm were particularly successful in revealing the core molecular clockwork that has subsequently been found to be mostly conserved in mammals. Analysis of mutants has also revealed key neural circuits regulating several behaviors.

One important choice to make when studying *Drosophila* behavior is whether to study larvae or adult flies. Several assays are available for either of these life stages, and examples are included here. Although larvae have a simpler nervous system, they have a less elaborate repertoire of behaviors and their behavior is unavoidably linked to ongoing development. Both the body and nervous system continuously grow as the animal transitions through the three instars and prepares for pupariation. It is therefore worthwhile to consider the goals of the research before choosing the life stage to study. Obviously, one should use adult flies to study certain behaviors, such as flight, courtship, and memory that lasts for several days.

Another important consideration for behavior and genetic analysis in general is the issue of "genetic background." Behavioral responses are highly polygenic and therefore, when analyzing strains from different sources, one must exercise caution that the comparisons are made between or among the appropriate groups. Ideally strains to be compared should be backcrossed for at least five generations to the appropriate wild-type flies with a defined genetic background before testing. Unfortunately, which "best" wild-type strain to choose depends on the assay to be used. Tully and Quinn (1985) found that some "wild-type" strains exhibit performance that is equivalent to mutant flies in their olfactory conditioning assay. Paradoxically, one of the strains that performs poorly in the olfactory task is the best performer in a visual learning assay (Wolf and Heisenberg 1991).

The chapters in this section introduce procedures to investigate and quantify a variety of *Drosophila* behaviors. Chapters 25–27 cover olfactory and visually guided behavior and conditioning paradigms to study learning and memory in adult and larval forms. Chapters 28 and 29 discuss analyses of the complex social behaviors of aggression and courtship. These innate behaviors can also be modified by experience, and Chapter 30 describes methods to measure courtship plasticity. Fly activity follows a circadian rhythm, and flies that are in an extended period of behavioral quiescence exhibit many hallmarks of sleep. Analysis of circadian rhythm and sleep is described in Chapter 31. Last, focusing on analyses in larvae, the other essential homeostatically regulated behavior of feeding is discussed in Chapter 32.

How far can the study of *Drosophila* behavior take us? It can certainly tell us a lot about why and how flies behave, but is this information useful to our understanding of behavior in general? There is no doubt that analysis of mutant flies can uncover conserved molecular mechanisms that are both sensory and central. Further, in recent years attention has shifted also toward understanding the neural circuit context in which these molecular mechanisms function. It seems likely that organizing principles of neural circuit function will emerge and that at least some of them will be useful to our understanding of circuit function in higher animals. Historically those studying animal behavior have tended to anthropomorphize—that is, to try to explain the behavior of other animals in very human terms. The extent to which this is necessary and helpful is the subject of ongoing debate, but it seems plausible that some of the physiological manifestations of thirst, hunger, lust, aggression, sleep, fear, and pain, but not the subjective "feelings" associated with these states, will exist in the fly brain. Some of us believe the investigator largely sets the boundaries and that flies will also enrich our understanding of how such apparently complex concepts such as motivation and attention can be represented in the brain.

REFERENCES

Tully T, Quinn WG. 1985. Classical conditioning and retention in normal and mutant *Drosophila melanogaster. J Comp Physiol A* **157:** 263–277.

Wolf R, Heisenberg M. 1991. Basic organization of operant behavior as revealed in *Drosophila* flight orientation. *J Comp Physiol A* **167:** 269–283.

25 | Visual Learning and Perception in *Drosophila*

Bruno van Swinderen

Queensland Brain Institute, The University of Queensland, Brisbane, Queensland 4072, Australia

ABSTRACT

The mechanisms of vision in *Drosophila* can now be studied in individuals and in populations of flies by using various paradigms. This chapter presents three basic strategies for conducting visual perception and learning studies: individual studies performed on single flies on solid supports (larvae on agar or adults in a T-maze) using a light/dark association paradigm; population studies performed on flies using a color/shaking paradigm; and studies of behavior or electrophysiology to track responses to visual stimuli in tethered individual flies. Some of these approaches require a substantial amount of sophisticated equipment and expertise compared with other behavioral paradigms in *Drosophila*. Other approaches are easier to implement but are fairly limited in their ability to address questions of visual perception. Nevertheless, the simpler approaches treating vision in one dimension (light, dark, color) do provide effective paradigms for genetic analysis. Finally, some recently developed paradigms allow for a level of sophisticated analysis of visual perception in *Drosophila* while still remaining simple and efficient.

INTRODUCTION

Vision is a major sensory modality in fly behavior: More than one-half of the *Drosophila* brain is devoted to visual processing (Borst 2009). Although the organization of early visual processing in the fly optic lobes is relatively well understood (Egelhaaf et al. 2002), mechanisms of visual perception and learning in *Drosophila* remain understudied and less well understood than, for example, olfactory learning. This is primarily because there are no simple assays for measuring fly behavior in response to the kind of complex visual stimuli (e.g., shapes within changing context) that have been most revealing in studies of learning and perception, as exemplified in honeybees (Giurfa and Menzel 1997). How can you know whether a fly can see something? How do flies discriminate between shapes? How do they extract a visual object from its background context? This chapter presents approaches that allow us to begin to answer these questions.

Visual learning occupies an odd position in the field of *Drosophila* behavior. On one hand, it involves some of the most elegant and sophisticated techniques for investigating perception in any animal model (for an overview, see van Swinderen 2005). On the other hand, visual paradigms have been less successful than olfactory paradigms for unraveling mechanisms of learning and memory, partially because the best approach for measuring visual learning uses individual flies in tethered flight (Heisenberg and Wolf 1984)—not a high-throughput paradigm.

Vision in *Drosophila* can be addressed at its most basic level by phototaxis-type assays (which have a long history, see Rockwell and Barr Seiger 1973), in which flies walk toward a light source. It is through such approaches that a number of visual mutants were isolated, which contributed to our understanding of processing in the fly eye (e.g., *sevenless, photophobe*; Ballinger and Benzer 1988). However, these one-dimensional assays have proven less useful for understanding visual perception and learning. Efforts to try to adapt olfactory-type T-mazes with light sources essentially replacing odor plumes have been largely unsuccessful, possibly because flies are so sensitive to visual context (discussed further below); any change from a training chamber to a test chamber is potentially a change in visual context. In paradigms where T-maze visual learning has been successful, context was carefully controlled by testing flies in the same chambers in which they were trained. These successes include Folkers and Spatz's color-shaking paradigms (Folkers and Spatz 1981) and Le Bourg and Buecher's light/quinine assay (Le Bourg and Buecher 2002), discussed below, where, in both cases, flies choose to walk into a chamber not previously associated with an aversive unconditioned stimulus (shaking or quinine). However, neither of these assays permits testing behavioral responses to visual objects. To date, that is the exclusive domain of flight arena paradigms.

The flight arena paradigm for visual learning and perception in *Drosophila* was invented, adapted, and perfected over several decades by Tom Poggio and Werner Reichardt (Poggio and Reichardt 1976), Karl Götz, Martin Heisenberg, and Reinhard Wolf (Heisenberg and Wolf 1984), and Michael Dickinson and others (Reiser and Dickinson 2008). This paradigm (described in more detail below) involves tethering a single fly and recording its behavior in flight (by a torque meter, wing-beat analyzer, or video camera) to monitor the fly as it directly controls, by negative biofeedback, the position of images displayed on a screen around or in front of it. Typically, images are displayed on the inside of a drum surrounding the fly, and the fly's left–right flight choices control the angular position of the rotating drum and images. Whereas most images were originally physical objects pasted on the inside of a mechanical drum, more recent approaches (Reiser and Dickinson 2008) have made use of virtual moving objects displayed on light-emitting diode (LED) arenas, which are now commercially available (www.mettrix.com). Because flies will fly toward (i.e., fixate) objects in this paradigm, thereby reporting their visual choices, the next steps for devising learning and memory protocols using the flight arena were fairly straightforward (e.g., see Liu et al. 1999). These typically involved a three-step procedure: First, distinct objects are presented to a tethered fly in flight to determine any naïve preferences; then, one of the objects is associated with an unconditioned stimulus (US), typically a focused beam of heat using infrared (IR) light, such that the fly is punished whenever the object is in front (i.e., it flies toward the object) and is relieved when it flies toward the alternate object. This sequence shows operant visual learning—the ability of the fly to modify its behavior to avoid punishment. Finally, the distinct objects are presented to the fly again in the absence of the US (heat) to test whether the animal has learned to avoid the visual previously associated with the punishment. Throughout the experiment, performance is measured by the amount of time spent fixating on one object versus the other, divided by total time flying. The appeal of this paradigm lies in its thoroughness. We determine naïve preference, operant learning, and memory phenotypes; and we also observe the fine-scale behavioral processes behind the performance indices (thoroughly described in Heisenberg and Wolf 1984), something completely absent from olfactory leaning assays: *How* did the animals learn? In addition, paradigms in the arena allow for a deep exploration of visual psychophysics, such as novelty recognition (Dill et al. 1993), context generalization (Brembs and Wiener 2006), responses to contradictory cues (Tang and Guo 2001), or compound stimuli (Brembs and Heisenberg 2001), and even attention-like behavior (van Swinderen and Greenspan 2003; Zhang et al. 2007).

To show visual choice in the flight arena, a fly must remain in flight for an extended period of time—at least a couple of minutes. A large number of mutants potentially useful for dissecting mechanisms of visual learning are simply incapable of sustained flight because of pleiotropic effects on flight muscles or on general health. Taken with the low-throughput nature of the flight arena, this physical demand has limited the use of the flight arena as a gene discovery device and even its uses for understanding neural mechanisms. The best attempts at using the tethered flight paradigm to understand neural mechanisms of visual learning have mostly focused on mutants previously shown in olfactory paradigms to be relevant to learning and memory (Liu et al. 2006; Brembs and Plendl 2008). It appears, therefore, that, to fully unravel visual learning processes, other paradigms must be developed that combine the versatility of the flight arena (i.e., in the images that can be displayed and the short-term processes involved in perception) with the efficiency of olfactory learning assays.

Recently, population assays have been described in which walking or flying behavior in response to visuals are used to measure visual perception (van Swinderen and Flores 2007; Fry et al. 2008; Katsov and Clandinin 2008). The behavior of flies responding to moving visuals (displayed on a computer monitor or an LED array) is filmed and is analyzed by motion-capture software (for walking, see Katsov and Clandinin 2008; for flight, see Fry et al. 2008). At the opposite extreme, single fly walking paradigms were adapted for genetic analysis (Seugnet et al. 2008, 2009), using only light/dark as stimuli. In this chapter, we will consider the increasing levels of sophistication of the various visual paradigms: from single fly (or larvae) walking (or crawling) paradigms to population assays measuring responses to colors or moving visuals and ending with single fly tethered paradigms for use in the flight arena or electrophysiology.

Individual Fly Assays: Visual Learning in *Drosophila* Larvae

In larval visual learning (Gerber et al. 2004; Kaun et al. 2007), larvae are transferred back and forth between well-lit or dark agarose plates that either do or do not contain fructose (which is appetitive) as the US. Such appetitive larval visual memory is tested by tallying the time spent in light or dark quadrants of a plate (without a US).

MATERIALS

Reagents

Agarose (electrophoresis grade)
Drosophila larvae, aged 90–115 h since a 24-h batch of egg-laying
US: fructose

Equipment

Cold-light boxes such as those used for visualizing radiographs (two)
Petri dishes, plastic, 9-cm diameter
Shielding devices
    For training: Light box shields for light and heat: black cardboard shielded from below by aluminum foil to prevent heating up of the cardboard
    For testing: X-mask made of foil/cardboard to shield two opposing quadrants of the Petri dish

METHOD

1. Prepare a 1% agarose solution in water, boil, and 10 min after boiling, add fructose (1 M final). Pour a thin layer into the Petri dishes, and let cool. Also, prepare plates with agarose only. For a sample size of one, prepare 20 fructose-containing (10 for each of the reciprocally trained larvae; see below) and 22 agarose-only containing plates (10 each for training, 1 each for testing), as fresh plates must be used for every trial.

2. Cover one of the light boxes with foil/cardboard for training. For testing, prepare an X-mask of foil/cardboard such that opposing quadrants let light through (see Fig. 1).

3. Using a fine brush, collect 90–115-h larvae from a food bottle, and gently wash them with tap water.

4. For training, ensure that the room is dark, with the only light coming from the noncovered light box. Below, a sample training regimen is detailed in which larvae associate light with reward and are placed in the dark without reward (light+/dark):

    i. Place a larva on a fructose plate.

    ii. Set the plate on the noncovered light box for 1 min.

    iii. Transfer the larva from the fructose plate onto a fresh plate with agarose only, and place it on the covered light box.

    iv. Repeat Steps i–iii 10 times.

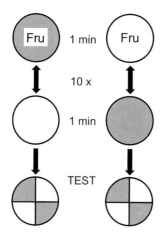

FIGURE 1. Larvae are transferred from darkened plates (gray) with fructose to lit plates (white) without fructose, or the converse. An average of both experiments is performed to yield a learning score.

5. For testing:

    i. Put the X-mask onto the noncovered light box.

    ii. Transfer the trained larva to test plates that contain agarose only, and place them onto the X-mask of the light box.

    iii. Score the position of the larva every 10 sec for 5 min, as in light or dark.

6. Calculate a first preference index (PI) as PREF1 = [(counts in light) – (counts in dark)]/(total counts).

7. Perform the reciprocal experiment (light/dark+) following Steps 1–5 and using another larva and fresh plates, and calculate a second PI as PREF2 = [(counts in light) – (counts in dark)]/(total counts).

8. Calculate a PI as PI = (PREF1 – PREF2)/2, and analyze the data with nonparametric statistics.

9. Perform the experiment for both reciprocal regimens following Steps 1–6 and using the respective other sequence of trials (i.e., dark/light+ and dark+/light, respectively).

DISCUSSION

The duration of the training trials may also be 3 min (Kaun et al. 2007). The experiment may be multiplexed by training a group of larvae, but testing them individually, if multiple light boxes are available for testing, or by performing the tests in smaller-diameter Petri dishes using 0.5 × 0.5-cm checkerboard masks. Whether a mass version of the assay is feasible and whether aversive learning may be uncovered by performing the test in the presence of an aversive reinforcer (as has been shown for olfactory learning Gerber and Hendel 2006) remain to be tested.

Protocol 2

Individual Fly Assays: Aversive Phototaxic Suppression

For adult flies, a paradigm called aversive phototaxic suppression (APS) can be used (Le Bourg and Buecher 2002). This method exploits flies' phototaxic reflex. By associating a lit chamber with quinine (which is aversive), repeated trials on a single animal result in a learned response to avoid (or to suppress) phototaxis. Of note, the US quinine must be present throughout the experiment for APS to work, unlike other memory assays in which the US is removed during testing. A full discussion of the protocol can be found in Le Bourg and Beucher (2002), but see also Seugnet et al. (2009).

MATERIALS

Reagents

Drosophila adult flies
Quinine hydrochloride solution (0.1 M), prepared in water

Equipment

Cylindrical vials (32) open on both ends (e.g., modified Falcon tubes)
Filter paper to line the inside of vials, cut to size; prepare several lined vials (e.g., 32)
Light shield for vials (e.g., opaque plastic tubes)
Light source, such as a dissecting microscope light on a gooseneck
Plastic block (opaque) with T-maze (2-mm- to 3-mm-wide) grooved in (see Fig. 2)
Plexiglas cover for maze; red plastic filter to place between cover and block

METHOD

1. Line a series of 16 vials with filter paper soaked with 320-μL quinine solution, and place in a lighted chamber. Prepare another set of 16 vials, lined with filter paper soaked in 320-μL water, and place in a dark chamber (i.e., inside the opaque plastic cover tubes).

2. Assemble the apparatus (see Fig. 2). The T-maze leads to the two vertical vials—one in the dark (e.g., covered), the other exposed to light.

3. Load a single fly (in the dark) at the entrance of the maze.

4. Immediately collect the fly as soon as it emerges in either vertical vial, and reload it into the maze. Note the outcome, whether the fly emerged from the lit/quinine vial or from the dark/water vial, in each case.

5. Repeat Steps 3 and 4 sixteen times, alternating the position of the lit/quinine vial and the dark/water vial each time.

6. Repeat the sequence in Steps 3–5 for at least eight flies. Then, divide the data obtained for each fly into four successive blocks of four runs, and calculate the proportion of dark vial choices for each block of four runs. Converted to percentages, this PI indicates APS if the results for block 4 yield significantly more dark choices compared with those for block 1.

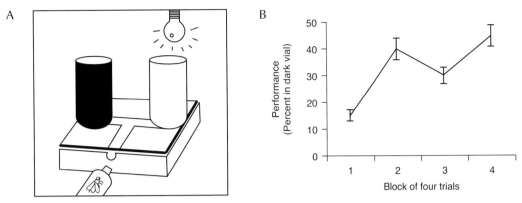

FIGURE 2. APS. (*A*) An opaque plastic block with a 2-mm-wide path leads to a choice point consisting of two alleys ending in two vertical chambers. One chamber is in the dark; the other is lighted. Quinine solution in the lit chamber produces APS after multiple reruns of the same fly. (*B*) Sample data. By the fourth block of four runs, half the flies are choosing the dark vial (i.e., suppressing phototaxis).

DISCUSSION

Although laborious (the assays described in Protocols 1 and 2 take ~10–20 min per animal, or hours to get some decent statistics), these paradigms are reliable and are easy to set up. However, they have not been shown to be useful for exploring other visual dimensions such as color or shapes. There have also been attempts to measure visual perception and forms of memory in single walking flies using more complex visual stimuli in arenas (e.g., see Wehner 1972; Strauss and Pichler 1998; Campbell and Strausfeld 2001; Neuser et al. 2008), which will not be discussed here.

Population Assays: Conditioning to Colors

Curiously, there has never been a widely used population assay for visual learning in *Drosophila*, although there have been a few brave attempts with clever devices (Götz 1970; Pflugfelder and Heisenberg 1995). Adapting the simple olfactory T-maze platform (see Chapter 26) for visual experiments has never worked reliably, even though vision is clearly important in that paradigm: Olfactory experiments are routinely performed in the dark (or red light) to prevent visual context cues. Nevertheless, some population paradigms have shown significant visual learning in *Drosophila*. These studies, first presented by Folkers and Spatz several decades ago (Folkers and Spatz 1981) but not repeated until recently (van Swinderen et al. 2009), used colors as conditioned stimuli (CS) and shaking as the US. There are several ways to attempt this strategy; a simple version of the paradigm, conditioning to colors using a shaking device, is described here.

A conditioning chamber, called a crab, shown in Figure 3, is designed to center the flies after shaking by having them tumble down to the lowest point between joined glass tubes forming a V. Thus, vibration should be just strong enough to center most flies. After shaking, flies display a geotactic response and climb up either side of the V, and their choice of which side to climb is influenced by color displays on either side. The proportion of flies on either side (calculated as PI = ([# visible flies on CS$^-$ side] − [# visible flies on CS$^+$ side])/[total # visible flies]) determines the flies' natural preference or their learned avoidance of a color associated with shaking. Each experiment thus yields four PIs per time point (one for each chamber), which are averaged. A full description of the protocol can be found in van Swinderen et al. (2009).

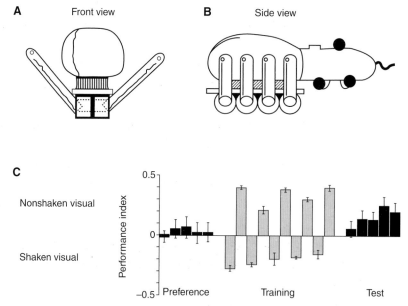

FIGURE 3. Visual learning paradigm. (*A*) Front view of the apparatus. Shaking causes the flies in the glass tubes to tumble down to the center; after which they can choose to climb either tube. The crab is the eight-tube device clipped on to the shaker. (*B*) Side view of the apparatus. A battery-operated skin-cleansing device (Neutrogena) clamped to a post is adequate. (*C*) Typical data showing visual learning after classical conditioning.

MATERIALS

Reagents

Drosophila adult flies

Equipment

Battery-operated shaking device, lightweight and adapted to be controlled with a switch (e.g., skincare products, from www.neutrogenawave.com).

Computer and computer monitor (cathode ray tube [CRT])

Glass test tubes, eight small (1 cm diameter, 6 cm length)

IR camera on a stand and recording equipment or software

Plastic sheeting (6 X 6-cm)

Silicon tubing, thick (1 cm/1.5 cm inner diameter/outer diameter X 2-cm)

Software for preparing visual images (visuals can be prepared with basic software such as Paint in Microsoft, or moving stimuli can be made with SWiSH)

Alternatively, the visuals and shaking stimulus can be controlled automatically with dedicated software and output devices (e.g., SWiSH to run Flash movies, and LabVIEW to coordinate visuals and stimulus via a National Instruments data acquisition [DAQ] board)

Super Glue

METHOD

Preparing the Shaker

See Figure 3.

1. Attach four 2-cm pieces of thick silicon tubing to a square of plastic sheeting.
2. Insert a test tube into each side of the four tube sections.
3. Angle all test tubes up so that they form a V with arms at 45° from vertical. This piece is called the crab.
4. Attach the crab to the shaker with the tubes pointing up.
5. Position the shaker (with crab attached) 5 cm above the middle of an upturned CRT monitor.
6. Position an IR camera above the shaker so that all eight tubes are visible.
7. Load flies (25–30) into each of the four chambers, making sure to reform the crab into its V shape afterward, positioned in the middle of the screen.
8. Initiate the program controlling visuals and shaking in a classical conditioning scenario (as described in the steps below, or do this manually).

Conditioning the Flies to Colors

A typical color-conditioning experiment is performed as follows, with a PI = ([# visible flies on CS⁻ side] − [# visible flies on CS⁺ side])/[total # visible flies] calculated for each chamber set, for each time point.

9. Natural preference: Carry out 5-sec shaking (to center flies) over a dark screen on the monitor, immediately followed by a 2-min exposure to the two different colors. The colors (e.g., blue vs. green) should be separated on the screen by a black bar approximately the width of the crab (10 cm). Count the flies every 20 sec in the replay of filmed experiments.

10. Training:

    i. Carry out five 1-min sessions of shaking associated with one color, alternating with five 1-min sessions without shaking associated with the other color (each color is presented alone on alternate sides of the crab). A shaking session consists of six 5-sec periods of vibration, each preceded by a 5-sec period without vibration. The entire training set is, thus, 10-min long.

    ii. Count the flies in each tube pair immediately before shaking in a freeze-frame replay of filmed experiments (or every 10 sec for the nonshaken visual).

    iii. Calculate an average PI for each of the consecutive 10 training epochs (five shaking sessions and five nonshaken sessions).

11. Test:

    i. On completion of training, keep the flies in the dark for 1 min (or more, to test memory), and then shake for 5 sec (in the dark) to center them.

    ii. Present both visual stimuli together, one on either side of the crab.

    iii. Count the flies every 20 sec, for 2 min, without any shaking sessions.

DISCUSSION

The color/shaking visual learning paradigm is quite easy to set up. There is substantial flexibility with regard to the visual stimuli that can be used, although brightness is key. As with other learning paradigms, it is important to balance training regimes: The final PI for a strain should be an average of separate experiments associating each color with shaking, in turn. Also, color presentations should alternate sides on the monitor, so that flies do not associate a tube side rather than a color. To best accomplish this, the entire paradigm should be automated, using systems such as LabVIEW.

Population Assays: Optomotor Maze

Recent publications have unveiled a novel approach for querying visual perception in fly populations, the optomotor maze (van Swinderen and Flores 2007). This paradigm can be adapted for visual learning by simply rerunning flies in the maze (habituation) or as a more sophisticated version of the APS paradigm described in Protocol 2. The optomotor maze, shown in Figure 4, provides an efficient paradigm to assay visual perception in *Drosophila*. The setup is very simple: An eight-choice maze consisting of 3-mm paths grooved into a transparent Plexiglas or acrylic slab is placed over an upturned CRT monitor on which visuals are displayed (Fig. 4A). The placement or movement of the visuals on the CRT (Fig. 4B), which the flies can see through the flat bottom of the maze, influences their turning behavior at each choice point. In addition, the distribution of a population at the end of the maze (Fig. 4C) determines the responsiveness of a strain to a particular visual scenario, such as a moving grating, but also to moving objects or a color gradient. A description of this paradigm can be found in van Swinderen and Flores (2007).

MATERIALS

Reagents

Drosophila adult flies

Equipment

Box, to cover the maze
Camera system and recording equipment to film the maze (optional)
Clamps, to secure the maze tightly to the glass plate
CRT with a computer to run the visual displays
Eight-point choice maze
> This piece of equipment is the most specialized and requires machine-shop expertise: 3-mm-wide paths are grooved into a transparent plastic slab (30 x 20 x 1-cm) such that one entrance point expands to nine collection points by way of eight interconnected T-junctions (see Fig. 4D).

Glass plate (30 x 20-cm)
> The slab is placed groove-face down on the glass.

One-way chambers for the nine collection points at the end of the maze
> These can be made with plastic pipette tips inserted into glass tubes.

Pedestals, such as 3-cm rubber stoppers (four)
> The maze is placed glass-side down on the rubber stoppers, which frame the image display on the CRT.

Polyethylene jumbo transfer pipettes, disposable
Software to display visuals (SWiSH, Vision Egg, MATLAB Psychtoolbox)

METHOD

1. Collect 25–30 flies (under CO_2) into disposable polyethylene jumbo transfer pipettes (four to eight sets of tubes will provide decent statistics for a strain or visual condition). Add 10-μL water to each tube, and stopper with a cotton plug. Store these overnight in an incubator set at room temperature.

2. Before an experiment (the next day), place transfer pipettes with flies in a dark chamber for 5 min.

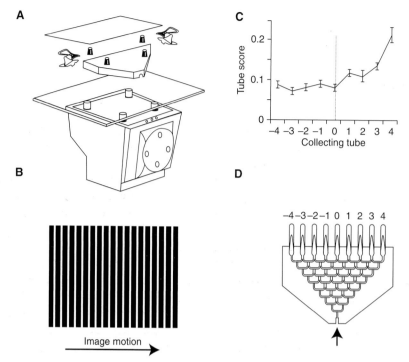

FIGURE 4. The optomotor maze. (*A*) Maze setup. The maze is placed on a glass plate over an upturned CRT monitor. The maze must be clamped tightly to the glass, and surrounding distracters should be blocked with a covering or box surrounding the maze. (*B*) Moving gratings are displayed on the computer screen. (*C*) A typical optomotor response to a moving grating. (*D*) Maze design. Flies enter at arrow.

3. Run the desired movie on the monitor (at 800 x 600-pixel resolution) with motion moving from right to left or left to right.

4. Insert the pipette with flies at the entrance of the maze, tapping gently to encourage the flies to enter. Allow 2–5 min for flies to complete the maze.

5. Collect and count the number of flies in each destination chamber.

6. Run additional batches of the same genotype, alternating the direction of the movie each time (e.g., right to left, then left to right).

7. Calculate the optomotor response for each run as a weighted average:

Optomotor index = $\Sigma(n^*[\text{\# flies in tube}]/[\text{total \# flies}])$,

where n is tube number $-4, -3, -2, -1, 0, 1, 2, 3, 4$. Remember to reverse the numbers for movie runs going in the opposite direction.

DISCUSSION

The maze paradigm is easily adaptable to different visual experiments and only limited by what can be displayed on a computer monitor. Yet, the optomotor maze can be a surprisingly fickle assay, probably because attention-like mechanisms are involved (van Swinderen 2007; van Swinderen and Flores 2007), especially with regard to the suppression of moving visual stimuli. Thus, anything that might prove distracting for flies might abolish the visual response. These issues include spacing that is too wide between the maze and the glass plate (clamp it down tighter) and geometrical or light asymmetries around the maze. Also, because well-fed flies perform poorly in the maze (as in other behavioral paradigms), starving them overnight before the experiment is recommended. It is possible to adapt this paradigm to visual learning assays, the easiest being simple tests of habituation (van Swinderen et al. 2009).

Protocol 5

Tethered Paradigms

The most successful approaches for studying visual perception and visual learning in *Drosophila* have been single fly paradigms in which tethered individuals respond to different visual stimuli. The equipment and protocols involved are quite sophisticated and differ depending on whether behavior or electrophysiology will be pursued. For either approach, flies must first be secured to a metal wire. This is typically performed by first cooling flies down to 4°C and then gluing them to a copper or tungsten wire with ultraviolet (UV)-activated cement. For electrophysiology, tethering requires a few extra steps to accommodate the placement of electrodes. Prepared individuals are then placed inside a cylindrical arena where images can be presented, or in front of a computer screen or even in front of a laptop. Flight dynamics or brain activity in response to visual stimuli is recorded by using a variety of specialized and/or commercially available electronic devices. Historically, most visual learning experiments were performed by measuring fly behavior with a torque meter and by displaying visuals on a mechanically rotating drum (Heisenberg and Wolf 1984). Because this approach is not easily adaptable to most laboratories, the focus here is on preparing the tethered flies for visual experiments and providing some references for subsequent protocols or equipment required for vision experiments.

The procedure followed for tethering individual flies to metal wires varies slightly, depending on whether the experiment will involve flight behavior or electrophysiology. In general, for flight behavior, the amount of glue contact to the fly is minimized. For electrophysiology, more structure is required to accommodate electrodes implanted in the brain. For experiments in the mechanical flight arena, Bjoern Brembs has provided a detailed movie available online at http://www.jove.com/index/details.stp?id=731doi:10.3791/731.

A useful manual for conducting tethered flight experiments (user's guide) is provided in Michael Dickinson's laboratory website http://www.dickinson.caltech.edu/PanelsPage, and see also Reiser and Dickinson (2008).

MATERIALS

Reagents

Drosophila adult flies

Equipment

Brass support rods and connectors (see Fig. 5)
Cold surface or cooling system, such as a cold block in an ice tray or on a Peltier system
Connectors (Omnetics Connector Corporation), alligator clips
Dental gun and UV-activated (or blue LED-activated) glue (Henry Schein Dental)
Dissecting microscope
Fly handling equipment (aspirator, brush)
Glue applicator (e.g., a length of tungsten wire)
Micromanipulator (e.g., MM3) and magnetic stand (e.g., GJ-8) (Narishige)
Tungsten or copper wires, thin (0.005-inch; A-M Systems)
Razor blade
Soldering equipment (soldering iron and solder)

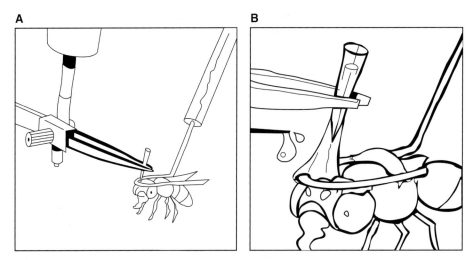

A **B**

FIGURE 5. Tethering and electrophysiology. (*A*) The fly is glued (under cold anesthesia) to a tungsten wire that is soldered to an electrical connector on a post. The post is attached to a micromanipulator. On a second micromanipulator, an attached pair of forceps holds a broken-off glass electrode tip. (*B*) Implanted glass electrode tips are glued in place by sculpting the semicured dental cement. After final curing with UV light, the secured electrodes are released by opening the attached forceps.

METHOD

Preparing the Tether

1. Cut 2- to 3-cm lengths of thin tungsten wire, and solder these onto the end of connectors (e.g., electrical pins that clip on to sockets).

2. Bend the end of the wire into the correct shape to attach easily to the fly: for flight, a small hook; for electrophysiology, a loop (see Fig. 5).

   To make the loop the right size (it must be smaller than the width of a fly's head), twist it around a syringe needle.

Tethering the Fly

3. Squeeze a drop of glue onto a microscope slide.

4. Aspirate a small number (two to five) of flies (females work best) into a glass test tube, and plunge the tube into an ice bucket for 30 sec to anesthetize flies.

5. Sprinkle flies onto a cold (4°C) surface.

6. Select a healthy-looking individual with intact wings, and position it on the block so as to best accommodate the tether.

7. Using the micromanipulator, practice lowering the tether onto the fly to see where contact will be made. Readjust the fly or tether angle to make sure that contact will be optimal (i.e., close to vertical). Contact points should span the head and thorax minimally for behavioral experiments and more extensively for electrophysiology (see Fig. 5). Then retract the tether with the micromanipulator along the vertical axis.

8. Apply a bead of glue to the contact points on the tungsten wire. For electrophysiology, a smaller bead on the left side is required to leave room for an electrode into the left eye.

9. Lower the tether, and watch the glue beads make contact with the required points, carefully readjusting the fly if necessary. For behavioral experiments, a minimal drop of glue should span

the back of the head on the front of the thorax. For electrophysiology, one bead should contact the dorsal part of the left eye, and the other bead should contact the dorsal rim of the right eye.

10. Point the activated UV gun for 20 sec to either side of the glue beads, ~1 cm from the fly.

11. For electrophysiology, now span the thorax with some more glue to secure it to the tungsten loop (see Fig. 5).

DISCUSSION

Electrophysiology

The tethered fly preparation as described above provides a good platform for implanting electrodes into an intact fly to record brain responses to visual stimuli (Nitz et al. 2002; van Swinderen and Greenspan 2003; van Swinderen 2007; van Swinderen et al. 2009). An alternative preparation used to record visual responses in the fly brain is to secure flies at the end of a pipette tip with just the head emerging (Joesch et al. 2008). The highly specialized nature of electrophysiological recordings is beyond the scope of this chapter (e.g., see Chapter 18), but appropriate tethering of the fly is a crucial first step.

Flight Arena and Visual Displays

The next step for either behavioral or brain-recording paradigms is the presentation of visual stimuli to the tethered fly. This can be accomplished in a variety of ways and depends largely on the equipment and expertise available in the laboratory. The simplest approach is to present one or two computer screens in front or to the sides of the animal (Yamaguchi et al. 2008). Moving images on the liquid-crystal displays can be programmed using any of a variety of software packages (SWiSH, Vision Egg, MATLAB Psychtoolbox). More traditionally, the tethered fly is placed inside a specially made LED arena (available from Mettrix Technology, www.mettrix.com, and see http://www.dickinson.caltech.edu/PanelsPage) onto which images can be programmed in MATLAB or LabVIEW. For behavioral approaches, a method of recording flight dynamics is required. This typically involves quite sophisticated equipment, either a torque meter to measure left–right thrusts of the animal or an IR wing-beat analyzer to measure wing-beat differentials. These devices are the real bottlenecks for conducting tethered visual paradigms, as few laboratories have access to them. Torque meters have typically been home-engineered devices, and wing-beat analyzers have been available from principally one source (David Smith at James Franck Institute [JFI] Electronics, University of Chicago; dr-smith@uchicago.edu). Signals from either device can then be fed back in real time to the programs controlling the image displays, and, in this way, the fly can report its visual choices in operant or in classical conditioning experiments. A simple alternative is to feed back signals from photodiodes pasted to a screen displaying close-up abdomen movements of a filmed tethered fly in flight (Strauss et al. 2001). Turns are usually associated with changes in abdomen position; these are detected on-screen by the photodiodes to directly control image flow.

For the electrophysiological experiment, similar visual displays can be used as for behavioral paradigms (Fig. 6), but flight is not necessary to gather data relevant to visual perception. To interpret brain local field potential (LFP) responses to visual stimuli, it is necessary to apply spectral calculations (e.g., a Fourier analysis in MATLAB) to the appropriate temporal segment of brain recording corresponding to a particular visual stimulus or visual transition. It is, therefore, crucial to always record a signal for the visual stimulus (e.g., a sine wave for an oscillating stimulus) alongside the LFP.

Special Considerations

The tethered paradigms (flight and electrophysiology) reveal a great deal about visual processes in *Drosophila* but are also the hardest to set up in a laboratory. This is because highly specialized equip-

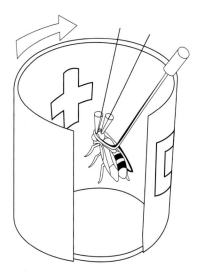

FIGURE 6. The recording arena. A prepared fly, with implanted electrodes, is placed in the center of an LED arena on which virtual objects are displayed such as a square and a cross. It is crucial to cover the inside of the arena with conductive mesh and to ground the entire preparation.

ment is required such as the wing-beat analyzer for flight paradigms (available from David Smith at JFI electronics, University of Chicago: dr-smith@uchicago.edu), or the field effect transistors for brain recording paradigms (available from Larry Andrews, www.nblablarry.com), or specialized LED arenas (available from Mettrix Technology, www.mettrix.com). Considerable expertise is required to put these together before even starting an experiment. The most adaptable visual display system available, a modular LED arena, is described by Reiser and Dickinson in a recent publication (Reiser and Dickinson 2008).

REFERENCES

Ballinger DG, Benzer S. 1988. *Photophobe (Ppb)*, a *Drosophila* mutant with a reversed sign of phototaxis; the mutation shows an allele-specific interaction with *sevenless. Proc Natl Acad Sci* **85**: 3960–3964.

Borst A. 2009. *Drosophila*'s view on insect vision. *Curr Biol* **19**: R36–R47.

Brembs B, Heisenberg M. 2001. Conditioning with compound stimuli in *Drosophila melanogaster* in the flight simulator. *J Exp Biol* **204**: 2849–2859.

Brembs B, Plendl W. 2008. Double dissociation of PKC and AC manipulations on operant and classical learning in *Drosophila. Curr Biol* **18**: 1168–1171.

Brembs B, Wiener J. 2006. Context and occasion setting in *Drosophila* visual learning. *Learn Mem* **13**: 618–628.

Campbell HR, Strausfeld NJ. 2001. Learned discrimination of pattern orientation in walking flies. *J Exp Biol* **204**: 1–14.

Dill M, Wolf R, Heisenberg M. 1993. Visual pattern recognition in *Drosophila* involves retinotopic matching. *Nature* **365**: 751–753.

Egelhaaf M, Kern R, Krapp HG, Kretzberg J, Kurtz R, Warzecha AK. 2002. Neural encoding of behaviourally relevant visual-motion information in the fly. *Trends Neurosci* **25**: 96–102.

Folkers E, Spatz H-C. 1981. Visual learning behavior in *Drosophila melanogaster* wild-type AS. *J Insect Physiol* **27**: 615–622.

Fry SN, Rohrseitz N, Straw AD, Dickinson MH. 2008. TrackFly: Virtual reality for a behavioral system analysis in free-flying fruit flies. *J Neurosci Methods* **171**: 110–117.

Gerber B, Hendel T. 2006. Outcome expectations drive learned behaviour in larval *Drosophila. Proc Biol Sci* **273**: 2965–2968.

Gerber B, Scherer S, Neuser K, Michels B, Hendel T, Stocker RF, Heisenberg M. 2004. Visual learning in individually assayed *Drosophila* larvae. *J Exp Biol* **207**: 179–188.

Giurfa M, Menzel R. 1997. Insect visual perception: Complex abilities of simple nervous systems. *Curr Opin Neurobiol* **7**: 505–513.

Götz KG. 1970. Fractionation of *Drosophila* populations according to optomotor traits. *J Exp Biol* **52**: 419–436.

Heisenberg M, Wolf R. 1984. *Vision in* Drosophila: *Genetics of microbehavior.* Springer-Verlag, Berlin.

Joesch M, Plett J, Borst A, Reiff DF. 2008. Response properties of motion-sensitive visual interneurons in the lobula plate of *Drosophila melanogaster. Curr Biol* **18:** 368–374.

Katsov AY, Clandinin TR. 2008. Motion processing streams in *Drosophila* are behaviorally specialized. *Neuron* **59:** 322–335.

Kaun, KR, Hendel T, Gerber B, Sokolowski MB. 2007. Natural variation in *Drosophila* larval reward learning and memory due to a cGMP-dependent protein kinase. *Learn Mem* **14:** 342–349.

Le Bourg E, Buecher C. 2002. Learned suppression of photopositive tendencies in *Drosophila melanogaster. Anim Learn Behav* **30:** 330–341.

Liu G, Seiler H, Wen A, Zars T, Ito K, Wolf R, Heisenberg M, Liu L. 2006. Distinct memory traces for two visual features in the *Drosophila* brain. *Nature* **439:** 551–556.

Liu L, Wolf R, Ernst R, Heisenberg M. 1999. Context generalization in *Drosophila* visual learning requires the mushroom bodies. *Nature* **400:** 753–756.

Neuser K, Triphan T, Mronz M, Poeck B, Strauss R. 2008. Analysis of a spatial orientation memory in *Drosophila. Nature* **453:** 1244–1247.

Nitz DA, van Swinderen B, Tononi G, Greenspan RJ. 2002. Electrophysiological correlates of rest and activity in *Drosophila melanogaster. Curr Biol* **12:** 1934–1940.

Pflugfelder GO, Heisenberg M. 1995. Optomotor-blind of *Drosophila melanogaster:* A neurogenetic approach to optic lobe development and optomotor behaviour. *Comp Biochem Physiol A Physiol* **110:** 185–202.

Poggio T, Reichardt W. 1976. Visual control of orientation behaviour in the fly. Part II. Towards the underlying neural interactions. *Q Rev Biophys* **9:** 377–438.

Reiser MB, Dickinson MH. 2008. A modular display system for insect behavioral neuroscience. *J Neurosci Methods* **167:** 127–139.

Rockwell RF, Barr Seiger M. 1973. Phototaxis in *Drosophila:* A critical evaluation. *Am Sci* **61:** 339–345.

Seugnet L, Suzuki Y, Vine L, Gottschalk L, Shaw PJ. 2008. D1 receptor activation in the mushroom bodies rescues sleep-loss-induced learning impairments in *Drosophila. Curr Biol* **18:** 1110–1117.

Seugnet L, Suzuki Y, Stidd R, Shaw PJ. 2009. Aversive phototaxic suppression: Evaluation of a short-term memory assay in *Drosophila melanogaster. Genes Brain Behav* **8:** 377–389.

Strauss R, Pichler J. 1998. Persistence of orientation toward a temporarily invisible landmark in *Drosophila melanogaster. J Comp Physiol A* **182:** 411–423.

Strauss R, Renner M, Gotz K. 2001. Task-specific association of photoreceptor systems and steering parameters in *Drosophila. J Comp Physiol A* **187:** 617–632.

Tang S, Guo A. 2001. Choice behavior of *Drosophila* facing contradictory visual cues. *Science* **294:** 1543–1547.

van Swinderen B. 2005. The remote roots of consciousness in fruit-fly selective attention? *Bioessays* **27:** 321–330.

van Swinderen B. 2007. Attention-like processes in *Drosophila* require short-term memory genes. *Science* **315:** 1590–1593.

van Swinderen B, Flores KA. 2007. Attention-like processes underlying optomotor performance in a *Drosophila* choice maze. *Dev Neurobiol* **67:** 129–145.

van Swinderen B, Greenspan RJ. 2003. Salience modulates 20–30 Hz brain activity in *Drosophila. Nat Neurosci* **6:** 579–586.

van Swinderen B, McCartney A, Kauffman S, Flores K, Agrawal K, Wagner J, Paulk A. 2009. Shared visual attention and memory systems in the *Drosophila* brain. *PLoS ONE* **4:** e5989. doi: 10.1371/journal.pone.0005989.

Wehner R. 1972. Spontaneous pattern preferences of *Drosophila melanogaster* to black areas in various parts of the visual field. *J Insect Physiol* **18:** 1531–1543.

Yamaguchi S, Wolf R, Desplan C, Heisenberg M. 2008. Motion vision is independent of color in *Drosophila. Proc Natl Acad Sci* **105:** 4910–4915.

Zhang K, Guo JZ, Peng Y, Xi W, Guo A. 2007. Dopamine-mushroom body circuit regulates saliency-based decision-making in *Drosophila. Science* **316:** 1901–1904.

26 Aversive and Appetitive Olfactory Conditioning

Michael J. Krashes and Scott Waddell

Department of Neurobiology, University of Massachusetts Medical School, Worcester, Massachusetts 01605

ABSTRACT

Olfactory memory paradigms have been extensively used to study memory in *Drosophila* since the late 1960s, and over the intervening years, investigators have made a series of worthwhile changes and "tweaks" to the procedures. Our intention in this chapter is to provide the reader with a detailed description of the aversive (odor-shock) and appetitive (odor-sugar) memory assays currently used in the Waddell laboratory. We also describe the essential control assays for sensory acuity and locomotor behavior. It should be emphasized that the assays we describe work adequately in our laboratory but are also amenable to, and may be further improved by, additional adjustments.

INTRODUCTION

The nature of memory has fascinated philosophers, psychologists, and biologists for centuries. However, the overwhelming complexity of the human brain and the intricacy of memory function and dysfunction might lead one to question whether memory is beyond a physical description. Yet studies of memory in animal models make the most optimistic of us believe that we are on the verge of finding relevant "traces" of memory in the brain and understanding the underlying neural manifestations. This "new dawn" of enquiry is most evident in genetically tractable animal models in which one can interrogate, with remarkable precision, the function of molecules within defined neural circuits in their relatively small brains.

The field of investigation into *Drosophila* memory was initiated in the late 1960s by William G. "Chip" Quinn and William A. Harris under the tutelage of Seymour Benzer (Quinn et al. 1974). These studies were initiated with the goal of identifying genes that encode critical elements of the learning machinery. The aversive and appetitive memory assays described in this chapter were originally devised in Chip Quinn's laboratory by Tim Tully and Bruce Tempel, and we direct readers to the original publications (Tempel et al. 1983; Tully and Quinn 1985). It is our opinion, at least, that the study of memory in the fruit fly is now able to span the traditional boundaries between molecular, cellular, systems, and behavioral neuroscience.

Aversive Olfactory Conditioning

In the standard protocol used for investigating memory in *Drosophila*, a population of approximately 100 fruit flies is trained to associate a particular odorant with electric shock punishment. When subsequently given a choice between the previously shock-paired odorant and another odorant the flies preferentially avoid the conditioned odorant. The persistence of memory in this paradigm depends on the training protocol. A single 1-min training session forms memory that is barely detectable 24 h later, but multiple training trials (5–10) with intervening rest intervals form long-term memory that lasts for days (Tully et al. 1994).

This aversive conditioning paradigm has been instrumental to progress in the field, having, with its predecessor (Quinn et al. 1974), been successfully used to screen for memory-defective mutant fly stocks. We refer readers to a recent review of the field for a list of the many genes that have been implicated in fly memory (Keene and Waddell 2007).

Odorant Choice and "Balance"

For mostly historical reasons 3-octanol (OCT) and 4-methylcyclohexanol (MCH) have been preferentially used as odorants in olfactory conditioning experiments. Although Quinn et al. (1974) reported that, "Not all odors work," many combinations have been used: menthol, geraniol, stearic acid, caproic acid, amyl acetate, and benzaldehyde in Dudai et al. (1976). Benzaldehyde has subsequently been used in several studies but we urge caution with its use, because it is also sensed via a nonolfactory, probably gustatory, route (Keene et al. 2004). Isoamyl acetate and ethyl acetate were used in studies described by Schwaerzel et al. (2003) and 6-methyl-5-hepten-2-one, methyl benzoate, methyl salicylate, methyl hexanoate, diethyl succinate, 4-methylphenol, and geranyl acetate in those of DasGupta and Waddell (2008).

A key feature of choosing an odorant pair is that one can identify concentrations of the two odorants that are "balanced" so that naïve flies distribute 50:50 when given the choice between the two odorants in the T-maze. These relative odorant concentrations must be determined empirically before performing olfactory conditioning experiments.

MATERIALS

CAUTION: See Appendix for proper handling of materials marked with <!>

Reagents

Drosophila adult flies
Fly food bottles and vials
MCH <!>
Mineral oil
OCT <!>

Equipment

Behavior room, temperature- and humidity-controlled
Bunsen burner
Clamp

Collection tubes (Falcon 14-mL polystyrene round-bottom tubes and cotton)
Diamond glass cutter
Electric drill with 5/32-inch bit
Flowmeter (to measure airflow)
Fly aspirator
Glass pipettes (10-mL)
Glass vials (25-mL) for diluting/delivering odorants
Nalgene 50 silicone tubing and connectors (3.2–5.5-mm)
Needle valves
Odor tops/stoppers (see Steps 2–7 for production instructions)
S48 stimulator (Grass Instruments, Astro-Med, Inc.)
Switchbox with electric leads
Timer
T-maze with two blank "testing" tubes (see Fig. 1)
Training tube, electric-grid-lined
Vacuum pump or house vacuum

METHOD

Day 1

Preparing the Flies

1. The day before conditioning, transfer all flies (separated by genotype) to fresh food bottles. Store overnight at the same conditions in which they were raised (typically at 25°C, relative humidity 60%, on a 12 h light/dark cycle). *Ideally, flies should be trained 3–6 d after eclosion.*

Assembling the Odor Tops

See Figure 1.

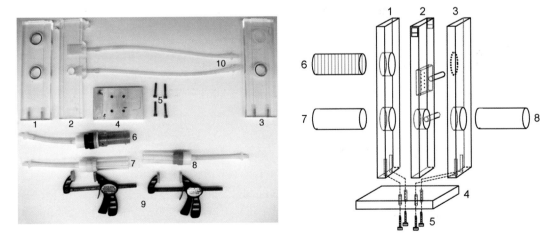

FIGURE 1. Parts and assembly of the T-maze apparatus. (1) Left-hand side plate of T-maze; (2) middle elevator section of T-maze; (3) right-hand side plate of T-maze; (4) aluminum base of T-maze; (5) screws to attach base to side plates of T-maze; (6) electrifiable training tube (can be replaced with a tube containing a sucrose impregnated filter paper for appetitive conditioning); (7) testing tube; (8) testing tube; (9) Quick-Grip clamps (see Fig. 2 for placement); and (10) silicone tubing to connect machine to vacuum source with male quick connect on one and female quick connect on the other. This allows for foolproof connectivity to the valve with the appropriate airspeed.

2. Drill two adjacent holes (5/32-in) in the silicone stopper with the electric drill.

3. Using the glass cutter, break a 10-mL glass pipette in two equal pieces and heat-polish the broken ends.

4. Heat one half pipette at its midpoint over a Bunsen burner and, after heating, bend it into an "L." Allow it to cool to room temperature.

5. Using ethanol for lubrication, slide the end of the other (straight) half pipette through one hole in the stopper until half the length protrudes. Push the end of the L-shaped pipette through the other hole so that the bottom protrudes slightly from the stopper.

6. Repeat Steps 2–5 to make another odor top. Label one odor top OCT and the other one MCH.

     These odor tops can be reused from experiment to experiment.

7. Attach a 12-in piece of Nalgene 50 silicone tubing with quick connectors to the L-shaped pipette.

     This tubing will connect to either the training tube or the testing tube.

Setting up the T-Maze

See Figure 1.

8. Ensure the T-maze is clean and free of dead flies. It can be washed with detergent, water, and ethanol, but be sure to rinse it very well with distilled water to eliminate any lingering odor.

9. Attach a 4-in piece of Nalgene 50 silicone tubing with quick connectors to each testing tube and to the training tube. Label one testing tube OCT and the other training tube MCH.

     These will be connected to the appropriate odorant vials during the experiment.

10. Make sure the machine has been assembled properly and use clamps to ensure that the machine is airtight.

     Note that the elevator should still slide up and down.

11. Last, attach a 12-in piece of Nalgene 50 silicone tubing and quick connectors to the two pipes on the back of the machine.

     This tubing will connect to the vacuum system in the behavior room and is required to draw air and odorants through the machine.

Preparing the Collection Vials

12. Plug a rack of collection tubes (14-mL polystyrene round-bottom tubes) with cotton. Label each collection tube with the genotype of flies, and the odorant they were exposed to with the electric shock (i.e., wild type + OCT).

     These tubes will be used to collect the flies from the T-maze tubes after testing memory.

Day 2

Diluting the Odorants

13. Pipette 10 mL of mineral oil into two glass 25-mL vials.

14. Add 10 μL of OCT to one vial and label it OCT and add 20 μL of MCH to the other vial and label it MCH. Note that naïve flies distribute 50:50 between these odorant dilutions when presented in our T-mazes with our odor delivery setup. One should empirically determine relative odorant concentrations producing equal distribution of naïve flies for their own apparatus and for each odorant pair chosen.

15. Cap both vials and vortex for 20 sec to mix.

16. Remove the vial caps and insert the odor tops made in Steps 2–7.

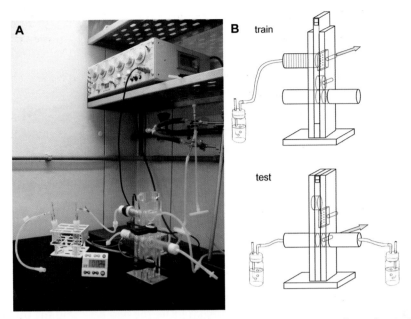

FIGURE 2. Aversive olfactory conditioning. (*A*) The complete apparatus in training mode. Airflow is tightly regulated using needle valves. Stimulator (on shelf) output is plugged into an intermediate switchbox (not visible). (*B*) Training: One odorant bubbler is connected to the top training tube and the elevator is in register so that air is pulled through the training tube. Testing: An odorant bubbler is attached to either test arm. The elevator is now in register with the two testing tubes so the two odorants used in training are being drawn into either arm of the T-maze and the flies can move into either arm. Odor bubblers are shown in these images. An odor top with two pieces of glass pipette pushed through a silicone stopper is placed into a 25-ml glass vial containing a dilution of odorant in 10 ml of mineral oil.

T-Maze Setup

See Figure 2.

17. Connect the tubing on the back of the T-maze to the vacuum system in the behavior room.

18. Position the T-maze elevator in the middle of the machine so that the vacuum pulls air through the training tube.

19. Attach the flowmeter to the tubing on the training tube and adjust the airflow, with an inline needle valve, so that the vacuum pulls air through the machine at a rate of 750 mL/min. Similarly set the airflow through the testing tubes (there are two tubes/inlets so the flow is 1500 mL/min, double that of the training tube).

Electric Shock Delivery-Stimulator Settings

20. Set the stimulator to the following: Delay 3.5 sec, Duration 1.5 sec, Volts 90 mV, Stimulus Mode "Repeat."

    This ensures that each population of flies will receive 12 electric shocks per 1 min.

21. Connect the stimulator to the switchbox and turn on the stimulator but *not* the switchbox.

Aversive Conditioning Experiment

See Figure 2.

22. Using the fly aspirator, transfer approximately 100 flies of the same genotype from a food bottle into the training tube and immediately attach it to the upper port of the T-maze.

23. Attach the electric leads from the switch box to the shock tube but do *not* switch it on yet.

24. Connect odorant A (OCT) to the training tube and then switch on the switch box. Start the timer and wait for 1 min.

> The flies in the shock tube will receive a series of 12 electric shocks in 1 min in the presence of the odorant A (OCT).

25. Simultaneously turn off the switch box and remove odorant A (OCT), to leave the flies inside the tube, receiving air from the room due to the vacuum pull.

26. Leave the flies in the tube for 30 sec.

27. Next, attach odorant B (MCH) to the training tube for 1 min but do *not* turn on the shock.

28. After 1 min of odorant B (MCH), disconnect the odorant and give the flies another 30 sec of fresh air.

Storing the Flies

If testing immediate memory or "learning," omit Steps 29 and 30 and go directly to Step 31.

29. If testing later memory, transfer the flies from the training tube to a fresh food vial, in which they will be stored until the time of testing.

30. When it is time for testing, transfer the flies back into the training tube and place in the upper port of the T-maze.

Testing

31. Slide the elevator of the T-maze so that the "lift" is in register with the training tube.

32. Knock the flies from the shock tube into the hole in the elevator by sharply tapping the T-maze on its side. Quickly push the elevator half down so that the flies are now trapped in the elevator.

33. Set the timer for 2 min. Connect odorant A (OCT) to the left testing tube and odorant B (MCH) to the right testing tube.

34. Switch off the lights, push the elevator all the way down so the hole is even with the two testing tubes, and start the timer.

> The vacuum will now pull the two odorants and the flies will distribute between the two testing tubes. This step should be performed in the dark to eliminate potential phototaxic effects from uneven lighting.

35. After a 2-min testing period, raise the elevator to trap the flies in either testing tube and switch the lights back on.

36. Carefully tap the machine on its side to knock the flies to the bottom of one testing tube, detach the testing tube, and transfer the flies to a collection tube.

37. Repeat Step 36 with flies in the opposite testing tube, transferring them to a separate collection tube. Label these collection tubes with CS+ (for the odorant that was paired with shock) or CS– (for the other odorant not paired with shock); label also the genotype tested and the odorant that was paired with the shock (in this case, OCT). Discard the few flies that remain in the elevator.

Reciprocal Training

38. Repeat the training, storage, and testing (Steps 22–37) with a different population of flies of the same genotype. But, with this group, pair the other odorant with the electric shock stimulus (in this example, MCH): pair odorant B (MCH) with shock for 1 min, give 30 sec of fresh air, and then odorant A (OCT) without shock for 1 min. Store, test, and collect the flies exactly as in Steps 29–37.

> Reciprocal training is an important feature of the experimental design. It averages odorant-specific differences if one is not interested in that detail—although some investigators are (see, e.g.,

DasGupta and Waddell 2008). More importantly, reciprocal training accounts for subtle skew in the distribution of flies that results from an imbalance in the relative concentrations of the odorants. For example, a skew away from odorant A toward B would be amplified if A was paired with shock but neutralized if odorant B is paired with shock. By averaging the two scores, the naïve skew is accounted for.

Calculating a Performance Index

39. Count the flies in the collection tubes under CO_2 anesthesia. Alternatively, the flies can be counted when dead, after placing them in a $-80°C$ freezer for 15 min.

    The flies in the CS+ collection tube are those that ran toward the odorant previously paired with the electric shock (wrong choice!). The flies in the CS– collection tube are those that ran toward the odorant that was not paired with the electric shock (correct choice).

40. Determine the performance index (PI).

    i. Calculate a half-score by subtracting the number of flies in the CS+ tube from the number of flies in the CS– tube and divide by the total number of flies:

       [(#flies CS–) – (#flies CS+)]/(total #flies).

    ii. Determine the half-score in the reciprocal experiment in which the other odorant was paired with the electric shock.

    iii. Average the two half-scores,

        [(half-score in which OCT was the CS+) + (half-score in which MCH was the CS+)]/2.

    to generate a single performance index or PI ($n = 1$).

    A score of 1 would represent perfect performance with all flies making the appropriate choice, whereas a score of 0 would represent the flies distributing evenly between the test tubes. Most published olfactory memory data is $n \geq 8$ per group.

DISCUSSION

This aversive shock-reinforced protocol can routinely yield a performance index ranging from 0.6 to 0.9 for "learning" or immediate (3-min) memory with the majority of flies avoiding the T-maze arm containing the odorant that was previously paired with electric shock. The performance index steadily decays as the time after training increases. We consistently obtain memory scores of 0.3 when measured 3 h after training. In our hands memory 1 d after training is rarely detected, and the flies distribute evenly between the two odorants. Others have observed significant 24-h scores after a single training session (Tully et al. 1994). However, repeated training sessions are required to form robust long-term aversive memory. A run of five to 10 repetitions of the protocol described above, with 15-min intertrial rest intervals, forms protein synthesis–dependent long-term memory that lasts for days. The requirement for spaced training to form long-term aversive memory mirrors that for many learning tasks in mammals, making this a good paradigm to understand the neural mechanisms that constrain long-term memory formation (Pagani et al. 2009).

Protocol 2

Appetitive Olfactory Conditioning

This protocol can be used to train populations of fruit flies to associate a specific odorant with sucrose reward rather than with punitive electric shock (Tempel et al. 1983; Schwaerzel et al. 2003; Krashes and Waddell 2008). When tested, flies approach the conditioned odorant. There are several advantages to using the appetitive paradigm. First, the reinforcer is physiologically relevant. Furthermore, because the neural circuits handling gustatory information are being identified, it is just a matter of time until the complete network from sensory neuron to association area is known. The appetitive paradigm also provides a unique entrée to consolidated memory because a single 2-min training session is sufficient to form long-term protein synthesis–dependent memory. This allows one to investigate the mechanics of consolidation in the absence of ongoing acquisition trials, such as with the spaced aversive conditioning protocol. It is critical to emphasize that flies have to be hungry to efficiently form appetitive memory and to retrieve it. Feeding them to satiety after training suppresses memory performance but robust performance returns if the flies are restarved (Krashes and Waddell 2008). This state dependence might fairly be viewed as a disadvantage by some investigators but it also implies integration in the fly brain between neural systems representing internal satiety state and those for memory retrieval (Krashes et al. 2009).

MATERIALS

CAUTION: See Appendix for proper handling of materials marked with <!>

Reagents

Drosophila adult flies
Glass milk bottles and vials
MCH <!>
Mineral oil
OCT <!>
Saturated sucrose solution

Equipment

Behavior room, temperature- and humidity-controlled
Clamp
Collection tubes (Falcon 14-mL polystyrene round-bottom tubes and cotton)
Flowmeter (to measure airflow)
Fly aspirator
Glass vials (25-mL) for diluting/delivering odorants
3MM Whatman filter paper
Nalgene 50 silicone tubing and connectors (3.2–5.5-mm)
Needle valves
Odor tops/stoppers
Timer
T-maze, with two blank "testing" tubes
"Training'"tubes, two blank
Vacuum pump or house vacuum

METHOD

Day 1

Food Deprivation

1. The day before conditioning, transfer flies of each genotype to labeled empty milk bottles capped with a foam stopper. Flies should be starved for 16–20 h before training the following day, but it is essential to provide a source of hydration for the flies by placing a dampened rectangle of filter paper (or a thin layer of 1% agarose) in each milk bottle. Store the flies overnight under the same conditions in which they were raised (typically at 25°C, relative humidity 60% on a 12 h light/dark cycle).

Setting Up the T-Maze and Its Components

2. Follow Steps 2–12 of Protocol 1 to prepare odor tops, T-maze, and collection vials.

Day 2

Preparing the Training Tubes and Storage Vials

3. Cut two 5.5 x 3.5-in (l x w) Whatman 3MM filter paper rectangles.

4. Soak the filter paper.

    i. Soak one rectangle in water, dry it completely with a blow-dryer, roll it lengthwise, and slide it into an empty training tube; label it "blank."

    > This training tube will be used when flies are exposed to a specific odorant without sucrose.

    ii. Soak the other filter paper in saturated sucrose solution, dry it completely with a blow-dryer, roll it lengthwise, and slide it into an empty training tube; label it "sucrose."

    > This training tube will be used when flies are to be exposed to a specific odorant in the presence of sucrose.

5. Prepare the storage vials.

    > These vials are required if testing later memory, otherwise flies will be tested immediately after training.

    i. For each group to be tested cut out two 4 x 4-in 3MM Whatman filter paper squares.

    ii. Soak each square in water and stuff the dampened filter paper to the bottom of an empty vial. Plug the vial with cotton.

    > The wet filter provides the flies with a source of water, prevents desiccation, and ensures survival.

6. Dilute the odorants, and assemble the T-maze as described in Steps 13–19 of Protocol 1.

Appetitive Conditioning Experiment

7. Using the fly aspirator, transfer approximately 100 flies of the same genotype from a labeled milk bottle into the "blank" training tube and immediately attach it to the upper port of the T-maze.

8. As quickly as possible, connect odorant A (OCT) to the "blank" training tube and start the timer.

    > The flies in the "blank" training tube will be exposed to odorant A (OCT) for 2 min without sucrose.

9. After 2 min, remove odorant A (OCT), to leave the flies inside the tube receiving air from the room due to the vacuum pull.

10. Leave the flies in the tube for 30 sec.

11. Gently knock the machine on its side to move the flies into the end of the "blank" tube. Remove the tube with flies and transfer them quickly to the "sucrose" training tube.

12. Immediately attach the "sucrose" tube with flies inside to the top port of the T-maze and quickly connect odorant B (MCH) to the "sucrose" tube and start the timer. Expose the flies in the sucrose training tube to odorant B (MCH) for 2 min in the presence of sucrose.

13. After 2 min of treatment with odorant B (MCH), disconnect the odorant and leave the flies in room air for 30 sec.

Storing the flies

If testing learning or short-term memory, omit Steps 14–15 and go directly to Step 16 for testing.

14. If testing later memory, promptly transfer the flies from the sucrose training tube to a previously made storage vial with damp filter paper, to store them until the time of testing. If testing memory beyond 24 h, transfer the flies to a vial containing food, but subsequently starve them again before testing.

15. When it is time to test memory, transfer the flies back into the "blank" training tube and place the tube in the upper port of the T-maze.

Testing

16. Slide the elevator of the T-maze so that the "lift" is in register with the training tube.

17. Knock the flies from the training tube into the hole in the elevator by sharply tapping the T-maze on its side. Quickly push the elevator half down to trap the flies in the elevator.

18. Set the timer for 2 min. Connect odorant A (OCT) to the left testing tube and odorant B (MCH) to the right testing tube.

19. Switch off the lights, push the elevator all the way down so the hole is even with the two testing tubes, and start the timer.

> The vacuum will now pull the two odorants and the flies will distribute between the two testing tubes. This step should be performed in the dark to eliminate potential phototaxic effects from uneven lighting.

20. After a 2-min testing period, raise the elevator to trap the flies in either testing tube and switch the lights back on.

21. Carefully tap the machine on its side to knock the flies to the bottom of one testing tube, detach the testing tube, and transfer the flies to a collection tube.

22. Repeat Step 21 with flies in the opposite testing tube, transferring them to a separate collection tube. Label the collection tubes with CS+ (for the odorant that was previously paired with sucrose) or CS– (for the other odorant not paired with sucrose); label also with the genotype tested and with the odorant that was paired with the sucrose (in this case, MCH). Discard the few flies that remain in the elevator.

Reciprocal Training

23. Repeat the training, storage, and testing (Steps 7–22) with a different population of flies of the same genotype. But, with this group, pair the other odorant with sucrose (in this example, OCT): pair odorant B (MCH) without sucrose for 2 min, give 30 sec of fresh air, and expose to odorant A (OCT) with sucrose for 2 min. Store, test, and collect the flies exactly as in Steps 14–22.

Calculating a Performance Index

24. Count the flies in the collection tubes under CO_2 anesthesia. Alternatively, the flies can be counted when dead, after placing them in a –80°C freezer for 15 min.

    The flies in the CS+ collection tube are those that ran toward the odorant previously paired with the sucrose (correct choice!). The flies in the CS– collection tube are those that ran toward the odorant that was not paired with the sucrose (wrong choice).

25. Determine the performance index (PI).

    i. Calculate a half-score by subtracting the number of flies that in the CS– tube from the number of flies in the CS+ tube and divide by the total number of flies:

       [(#flies CS+) – (#flies CS–)]/(total #flies).

    ii. Determine the half-score in the reciprocal experiment in which the other odorant was paired with the sucrose.

    iii. Average the two half-scores,

       [(half score in which OCT was the CS+) + (half score in which MCH was the CS+)]/2

       to generate a single performance index ($n = 1$).

DISCUSSION

This appetitive memory paradigm reliably produces a performance index around 0.3 either immediately or several days after training and we have recently achieved learning scores in the 0.6–0.7 range. The difficulty using this assay comes when assaying very extended memory. An extended period of food deprivation is lethal and therefore one can only test memory beyond 24 h in the few flies that remain alive (following 16 h of food deprivation before and 24 h of deprivation after training). This hurdle can, however, easily be overcome, and the utility of the assay extended, by feeding flies after training (Krashes and Waddell 2008). Note that flies fed for as little as 1 h after training will live for several days, but they show very little appetitive memory performance unless they are starved again. Memory therefore persists in satiated flies, but the flies must be remotivated by food deprivation to express the memory (Krashes et al. 2009). Sixteen hours of food deprivation before training and testing produces maximal learning and memory scores, respectively. It is critical to be consistent with feeding and starving regimens, because the satiety state has a profound effect on learning and memory performance. It is therefore important to distinguish a partially satiated fly from one with poor memory.

Control Assays for Sensory Acuity and Locomotion

Olfactory conditioning involves the flies associating odorant information with either electric-shock punishment or sugar reward. In both of these approaches, memory is measured as a behavioral performance: whether the flies avoid or approach the arm of the T-maze that is suffused with the appropriate odorant. It is therefore critical to determine that the memory defect does not simply result from an inability to sense odorants, sugar, or shock or to move well enough to perform the memory task. The acuity assays are straightforward and quick to perform. Olfactory and shock acuity are measured in the T-maze in the same way as memory and are therefore also good "task-relevant" controls in addition to reporting the appropriate locomotor capability.

Olfactory Acuity

To test olfactory acuity untrained flies are given the choice between a diluted odorant (as used in conditioning) and air bubbled through mineral oil in the T-maze. Flies are loaded into the machine as if testing memory and transported in the elevator to the T-maze choice point. After a 2-min choice period they are trapped in either arm, removed, and counted. A performance index is calculated as the number of flies in the air arm minus those in the odorant arm divided by the total number of flies. Most odorants used in conditioning are naturally aversive at the concentrations used. One can also vary odorant concentration to test relative acuity more rigorously.

Shock Reactivity

The shock acuity assay is performed and quantified similarly. Untrained flies chose between a tube containing an electrified grid and a tube containing a nonelectrified grid. After a 2-min choice period, they are trapped in either arm, removed, and counted. A performance index is calculated as the number of flies in the nonelectrified arm minus those in the electrified arm divided by the total number of flies. One can also vary the intensity of the electric shock to test relative acuity more rigorously.

Sugar Acuity

Sucrose acuity has been measured in different ways over the years (Tempel et al. 1983; Schwaerzel et al. 2003), and we now use a variant of the taste preference test used frequently in studies of gustation (Marella et al. 2006). Flies are starved overnight and taste preference is assayed on quadrant plates, two of each containing 1% agarose plus or minus 100 mM sucrose. Approximately 60 flies are placed on the plate and allowed to explore the agarose quadrants for 5 min, at which time they are recorded using a digital camera and BTV Pro software. The number of flies on each quadrant is manually counted at the 5 min time point. A sucrose preference index is calculated as

$$PI = [(\#flies\ on\ sucrose\ quadrants) - (\#flies\ on\ agarose)]/(total\ \#flies).$$

Data Analysis and Statistics

It is critical that the data is subjected to the appropriate statistical analysis. If performed incorrectly, irrelevant differences might be judged to be important and important differences might be missed. Because of the inherent variability of behavior, small differences are unlikely to be judged as significant if the number of samples is too low. Likewise low sample number can sometimes yield statistical differences that disappear as the number of samples increases. Most published olfactory memory data is at least $n \geq 8$ for each genotype and the scores are displayed as the mean PI ± standard error of the mean (S.E.M.). We routinely perform statistical analyses using KaleidaGraph (Synergy Software). Overall analyses of variance (ANOVAs) are followed by planned pairwise comparisons between the relevant groups with a Tukey HSD (Honestly Significant Difference) post hoc test. As we cannot possibly do justice to a discussion of statistics in biology here, we refer readers to Zar (2007) for a detailed account of the subject.

ACKNOWLEDGMENTS

Work in our laboratory is supported by National Institutes of Health grants MH09883, MH081982, and GM085788 to S.W. and a National Research Service Award DA024499 to M.J.K.

REFERENCES

DasGupta S, Waddell S. 2008. Learned odor discrimination in *Drosophila* without combinatorial odor maps in the antennal lobe. *Curr Biol* **18**: 1668–1674.

Dudai Y, Jan YN, Byers D, Quinn WG, Benzer S. 1976. *dunce*, a mutant of *Drosophila* deficient in learning. *Proc Natl Acad Sci* **73**: 1684–1688.

Keene AC, Waddell S. 2007. *Drosophila* olfactory memory: Single genes to complex neural circuits. *Nat Rev Neurosci* **8:** 341–354.

Keene AC, Stratmann M, Keller A, Perrat PN, Vosshall LB, Waddell S. 2004. Diverse odor-conditioned memories require uniquely timed dorsal paired medial neuron output. *Neuron* **44:** 521–533.

Krashes MJ, Waddell S. 2008. Rapid consolidation to a radish and protein synthesis-dependent long-term memory after single-session appetitive olfactory conditioning in *Drosophila*. *J Neurosci* **28:** 3103–3113.

Krashes MJ, DasGupta S, Vreede A, White B, Armstrong JD, Waddell S. 2009. A neural circuit mechanism integrating motivational state with memory expression in *Drosophila*. *Cell* **139:** 416–427.

Marella S, Fischler W, Kong P, Asgarian S, Rueckert E, Scott K. 2006. Imaging taste responses in the fly brain reveals a functional map of taste category and behavior. *Neuron* **49:** 285–295.

Pagani MR, Oishi K, Gelb BD, Zhong Y. 2009. The phosphate SHP2 regulates the spacing effect for long-term memory induction. *Cell* **139:** 186–198.

Quinn WG, Harris WA, Benzer S. 1974. Conditioned behavior in *Drosophila melanogaster*. *Proc Natl Acad Sci* **71:** 708–712.

Schwaerzel M, Monastirioti M, Scholz H, Friggi-Grelin F, Birman S, Heisenberg M. 2003. Dopamine and octopamine differentiate between aversive and appetitive olfactory memories in *Drosophila*. *J Neurosci* **23:** 10495–10502.

Tempel BL, Bonini N, Dawson DR, Quinn WG. 1983. Reward learning in normal and mutant *Drosophila*. *Proc Natl Acad Sci* **80:** 1482–1486.

Tully T, Quinn WG. 1985. Classical conditioning and retention in normal and mutant *Drosophila melanogaster*. *J Comp Physiol [A]* **157:** 263–277.

Tully T, Preat T, Boynton SC, Del VM. 1994. Genetic dissection of consolidated memory in *Drosophila*. *Cell* **79:** 35–47.

Zar JH. 2007. *Biostatistical analysis*. Prentice-Hall, Upper Saddle River, NJ.

27 | Odor–Taste Learning in Larval *Drosophila*

Bertram Gerber, Roland Biernacki, and Jeannette Thum

Universität Würzburg, Biozentrum Am Hubland, 97074 Würzburg, Germany

ABSTRACT

The *Drosophila* larva is an emerging model for studies in behavioral neurogenetics because of its simplicity in terms of cell number. Despite this simplicity, basic features of neuronal organization and key behavior faculties are shared with adult flies and with mammals. Here, we describe a Pavlovian-type learning assay in fruit fly larvae. A group of larvae is sequentially exposed to specific odors in the presence or respectively the absence of sugar, and then tested to determine whether they prefer the odor previously experienced with the reward. This protocol offers a robust, simple, cheap, and reasonably quick test for learning ability (an aversive version is available as well, using either high-concentration salt or quinine as punishment). With the concerted efforts of the *Drosophila* research community, we anticipate it will allow us to unravel the full circuitry underlying odor–taste learning on a single-cell level.

INTRODUCTION

What shall I do? This question is one that confronts humans and animals alike, all the time. To answer it, our needs, sensory impressions, and the possible ways to behave must be adaptively integrated. How this triad of sensory impressions, needs, and behavior is organized is a core scientific problem of neuroscience. As the biological needs of man and animals are in principle similar (to reproduce, to eat, to avoid being eaten, etc.), they provide a reasonable starting point for studying the basic principles of behavior, including its learned alterations. Here, we describe a behavioral tool for such research, focusing on associative odor–taste learning in *Drosophila* larvae (Scherer et al. 2003, and references below). Studies using odor–electric shock learning (Aceves-Piña and Quinn 1979) are not covered in this chapter.

We reasoned that larvae are always hungry, and therefore food should be a particularly powerful reward. A group of larvae is placed onto a sweet-tasting, rewarding sugar substrate (+) and presented with a specific odor A (experience denoted as A+). After this exposure, the larvae are transferred to a sugar-free situation, characterized by another odor (B). After three such A+/B experiences, we test the larvae by offering them a choice between the two odors in the absence of sugar. If they have associated odor A with the reward, they should track down this odor in search of sugar. Importantly, this effect is measured relative to reciprocally (A/B+) trained larvae; the experimental setup is shown in

443

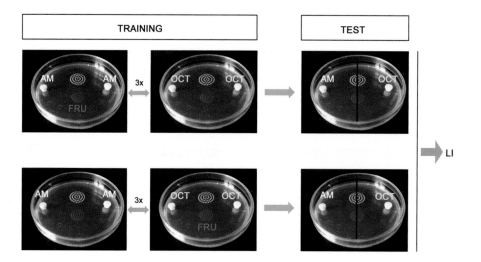

FIGURE 1. (*A*) The reciprocal principle. A two-group, reciprocal training design is used: In one of the groups, *n*-amyl-acetate (AM) is presented with a sugar reward (+) and subsequently 1-octanol (OCT) is presented without reward (AM+/OCT). The other group receives reciprocal training (AM/OCT+). After training exposures, animals are tested for their choice between AM and OCT. Relatively higher preferences for AM after AM+/OCT training than after AM/OCT+ training reflect associative learning and are quantified by the learning index (LI). (Image courtesy of B. Michels.)

Figure 1 (note that experiments that do not use such a reciprocal design are not considered in this chapter; see Gerber and Stocker 2007 for discussion).

Thus, the odor memory promotes the search for sugar. That is, after training with sugar, the test offers the larva a choice with one odor suggesting "over there you will find sugar" and the other "over there you will not find sugar." In the absence of sugar, the larva thus searches for the reward. If sugar actually is present, however, such search is not warranted. Indeed, if the sugar is presented during the test, conditioned behavior is abolished. In contrast, after aversive training (e.g., using quinine as punishment) one odor suggests "over there you will suffer from quinine" whereas the alternative suggests "over there you will not suffer from quinine." In the presence of quinine, therefore, the olfactory quinine associations can give direction to the escape from the aversive reinforcer, whereas if quinine actually is absent, such flight behavior is not warranted. Thus, it is the expected outcome (finding food; escape from unpleasant situations), rather than the "value" of the odor per se, which fuels conditioned behavior (for a more detailed discussion, see Gerber and Hendel 2006). In other words, learned behavior is viewed as an action in pursuit of its outcome, rather than a response triggered by the learned odor.

In the *Drosophila* larva, these kinds of learning are slowly beginning to be understood neurogenetically (see Gerber and Stocker 2007; Gerber et al. 2009; and references therein for more detail): All olfactory sensory neurons originate in the dorsal organ, pass above the pharynx into the brain, and project into the antennal lobe (Fig. 2A,B). Downstream from the antennal lobe, the olfactory pathway bifurcates: One branch of the projection neurons leads to the lateral protocerebrum, comprising premotor centers that support innate olfactory behavior; the other branch takes a "detour" via the mushroom bodies where memory traces for learned odorants can be localized. For learned odor responses, it is via this route that the respective motor centers are thought to be activated. Notably, the lateral protocerebrum thus receives both direct and indirect olfactory input, directly from the projection neurons and indirectly via the mushroom body output neurons, to orchestrate innate and learned behavioral tendencies.

The gustatory pathways, in contrast, originate from multiple external and internal sense organs and run underneath the pharynx, bypass the brain proper, and run toward multiple target areas in the subesophageal ganglion and premotor centers (Fig. 2C). Thus, taste pathways are linked rela-

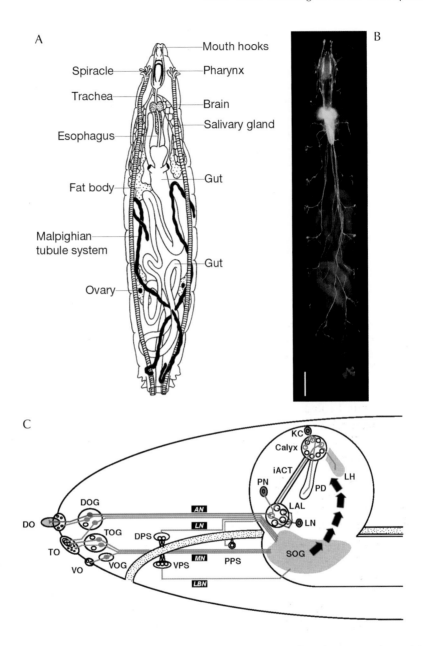

FIGURE 2. (A) Major body parts of a *Drosophila* larva. (B) Larval nervous system. Fluorescent image of the larval nervous system, clearly showing the brain with the ventral nerve cord and the major nerve cords. Scale bar, 100 μm. (C) Circuitry. Overview of the chemosensory pathways of the *Drosophila* larva. Olfactory pathways (blue) project into the brain proper, whereas gustatory input (brown) is collected in various regions of the subesophageal ganglion. The bold arrows indicate the proposed octopaminergic pathway to short circuit a taste-driven "value" signal carried by neurons from the subesophageal ganglion toward the brain. (Reprinted, with permission of Landes Bioscience, from Stocker 2007.) AN, antennal nerve; DO/DOG, dorsal organ/ganglion; DPS, dorsal pharyngeal sense organ; iACT, inner antennocerebral tract; KC, Kenyon cells; LAL, larval antennal lobe; LBN, labial nerve; LH, lateral horn; LN, local (inter)neurons; *LN*, labral nerve; MN, maxillary nerve; PD, pedunculus; PN, projection neuron; PPS, posterior pharyngeal sense organ; SOG, subesophageal ganglion; TO/TOG, terminal organ/ganglion; VO/VOG, ventral organ/ganglion; VPS, ventral pharyngeal sense organ. (A) Courtesy of The Carnegie Institution; (B) reprinted, with permission, from Sun et al. 1999 (© National Academy of Sciences, USA); (C) modified, with permission from Landes Bioscience, from Stocker 2006.

tively closely to the motor system. Notably, from the subesophageal ganglion, fibers split off from the gustatory pathway to send information about the value of the food to the brain ("good" or "bad": "valuation" neurons).

The mushroom bodies thus bring together the three core aspects of behavioral control: the sensory signal of the odor in terms of the pattern of activated mushroom body cells, the needs of the animal in the form of the "good" signal, and the motor program in the form of the activation of the mushroom body output neurons. Effectively, therefore, within the mushroom bodies a sensory signal ("Which odor?") is reformatted into a motor command ("Go there!") on the basis of the learned "value" of the odor.

In principle the pathways for smelling, tasting, and learning in the larva are similar to those found in adult flies, but in the larvae the numerical simplicity is striking: For example, in the larvae there are only 21 olfactory sensory neurons, 21 antennal lobe glomeruli, less than 40 projection neurons, a not yet fully described small set of local antennal lobe interneurons, a glomerularly defined input region to the mushroom bodies with less than 40 glomeruli, and only about 600 mushroom body neurons. This numerical simplicity should allow a validation of our admittedly tentative working model of odor–taste learning on the level of single, identified neurons.

Learning Assays in *Drosophila* Larvae

This protocol describes a two-group, reciprocal training design: One group of *Drosophila* larvae is exposed to *n*-amylacetate (AM) with a sugar reward (+), then subsequently exposed to 1-octanol (OCT) with no reward (denoted AM+/OCT). The other group receives the reciprocal training (AM/OCT+). The two groups of larvae are then tested for their choices between AM and OCT. Relatively higher preferences for AM after AM+/OCT training than after AM/OCT+ training reflect associative learning and are quantified by the learning index (LI), as discussed in the Data Analysis section.

To ensure that the experimenters' expectations cannot have an impact on the way the experiment is performed and/or data are scored, ask a colleague to "code" the Petri dishes for you; that is, the colleague should label, for example, the FRU-containing Petri dishes as "X" and the PUR Petri dishes as "Y." This code should be revealed to the experimenter only after all data have been recorded. The same kind of procedure should be used when comparing different genotypes.

The four possible experimental training arrangements are shown in Table 1: AM is presented on the X-Petri dishes and OCT on the Y-Petri dishes (arrangements a and b) or, for the reciprocally trained groups, AM is presented on the Y- and OCT on the X-Petri dishes (arrangements c and d).

With regard to the sequence of training trials, the first trial is performed either with the X-Petri dishes (arrangements a and d) or with the Y-Petri dishes (arrangements b and c).

From our data, the sequence of training trials had no significant effect on test behavior; for example, the larvae distribute between AM and OCT in the same way, regardless of whether they were trained as *AM-X first then OCT* or *OCT first then AM-X*. This observation should be confirmed, however, for each respective data set.

To prevent systematic effects of stimuli in the surrounding experimental environment, one should perform the test in one-half of the cases such that AM is presented to the left and OCT to the right (depicted in Table 1); in the other set of the cases, AM should be placed to the right and OCT to the left (Table 2). This may be important, for example, in circumstances in which the larvae can use a light source (window) to orient.

TABLE 1. The four possible training arrangements

Arrangement	a[a]	b	c	d
1. trial	AM-X	OCT-Y	AM-Y	OCT-X
2. trial	OCT-Y	AM-X	OCT-Y	AM-Y
3. trial	AM-X	OCT-Y	AM-Y	OCT-X
4. trial	OCT-Y	AM-X	OCT-X	AM-Y
5. trial	AM-X	OCT-Y	AM-Y	OCT-X
6. trial	OCT-Y	AM-X	OCT-X	AM-Y
TEST	AM vs. OCT	AM vs. OCT	AM vs. OCT	AM vs. OCT

[a]Arrangement a is used as the example in the protocol.

TABLE 2. Versions of the experimental arrangements in which during the test AM is presented to the left and OCT to the right

Arrangement	α	β	χ	δ
1. trial	AM-X	OCT-Y	AM-Y	OCT-X
2. trial	OCT-Y	AM-X	OCT-X	AM-Y
3. trial	AM-X	OCT-Y	AM-Y	OCT-X
4. trial	OCT-Y	AM-X	OCT-X	AM-Y
5. trial	AM-X	OCT-Y	AM-Y	OCT-X
6. trial	OCT-Y	AM-X	OCT-X	AM-Y
TEST	OCT vs. AM	OCT vs. AM	OCT vs. AM	OCT vs. AM

MATERIALS

CAUTION: See Appendix for proper handling of materials marked with <!>.

Reagents

Agarose, 1% solution in distilled water
Agarose, 1% solution, supplemented with 2 M fructose or sucrose
Drosophila larvae

> Flies are reared in culture bottles at 25°C, with a relative atmospheric humidity of 60%–70% and with a 14/10-h light/ dark cycle. Five-day-old larvae are used (90–120 h after egg laying); in particular, only larvae that are still in the food paste, and not larvae that may have begun to crawl up the side of the glass (known as the "wandering stage"), should be used for the learning assay.

n-amylacetate (AM; Chemical Abstract Services [CAS] 628-63-7)
1-octanol (OCT; CAS, no. 111-87-5) <!>
Paraffin (CAS, no. 8012-95-1)

Equipment

Odor containers

> Eight odor containers, a sample of which can be obtained free of charge from the authors, are needed. All containers, instruments, and any item that comes into contact with the containers must always be used for exclusively AM or for exclusively OCT to avoid contamination.

Petri dishes (90-mm)
Petri dish lids, perforated

> A sample of the perforated lids can be obtained from the authors free of charge.

Stopwatch

METHOD

Preparing the Assay Plates and the Larvae to Be Trained

See Figure 3.

1. On the day before the experiment, prepare eight 90-mm Petri dishes with a thin layer of 1% agarose only (PUR) and six Petri dishes with 1% agar containing 2 M fructose (FRU) (alternatively, sucrose can be used). Label the Petri dishes and their lids as PUR or FRU.

   > Ask a colleague to "code" the Petri dishes for you; for example, the colleague may label the FRU-containing Petri dishes as "X" and the PUR Petri dishes as "Y." This code should be revealed to the experimenter only after all data have been recorded. Such a procedure ensures that the experimenters´ expectations cannot have an impact on how the experiment is performed and/or how data are scored, and should be used when comparing different genotypes.

2. On the morning of the day of the experiment, fill eight odor containers with odorant.

   i. Dilute AM 1:50 with paraffin and fill each of four containers with 10 µL of the diluted AM.

   > Diluting AM is important for practical reasons: It ensures that AM and OCT are *about* equally attractive in experimentally naïve larvae (this may need to be confirmed for your wild-type fly strain of choice), so that learning-induced changes in the *relative* preference between both odors are easier to detect.

   ii. Fill each of the other four containers with 10 µL OCT.

   iii. Store all containers filled with AM together in one small Petri dish, and use another small Petri dish to store the OCT containers, to prevent evaporation of odor into the laboratory.

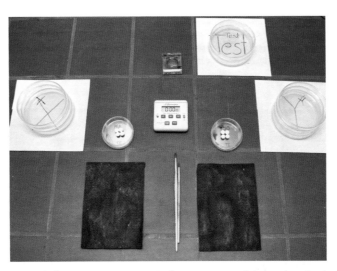

FIGURE 3. Workplace. Equipment required to carry out training and testing is shown.

Harvesting and Washing of Larvae from the Culture Vials

3. Using a spatula, remove some food paste containing larvae, and transfer it to a Petri dish with some drops of tap water in it. Wash the larvae by gently stirring.

4. Use a brush to "harvest" approximately 30 larvae, and place them into another Petri dish; then draw the larvae together into a "drop."

   It is important that none of the food paste remains within that "drop." Otherwise the food and the food odor may be present during the subsequent training and obscure the learning process.

Training and Testing

See Tables 1 and 2.

5. Carry out the training trial.

   i. Using the X-marked Petri dishes, place two AM-filled odor containers opposite one another, ~7 mm from the edge of the Petri dish.

   ii. With the brush, gently place the 30 larvae onto the middle of the X dish. Make sure the larvae are not "trapped" within a drop of water, because they may otherwise have difficulty overcoming the surface tension of the drop. This ensures both that their free wandering on the Petri dish and their ability to smell and/or to taste remains uncompromised.

   iii. Place a perforated lid on the dish and start the stopwatch.

   The perforation of the lid is meant to improve aeration; to avoid contamination, any given perforated lid should be used exclusively either for AM trials, for OCT trials, or for the test.

   iv. At 4 min, prepare the next trial by placing two odor containers filled with OCT into the Y dish. Place a perforated lid on the dish.

   v. After a total of 5 min has passed, transfer the larvae from the X dish onto the Y dish and place a fresh perforated lid on the Y dish. Start the stopwatch and track exposure time for X min with a stopwatch.

   vi. Put the AM containers back in the small Petri dish designated for them, close its lid, and dispose of the used X dish.

   vii. Follow the third to sixth training trials according to Table 1, arrangement a.

6. Carry out the testing.

   i. Prepare the test dish during the sixth training trial. At 4 min during the last training trial, place an AM-filled odor container on the left side and an OCT-filled odor container on the right side of the test Petri dish.

      Be sure to carry out the test in one-half of the cases such that AM is presented to the left and OCT to the right (as in this example); in the other one-half of the cases, AM should be placed to the right and OCT to the left. This is important in cases in which the larvae can potentially use any directional cues for their orientation.

   ii. After a total of 5 min has passed in the last training trial, transfer the larvae from the last Y dish with the OCT containers onto the test dish. The larvae should be placed in the middle, roughly aligned along the midline; cover the test dish with a perforated lid.

      Here also it is important that the larvae are not "trapped" in a drop of water.

   iii. Place the two OCT-filled odor containers from the last training dish back into their designated small Petri dish, and close its lid. Dispose of the last training Petri dish.

   iv. At 3 min, count the number of animals on the AM side and the number of animals on the OCT side of the dish.

      Any animals that at this time point are found on the Petri dish lid are not counted.

   v. Put the odor containers back into their respective small Petri dishes, and dispose of the test Petri dish with the larvae in it.

   vi. Calculate the PREF score as described in the Data Analysis section.

If you want to finish your experimental work and resume it on the next day, *separately* wash the AM and the OCT containers. Add odorless detergent to warm water, add the odor containers, and gently stir on a magnetic stirrer for at least 1 h. Then, remove the odor containers and dry them *separately* overnight in an incubator.

   Important: Never use the same odor containers for different odors.

7. For the next round of experiments, use a new group of larvae for reciprocal training (e.g., arrangement c from Table 1) and a fresh set of Petri dishes. That is, in this round, combine AM with the Y-marked and OCT with the X-marked Petri dishes.

   i. Remove approximately 30 larvae from the culture bottle.

   ii. Carry out the training as described in Step 5: The first and second training trials are now AM-Y and OCT-X, respectively.

   iii. Carry out the test as described in Step 6 and calculate the PREF and LI values as detailed in the Data Analysis section.

   iv. Now carry out the experiment according to b, and then according to d as described in Table 1.

   v. Carry out the experiment according to α, χ, β, δ, as described in Table 2. Ask a colleague to decode the identities of the X- and Y-marked Petri dishes. For the example discussed below, we assume X corresponds to a sugar-containing Petri dish, and Y to the Petri dish containing agarose only.

DISCUSSION

If you are interested in comparing learning abilities between genotypes, it is important to test for the behavioral specificity of the defect. You therefore need to compare experimentally naïve animals of the different genotypes in terms of (1) their preference between fructose and plain agarose (see Hendel et al. 2005; Niewalda et al. 2008; or Schipanski et al. 2008 for how to do this), (2) their preference between an AM-filled and an empty container, as well as (3) their preference between an OCT-filled and an empty container.

The rationale for not testing the relative preference between the two odors is that, for practical reasons, odor concentrations are chosen such that naïve, wild-type animals show about zero preference between them; therefore, one may expect both naïve wild-type and naïve mutant larvae to be indifferent with respect to the two odors. This indifference, however, may have different causes in wild-type and mutant larvae: The wild-type may be indifferent, whereas the mutant may be anosmic. This problem is avoided if odor detection, rather than relative preference, is tested.

Still, any mutant learning defect can be seen only after training (i.e., after animals have undergone extensive handling), exposure to the reward, and exposure to the odors. You should therefore test whether a given mutant still is able to (1) detect AM versus an empty odor container if you treat the larvae exactly as during training except that you omit the reward and merely expose to both odors; (2) detect OCT after that same regimen; (3) detect AM versus an empty odor container if you treat the larvae in a training-like way except that you omit the odors and merely expose to the reward; and (4) detect OCT after that same regimen (see Michels et al. 2005). Indeed, handling may stress the animals, change motivation, and/or induce fatigue; repeated odor exposure may lead to sensory adaptation, habituation, and/or latent inhibition and sugar exposure and/or uptake to contextual learning and/or changes in satiety (for discussion, see Gerber and Stocker 2007). Therefore, these kinds of control procedures, introduced in Michels et al. (2005), to us do seem warranted when describing a "learning mutant," and as we anticipate should become standard in the field.

If you wish to extend your analyses to aversive learning (using, e.g., high salt or quinine as punishment), please note that to uncover conditioned aversive behavior one must test in the presence of the respective aversive reinforcer (see above as well as the discussion in Gerber and Hendel 2006).

Finally, one can in principle perform the learning experiment in exactly the same way as detailed here, but omitting one of the two odors from its respective odor container. For example, you can measure the behavior toward AM of larvae that have received paired presentations of AM with the reward during one kind of trial and exposure to an unrewarded Petri dish with an odorless container during the other type of trial (AM+/empty); these larvae then must be compared with larvae that have received unpaired presentations of AM and reward (AM/empty+). Clearly, the test involves an AM-filled container on the one side and an empty container on the other (Selcho et al. 2009).

Note that in an academic setting, experiments are typically performed in a fume hood and without "landmarks" that could allow larval orientation (windows, heaters, etc.). However, these experiments are robust enough to be performed in a nonacademic setting as well—for example, by a group of 20 eighth-grade schoolchildren in a regular classroom (Fig. 4).

FIGURE 4. Robustness of the assay. The experiment is robust enough for nonacademic settings. Students of the eighth grade at the Gymnasium Stettensches Institut Augsburg are shown performing a one-day course in larval learning.

DATA ANALYSIS

Regarding the appetitive learning effect in the example presented here, we want to test whether the joint presentation of an odor with a sugar reward leads to an association between the two. Thus, if larvae are brought to associate different odors with the sugar—as in our case by AM+/OCT training in one case and OCT+/AM training in the other—the larvae should behave differently in the test: Those larvae that were rewarded in the presence of AM should be found more frequently on the AM side relative to those that were rewarded in the presence of OCT. Therefore, the preferences of these two groups are calculated as

$$PREF = (\#AM - \#OCT)/\#Total. \qquad (1)$$

In this equation, # indicates the number of larvae observed on the respective half of the test dish. Consequently, a PREF score of 1 means that all the animals were located on the AM side, whereas a PREF score of –1 means that all the animals were found on the OCT side.

These PREF scores are determined for the reciprocally trained groups (arrangements a and c; arrangements b and d; as well as for α–χ and β–δ). They should then be displayed, separated by the odor–sugar contingency (i.e., separately for all AM-rewarded and all OCT-rewarded groups), as box-and-whisker plots with the median as the middle line, the 25% and 75% quartiles as box boundaries, and the 10% and 90% quantiles as whiskers (Fig. 5A). The median can be determined by dividing an ascending series of the PREF scores by 2 and taking the value right at the midway point, so that 50% of the PREF scores are larger than this value and 50% are smaller. Correspondingly, the 25% quartile

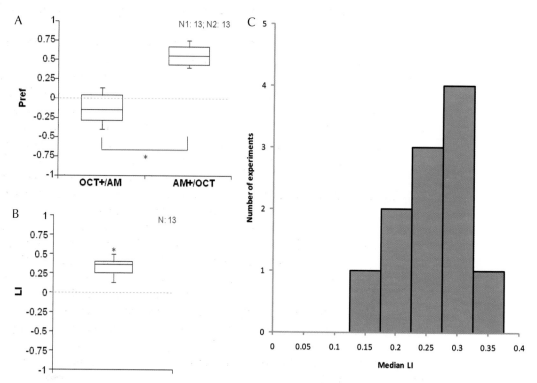

FIGURE 5. Example of results from a nonacademic setting (see Fig. 4). (*A*) PREF scores of animals rewarded in the presence of OCT, but not of AM (*left*), or rewarded in a reciprocal manner—in the presence of AM, but not in the presence of OCT (*right*). Data are presented as median (middle line), and 25%, 75% quartiles (box boundaries), and 10%, 90% (whiskers) quantiles. Sample size is 13 for both groups. The asterisk between the plots refers to a Mann–Whitney U-test. (*B*) Learning indices (LIs) as calculated from the data in *A*, displayed as box-and-whisker plot. Sample size is 13. (Asterisk) $P < 0.05$ in a one-sample sign test. (*C*) From a series of 11 experiments in various nonacademic settings, the number of experiments finding median learning indices falling into the respective class is presented. Thus, typically the experiments in these kinds of settings usually yield learning indices of between 0.2 and 0.3.

marks a PREF value such that 25% of the PREF scores are larger and 75% are smaller; the 75% quartile marks a PREF value such that 75% of the PREF scores are larger and 25% smaller. The same logic applies to the 10% and 90% quantiles as well. These box-and-whisker plots, rather than a display of the mean and standard error of the mean (SEM), are warranted because typically the kinds of behavioral data from this experiment are not distributed in a Gaussian (bell-shaped) way.

Our expectation is that the PREF scores of larvae that were rewarded in the presence of AM ($PREF_{AM+}$) should be higher relative to those that were rewarded in the presence of OCT ($PREF_{OCT+}$). Note that the comparison between $PREF_{AM+}$ and $PREF_{OCT+}$ is "buffered" against potential overall preferences for one or the other of the odors. That is, we can conclude that the larvae have learned if either $PREF_{AM+} = 1$ and $PREF_{OCT+} = 0$, or if $PREF_{AM+} = 0.5$ and $PREF_{OCT+} = -0.5$ or if $PREF_{AM+} = 0$ and $PREF_{OCT+} = -1$. Whether a learning effect can be shown should be tested by means of a Mann–Whitney U-test comparing the $PREF_{AM+}$ versus the $PREF_{OCT+}$ values (significance level 5%, i.e., $P < 0.05$). This type of test is warranted because the data are not typically distributed normally, precluding the use of parametric tests such as t-tests.

However, we may be interested not only in whether the larvae have learned at all, but also in how much they have learned. This is necessary, for example, for comparing the learning ability of wild-type larvae to that of mutant animals. To this end, we calculate the learning index (Equation 2), to quantify how much the PREF scores of AM-rewarded animals differ from the PREF scores of OCT-rewarded animals (division by 2 ensures that LI scores vary between –1 and 1). As we noted above, the sequence of training trials is typically without effect; therefore LIs can be computed either from arrangements a and c in Table 1 or from arrangements a and d in Table 1:

$$LI = (PREF_{AM+} - PREF_{OCT+})/2. \qquad (2)$$

Thus, LI scores systematically greater than 0 show appetitive learning; if the LI scores are ~0, there is no evidence of learning; if they are systematically negative, one would need to conclude that aversive learning has occurred (for a discussion of the general utility of the LI, see the Appendix to Hendel et al. 2005). To display the LI values, a box-and-whisker plot should again be used (Fig. 5B; the median learning indices are shown in Fig. 5C). To test statistically for whether the LI scores differ from 0, a one-sample sign test should be used. This test can be approximated fairly easily: Count how many LI values are >0, and how many LI values are <0 (e.g., 13 LIs are greater than 0, and three LIs are less than 0). The lesser of these two numbers is then subtracted from the greater; this difference is known as test statistic T (in our example $T = 13 - 3 = 10$). If T is greater than $2 \cdot \sqrt{n}$ (where n is the total number of learning indices different from 0), the probability of error is <5%. In our example, therefore, $2 \cdot \sqrt{n} = 2 \cdot \sqrt{16} = 8$. Because $T > 8$, the LIs across the whole experiment differ significantly from 0: The larvae have learned. In cases where two genotypes are to be compared in terms of their LI scores, a Mann–Whitney U-test is applicable.

In cases of multiple-group comparisons (e.g., when performing rescue experiments or a mutant screen), a Kruskal–Wallis test should be performed first. In case of significance, this should be followed by pairwise U-tests that should be properly corrected for such multiple comparisons to keep the experiment-wide error rate at 5%. One such correction, a very conservative one, is known as the Bonferroni correction. This can be performed by dividing the critical P value by the number of pairwise tests performed. Suppose the Kruskal–Wallis test has yielded significance regarding a four-genotype experiment, and you want to do three pairwise comparisons (e.g., wild-type vs. rescue strain, wild-type vs. driver control, and wild-type vs. effector control). Then, the critical P value for each individual U-test should be set as $P < (0.05/3) = 0.017$, such that the experiment-wide error remains at $P < 0.05$.

ACKNOWLEDGMENTS

The development of this learning paradigm was made possible by the Volkswagen Foundation and the support by R.F. Stocker, Université Fribourg, Switzerland.

B.G. is a Heisenberg Fellow of the Deutsche Forschungsgemeinschaft (DFG), which supports our ongoing research via grants SFB 554 (Arthropod Behavior), SFB TR 58 (Fear, Anxiety, Anxiety Disorders), and GK 1156 (Synaptic and Behavioral Plasticity).

The efforts to make this kind of learning experiment accessible in nonacademic settings were supported by the Robert Bosch Foundation via the Bio-logisch! program. We thank numerous high school and undergraduate students from our laboratory courses, as well as the graduate student teaching assistants advising them, for patience, tireless experimentation, and feedback. All images are courtesy of J. Thum and R. Biernacki, unless noted otherwise.

Translated from the German by R.D.V. Glasgow, Zaragoza, Spain.

We thank J. Wessnitzer and J. Young, University of Edinburgh, for comments.

REFERENCES

Selection of Review Articles

Davis RL. 2004. Olfactory learning. *Neuron* **44**: 31–48.

Demerec M, Kaufmann BP. 1972. Drosophila *guide. Introduction to the genetics and cytology of* Drosophila melanogaster. Carnegie Institution, Washington, DC.

Gerber B, Stocker RF. 2007. The *Drosophila* larva as a model for studying chemosensation and chemosensory learning: A review. *Chem Senses* **32**: 65–89.

Gerber B, Tanimoto H, Heisenberg M. 2004. An engram found? Evaluating the evidence from fruit flies. *Curr Opin Neurobiol* **14**: 737–744.

Gerber B, Stocker RF, Tanimura T, Thum AS. 2009. Smelling, tasting, learning: *Drosophila* as a study case. *Results Probl Cell Differ* **47**: 139–185.

Hallem EA, Dahanukar A, Carlson JR. 2006. Insect odor and taste receptors. *Annu Rev Entomol* **51**: 113–135.

Heisenberg M. 2003. Mushroom body memoir: From maps to models. *Nat Rev Neurosci* **4**: 266–275.

Hildebrand JG, Shepherd GM. 1997. Mechanisms of olfactory discrimination: Converging evidence for common principles across phyla. *Ann Rev Neurosci* **20**: 595–631.

Keene AC, Waddell S. 2007. *Drosophila* olfactory memory: Single genes to complex neural circuits. *Nat Rev Neurosci* **8**: 341–354.

Scott K. 2005. Taste recognition: Food for thought. *Neuron* **48**: 455–464.

Strausfeld NJ, Hildebrand JG. 1999. Olfactory systems: Common design, uncommon origins? *Curr Opin Neurobiol* **9**: 634–639.

Stocker RF. 1994. The organization of the chemosensory system in *Drosophila melanogaster*: A review. *Cell Tissue Res* **275**: 3–26.

Stocker RF. 2001. *Drosophila* as a focus in olfactory research: Mapping of olfactory sensilla by fine structure, odor specificity, odorant receptor expression, and central connectivity. *Microsc Res Tech* **55**: 284–296.

Sun B, Xu P, Salvaterra PM. 1999. Dynamic visualization of nervous system in live *Drosophila*. *Proc Natl Acad Sci* **96**: 10438–10443.

Vosshall LB, Stocker RF. 2007. Molecular architecture of smell and taste in *Drosophila*. *Annu Rev Neurosci* **30**: 505–533.

Zars T. 2000. Behavioral functions of the insect mushroom bodies. *Curr Opin Neurobiol* **10**: 790–795.

Original Articles Concerning Learning in Larval *Drosophila*

Gerber B, Hendel T. 2006. Outcome expectations drive learned behaviour in larval *Drosophila*. *Proc Biol Sci* **273**: 2965–2968.

Gerber B, Scherer S, Neuser K, Michels B, Hendel T, Stocker RF, Heisenberg M. 2004. Visual learning in individually assayed *Drosophila* larvae. *J Exp Biol* **207**: 179–188.

Hendel T, Michels B, Neuser K, Schipanski A, Kaun K, Sokolowski MB, Marohn F, Michel R, Heisenberg M, Gerber B. 2005. The carrot, not the stick: Appetitive rather than aversive gustatory stimuli support associative olfactory learning in individually assayed *Drosophila* larvae. *J Comp Physiol A* **191**: 265–279.

Kaun KR, Hendel T, Gerber B, Sokolowski MB. 2007. Natural variation in *Drosophila* larval reward learning and memory due to a cGMP-dependent protein kinase. *Learn Mem* **14**: 342–349.

Knight D, Iliadi K, Charlton MP, Atwood HL, Boulianne GL. 2007. Presynaptic plasticity and associative learning are impaired in a *Drosophila* presenilin null mutant. *Dev Neurobiol* **67**: 1598–1613.

Michels B, Diegelmann S, Tanimoto H, Schwenkert I, Buchner E, Gerber B. 2005. A role of synapsin for associative learning: The *Drosophila* larva as a study case. *Learn Mem* **12:** 224–231.

Neuser K, Husse J, Stock P, Gerber B. 2005. Appetitive olfactory learning in *Drosophila* larvae: Effects of repetition, reward strength, age, gender, assay type and memory span. *Anim Behav* **69:** 891–898.

Niewalda T, Singhal N, Fiala A, Saumweber T, Wegener S, Gerber B. 2008. Salt processing in larval *Drosophila:* Choice, feeding, and learning shift from appetitive to aversive in a concentration-dependent way. *Chem Senses* **33:** 685–692.

Scherer S, Stocker RF, Gerber B. 2003. Olfactory learning in individually assayed *Drosophila* larvae. *Learn Mem* **10:** 217–225.

Schipanski A, Yarali A, Niewalda T, Gerber B. 2008. Behavioral analyses of sugar processing in choice, feeding, and learning in larval *Drosophila*. *Chem Senses* **33:** 563–573.

Schroll C, Riemensperger T, Bucher D, Ehmer J, Voller T, Erbguth K, Gerber B, Hendel T, Nagel G, Buchner E, et al. 2006. Light-induced activation of distinct modulatory neurons triggers appetitive or aversive learning in *Drosophila* larvae. *Curr Biol* **16:** 1741–1747.

Selcho M, Pauls D, Han KA, Stocker RF, Thum AS. 2009. The role of dopamine in *Drosophila* larval classical olfactory conditioning. *PLoS One* **4:** e5897. doi: 10.1371/journal.pone.0005897.

Yarali A, Hendel T, Gerber B. 2006. Olfactory learning and behaviour are "insulated" against visual processing in larval *Drosophila*. *J Comp Physiol A* **192:** 1133–1145.

Zeng X, Sun M, Liu L, Chen F, Wei L, Xie W. 2007. Neurexin-1 is required for synapse formation and larvae associative learning in *Drosophila*. *FEBS Lett* **581:** 2509-2516.

Studies Concerning Odor–Electric Shock Learning in Larval *Drosophila*

Aceves-Pina EO, Quinn WG. 1979. Learning in normal and mutant *Drosophila* larvae. Science **206:** 93–96.

Forbes B. 1993. "Larval learning and memory in *Drosophila melanogaster*." Thesis, Institut für Genetik, Universität Würzburg. (This study reports a failure to observe odor–electric shock learning.)

Heisenberg M, Borst A, Wagner S, Byers D. 1985. *Drosophila* mushroom body mutants are deficient in olfactory learning. *J Neurogenet* **2:** 1–30.

Khurana S, Abu Baker MB, Siddiqi O. 2009. Odour avoidance learning in the larva of *Drosophila melanogaster*. *J Biosci* **34:** 621–631.

Tully T, Cambiazo V, Kruse L. 1994. Memory through metamorphosis in normal and mutant *Drosophila*. *J Neurosci* **14:** 68–74.

28 | Studying Aggression in *Drosophila*

Sarah J. Certel[1] and Edward A. Kravitz[2]

[1]*Division of Biological Sciences, COBRE Center for Structural and Functional Neuroscience, University of Montana, Missoula, Montana 59812-1552;* [2]*Department of Neurobiology, Harvard Medical School, Boston, Massachusetts 02115*

ABSTRACT

Aggression is an innate behavior that likely evolved in the framework of defending or obtaining resources. This complex social behavior is influenced by genetic, hormonal, and environmental factors. In many organisms, aggression is critical to survival, but the ability to control and suppress aggression in distinct contexts also is necessary. Invertebrate organisms, with their relatively simple nervous systems and a multiplicity of powerful tools available to examine their often elaborate and complex behavioral displays, have become increasingly valuable models for investigating the genetic and systems biological roots of social behavior. In this chapter, we outline methods for analyzing aggression in *Drosophila*: The design encompasses eco-ethological constraints that emphasize an understanding of normal aggression. The details include steps for constructing a fight arena, isolating and painting flies, introducing flies to an arena, and videotaping and scoring fights. These experimental protocols are in current use to identify candidate genes important in aggression and to elaborate the neuronal circuitry underlying the display of aggression and other social behaviors.

INTRODUCTION

Animals are born with sets of innate behaviors. Hardwired into nervous systems, these allow animals to react instinctively to environmental stimuli and thereby enhance their prospects for survival and reproduction. Examples of aggression are found throughout the animal kingdom and like other behavioral phenotypes, aggressive traits have been shaped by evolution (Huntingford and Turner 1987). Even though hardwired, aggression can be molded by hormones, experience, and environmental factors (Brain 1979; Monaghan and Glickman 1992; Lambert et al. 2006; Yurkovic et al. 2006; Koolhaas et al. 2007; Hines 2008; Kinsley et al. 2008). Plasticity increases the utility of this important social behavior, but in turn, increases the challenge of unraveling the molecular biological causes and modifiers of aggression.

Drosophila as a model system brings powerful genetic tools and capabilities to the study of aggression. The first description in the literature of this behavior in *Drosophila* was likely that reported by Sturtevant, nearly 100 years ago (Sturtevant 1915). Decades later, several papers described aggression in fruit flies in greater detail (Dow and von Schilcher 1975; Jacobs 1978; Hoffmann 1987a,b). Despite these publications, aggression in fruit flies was not well known until the publication of recent papers demonstrating that (a) fighting behavior could be observed between pairs of male and pairs of female flies in simplified fight arenas; (b) some behavioral patterns seen in male and female fights were similar whereas others were sex selective; (c) males formed hierarchical relationships, whereas females did not; (d) quantitative analyses of the behavior could be performed using traditional methods of behavioral analysis to reveal the structure of average fights including the construction of ethograms, the generation of transition matrices, and the calculation of first-order Markov chains (Chen et al. 2002; Nilsen et al. 2004); and (e) genes could be identified that were selectively associated with aggression (Dierick and Greenspan 2006; Edwards et al. 2006). *Drosophila* fights resemble those examined in other species in that they are broken up into brief encounters (meetings between the flies) where flies come together, interact for a period of time, and then break apart again. During each encounter a variety of behavioral patterns are seen: These show statistically varying transitions between them. In male fights, the patterns include approach, fencing, wing threats (of >2 sec duration), lunging, holding, boxing, tussling, chasing, and retreat. In female fights, approach, fencing, and brief wing threats that resemble the patterns seen in males are observed. At the mid-intensity level, however, instead of lunging, females shove each other and head butt the opponent in competition for resources. These sex-specific aggressive patterns are robust and used in complex social situations involving multiple male and female flies as reported in previous studies (Dow and von Schilcher 1975; Hoffmann 1987a,b; Ueda and Kidokoro 2002; Dierick and Greenspan 2006). Studies from the Hoffmann laboratory (Hoffmann 1987a), in particular, carefully characterized and identified many of the behavioral patterns that can also be observed in the one-on-one fight situation described here.

The field of aggression research has been advanced by studies in *Drosophila* through the use of highly informative combinations of genetic, genomic, and anatomical tools, illustrated by the following examples. (1) Two laboratories performed informative microarray analyses on independently selected and inbred highly aggressive lines derived from wild-type Canton-S flies (Dierick and Greenspan 2006; Edwards et al. 2006). (2) To address the question of how sex-specific patterns of behavior are wired into nervous systems, genetic tools were used in multiple studies to masculinize all or parts of the brains of females or feminize all or parts of the brains of males (Vrontou et al. 2006; Chan and Kravitz 2007). The results showed that male patterns of aggression could be transferred to females and female patterns of aggression to males by manipulation of genes of the sex determination hierarchy. (3) Several laboratories used traditional mutants to eliminate the function of well-characterized amine neuromodulators like serotonin, octopamine, and the neuropeptide F; and such manipulations led to decreases in aggression (Dierick and Greenspan 2007; Hoyer et al. 2008; Zhou et al. 2008). (4) Octopamine also modulated the choice between aggression and courtship in a particular segment of the fight ritual of male flies and led to enhanced male–male courtship behavior (Certel et al. 2007). In addition to identifying factors that regulate the expression of aggressive behavior, results from these studies also provide an entryway into identifying specific neurons and neuronal circuits that regulate aggression. Ultimately, the goal of most of these studies is to understand aggression in terms of network function and organization.

Designing an Experimental Arena for the Study of Behavior

Because aggression is used in the wild to defend or to obtain resources, it seems reasonable to include resources in the design of laboratory fight arenas. Thus, conditions can be established that resemble behavioral situations that animals normally encounter in the wild (Blanchard et al. 2003; Branchi and Ricceri 2004; Crabbe and Morris 2004). We therefore considered two key areas in designing our experimental paradigm: First, we attempted to include ethological constraints within our design (e.g., in the wild, losers commonly leave a resource or territory after social defeat; therefore space for retreat was included in our arenas); second, it is important to record as much of the behavior shown by the animals as possible for use in later quantitative analyses. Both considerations are essential in attempting to relate molecular findings to normal and abnormal behavioral observations.

Designing behavioral paradigms that include ethological considerations can be difficult: It is necessary to be able to constantly and consistently observe the behavior and, at the same time, provide laboratory conditions that resemble the real world. Fruit flies are, after all, real animals that respond to their environment as strongly as they do to the presence of another animal. In the paradigm outlined below, behavior is viewed with video cameras positioned at an oblique angle to the side of the fight arena. This angle allows accurate video images of the complex patterns and fight sequences shown by both animals viewed in three-dimensional space.

Routine analysis of the behavior involves tallying all of the behavioral patterns seen in individual encounters throughout fights. Other events that can be scored include latency to initiate fighting, number of encounters, winners and losers of individual encounters, and whether or not a hierarchical relationship is established. It also is important to consider the pairings of flies for fights. For example, pairing mutant and control flies may offer a relevant ethological situation, addressing questions of whether the mutants can win fights against wild-type opponents. However, the outcome of this particular pairing may be that mutant flies continually retreat from control flies, thereby always losing fights. To observe what aggressive responses a mutant fly is actually capable of performing, two mutant flies can be paired. The analysis of both types of pairings might provide different but useful data in understanding how a particular gene functions in a behavioral context. Finally, females fight for resources in same-sex pairing too. Thus studies that use genetic or engineered mutants and show interesting behavioral phenotypes in male flies also should be examined with pairs of female flies.

Observing and Analyzing Aggressive Behavior

For a behavior to evolve by natural selection, the reproductive benefits must exceed the reproductive costs of displaying that behavior. During aggression, the costs include risk of injury and death in combat between animals possessing dangerous weapons. As a consequence, sequences of interactions at different levels of intensity have evolved within most species that can lead to conflict resolution without escalation to all-out fighting behavior (Archer 1988; Wingfield et al. 1993). In defining aggressive status, investigators often ignore intermediate steps and score only the terminal phases of agonistic conflicts, the escalated fights. There is much to be gained, however, in analyses that include observation of all phases of the agonistic meetings between pairs of organisms. Included would be information on what triggers the aggression, the motivation of animals to fight, the progression of fights, changes in the motor patterns used by animals during fights, and, ultimately, answering the question of whether the usage of particular patterns defines how a conflict is resolved. In the fight chambers we use, for example, an ultimate winner might lunge more than 100 times before an ultimate loser retreats from the food surface. The key to the outcome of a fight of this type might not be increased aggressiveness by the ultimate winner, but the refusal of the ultimate loser to back down.

Use of the experimental paradigm described here and subjecting the gathered data to a comprehensive analysis of the patterns of behavior seen during fights is labor intensive. For high-throughput screening of mutants, it may not be practical to perform primary screens in this way. In these situations, it might be better to use smaller size chambers in a primary screen, followed by examination of any interesting mutants using a secondary screen allowing a thorough behavioral analysis to be performed. If interested in aggression as it relates to the normal usage of this behavior in the wild, whatever chamber is used should contain a defined territory and/or resources that animals might ordinarily be willing to compete over. Key fight components to examine in simplified scoring protocols might include latency to first encounter, latency to first lunge, and numbers of lunges performed by ultimate winners and losers. It is also important to ask whether a hierarchical relationship has been established within the time window of the fight. Additional parameters that can be scored in male fights include whether high-intensity aggressive components like boxing and tussling are seen and/or whether courtship components like wing extensions and abdomen bending are observed during fights.

As *Drosophila* geneticists, we understand that observing divergence from normality in the case of traditional or engineered mutants offers insights into the normal function of genes, neurons, and neuronal circuits. In extrapolating these techniques to the study of behavior in *Drosophila*, it is important to remember that the behavioral patterns observed during agonistic encounters are part of a complex repertoire that includes many behavioral patterns, each one of which bears a certain statistical likelihood of transitioning to another. Ignoring the complexity of the social situation may lead to false conclusions as to the consequences of created mutations. On the other hand, with careful examination of the behavior, valuable information can be gained toward understanding what aspects of aggression are fundamental to fights in all species of animals and which are selective for the species under examination.

Protocol

Scoring and Analyzing Aggression in *Drosophila*

This protocol describes the various steps for carrying out studies of aggression in *Drosophila*: creating a chamber to serve as a fight arena, isolating and identification of flies, introducing flies to an arena, and videotaping and scoring fights. Examples of videotaping these studies are shown in Movies 28.1, 28.2, and 28.3.

A complete video presentation of this protocol can be found at the JoVE online journal (see Mundiyanapurath et al. 2007).

MATERIALS

Reagents

Acrylic paint

Agarose (2%; molecular biology grade)
> Prepare by mixing the agarose with deionized water and microwaving until it fully dissolves.

Drosophila pupae and adult flies
> Keep all flies (stock and cross vials, isolation tubes with pupae) in an environmentally controlled 12-h light/12-h dark cycle incubator at the desired temperature (e.g., 25°C). Maintain eclosed, aging flies in the same incubator to control their activity periods.

Dry active yeast

Fly food, standard

Equipment

Black filter paper circles (8.5-cm) with 2-cm hole cut in center

Camera (Sony DCR-HC62 [or model with good optics])

Diamond glass cutter

Glue (Dow Corning DAP 100% silicon household adhesive for ceramics and china or Devcon High Strength 5 Minute Epoxy)

Incubator, environmentally controlled 12-h light/12-h dark cycle, pre-set to the desired temperature (e.g., 25°C)

Incubator, high-humidity environment (45%–55%), set at a temperature of 25°C

Isolation vials, borosilicate glass culture tubes (16 × 100-mm) (VWR International, no. 47729-576)

Laboratory label tape, colored

Microscope slides, plain (Fisher Scientific brand)

Mortar and pestle, small

Pasteur pipette

Polystyrene disposable sterile Petri dishes (100 × 15-mm) (VWR International)

Ruler

Scintillation vial caps (15 × 10-mm) (Kimble)

Straight dissecting needle with birch handle

Wooden toothpicks

METHOD

Assembling the Chamber Walls

See Figure 1.

1. Take two plain glass slides, mark the middle at the two edges, and, using a straight-edged ruler as a guide, score the glass with the diamond cutter. Slide the diamond cutter back and forth a few times generating a scraping noise.

2. Gently apply pressure to the scored line to break the slide into two halves. Face the scored side away from you for safety.

3. Generously apply the glass adhesive along an edge of one piece of the broken slide. Then align and firmly press the second slide piece onto the adhesive. Place onto a flat surface and allow to set for at least an hour.

4. After an hour, repeat the application of the adhesive step to the other edges and form a square chamber with the second two-slide piece. Allow the glue to harden overnight.

Placing the Walls in a Petri Dish

See Figure 2.

5. Fill the top of a Petri dish to a depth of at least 5 mm with a heated solution of 2% agarose.

6. Wait a few minutes, but while the agarose is still molten, place the walls of the chamber in the center of the dish.

7. Using a flame-heated straight dissecting needle, make a set of small holes (>1 mm) in the center of the bottom of the Petri dish for ventilation, and make a larger hole (~4 mm diameter) slightly to the side for introducing flies to the chamber. Place a piece of removable label tape over the larger hole.

Making the Food Cup

See Figures 2 and 3.

8. Heat a vial of fly food to the point at which the food is melted. Use a pasteur pipette to transfer the food to a scintillation vial cap, filling it to form a flat surface at the top.
   Be careful to avoid air bubbles and allow the food to cool.

FIGURE 1. Two sides of the fighting chamber glued together. The two halves of a marked, broken standard microscope slide were glued together at right angles.

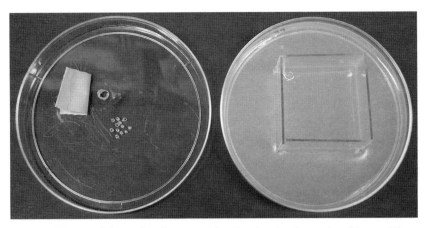

FIGURE 2. Top view of the two fighting chamber parts. The glass box has been placed into a 1% agarose solution. Ventilation holes and a single larger aspiration hole have been made in the cover.

9. Make fresh yeast paste by grinding a small amount of dry yeast in a mortar and pestle and adding a few drops of water. Add more yeast paste or water until the consistency is thick enough to allow a small visible drop to be picked up by a toothpick and applied to the center of the scintillation vial cap food surface.

10. Place the vial cap containing food in the center of the chamber and place the bottom of the Petri dish containing the small holes, inverted so that the sides are up and not obscuring the view, on top of the chamber.

11. Introduce two 3–5-d-old flies by aspiration into the chamber through the larger hole in the top, cover that hole with tape, and place a piece of black filter paper with a 2-cm hole cut in it into the top dish.

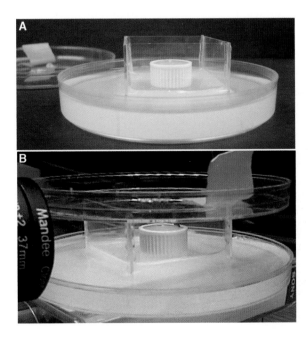

FIGURE 3. Side view of a finished chamber showing placement of the scintillation vial/food cup (*A*) and videotaping angle (*B*).

12. Place a light source above the chamber, far enough away that it will not heat the chamber, and positioned to effectively illuminate the entire food surface.

13. Move the entire assembly into a high-humidity environment (45%–55%) at a temperature of 25°C.

Social Isolation and Painting

Select flies maintained in an environmentally controlled 12-h light/12-h dark cycle incubator at the desired temperature (e.g., 25°C). Begin fights within the first hour after the onset of the light cycle to maximize activity and to reduce possible circadian rhythm effects.

14. Prepare isolation vials by adding 1.5 mL of boiled fly food to individual glass tubes. Allow the food to solidify.

15. Gently remove a late-stage male or female pupa from the cross or stock vial and place it on the side of the isolation vial.

16. Return isolation vials plus pupae to the light/dark cycle incubator.

17. Eclosed flies can be painted for individual identification, but this must be performed at least 24 h before fighting to eliminate any possible effects of the anesthesia.

    i. Anesthetize the flies under CO_2.

    ii. Place a tiny dab of acrylic paint on the thorax with a toothpick.
        Avoid contact of paint with head, wings, or abdomen.

    iii. Allow the paint to dry briefly, then gently transfer the fly back into the isolation vial.

Videotaping and Behavioral Analysis

As shown in Figure 3B, video cameras are positioned to record an oblique view of the fighting territory and the pair of flies.

18. Record the fights (usually for up to 1.5 h), beginning the video recording when flies are first placed in the arena.

    i.   To view encounters, attach the camera to a Mac computer.

    ii. Open the iMovie program and begin playing the tape.

    iii. Record the start time, the time when both flies are first together on the food surface, and the time of the first encounter (meeting between the flies when they start to interact and are within one body length of each other).

    iv. Score the time period of latency to the first encounter.

19. Create and observe movie clips of the individual encounters.

    i.   Use the space bar on the iMovie program (see iMovie Help) to generate clips of individual encounters.

    ii. Examine each clip or encounter either at a single frame level or at reduced speeds, and observe the behavioral patterns during each encounter can be scored.

    iii. If desired, score also the encounter duration, time between encounters, as well as any other information of interest, either from the original tapes or from the clipped encounters.

20. Enter the data from individual fights into Excel spreadsheets. Gather and record also any other pertinent data, including the genotypes of the paired animals, the calendar date, numbers of encounters in the fight, the patterns observed, and whether or not a hierarchical relationship was established.

ACKNOWLEDGMENTS

We thank members of the Kravitz Laboratory for many helpful discussions. This work has been supported by research grants from the National Institute of General Medical Sciences (GM074675 and GM067645 to E.A.K) and the National Science Foundation (IDS-075165 to E.A.K.).

REFERENCES

Archer J. 1988. *The behavioural biology of aggression.* Cambridge University Press, New York.

Blanchard RJ, Wall PM, Blanchard DC. 2003. Problems in the study of rodent aggression. *Horm Behav* **44:** 161–170.

Brain P. 1979. Hormones, drugs and aggression. *Ann Res Rev* **3:** 1–38.

Branchi I, Ricceri L. 2004. Refining learning and memory assessment in laboratory rodents. An ethological perspective. *Ann Ist Super Sanita* **40:** 231–236.

Certel SJ, Savella MG, Schlegel DC, Kravitz EA. 2007. Modulation of *Drosophila* male behavioral choice. *Proc Natl Acad Sci* **104:** 4706–4711.

Chan YB, Kravitz EA. 2007. Specific subgroups of FruM neurons control sexually dimorphic patterns of aggression in *Drosophila melanogaster. Proc Natl Acad Sci* **104:** 19577–19582.

Chen S, Lee AY, Bowens NM, Huber R, Kravitz EA. 2002. Fighting fruit flies: A model system for the study of aggression. *Proc Natl Acad Sci* **99:** 5664–5668.

Crabbe JC, Morris RG. 2004. *Festina lente*: Late-night thoughts on high-throughput screening of mouse behavior. *Nat Neurosci* **7:** 1175–1179.

Dierick HA. Greenspan RJ. 2006. Molecular analysis of flies selected for aggressive behavior. *Nat Genet* **38:** 1023–1031.

Dierick HA. Greenspan RJ. 2007. Serotonin and neuropeptide F have opposite modulatory effects on fly aggression. *Nat Genet* **39:** 678–682.

Dow MA, von Schilcher F. 1975. Aggression and mating success in *Drosophila melanogaster. Nature* **254:** 511–512.

Edwards AC, Rollmann SM, Morgan TJ, Mackay TF. 2006. Quantitative genomics of aggressive behavior in *Drosophila melanogaster. PLoS Genet* **2:** e154. doi: 10.1371/journal.pgen.0020154.

Hines M. 2008. Early androgen influences on human neural and behavioural development. *Early Hum Dev* **84:** 805–807.

Hoffmann AA. 1987a. A laboratory study of male territoriality in the sibling species *Drosophila melanogaster* and *D. simulans. Anim Behav* **35:** 807–818.

Hoffmann AA. 1987b. Territorial encounters between *Drosophila* males of different sizes. *Anim Behav* **35:** 1899–1901.

Hoyer SC, Eckart A, Herrel A, Zars, T, Fischer SA, Hardie SL, Heisenberg M. 2008. Octopamine in male aggression of *Drosophila. Curr Biol* **18:** 159–167.

Huntingford FA, Turner A. 1987. *Animal conflict.* Chapman Hall, New York.

Jacobs ME. 1978. Influence of β-alanine on mating and territorialism in *Drosophila melanogaster. Behav Genet* **8:** 487–502.

Kinsley CH, Bardi M, Karelina K, Rima B, Christon L, Friedenberg J, Griffin G. 2008. Motherhood induces and maintains behavioral and neural plasticity across the lifespan in the rat. *Arch Sex Behav* **37:** 43–56.

Koolhaas JM, de Boer SF, Buwalda B, van Reenen K. 2007. Individual variation in coping with stress: A multidimensional approach of ultimate and proximate mechanisms. *Brain Behav Evol* **70:** 218–226.

Lambert KG, Tu K, Everette A, Love G, McNamara I, Bardi M, Kinsley CH. 2006. Explorations of coping strategies, learned persistence, and resilience in Long–Evans rats: Innate versus acquired characteristics. *Ann NY Acad Sci* **1094:** 319–324.

Monaghan EP, Glickman SE. 1992. *Hormones and aggressive behavior.* MIT Press, Cambridge, MA.

Mundiyanapurath S, Certel S, Kravitz EA. 2007. Studying aggression in *Drosophila* (fruit flies). *J Vis Exp* **25:** 155.

Nilsen SP, Chan YB, Huber R, Kravitz EA. 2004. Gender-selective patterns of aggressive behavior in *Drosophila melanogaster. Proc Natl Acad Sci* **101:** 12342–12347.

Sturtevant AH. 1915. Experiments on sex recognition and the problem of sexual selection in *Drosophila. J Anim Behav* **5:** 352–366.

Ueda A, Kidokoro Y. 2002. Aggressive behaviours of female *Drosophila melanogaster* are influenced by their social experience and food resources. *Physiolog Entomol* **27:** 21–28.

Vrontou E, Nilsen SP, Demir E, Kravitz EA, Dickson BJ. 2006. *fruitless* regulates aggression and dominance in *Drosophila. Nat Neurosci* **9:** 1469–1471.

Wingfield JC, Doak D, Hahn TP. 1993. *Integration of environmental cues regulating transitions of physiological state, morphology and behavior.* Academic, New York.

Yurkovic A, Wang O, Basu AC, Kravitz EA. 2006. Learning and memory associated with aggression in *Drosophila melanogaster. Proc Natl Acad Sci* **103:** 17519–17524.

Zhou C, Rao Y, Rao Y. 2008. A subset of octopaminergic neurons are important for *Drosophila* aggression. *Nat Neurosci* **11:** 989–990.

MOVIE LEGENDS

Movies are freely available online at www.cshprotocols.org/drosophilaneurobiology.

MOVIE 28.1. One male fly approaches the other and lifts his wings in a threatening pose. This wing threat extends for >10 sec although the movie is clipped before the end. Movie kindly provided by Olga Alekseyenko.

MOVIE 28.2. The male fly on the left lunges three times in quick succession at the male on the right. Lunging is when one male fly rears up on his hind legs and snaps down on the other fly. Movie kindly provided by Olga Alekseyenko.

MOVIE 28.3. This extended encounter starts with the male fly on the left approaching a second male and lunging twice in quick succession. The second male turns and retaliates with lunges that escalate to boxing, tussling, holding, and many lunges. Movie kindly provided by Olga Alekseyenko.

29 | Courtship

Stephen F. Goodwin[1] and Kevin M.C. O'Dell[2]

[1]Department of Physiology, Anatomy and Genetics, University of Oxford, Oxford OX1 3PT, United Kingdom; [2]Molecular Genetics, Faculty of Biomedical & Life Sciences, University of Glasgow, Glasgow G12 8QQ, United Kingdom

ABSTRACT

Courtship can be defined as behavioral interactions between males and females, the evolutionary objective of which is copulation and the ultimate perpetuation of the species. To enable courting individuals to evaluate conspecificity and aspects of evolutionary or sexual fitness, courtship involves multiple sensory perceptions, actions, and responses that must be processed in an appropriate fashion. Therefore wild-type *Drosophila* need to show, perceive, and evaluate visual, chemosensory (taste and smell), tactile, and auditory cues, to enable courtship. In this chapter we review a wide range of protocols used to investigate courtship behavior in *Drosophila*. In addition, we evaluate the relative merits of specific experimental strategies and, in particular, discuss the nature of appropriate controls.

INTRODUCTION: SETTING THE GROUND RULES

To the naïve, courtship may appear to be the easiest research program to undertake. If we put a male and female fly together they will surely interact, and even the laziest or most inept researcher will be able to collect data of some sort. Indeed, even if the flies do nothing we can measure the fact that the flies have an absence of behavior, the underlying cause of which one can speculate on. However, even the slightest variation in the genetic status of the flies or the environment within which the experiments are conducted can lead to profound changes in behavior. Unfortunately, there is little consensus as to the "ideal" experimental design or strategy, and it is often difficult to compare data between laboratories.

Historically, courtship has been studied by traditional behavioral techniques. This requires direct observation and data collection, although video and digital analyses are now more widespread. For researchers whose primary interest is in evolution, function, or sex-specificity of courtship behavior, it is absolutely critical that experimental design is robust, that all realistic attempts are made to control for genetic and environmental variation, and that appropriate courtship controls are used. Courtship analyses are further complicated by the fact that both sexes modify their behavior in light of prior life history experiences (see Chapter 30). For researchers whose interest is in some other aspect of *Drosophila* biology, such as neural function, studies of courtship behavior can prove invaluable. Given that wild-type courtship requires such a high level of sensory and general fitness, it is dif-

467

ficult to imagine how a mutant or transgenic fly can perform normal courtship if any aspect of its biology is in some way perturbed. So courtship provides a simple tool by which the general fitness of any strain can be rapidly evaluated.

So perhaps the most critical advice is to control for everything. Indeed, as Donald Rumsfeld once said (February 12, 2002, Department of Defense news briefing): "[A]s we know, there are known knowns; there are things we know we know. We also know there are known unknowns; that is to say we know there are some things we do not know. …But there are also unknown unknowns—the ones we don't know we don't know." Critical environmental variables that we know need to be controlled include temperature (partly as a secondary consequence of locomotor behavior: Gilbert et al. 2001), humidity (at least regarding its effect on courtship conditioning: Gailey et al. 1984), time of day (reviewed by Ishida et al. 2001), quality of food, and mating chamber size. Flies should also be matched by age and genetic background. Mating frequencies are also affected by the numbers and ratio of each sex used and as sexually experienced flies modify their behavior, the sexual history of each contributing fly must be known and controlled for (reviewed by O'Dell 2003).

From a courtship perspective all wild-type strains are not the same (Eastwood and Burnet 1977). Indeed, using a genomic approach Baker et al. (2007) have shown that transcriptional variation among common laboratory strains of *Drosophila* can differ dramatically and they conclude that much of this variation reflects sex-specific challenges associated with the divergent morphological and regulatory pathways operating within males and females. Therefore a comparative courtship analysis between a wild-type strain and mutant or transgenic line may well reveal differences associated with genetic background rather than those of the mutation or transgene of interest. Clearly, it is essential to address problems of genetic background and inbreeding as well as being consistent with respect to other variables including age, prior social (sexual) experiences, and sex ratio (O'Dell 2003).

The nature and frequency of sexual interactions is profoundly affected by the size of the mating arena, yet there is little consensus as to the size or design of an optimal mating chamber (O'Dell 2003). As a rule the larger the chamber the longer it takes the courting couple to meet. One might argue that this is more reflective of the "natural" environment, but it is harder to observe, and especially record in detail, courtship in a large arena. Critically, stressed flies are likely to perform less courtship than unstressed flies. Therefore, whether the flies are transferred to the mating chamber by anesthesia, an aspirator, or some other less traumatic route, it is highly desirable that the flies have a 5–10 min recuperation period in their new environment. Therefore mating chambers can be designed with removable barriers that keep males and females apart before the experimental period.

Any investigation of animal behavior is complicated but studies of *Drosophila* courtship are particularly fraught with difficulties as flies show a rich ensemble of interactive behaviors that at least in some respects are open to the observer's interpretation (O'Dell 2003). One way of addressing this would be to use automated monitoring systems that remove the subjectivity from the process. Dankert et al. (2009) have recently introduced just such a system that monitors interacting pairs of flies and computes their location, orientation, and wing posture. They claim their system detects and distinguishes circling (orientation), wing extension (courtship song), and copulation and enables the automated construction of ethograms, saving considerable time and effort. It will be interesting to see whether the reported efficiency and accuracy of their automated system is adopted by laboratories that have previously relied on observations by experimenters.

The Best Laid Plans

The following protocol allows determination of two aspects of courtship in *Drosophila*: to assess whether there is a deficiency in mating frequency and, if this is the case, to resolve the nature of the specific problem. If the objective is simply to know whether a particular mutant or transgenic strain has a mating deficiency, this first part of the approach provides an objective, high-throughput strategy for the minimum of effort that is ideal for determining whether a specific strain has any courtship defect. Any strain that mates at a frequency comparable to that of wild-type flies must be considered reasonably fit in an evolutionary sense. Having established that a specific strain has an abnormal mating frequency, we are then interested in determining whether there is a specific courtship defect, described in the second half of the protocol. This determination will require the direct live observation or digital recording of courtship.

It is assumed that the researcher is comparing the courtship behavior of their mutant or transgenic line to that of an appropriate wild-type control. Every attempt should be made to control for genetic background, as mutant strains frequently acquire "modifiers" as a result of inadvertent selection in homozygous fly stocks (de Belle and Heisenberg 1996). As a rule, mutant genes or transgenes should be backcrossed into different wild-type strains (O'Dell 2003) to avoid phenotype suppression. Clearly experiments involving several mutant or transgenic strains can be investigated in parallel and this can be factored in as appropriate. Note that every time a new variable is introduced (mutant vs. wild type; young vs. old; food X vs. food Y) the effective experimental size is doubled.

MATERIALS

CAUTION: See Appendix for proper handling of materials marked with <!>.
See the end of the chapter for recipes for reagents marked with <R>.

Reagents

CO_2 <!> for anesthesia (foreplay)
   The use of CO_2 anesthesia can affect the quality of any behavior for a period of time (Joachim and Curtsinger 1990; Seiger and Kink 1993). Therefore, if at all possible, we recommend transferring the flies with a mouth aspirator (see Chapter 30, Fig. 1). The flies should then be given an opportunity to acclimatize to their new environment before commencing courtship.

Drosophila strains to be analyzed (healthy flies are happy flies)
   Flies should be of optimal size, well fed and watered, and stress-free. It is intuitively obvious that sick, stressed, or scared flies will not perform optimally in a behavioral task. Good fly husbandry will generate responsive healthy animals. Flies must not be maintained at such a density that impairs their fitness and well-being, and they must have high levels of both short-term and long-term activity. The optimal age for sexual activity is between 2 and 7 d posteclosion (for populations kept at 25°C).

Food media <R> (healthy body, healthy mind)
   Standard fly food media contains agar, glucose, sucrose, yeast, cornmeal, wheat germ, soya flour, molasses, and the mold inhibitor Nipagen M (0.1%). However, in learning and memory paradigms some authors use supplemented food media (Tully and Quinn 1985).

Equipment

Digital recording device (voyeurism)

The quality of the digital recording necessary will depend on the detail of the questions being asked. For example, if the question is simply whether a target female is attractive or not, then any reasonable recording that allows the observer to calculate the frequency of any courtship directed by a male to the female, or enables the observer to score the frequency of wing extension (courtship song), will suffice. Questions regarding more complex or specific behaviors or interactions, such as tapping or genital contact, may require a higher quality and/or higher magnification of digital recording.

General environment (the house of fun)

It is essential that flies be housed in a specialized room that can manage constant temperature (usually 25°C) and humidity (50%–70%). If an incubator is being used to house fly stocks, then a modification is required because living flies produce ammonia, which corrodes the copper wiring. As courtship behavior shows a strong circadian effect, the fly room or incubator should be on a 12 h:12 h light/dark (LD) cycle.

Mating chambers (setting the mood)

Mating chambers come in a variety of shapes and sizes, and there is little consensus as to the optimal mating chamber design (O'Dell 2003). However, a mating chamber must fulfill several functions: It must provide the correct environment for unstressed flies to acclimatize, meet, and have the opportunity to mate, while allowing the experimenter to observe the behavior either directly or via a digital recording device. In our laboratories we use circular mating chambers milled from a block of polythene and fitted with a clear Perspex lid. The cells are 21.5 mm in diameter and 4 mm deep and can be divided in two by a retractable divider. On opposite sides of the arena are 3-mm entrance holes stoppered with a removable Perspex plug through which individual flies can be aspirated (O'Dell et al. 1989).

METHOD

Estimation of Mating Frequency: Quantity or Quality?

1. Set up fly crosses in temperature-controlled, humidity-controlled, 12:12 LD circadian cycle environment. Avoid overcrowding.

2. Collect males and females from appropriate strains. Isolate the flies within 3 or 4 h of eclosion to ensure they have not mated. Keep the flies in sex-specific groups in food vials, and transfer to fresh vials on the day before experimentation.

   Retain the original holding tubes in which the allegedly virgin females were stored before experimentation and check for signs of offspring. The retention of the holding tubes will confirm whether the original females were virgins or not.

3. Without anesthetic, transfer 25 males and 20 females into a food bottle for 1 h. All classes should be set up in parallel.

   As every host laboratory is different it is best to "optimize" the experimental process. As a rule wild-type pairs should mate at a frequency of about 90%, so the host laboratory should alter the timescale of the "opportunity-to-mate" period such that wild-type pairings essentially mate at this frequency. If a strain is expected to mate more quickly or efficiently than wild type, then the "opportunity-to-mate" period can obviously be shortened to perhaps 25 or 30 min to calculate this.

4. After the mating period, transfer individual females to food vials. Discard males.

5. Revisit a few days later (e.g., 7 d) and examine the vials to determine whether females have produced offspring (in which case they mated) or not (in which case they did not mate).

Analysis of Specific Courtship Defects: What Am I Doing Wrong?

6. Set up fly crosses in temperature-controlled, humidity-controlled, 12:12 LD circadian cycle environment. Avoid overcrowding.

7. Collect males and females from appropriate strains within 4 h of eclosion. Keep in sex-specific groups or individually in fresh food vials. Transfer to fresh vials on the day before experimentation.

> Retain the original holding tubes in which the allegedly virgin females were stored before experimentation and check for signs of offspring. Here also, the retention of the holding tubes will confirm whether the original females were virgins or not.

8. Using an aspirator or some other nonanesthetic route, transfer male and female flies into the mating chamber for a defined period of time.

> See note to Step 3.

9. Score specific courtship behaviors live or record digitally.

> Whenever an observer is used to record data, it is essential that all courtships be scored blind such that the experimenter does not know which strain they are observing at any one time.

10. Determine the frequency and/or duration of specific courtship behaviors, as well as the sequence of courtship behaviors via ethograms.

BASIC STATISTICAL ANALYSES: IT'S ALL ABOUT TIMING

A variety of standard statistical analyses can be applied to courtship data. Clearly the simplest measure is whether pairs of flies perform a specific task, such as courting or mating, and how long it takes for such an event to occur. But courtship can also be broken down into the frequency, duration, and sequence of specific behaviors (O'Dell 2003).

In recordings of male–female pairs the most common and simplest analysis is the courtship index (CI), which was introduced by Siegel and Hall (1979). The CI is a measurement of the fraction of the observed trial during which time the male actively courts the female (or any other target). The CI is a cumulative score of all the courtship behaviors, including tapping, orientation, following, wing extension, and attempted copulation. Sequence analyses of courtship and the generation of ethograms, in which the order of events as well as their frequency and duration is recorded, clearly have the potential to explore aberrant courtship in great detail. However, only with the advent of more automated data capture techniques will it be possible to collect data on the scale and efficiency that will allow the potential of ethograms to be realized (Dankert et al. 2009). Ethograms will be invaluable for asking questions about interactions between courting flies and, in particular, investigating the consequences of abnormal behavior.

DISCUSSION: HOW WAS IT FOR YOU?

Historically most analyses of *Drosophila* courtship were broadly descriptive in nature, asking questions regarding the function of described sex-specific behaviors, especially their role in evolution and sexual isolation, and it is not the objective of this chapter to discuss an area of research that has been extensively reviewed (Dickson 2008; Villella and Hall 2008).

Irrespective of the biological question being addressed, predominantly behavioral assays still provide an invaluable tool with which to investigate the significance of specific courtship behaviors. Indeed Krstic et al. (2009) have exploited molecular tools to revisit questions regarding the relative importance of different aspects of sensory perception and processing to *Drosophila* courtship. In addition Nickel and Civetta (2009) recently used a strategy that is broadly in keeping with the method described in Steps 1–5 with a view to identifying genes responsible for asymmetric reproductive isolation between closely related species. A similar approach was used by Krupp et al. (2008) to determine how prior social experience modified pheromone expression and its consequence for male mating behavior.

Liu et al. (2008a) used genetic and pharmacological strategies to manipulate dopamine levels to study male–male courtship using assays of multiple males broadly following Steps 6–10 described above. Similarly, courtship can be used as a screening tool as part of broader studies, in which the exceptional sensitivity of sex-specific behavior assays can reveal perturbations or modifications of the nervous system. For example, Chang et al. (2008) analyzed the effect of glutamate on courtship (broadly in line with Steps 6–10 above) in FMR1 (fragile X ortholog) deletion strains with a view to identifying molecules that rescue the mutant phenotype.

Studies of the molecular basis of courtship often require a more detailed analysis of courtship. For example, Liu et al. (2008b) generated flies overexpressing juvenile hormone esterase and used analyses of mating frequencies (Steps 1–5) and courtship behavior (Steps 6–10) to help reveal consequent reduced pheromone abundance. Again, the success or otherwise of sophisticated transgenic approaches to our understanding of the molecular basis of sex-specific behavior relies heavily on the quality of the behavioral techniques and strategies used.

Ultimately, however sophisticated our experimental analyses may be, we should be cautious as to the ecological and ethological reality of the behavior we are observing and interpreting. Whereas there is little doubt that a revolution is occurring in our understanding of the molecular basis of sex-specific behavior in *Drosophila*, some behavioral sophistication must be lost as we impose a courtship environment on our flexible host organism. Clearly the next stage must be to investigate courtship in the natural environment, and we await the development of compost-proof digital cameras with great anticipation.

RECIPE

CAUTION: See Appendix for proper handling of materials marked with <!>.
Recipes for reagents marked with <R> are included in this list.

Food Media

Reagent	Quantity
Tayo agar	10 g
Sucrose	15 g
Glucose	33 g
Yeast	35 g
Maize meal	15 g
Wheat germ	10 g
Treacle	30 g
Soya flour	1 tbsp
Water to	1 L
Propionic acid <!>	5 mL
Nipagin solution <!>	10 mL

Nipagin solution: 25 g Nipagin M (Tegosept M, p-hydroxybenzoic acid methyl ester) in 250 mL ethanol.

Mix all of the solid reagents into 1 L of water (do *not* include nipagin or propionic acid at this point). Bring the solution to a boil, stirring constantly, then simmer for 10 min. Allow the solution to cool slightly (to about 70°C); then let it sit for 20 min. Add in the nipagin solution and propionic acid and dispense the media into aliquots: 8 mL into each fly vial and 70 mL into each fly bottle.

REFERENCES

Baker DA, Meadows LA, Wang J, Dow JAT, Russell S. 2007. Variable sexually dimorphic gene expression in laboratory strains of *Drosophila melanogaster*. *BMC Genomics* **8:** 454.

Chang S, Bray SM, Li Z, Zarnescu DC, He C, Jin P, Warren ST. 2008 Identification of small molecules rescuing fragile X syndrome phenotypes in *Drosophila*. *Nat Chem Biol* **4:** 256–263.

Dankert H, Wang L, Hoopfer ED, Anderson DJ, Perone P. 2009. Automated monitoring and analysis of social behavior in *Drosophila*. *Nat Methods* **6:** 297–303.

de Belle JS, Heisenberg M. 1996. Expression of *Drosophila* mushroom body mutations in alternative genetic backgrounds: A case study of the mushroom body miniature gene (*mbm*). *Science* **93:** 9875–9880.

Dickson BJ. 2008 Wired for sex: The neurobiology of mating decisions. *Science* **322:** 904–909.

Eastwood L, Burnet B. 1977. Courtship latency in male *Drosophila melanogaster*. *Behav Genet* **7:** 359–372.

Gailey DA, Jackson FR, Siegel RW. 1984. Conditioning mutations in *Drosophila* affect an experience-dependent courtship modification in courting males. *Genetics* **106:** 613–623.

Gilbert P, Huey RB, Gilchrist GW. 2001. Locomotor performance of *Drosophila melanogaster*: Interactions among developmental and adult temperatures, age and geography. *Evolution* **55:** 205–209.

Ishida N, Miyazaki K, Sakai T. 2001. Circadian rhythm biochemistry: From protein degradation to sleep and mating. *Biochem Biophys Res Comm* **286:** 1–5.

Joachim D, Curtsinger JW. 1990. Genotype and anesthetic determine mate choice in *Drosophila melanogaster*. *Behav Genet* **20:** 73–79.

Krstic D, Boll W, Noll M. 2009. Sensory integration regulating male courtship behavior in *Drosophila*. *PLoS One* **4:** e4457.

Krupp JJ, Kent C, Bileter J-C, Azanchi R, So AKC, Schonfeld JA, Smith BP, Lucas C, Levine JD. 2008. Social experience modifies pheromone expression and mating behavior in male *Drosophila melanogaster*. *Curr Biol* **18:** 1373–1383.

Liu T, Dartevelle L, Yuan C, Wei H, Wang Y, Ferveur J-F, Gu A. 2008a. Increased dopamine levels enhances male-male courtship in *Drosophila*. *J Neurosci* **28:** 5539–5546.

Liu Z, Li X, Prsifka JR, Jurenka R, Bonning BC. 2008b. Overexpression of *Drosophila* juvenile hormone esterase binding protein results in anti-JH effects and reduced pheromone abundance. *Gen Comp Endocrinol* **156:** 164–172.

Nickel D, Civetta A. 2009. An X chromosome effect responsible for asymmetric reproductive isolation between *Drosophila virilis* and heterospecific females. *Genome* **52:** 49–56.

O'Dell KMC. 2003. The voyeurs' guide to *Drosophila melanogaster* courtship. *Behav Processes* **64:** 211–223.

O'Dell KMC, Burnet B, Jallon J-M. 1989. Effects of the *hypoactive* and *inactive* mutations on mating success in *Drosophila melanogaster*. *Heredity* **62:** 373–381.

Seiger MB, Kink JF. 1993. The effect of anesthesia on the photoresponses of four sympatric species of *Drosophila*. *Behav Genet* **23:** 99–104.

Siegel RW, Hall JC. 1979. Conditioned responses in courtship behavior of normal and mutant *Drosophila*. *Proc Natl Acad Sci* **76:** 3430–3434.

Tully T, Quinn WG. 1985. Classical conditioning and retention in normal and mutant *Drosophila melanogaster*. *J Comp Physiol A* **157:** 263–277.

Villella A, Hall JC. 2008. Neurogenetics of courtship and mating in *Drosophila*. *Adv Genet* **62:** 67–184.

30 | Measurement of Courtship Plasticity in *Drosophila*

Aki Ejima[1,2] and Leslie C. Griffith[1]

[1]*Department of Biology, National Center for Behavioral Genomics and Volen Center for Complex Systems, Brandeis University, Waltham, Massachusetts 02454;* [2]*Career-Path Promotion Unit for Young Scientists, Kyoto University, Kyoto 606-8501, Japan*

ABSTRACT

Unsuccessful courtship reduces subsequent courtship behavior of male *Drosophila*. This experience-dependent behavior modification is called courtship conditioning and has been one of the major paradigms used to study learning and memory in *Drosophila*. This

chapter describes a protocol for quantifying courtship conditioning using a mated female trainer and discusses how the use of different trainer and tester flies and experimental conditions affect behavioral results.

INTRODUCTION

In *Drosophila*, courtship is an innate behavior that is genetically encoded. A naïve male who has been isolated after eclosion and has no social experience can perform normal courtship (see Chapter 29; Ejima and Griffith 2007). Although the basic behavior appears to be hardwired, prior sexual experience can allow a male to adjust the intensity of courtship to avoid fruitless energy expenditure courting unreceptive females.

In 1979, Siegel and Hall found that following unsuccessful courtship toward an unreceptive mated female, a male reduces his level of courtship (Siegel and Hall 1979). After 60 min of exposure to the mated female, the male showed reduced sexual motivation even toward a receptive virgin female. The effect lasted for >2 h. This experience-dependent behavior modification "courtship conditioning" has been one of the major paradigms for the study of learning and memory in *Drosophila* since then. The persistence of the memory can be modulated by altering training conditions. Courtship memory can last as long as 9 d if male flies are subjected to 5 h of spaced conditioning trials with a mated female trainer (McBride et al. 1999).

Courtship conditioning is considered to be a form of associative learning in which some negative cue(s) from the unreceptive mated female (unconditioned stimulus [US]) gets associated with positive courtship-stimulating cues (conditioned stimulus [CS]) (e.g., the courtship-stimulatory female pheromone [Table 1; Ackerman and Siegel 1986]). Tompkins et al. (1983) reported that olfac-

475

TABLE 1. Courtship learning assays

Trainer	Tester	Modification	Reference
Mated female	Virgin female	Associative memory	Siegel and Hall 1979
Virgin female + quinine	Virgin female	Associative memory	Ackerman and Siegel 1986
Virgin female + no copulation	Virgin female	Associative memory	Ejima et al. 2005
Mated female	Mated female	Repeated reinforcement	Gailey et al. 1984
Immature male	Immature male	Habituation	Gailey et al. 1982

tion-defective mutant males failed to exhibit courtship conditioning, indicating that the US and/or CS had an olfactory component. What are the olfactory cues? Mass spectrometry analysis of cuticular hydrocarbons revealed that mated females have a courtship-inhibitory pheromone, *cis*-vaccenyl acetate (cVA; Ejima et al. 2007), which is synthesized in the male ejaculatory bulb and transferred to the female during copulation. Is cVA an obligate US and essential for courtship learning? When a male is paired with a virgin female that contains no cVA, the pair copulate and the male shows no courtship conditioning (Siegel and Hall 1979). However, when copulation is artificially prevented, the male shows courtship suppression, suggesting that an unsuccessful experience can work as a US and an inhibitory pheromone is not essential for learning to suppress courtship (Ejima et al. 2005). In courtship conditioning using virgin trainers, the male associates the negative experience (US) with the pheromone profile specific to the age of the trainer females (CS) and reduces behavioral sensitivity to that pheromonal stimulus only. Thus, learning to suppress courtship can be driven by multiple classes of cues and the memory formed can be either general (decreased courtship toward all females) or selective (decreased courtship of only that type of female used for training).

Sexual experience with immature males also induces behavioral modification (Gailey et al. 1982). Mature males are attracted to immature males, who have not yet developed courtship-inhibitory male pheromones, and show high levels of courtship. However, during 30–60 min of exposure to an immature male, the mature male gradually reduces his courtship level and produces immature male-specific courtship suppression. This experience-dependent behavior modification is habituation, in which the chemosensory pathway for the immature male pheromones becomes desensitized by the prolonged exposure to the stimulus. This modification can be reproduced by pretest exposure to a pheromone extract alone (Gailey et al. 1982) and is sensitive to dishabituation, unlike courtship conditioning produced by female training (Ejima et al. 2005).

Protocol

Assay for Courtship Suppression

In this protocol we describe the basic behavioral assay for measuring short-term memory of courtship suppression induced by training with a mated female.

MATERIALS

CAUTION: See Appendix for proper handling of materials marked with <!>.
See the end of the chapter for recipes for reagents marked with <R>.

Reagents

CO_2 <!> (optional; see Step 6)
Drosophila melanogaster, naïve males and females (wild-type; e.g., Canton-S or Oregon-R)
Ether <!> (optional; see Step 6)
Food media (autoclaved) <R>

Equipment

Aspirator, mouth (see Fig. 1)
Bunsen burner (for mouth aspirator; see Fig. 1)
Courtship chamber
> A dissection plate plus slide glass (Fig. 1B) can be used as a courtship chamber. Slide the slide glass lid on the dissection plate and introduce animals through the gap into the hollow. A wheel chamber made with Plexiglas is suitable for observation of multiple animals. Rotate the top layer and introduce animals through an entrance hole into each cell (see Movie 30.1).

Filter paper, Whatman ashless no. 42
> Place water-soaked filter paper in each cell of the courtship chamber.

Incubator, set to 25°C, with a light/dark cycle
Light, white (for bright conditions) or dim-red (>700 nm, for dark conditions; see Discussion)
Mesh (e.g., Tetko 3-180/43, nylon stockings or cheese cloth) (for mouth aspirator, see Fig. 1)
Pasteur pipette, disposable, glass (e.g., Corning 7095D5X) (for mouth aspirator, see Fig. 1)
Scissors, fine (optional; see Step 6)
Test tube
Tube, PVC (polyvinyl chloride; e.g., Nalgene, premium tubing, nontoxic) (for mouth aspirator, see Fig. 1)
Vials
Video recording equipment

METHOD

Preparing Adult Flies

1. Raise flies at 25°C and 70% relative humidity under a 12 h:12 h light/dark (L/D) cycle.

2. Collect the flies on the day of eclosion and keep them in a 25°C incubator with an L/D cycle. Use autoclaved food with no sprinkled yeast granules (to avoid ethanol exposure).

A

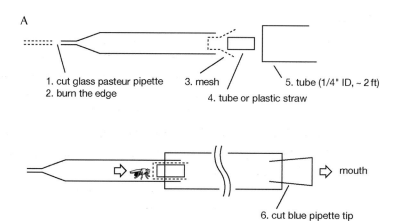

1. cut glass pasteur pipette   3. mesh   5. tube (1/4" ID, ~ 2 ft)
2. burn the edge
                                   4. tube or plastic straw

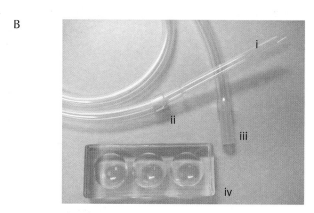

⇨ mouth

6. cut blue pipette tip

B

FIGURE 1. Constructing a mouth aspirator. (*A*) (1) Cut the tip of a glass pasteur pipette. (2) Burn the tip to heat-polish the sharp cutting edge and adjust the hole size. (3) Insert a piece of mesh cloth. (4) Insert a piece of plastic tube or straw to hold the mesh against the inside wall of the pasteur pipette. (5) Cover the tail end of the glass pasteur pipette with 1/4" ID (inner diameter) PVC tube. (6) Cut tapered tip of a blue pipette tip (mouthpiece) and insert it into the tail end of tube. A fly aspirated into the pasteur pipette will be trapped in front of the mesh. Arrows represent airflow on suction. (*B*) Photo of a mouth aspirator (from *A*) and a dissection-plate chamber. (i) Pasteur pipette; (ii) mesh trap; (iii) mouthpart; (iv) dissection plate and slide glass. If the PVC tube is long enough (~2 ft), the mouth aspirator can be carried hanging around the neck as a fashion statement.

      i. Collect naïve males and keep them in individual test tubes (to prevent pretest social experience).
     ii. Collect target females and males for mating and store in groups (e.g., 30 animals per vial).

3. Maintain the collected flies for 4 or 5 d until they are sexually mature.

Preparing Mated Female Trainers

4. On the day before courtship training, pair a virgin female and a male (both group-stored, 3 or 4 d old) and wait for copulation.

5. Collect and store the mated females whose copulations lasted longer than 14 min (Canton-S).

    The average "full-copulation" duration is variable, depending on the genetic background. Determine the cutoff duration by checking fertility and remating rate in advance.

Courtship Training and Test

6. To prepare immobilized tester females, anesthetize them with CO_2 or ether, and freeze-kill or decapitate them with fine scissors immediately before the experiment.

   Mild CO_2 exposure for 50 min causes immobilization for the following 10 min.

   Apply strong ether vapor to kill the females. Some investigators prefer to avoid the use of ether, because ether fumes may be emitted by the female.

7. Assemble a courtship chamber containing wet filter paper in each cell to maintain humidity.

8. Label the chamber with the date and fly identification numbers (do not include genotype or specific manipulation).

9. Transfer a test male (individual-stored, naïve) to each cell of the chamber using a mouth aspirator without anesthesia (see Movie 30.1).

10. Allow the males to acclimate to the chamber for 5 min.

11. Using a mouth aspirator, introduce a trainer female (live and mated the day before) to each cell and immediately start video recording (10 min; initial courtship, Fig. 2).

12. Observe the pair together for a total of 60 min. Video record the last 10 min of the 60 min training (final courtship).

13. Transfer the trained male to a clean courtship chamber. Add an immobilized virgin tester female (from Step 6) immediately after training (0 min memory) or after a certain period of time (1–3 h, short-term memory) and start video recording (10 min; test courtship).

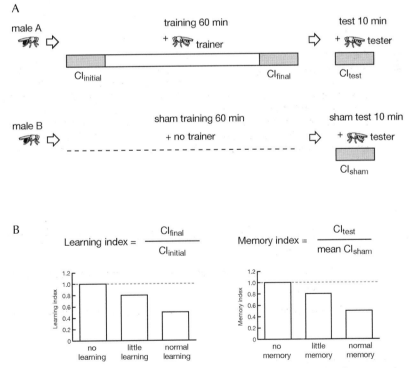

FIGURE 2. Schematic figure of training and test process. (*A*) A male (male A) is paired with a trainer female for 60 min. The courtship response is video recorded during the first and last 10 min (shadowed box). After the training, the trained male is transferred to a clean chamber and paired with a tester female (test courtship). A sham control (male B) is kept alone in a chamber and then receives the same test (sham test). (*B*) Representative learning and memory indices. If the courtship level in the final 10 min of training is equivalent to that of the first (initial) 10 min (learning index ≈ 1.0), the male did not reduce the courtship intensity during the training period. A memory index ≈ 1.0 means that courtship after training is at the same level as sham, indicating that there was no behavioral modification by the training.

14. As a sham-trained control, isolate a male in a courtship chamber for 60 min and then test for courtship toward the tester female.

Analysis

15. Analyze the video recordings. During the behavioral analysis, the observer should refer to only the fly identification numbers and must be blind to the other experimental information (e.g., genotype or specific manipulation).

16. For each 10-min recording, calculate courtship index (CI), $CI_{initial}$, CI_{final}, CI_{test}, and CI_{sham}, which is the fraction of time a male spent in courtship activity in the 10-min observation period (see Chapter 29; Ejima and Griffith 2007).

    If pairs copulate during the training period or when male flies show low $CI_{initial}$ (<0.1), then the data are eliminated from future analysis.

17. Determine the learning index, which is a ratio of the courtship level in the final 10 min of the training (CI_{final}) to that of the initial 10 min ($CI_{initial}$), and which shows how the courtship level is changed during training (Fig. 2B).

18. Calculate the memory index by dividing CI_{test} by the mean of the sham control courtship levels (CI_{sham}). A memory index ≈ 1 indicates that there is no memory because the courtship level of the trained males is equivalent to that of sham-trained males.

DISCUSSION

The courtship conditioning protocol we described in this chapter is the most conventional one used and involves a mated female trainer and a virgin female tester. Courtship memory produced by this protocol is considered to be associative, but it requires active performance of the male during training for the effective modification (Ackerman and Siegel 1986; Ejima et al. 2005). A potentially damaged or sick male that has shown low courtship at the initial recording is considered to lack the training experience, therefore data from such males should be excluded from learning and memory analysis (see Step 16). In contrast, courtship suppression toward immature males can be reproduced without the male's performance of courtship during training. Exposure to the pheromone extract of immature males alone can induce behavioral modification (Gailey et al. 1982). For such a pheromone experiment, a double-layered courtship chamber with a mesh barrier can be used (see Ejima and Griffith 2007). Another feature that separates courtship habituation from associative courtship conditioning is sensitivity to a dishabituation stimulus. The courtship suppression toward immature males can be dishabituated (e.g., by vortexing for 1 min), whereas the courtship memory produced by conditioning using female targets is resistant to such mechanical disturbance (Ejima et al. 2005).

The persistence of the courtship modification largely depends on the experimental conditions during training. In a small chamber the male gets continuous exposure to the US and CS cues during training (for review, see Griffith and Ejima 2009) somewhat analogous to a "massed" training procedure. This type of small chamber training produces memory that lasts for hours. If training is carried out in a large chamber that allows relatively frequent breaks in between courtship bouts (a "spaced" training paradigm), memory can last for days (McBride et al. 1999). Lighting conditions are also critical for courtship experiments. Because visual cues are strong positive cues for male courtship, the availability of visual cues can sometimes mask subtle changes in sexual motivation of the trained male or in changes in the salience of nonvisual cues (Joiner and Griffith 2000). Dim red light with wavelength >700 nm can be used to substantially reduce visual information during the assay and increase emphasis on nonvisual cues. The effects of these and other sensory input on courtship are discussed in Ejima and Griffith (2007).

RECIPE

CAUTION: See Appendix for proper handling of materials marked with <!>.
Recipes for reagents marked with <R> are included in this list.

Food Media

Reagent	Quantity
H_2O	60 L
Agar	600 g
Yeast	1950 g
Cornmeal	4571 g
Dextrose	63000 g
NaKT (sodium potassium tartrate)	480 g
$CaCl_2$ <!>	60 g
Lexgard in EtOH <!>	169 g

Autoclave and distribute into individual test tubes.

REFERENCES

Ackerman SL, Siegel RW. 1986. Chemically reinforced conditioned courtship in *Drosophila*: Responses of wild-type and the *dunce, amnesiac* and *don giovanni* mutants. *J Neurogenet* **3:** 111–123.

Ejima A, Griffith LC. 2007. Measurement of courtship behavior in *Drosophila melanogaster*. *Cold Spring Harb Proto* doi:10.1101/ pdb.prot.4847.

Ejima A, Smith BP, Lucas C, Levine JD, Griffith C. 2005. Sequential learning of pheromonal cues modulates memory consolidation in trainer-specific associative courtship conditioning. *Curr Biol* **15:** 194–206.

Ejima A, Smith BP, Lucas C, van der Goes van Naters W, Miller CJ, Carlson JR, Levine JD, Griffith LC. 2007. Generalization of courtship learning in *Drosophila* is mediated by *cis*-vaccenyl acetate. *Curr Biol* **17:** 599–605.

Gailey DA, Jackson FR, Siegel RW. 1982. Male courtship in *Drosophila*: The conditioned response to immature males and its genetic control. *Genetics* **102:** 771–782.

Gailey DA, Jackson FR, Siegel RW. 1984. Conditioning mutations in *Drosophila melanogaster* affect an experience-dependent behavioral modification in courting males. *Genetics* **106:** 613–623.

Griffith LC, Ejima A. 2009. Courtship learning in *Drosophila melanogaster*: Diverse plasticity of a reproductive behavior. *Learn Mem* **16:** 743–750.

Joiner MA, Griffith LC. 2000. Visual input regulates circuit configuration in courtship conditioning of *Drosophila melanogaster*. *Learn Mem* **7:** 32–42.

McBride SM, Giuliani G, Choi C, Krause P, Correale D, Watson K, Baker G, Siwicki KK. 1999. Mushroom body ablation impairs short-term memory and long-term memory of courtship conditioning in *Drosophila melanogaster*. *Neuron* **24:** 967–977.

Siegel RW, Hall C. 1979. Conditioned responses in courtship behavior of normal and mutant *Drosophila*. *Proc Natl Acad Sci* **76:** 3430– 3434.

Tompkins L, Siegel RW, Gailey DA, Hall JC. 1983. Conditioned courtship in *Drosophila* and its mediation by association of chemical cues. *Behav Genet* **13:** 565–578.

MOVIE LEGEND

Movie is freely available online at www.cshprotocols.org/drosophilaneurobiology.

MOVIE 30.1. Introduction of flies into a wheel chamber using a mouth aspirator.

31 Sleep and Circadian Behavior Monitoring in Adult *Drosophila*

Cory Pfeiffenberger, Bridget C. Lear, Kevin P. Keegan, and Ravi Allada

Department of Neurobiology and Physiology, Northwestern University, Evanston, Illinois 60208

ABSTRACT

This section details how to set up and interpret experiments of sleep and circadian behavior in adult *Drosophila*, as measured using a locomotor activity assay. In particular, the following protocols describe how to adapt TriKinetics' *Drosophila* Activity Monitoring System to analyze circadian rhythmicity, period length, behavioral phase, sleep duration, consolidation of sleep, and latency to sleep. The general approaches described here can be applied to a wide range of behavioral activity experiments, including sleep deprivation analyses and general studies of hypoactivity and hyperactivity.

INTRODUCTION

Adult behavioral assays have been used with great success in *Drosophila melanogaster* for gene discovery and characterization. The locomotor activity assay has proven to be a particularly useful behavioral paradigm, as the robust nature of this assay makes it possible to identify altered behavior patterns in small populations or even individual flies. Locomotor activity assays have been used for several decades in *Drosophila* to study circadian rhythms, the daily patterns of behavior driven by internal clocks. More recently, this system has also been used to investigate *Drosophila* sleep. The locomotor assay that we describe can be used to simultaneously assess both circadian and sleep behavior; several methods can be used to analyze the data generated from such assays.

Circadian Rhythms in *Drosophila*

Circadian pacemakers drive daily physiological and behavioral rhythms in organisms ranging from cyanobacteria to humans (Bell-Pedersen et al. 2005). In *Drosophila melanogaster*, several aspects of behavior and development are subject to circadian regulation, including the emergence of adult *Drosophila* from pupal cases (eclosion). Among the most robust of these rhythms are the daily patterns of rest and activity showed by *Drosophila* adults (Cirelli and Bushey 2008). Distinct activity patterns are observed when *Drosophila* are exposed to daily light cycles that approximate a typical outdoor environment (12 h light:12 h dark). Under these conditions, activity levels peak near "dawn" and "dusk" (Fig. 1A, arrows). As daily activity rhythms are driven by an internal circadian clock, they persist strongly in flies shifted from cycling light/dark conditions (also known as light entrainment)

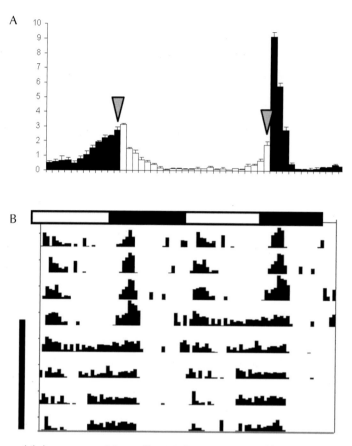

FIGURE 1. Adult *Drosophila* locomotor activity profiles. (A) Group activity profile for wild-type *Drosophila* maintained in 12 h light:12 h dark conditions (LD) over 5 d. Average, normalized activity levels are determined for the group over the last four LD days. Black bars represent activity levels during dark phase, and white bars indicate activity levels during light phase. Error bars indicate standard error of the mean. Gray arrowheads indicate activity increases before light transitions. (B) Activity profile of an individual wild-type fly subject to 4 d of LD followed by 5 d of constant darkness (DD). Vertical marks indicate scaled activity levels. This profile is double plotted, such that each day is represented twice on the graph. In LD, white horizontal bars indicate light phase; in DD, white bars indicate subjective light phase. Black horizontal bars indicate dark phase (LD) or subjective dark phase (DD). This fly maintains a rhythmic activity pattern during DD, with ~24-h periodicity.

to constant-dark conditions. In wild-type *Drosophila*, these free-running activity rhythms are maintained at near-24-h periodicity for many days (Fig. 1B).

Drosophila melanogaster was one of the first model organisms used to identify circadian rhythm genes. The first successful circadian screen in *Drosophila* examined pupal eclosion rhythms, identifying populations of flies in which adults emerged from pupal cases with a different daily distribution than wild-type flies (Konopka and Benzer 1971). In the decades that have followed, locomotor activity assays have been used much more prominently for screens and other circadian experiments. Several features of the locomotor activity assay are advantageous for high-throughput analyses, leading to the rapid discovery and functional characterization of many *Drosophila* circadian rhythm genes. Whereas examinations of pupal eclosion yield just a single datum for each fly, locomotor activity assays provide several days of data for each individual. This makes it possible to detect a reproducibly normal or aberrant rhythm from a single fly. Even the primary progeny from a mutagenesis experiment can be individually screened. Another important feature of locomotor activity assays is the commercial availability of a highly efficient, automated system for data collection. The *Drosophila* Activity Monitoring (DAM) System (TriKinetics, Waltham, MA) allows for continuous data collection from large numbers of individuals. Although the typical experiments last ~2 wk, only the initial setup and postrun analyses require significant effort on the part of the researcher, thus allowing multiple experiments to be run at any given time. Finally, the use of analysis tools such as those we will describe (ClockLab, Actimetrics, Wilmette, IL, and the Counting Macro [written by K. Keegan and T. Lee; available from the Allada Laboratory on request]) make it possible to quickly analyze multiple aspects of circadian behavior from each experiment.

Standard Circadian Locomotor Assay and Analyses

For the reasons described above, we focus on the use of the adult locomotor activity assays to assess circadian behavior. The TriKinetics DAM System records activity from individual flies maintained within 5 × 65-mm (diameter × length) glass or plastic tubes containing food at one end and an air-permeable plug at the other (Fig. 2A). The tubes are placed in activity monitors that direct an infrared beam through the midpoint of each tube (Fig. 2B). When a fly crosses the infrared beam, it is counted as an activity event. Events detected over the course of each consecutive sampling interval, or bin, are summed and recorded through the duration of an experiment for each fly. We normally assay circadian behavior during an entrainment period followed by constant conditions. A typical run consists of 5 d of 12 h light:12 h dark cycling (LD entrainment), followed by 7 d of constant darkness (DD). Circadian analyses are performed on both the LD and DD portions of the run, as described below.

LD Circadian Activity Profiles (aka LD Eductions)

Locomotor activity profiles obtained during LD entrainment can provide valuable circadian information. Wild-type flies normally show an increase in activity just before the lights-on and lights-off transitions (Fig. 1A, arrowheads). These so-called "morning" and "evening" activity increases are driven by particular groups of circadian pacemaker neurons, and many circadian mutants show a loss or phase shift in these activity increases (Nitabach and Taghert 2008). LD activity profiles provide a qualitative image of daily activity bouts, and the data used to generate these profiles can be used to quantitatively assess the phase and/or amplitude of particular activity bouts. For a run con-

FIGURE 2. TriKinetics *Drosophila* Activity Monitoring System. (*A*) 5 × 65-mm glass tube containing an individual *Drosophila melanogaster* adult. Agar-based food is present on the right side of the tube, while the opposite end contains a plastic cap with air hole. (*B*) Tubes are loaded into a TriKinetics *Drosophila* Acitivity Monitor. Two models are shown here. In each model, an infrared emitter/detector reports an activity count each time a fly crosses the center of the tube. Activity counts are sent through a TriKinetics Power Supply Interface Unit to a data collection computer (not pictured).

sisting of 5 d of LD entrainment, we normally assess circadian behavior on days 2–5. Day 1 of the experiment is devoted to loading flies and allowing them to entrain to the light conditions. LD activity profiles are assessed from groups of flies, usually corresponding to a particular genotype. Activity levels among individual flies are normalized, and the average normalized activity is determined within the population for any given time. For the eduction graph shown (Fig. 1A), the population activity from days 2–5 is averaged into a single 24-h day.

DD Period and Rhythmicity Assessment

Additional circadian information can be obtained by analyzing locomotor activity during constant conditions. Most circadian rhythm mutants maintained in LD entrainment will show some type of ~24-h activity pattern, a response caused by light transitions. Activity assessments made from entrained flies that have been shifted to constant darkness can provide insight into the state of internal clocks and the ability of these clocks to drive rhythmic outputs. From our typical LD DD run, we assess both free-running rhythmicity and period length with chi-squared periodogram analysis. Defects in rhythmicity can reflect defects in clock function, entrainment/input to the clock, or output from the clock. Changes in free-running period length are normally associated with alterations to the core circadian clock itself but can also reflect changes in intercellular communication between circadian pacemakers.

Drosophila Show a Sleep State

Sleep is a highly conserved behavior that is required for optimal performance and, in many cases, life of an organism (Everson et al. 1989). Although the sleep state does not permit an animal to procreate, feed, or defend itself, making it seemingly disadvantageous, sleeplike behavior has been observed in all animals studied to date, including the nematode *Caenorhabditis elegans* (Raizen et al. 2008) and the fruit fly *Drosophila melanogaster* (Hendricks et al. 2000; Shaw et al. 2000). High conservation of this behavior across a broad range of evolutionary diversity implies a very important function; consistent with this assumption, it has been observed that rats deprived of sleep show reductions in longevity similar to those deprived of food (Everson et al. 1989). Despite the clear importance of sleep, an explanation for why we and other animals enter this vulnerable and seemingly unproductive state remains lacking.

Drosophila demonstrate a behavioral state that shows traits consistent with sleep: They undergo periods of relative behavioral immobility that coincide with an increased arousal threshold after ~5 min of inactivity (Hendricks et al. 2000; Shaw et al. 2000). These bouts of inactivity are regulated by circadian (flies tend to sleep at the same time[s] each day) and homeostatic (flies deprived of sleep will show a sleep rebound when the deprivation is released) mechanisms (Hendricks et al. 2000; Shaw et al. 2000). Intriguingly, the *Drosophila* sleep state can be modified pharmacologically with many of the same agents that alter human sleep, suggesting similar molecular mechanisms underlie sleep in both fruit flies and humans. For example, caffeine decreases sleep duration, whereas the antihistamine hydroxyzine increases sleep in both mammals and fruit flies (Hendricks et al. 2000; Shaw et al. 2000). Finally, although fruit flies lack the necessary brain architecture to produce electroencephalograph (EEG)-recordable brain waves, fly sleep correlates with a change in brain activity as determined by measuring local field potentials (van Swinderen et al. 2004).

Sleep in *Drosophila* is Measured Behaviorally

To overcome the absence of an observable EEG, *Drosophila* sleep researchers have developed behavior-based paradigms to infer when a fly is asleep, as opposed to immobile but awake. It has been reported that periods of behavioral quiescence lasting ≥5 min correlate well with an increase in arousal threshold; these are categorized as sleep (Hendricks et al. 2000; Shaw et al. 2000). Thus, behavioral quiescence, detected with automated apparatuses like TriKinetics' DAM System, is the most commonly reported measure of *Drosophila* sleep. These behavior-based analyses show three desirable traits: (1) high throughput—thousands of flies can be monitored simultaneously, (2) efficient use of resources—DAM System–based studies are relatively easy and inexpensive to conduct, and (3) data uniformity—the DAM System provides a standardized data format as well as a central source to which technical problems and/or questions can be directed (www.trikinetics.com).

Video monitoring, another approach for examination of *Drosophila* sleep, was developed in large part to overcome a known weakness of behavior-based measures of *Drosophila* sleep, overestimation of sleep (flies may move without detection by the infrared beam technology of the DAM System; such active bouts are erroneously recorded as sleep). Indeed, video monitoring has been used to show that the DAM System provides an accurate representation of sleep duration during the dark phase, but can overrepresent sleep in the light (Hendricks et al. 2000). Whereas video monitoring provides maximum fidelity with respect to determination of fly activity and/or sleep, physical space requirements, initial hardware costs, data storage needs, and the relative difficulty of assay setup make it an impractical solution to high-throughput analyses (Zimmerman et al. 2008). Thus, the DAM System is the preferred method for data collection in which high-throughput analyses of *Drosophila* behavioral data are concerned, particularly those related to circadian rhythms and sleep.

Standard Sleep Analysis in *Drosophila*

For the vast majority of our experiments we monitor the flies for 5 d of 12 h light:12 h dark (LD) followed by 7 d of 24 h darkness (DD). Light is a powerful stimulus that can mask or otherwise alter many behaviors, especially sleep; this paradigm allows us to analyze sleep both in the presence and in the absence of this stimulus. Five days of LD allow the flies to recover from the process of being loaded into the tubes and to adjust to their new environment while still providing 4 d of "normal" activity under LD conditions. Monitoring 7 d of DD provides ample time for the flies to adjust to life in the absence of an external light stimulus, especially highlighting the influence of the internal molecular clock and the light-independent regulation of sleep. In addition to these explicit benefits, our 12-d schedule is practical because it (1) is long enough to provide data sufficient for multiple LD and DD analyses, but not so long that fly mortality becomes a hindrance; (2) matches the generation time of *Drosophila*, allowing the researcher to set up crosses and run behavior experiments with the same period; and (3) is amenable to a 2-wk schedule, allowing behavior experiments to always be started on the same day of the week.

There are many creative ways to analyze binary beam-crossing data to provide information about sleep. The protocols detail the collection of data, processing, and interpretation of three aspects of sleep: duration, consolidation (alternatively referred to elsewhere as fragmentation), and latency to sleep. Although this chapter does not provide a protocol for conducting a sleep deprivation experiment, it does provide the essential groundwork. Additional considerations and relevant references are described below.

Sleep Duration: Sleep duration can be expressed as either minutes of sleep per day or percent time spent asleep. In addition, one can determine how the flies' sleep is distributed throughout the day. Although short-sleeping fly data are generally trustworthy, data collected from long-sleeping flies should be scrutinized because it is possible that a reduced number of beam interruptions could be attributed to illness or lethargy rather than sleep.

Consolidation: We determine consolidation of sleep by three measurements. First is the average sleep bout length (ABL). Second, to counter the influence of very short sleep bouts (5 min as compared with 45 min), we also determine a consolidation index (CI), which is a weighted ABL. The individual sleep bout durations are squared before averaging, thus diminishing the influence of potentially dubious short sleep bouts. The third measure of consolidation is to simply count the total number of sleep bouts. Based on these measurements one can form hypotheses about why a mutant is sleeping more or less. For example, if a fly sleeps less because it is defective in maintaining sleep, one might expect the CI to be reduced while the number of sleep bouts would show a compensatory increase.

Latency to Sleep: Latency is a measurement that reveals the flies' ability to fall asleep. It is defined as the time lapse between lights-off and the first sleep bout. Latency to sleep, ABL, and/or CI can be used to provide a simple measure of sleep intensity in flies. If a fly has a very high sleep need, one would predict latency to sleep to be short and the CI to be large.

Sleep Deprivation and Rebound: Rebound after deprivation provides a measurement of the homeostatic regulation of sleep. There are several published and unpublished approaches. The most common involve mechanical sleep deprivation, either by manually tapping the fly housing or using a motorized apparatus to rotate the DAM boards, jostling the flies to prevent them from sleeping (Hendricks et al. 2000; Huber et al. 2004). As an alternative to mechanical deprivation, one can use a pharmacological approach to disrupt sleep; it is accepted that flies respond to several stimulants known to disrupt human sleep, such as caffeine and modafinil (Hendricks et al. 2000; Shaw et al. 2000). The antiepileptic carbamazepine has been shown to robustly deprive flies of sleep (Agosto et al. 2008). A caveat associated with any form of sleep deprivation is that it can be difficult to separate sleep-dependent effects from those of general stress. Using multiple sleep deprivation paradigms (mechanical and pharmacological) can help to address this issue.

Activity Monitoring with the DAM System

The following protocol details how to monitor locomotor activity levels in *Drosophila* using the *Drosophila* Activity Monitoring (DAM) System from TriKinetics. This protocol can be used to study circadian behavior, sleep/wake, and general activity levels (hypoactivity vs. hyperactivity).

MATERIALS

Reagents

Behavior food (5% sucrose [w/v] and 2% bacteriological grade agar; see Step 1)
Drosophila (wild-type) to be monitored (see notes to Step 8 for selection of flies)

Equipment

Behavior tubes (TriKinetics; 5 x 65-mm for *D. melanogaster*)
Caps for both ends of tubes (TriKinetics)
> We use TriKinetics black vinyl caps on the "food" side of behavior tubes and yellow polyvinyl chloride (PVC) caps on the "fly" side of the tubes. Before first use, yellow caps must be prepared by cutting the length of the cap (to 2–3 mm) and poking a small hole for air. Alternatively, small cotton plugs can be placed in the "fly" side of the tube, and the "food" side of the tube can be dipped in wax to seal.

DAMFileScan software (download from www.trikinetics.com)
DAM System monitoring software (download from www.trikinetics.com), run continuously on the PC/Mac
Dark room
Drosophila Activity Monitors (TriKinetics; Fig. 2)
Power Supply Interface Unit (PSIU; TriKinetics), connected to a PC or Mac via USB
Programmable incubator, temperature- and light-controlled (Percival Scientific)
Telephone cords (RJ-11) and connectors (standard, usually supplied by TriKinetics)

METHOD

Preparing Behavior Food

1. Mix 5% sucrose (w/v) and 2% bacteriological grade agar (w/v) with distilled water in an autoclave-safe container.

2. Autoclave the mixture to dissolve the ingredients and sterilize the solution.

3. Cool the food and store in a sealed microwave-safe container at 4°C for ~2 wk.

4. Fill a 500-mL beaker with the glass behavior tubes, positioned vertically (approximately 160 tubes).
   > Be sure to use clean and sterile glass behavior tubes. After use, we soak the tubes in a low-percentage bleach solution overnight, wash and rinse thoroughly, and autoclave on dry cycle.

5. Pour the liquid behavior food into the beaker using a funnel or metal spatula to direct the food to bottom of the beaker. Fill slowly until the food is ~2–3 cm up the behavior tubes.

6. When the food has solidified, remove the tubes, wipe off residual food from the outside of the tube, and place a black cap on the food side of the tube. Store the behavior tubes in a sealed container at 4°C for up to 2 wk.

Loading Flies into Activity Monitors

7. Warm prepared behavior tubes to room temperature

8. Place one fly in each tube.

> Anesthetics (e.g., CO_2) can have long-term effects on sleep. Avoid keeping the flies on a CO_2 pad for >5 min.
>
> Males and females show distinct sleep/activity patterns; the sex of each fly should be noted.
>
> Older flies have sleep patterns that are distinct from those of young flies. Avoid analyzing flies that are not age matched.
>
> Be careful not to allow the fly to touch the food while anesthetized, as they can easily become stuck.

9. Cap the tube and load it into the activity monitor (Fig. 2); each monitor can hold 32 flies.

> Verify that the food does not block the infrared beam.

Starting the Behavior Run

10. Connect the loaded activity monitor to the PSIU using a telephone cable. Additional telephone jacks/cables can be used to increase the number of monitors connected to a single PSIU.

11. Verify correct connection with DAM System software. Note that we typically set the DAM System software to record activity at 1-min intervals. In this case, no more than 60 boards should be assigned to any PSIU/DAM System, as the DAM System generally reads data at a rate of ~1 board/sec.

> If a monitor is correctly connected, the DAM System will temporarily display a yellow signal next to that monitor number, followed by a green signal to indicate full connectivity. If a monitor is not connected, the signal should read red. In case of a general connectivity problem or a loop in the system, the signal will read black. In this case, check all connections immediately to avoid data loss.

12. Program the incubator with the temperature and light cycle of interest.

> A typical run consists of 5 d of 12 h light:12 h dark (5LD), followed by 7 d of 24 h dark (7DD) at 25°C. Monitors are loaded and connected during the light phase of the first LD day. To minimize the effects of light coming from outside the incubator, we run all of our behavior experiments in incubators kept in a dark room.

13. Once a behavior run has started, do not open the incubator until the run is over.

Collecting and Preprocessing Data (DAM System/DAMFileScan)

14. The DAM System collects locomotor activity data as a single .txt file for each monitor. After the run has ended, copy each monitor file of interest into a new folder.

> We perform the data preprocessing and all subsequent analyses on a separate computer, as we typically run only the DAM System software on the data collection computers.

15. Open DAMFileScan (download from www.trikinetics.com).

> DAMFileScan allows the user to specify experiment start and end dates/times, consolidate recorded bins into longer bin lengths, and create individual .txt files for each fly, as required by ClockLab and the Counting Macro.

16. Import the data of interest by clicking the "Select Input Data Folder" button, and choose the folder containing the DAM System monitor .txt files.

17. Select the range of Monitors to be analyzed, and click the "Scan" button to verify the files are correct.

18. Choose a Bin Length. For circadian analyses, bin data to 30 min. For sleep analyses, use 1-min bins.

    To perform both sleep and circadian analyses, extract data separately as 30-min and 1-min bins.

19. In the "First Bin to Save" region set the Date to the day the monitors were loaded. Set the Time to the first bin after midnight (e.g., for 30-min bins, set this to 00:30:00; for 1-min bins, set to 00:01:00).

    Note that the saved data will include several hours of data before the monitors were actually loaded/connected. However, the first day of data collection can be excluded from all analyses.

20. In the "Last Bin to Save" region set the Date to the last day of your experiment, and the Time to the latest possible time (e.g., if you collected data at 2:00 p.m., set this to 14:00:00).

21. In the drop-down menu for Output File Type choose "Channel Files." This will generate an individual file for each of the 32 flies ("channels") from a given monitor.

22. In the Extra Readings menu, make sure that "Sum into bin" is selected. This is relevant if the Bin Length exceeds the DAM System Reading Interval.

23. Enter a run name for this experiment. All files created by DAMFileScan will contain this name.

24. Click the "Save" button to process the data.

    This will create a new folder with the same name as your run name, containing fly-specific .txt files. The folder will be in the same directory as the folder containing the original DAM System .txt files.

TROUBLESHOOTING

Problem: There is a high rate of mortality.

Solution: If a significant fraction of wild-type flies (>10%–20%) die during a 12-d behavior run, one should investigate potential causes. There are generally three causes.

1. Quality of the flies entering behavior. Verify the flies are healthy when they are placed in the tubes. Confirm that the length of the experiment does not exceed the average lifespan of the genotype(s) under investigation.

2. Sterility and cleanliness of the tubes and food. Verify the tubes and caps are thoroughly cleaned, and the food and tubes are autoclaved before use.

3. Desiccation. We have observed that, especially during the winter, food can dry out quickly. To help combat this, a tray of water is kept in the behavior incubator to maintain relatively constant humidity during dry months. One can also use incubators with controlled humidification and/or monitored humidity.

Problem: DAMFileScan "First Bin to Save" or "Last Bin to Save" does not include the entire behavior run.

Solution: If Monitor files are removed from the DAM System read folder, then they will be replaced by new files that lack all previous data. To resolve this, the data from the newer monitor file can be copied/pasted at the end of the older file. See DAM System Notes (www.trikinetics.com) for more information.

Processing Circadian Data

This section describes the use of ClockLab (Actimetrics, Wilmette, IL), a MATLAB-based program, and the Counting Macro (K. Keegan and T. Lee, Evanston, IL), an Excel-based program, to analyze circadian locomotor activity data collected from the DAM System. This section provides a detailed protocol for analyzing free-running rhythmicity and period length for individual flies and assessing group activity plots during both entrainment and constant conditions.

MATERIALS

Equipment

> ClockLab software (Actimetrics, http://www.actimetrics.com/)
> Counting Macro (written by K. Keegan and T. Lee; available from the Allada Laboratory on request)
> MATLAB software (The Mathworks; http://www.mathworks.com/)
> Microsoft Excel
> Run folder generated by DAMFileScan (from Protocol 1)

METHOD

Periodogram Analyses: Assessing Rhythmicity and Period Length

1. Open the ClockLab program within MATLAB.

2. Click the "Open File" button in the top left corner, and select an individual .txt file from within the run folder generated by DAMFileScan. If the data have been extracted correctly, each .txt file should correspond to an individual fly, and the file name will indicate the monitor (M001–M120) and position number (C01–C32) of that fly.

   > For Run Name 500a, Monitor 5, Fly 31, the corresponding file should be named 500aM005C31.txt.

3. In the bottom right panel, select a Block Size that corresponds to the Bin Length chosen in DAMFileScan (typically 30 min for ClockLab circadian analyses).

4. Set the Start Time in military format that corresponds to Zeitgeber/circadian time 0 (ZT/CT 0). For a typical LD DD behavior run, this refers to the time that the lights turn on each day during entrainment.

5. Set the End Time for the analysis. This should precede or equal the last data collection time that was set in DAMFileScan.

   > For consistency, we often set this to CT 0-1, typically the earliest time at which data are collected at the end of a behavior run.

6. At this point, the Start/End Dates (in bottom right panel) likely encompass the entire behavior run. (Optional) To examine the activity profiles from individual flies, click the Actogram button in the top right panel.

   > We normally view "scaled" actograms (gray drop-down menu under Actogram button). Once an actogram is open, the activity profile can be double-plotted by selecting the Actogram tab and

then checking "Double Plot." Start/End dates to be viewed can also be modified in the main ClockLab window.

7. Before proceeding with rhythmicity analyses, set analysis preferences by clicking on the "Preferences" button in the top right panel of the main ClockLab window.

   i. Under Periodogram (top right), set the desired confidence interval.

      We typically use a confidence interval of 0.01.

   ii. Set the range of period lengths to be considered.

      We usually consider period lengths ranging from 14 h to 34 h. For assessing ultradian rhythms or extremely long periods, these start and end values can be adjusted.

   iii. Close the Preferences window.

8. Select the Start/End dates for rhythmicity analyses in the main ClockLab window. The start date should be the first day of constant conditions, whereas the end date is usually the date on which the data were collected.

   For consistency, we normally analyze exactly 7 d of constant conditions, even if the available data extend beyond that day.

9. Click on "Batch Analysis" (bottom left button). In the left half of the Batch Analysis window, select the individual flies to be analyzed, or click the "Select All" button. In the right half of the window, check the boxes corresponding to the analyses to be performed.

   i. Check the "P'gram peak1" box to perform chi-squared periodogram analyses on all of the selected flies. The start date of analysis should again be set to the first day of constant conditions. However, owing to a bug in the ClockLab software, we normally set the end date *one day before* the last day of the run.

      Example: A 5LD 7DD run started on 01/09/2009 and ended after 9:00 a.m. on 01/21/09, with a lights-on time of 8:00 a.m. and a lights-off time of 8:00 p.m. The DD portion of this run began on 1/14/09. In the main ClockLab window (Step 8), Start Date would be set to 1/14/09 (Day #6) and End Date would be set to 1/21/09 (Day #13). The Start time should be set to 8, whereas the End time should be set to either 8 or 9. In the Batch Analysis window, P'gram peak should be assessed from 1/14 (Day #6) to 1/20 (Day #12). This will analyze 7 d of DD data.

   ii. Check the "Avg Counts" box to measure activity levels of each fly on the last day of the run. This will assess whether individual flies survive throughout the duration of the run. The start date of this analysis is normally the day before the end of the run (*Day #12 in the example above*). As described above, the end date should also be set to the day before the end of the run (*Day #12 in the previous example*).

   iii. Several additional analyses can also be performed using Batch Analysis. See ClockLab documentation for more information. Once all desired analyses have been checked and set up, click the "Go" button (bottom right corner).

10. The batch analysis will generate a text file called "BatchAnal" within the run folder. This file can be opened using Microsoft Excel. If difficulties are encountered opening this file, make sure that Excel is set to enable "All Documents" and not just "All Readable Documents." Excel should then use the "Text Import Wizard" to open the BatchAnal file (simply hit "Finish" or Return to open the file).

11. Within the batch analysis file, several columns of data are reported for each fly. A chi-squared periodogram assessment for a rhythmic fly will provide a Tau value, a Power value, and a Significance value. The Tau value is the period length (within the range specified) at which rhythmicity is calculated to be strongest. The Significance value represents the minimum measurement that would be considered rhythmic based on the confidence interval and the Tau value. The Power value is the observed rhythmicity measurement for that fly at the Tau value

indicated. To determine rhythmicity for an individual fly, we subtract the Significance measurement from the Power measurement. Flies that do not meet the minimum threshold for rhythmicity will display "NaN" within the Tau, Power, and Significance columns. We consider the rhythmicity value of these flies to be equal to 0.

12. Remove any flies that did not survive the entire behavioral run from all data analysis. Activity measurements on the last day of the run are reported within the "Avg_Counts/min" column. Any fly with a "0" value can be excluded from rhythmicity analysis. This value can also be used to exclude extremely low activity flies.

> Example: A fly with an Avg_Count/ min reading <0.01 had less than 15 total activity counts on the last day of the behavior run.

13. Compile the period values and rhythmicity data for all living flies. We normally report the following parameters for a given genotype: number of flies analyzed, percentage of rhythmic flies, average period (Tau) among rhythmic flies, and average rhythmic power (Power − Significance). The periodogram analysis would consider any fly with a Power − Significance (P − S) value >0 to be rhythmic. However, based on blind assessments of activity profiles from circadian mutant flies, we observe that very low P − S values often do not represent an obvious rhythm. Therefore, we consider flies rhythmic only if the P − S value is ≥10, and we exclude Tau values for flies in which P − S is <10. All P − S values from living flies (including "0" for "NaN" flies) are included when calculating average rhythmic power.

The Counting Macro: Importing Data and Setting Parameters for Circadian Analysis

Version 5.19.5 of the Counting Macro (CM) can be obtained from the Allada Laboratory. For installation instructions, please see the readme files included with the current distribution of the CM. The CM is designed to run on a Windows PC running Microsoft Excel versions 2000, 2002, and 2003. Some features of CM 5.19.5 are not compatible with Excel 2007, but this issue will be addressed in a future revision. The CM is not compatible with Excel for Mac. Unfortunately, it will not operate properly under any version of Excel for OSX. CM operation on a Mac running Excel in an emulatd Windows environment has not been tested.

14. Import your data into the CM.

    i. Open the CM Excel file.

> Excel will provide a warning that the file contains macros. Choose to "Enable Macros."

    ii. Click on the tab farthest to the right entitled "Main."

    iii. Click on the main button entitled "Drosophila Activity Monitor Counting Macro."

    iv. In the Main Window, click on the "Clear All Sheets" button to clear any preexisting data from the file.

    v. Click the "Import Data" button.

    vi. Click the "Choose Directory" button to select the run folder containing the fly-specific .txt files created by DAMFileScan; double-click any .txt file within that folder to open.

    vii. In Section 4, highlight the board numbers you would like to import from the folder selected.

> If the txt files have the ".txt" extension in the file name, check the corresponding box.

    viii. In Section 5, "Select Import Settings," choose the "Standard" option.

    ix. Click the "Import" button.

> This will create board-specific tabs, containing fly-specific columns of data. The data are arranged as number of beam breaks per bin, in which each cell is a bin.

    x. After import has finished, click the "Main Menu" button at the bottom of the window.

15. Provide run specifications for processing.

i. Click the "Run Specs" button in the Main Window.

ii. Choose a bin length at the top of the window; this should match the bin length specified in the DAMFileScan.

iii. Choose a time for lights-on. We normally use an 8:00 a.m. lights-on for all standard runs.

iv. To provide specifications for LD analysis, check the corresponding box.

v. For a normal 5LD analysis (see Fig. 1A), choose the "LD Eduction" option within "Standard Parameters," and skip to Step 15ix. Alternatively, input settings manually below.

vi. Choose a "Run day to start analysis." We generally use Day 2 (in which Day 1 is the day the boards were loaded).

> Circadian behavior normally entrains to LD conditions by Day 2. For temperature entrainment or unusual lighting regimes, additional entrainment days can be added before the start of data analysis.

vii. Choose the "# Days of Analysis." For 5LD, we normally analyze 4 d of data (Days 2–5).

viii. Set the "Offset Hours from Lights On." A typical circadian LD group activity profile (also known as eduction) plots data beginning with 6 h before the lights-on time, followed by the 12-h light phase, followed by the 6 h after lights-off. This corresponds to an Offset setting of –6 h.

ix. To perform a second circadian analysis, check the DD analysis box (or skip to Step 15xii). You can then choose one of the standard options: a 2LD 7DD daily activity profile, or a profile of DD Day 1.

x. If you would like to perform a nonstandard analysis, choose the "Run date to start analysis," "# Days of Analysis," and "Offset Hours from Lights On."

xi. Enter the "Time of Last Data Collection." This is typically >CT 1 on the last day of the run (≥9:00 a.m. for a run in which lights-on was set to 8:00 a.m.). Note that this parameter is not relevant for a typical circadian analysis.

xii. Click the "OK" button.

16. Remove dead and (optional) low-activity flies from analysis.

i. Click the "Fly Triage" button in the Main Window.

ii. To triage flies from LD analysis, click the "Label as "LD," don't delete" option button.

iii. In Section 2, choose the criteria by which the flies will be triaged. To remove only the flies that are dead/low activity at the end of the run, choose "Use End Range Selection Criteria Only." To remove flies that show low activity throughout the run, choose "Use Total and End Range Selection Criteria."

iv. In Section 3, choose the bin length used.

v. In the neighboring drop-down menu, choose "LD Eduction (Based on LD Run Specs)" and click the "Import Preset" button.

> This will fill in the bin ranges below based on the criteria specified in the Run Specs window (see Step 15).

vi. In Section 4, if "Total Range" criteria were included, select "Activity/Bin." Enter the minimum number of activity counts per bin to avoid triage.

vii. In the "End Range" section, select "Activity/Bin" and enter 0 in the field.

> Alternatively, to remove flies with extremely low activity on the last day of LD, select "Total Activity" and choose a minimum # of total activity counts that the fly must exceed to be included in the analysis.

viii. In Section 5, click the "…" button to specify a directory and file name for a fly triage summary file.

ix. Click the "GO" button. In the next window, verify the specs and click "OK."

x. To triage flies from DD analysis (as set in Run Specs), repeat the LD steps with the following modifications.

xi. In Section 1, choose "Label as "DD," don't delete," and be sure the "Preserve previous triage labels" box is checked.

> If left unchecked, the LD triage information will be lost and must be repeated.

xii. In Section 3, choose a preset based on the type of analysis set up in Run Specs. Preset options for circadian analysis include 2LD 7DD or DD Day 1 analysis. Either of these circadian options should also work for a manual entry made in the DD analysis Run Specs.

xiii. In Section 5, enter a new file name and click the "GO" button.

> If the same file name is used the LD triage file will be replaced.

xiv. Click the "Quit to Excel" button.

17. Enter names and location information for fly groups to be analyzed.

i. Click on the "MultiEducer" tab.

ii. In the left column ("Eduction Name"), enter the genotype or other defining characteristic for each fly population.

> Avoid using any characters (such as "/") that are not allowed in PC file names.

iii. In the left column ("Location"), enter the monitor number and positions associated with the name in the neighboring cell. Note that monitors (also known as boards) are named as indicated in the corresponding Excel worksheets: RunName|Board #.

> Follow the example presented in cell C7. As a shortcut, selection a Location cell and choose the "Add Board" button. Select the board of interest and click "OK" and the location cell will be partially filled. Note that commas are only needed *after* fly positions if another board location will be added.

The Counting Macro: Generating Circadian Activity Profiles

18. In the Main Window, click the "Activity Educer" button.

19. For LD analysis, click on "Import Analysis 1" (aka LD Run Specs). For DD analysis, click on "Import Analysis 2" (aka DD Run Specs). Note that LD and DD analyses must be run separately.

> The import option will use the parameters previously set up in "Run Specs." Alternatively, you can input values manually.

20. In Section 2, "Ignore flies that have been labeled in Triage with:," choose LD or DD based on the analysis to be performed.

21. Check the "Include profile graph" box.

22. Under "Multiple (from "MultiEducer" sheet)," Section 3, click the "..." button. Select a directory in which to save the graphs.

> When performing both LD and DD analyses, make sure to choose different directories, or files will be written over.

23. In Section 4, "Educe:," click the "Produce Eduction" button. An Excel file should be generated for each of the fly groups specified in the MultiEducer.

Interpreting Circadian Activity Profile Files

24. Open an Excel file generated by the Activity Educer to examine the activity profiles and the data used to generate these profiles.

25. Click on the "Eduction Workbook" tab. This contains the raw data for all flies that were included in the analysis. Each column represents an individual fly, and each row (starting with

Row 5) represents the number of activity counts (beam breaks) within a bin. Note that any fly marked as LD, DD, or LD/DD in Row 4 would be excluded from the corresponding analysis.

26. Click on the "Normalized Data" tab. This displays the normalized activity levels for each fly that was included in the analysis. Only the data designated for analysis is normalized, so you likely have to scroll down to see this data. To normalize, the raw activity measurement from a given bin is divided by the average activity/bin over the entire analysis period. Thus, the average normalized activity for any individual fly over the length of the analysis is equal to 1. Normalization prevents high-activity flies from having greater influence on the activity profiles than low-activity flies.

27. Click on the "Graphs" tab. This should contain two graphs and additional data.

    i. The top graph plots the normalized activity levels over 24 h and assumes an offset of –6 h, which is a typical plot used to display the LD or DD Day 1 activity profile. The data used to generate this profile are shown beginning in Column C, starting at the same row in which the Normalized Data are displayed. Column C reports the average normalized activity values over the first 24 h of analysis (Day 1). Depending on the number of days analyzed, subsequent columns report the average normalized activity during those days. Next, the "Average" column reports the normalized activity for a particular time period as averaged over all days. The "Std Error" column next to the "Average" column indicates the variability among different days. The values in the "Average" column are used to generate the top (Eduction) graph, and the values from the "Std Error" column are used to set the error bars in that graph.

    ii. Use and limitation of the top (Eduction) graph. The eduction graph compiles data from multiple days into a single day. This graph assumes a 24-h period and an offset of –6 h. It should only be used to compile data from multiple days during 24-h-based entrainment (e.g., LD eduction), not during constant conditions. It can be used to assess single 24-h periods of constant conditions (such as DD Day 1), but it will not generate error bars for single-day graphs. The bars on this graph are filled black or white to indicate 6 h of dark, followed by 12 h of light, followed by 6 h of dark. If the analysis performed does not conform to these conditions, then the bars can be reformatted as needed.

    iii. The lower graph plots the normalized activity levels over the entire length of the analysis (X axis = Bin #). It does not compress the data into a single day. Note that for 30-min bin data, each 24-h day contains 48 bins. The data used to generate this plot are shown in Column A ("Profile"), beginning in the same row as the Normalized data/Eduction data. The "Profile" value is simply the average normalized activity for all flies in the group. The standard error for this measurement is shown in Column B but is not plotted on the graph.

    iv. Use of the lower (Profile) graph. The profile graph can be used to examine group activity over multiple days. This is typically used to display data that include multiple days of constant conditions (such as a 2LD 7DD profile). It can also be used to compare different days of LD entrainment to each other.

    v. Phase assessments. The group or individual fly data in these files can be used to assess phase of activity. For example, you may want to determine the time at which "evening" activity peaked, in which "evening" can be any time from midday up until the time the lights turn off (between ZT 6 and ZT 12, or bins 25–36 in a typical LD eduction plot). To determine this for a group of flies, simply look for the maximum value of the normalized activity within that time range. Often, we look for the maximum increase in activity to assess the phase of an activity peak. Such an activity increase can be looked for over a fixed or variable time period. For a fixed time period, simply subtract one bin from another over a specified time range (e.g., Bin 36 – Bin 32 for a 2-h increase), and look for the largest values. To examine the largest increase over a variable time period (such as 2–4 h), use the MIN command in Excel (e.g., Bin36 − (MIN(BIN29:BIN32))). Many other possibilities exist for analyzing the data within Excel.

TROUBLESHOOTING

Problem: Error Message "Run-time error "9": Subscript out of range."

Solution: This is a general error message associated with the visual basic programming language. Essentially, the program is looking for a variable that does not exist. During Import, this error is obtained if the ".txt extension" box is not checked when needed. When running the Activity Educer, this error can occur if the Location cells in the MultiEducer tab are formatted incorrectly. Verify the user input is correct.

Problem: Error Message "Run-time error "1004.""

Solution: This error occurs when the Counting Macro attempts to create a file with a name forbidden by Windows owing to an unacceptable character (e.g., "/"), as entered in the MultiEducer section. Remove any characters that cannot be used in a file name by Windows.

Problem: Error Message "could not load an object because it is not available on this machine" or "compile error, cannot find project or library."

Solution: Install COMDLG32.OCX using the following instructions:

COMDLG32.OCX is included with the current distribution of the Counting Macro.

Windows XP:

Place COMDLG32.OCX in your Windows\System folder on your C: (or boot) drive.

Then you need to register it with Windows Registry. To do that, go to the Start Menu   Run. Enter this command in the box:

    regsvr32 \windows\system\COMDLG32.OCX

(Note: There is a space after "regsvr32." Just copy the whole command into the box. "windows\system" refers to the location of your file.)

Press OK and you should get a notification that it succeeded.

Windows Vista:

Run cmd.exe as administrator (right-click on cmd.exe and "Run as administrator").

Type in "regsvr32 \windows\system\COMDLG32.OCX" at the command prompt (any directory will work).

Processing Sleep Data

This section details how to use the Counting Macro, an Excel-based program, to process data created with the *Drosophila* Activity Monitoring System for sleep analyses. Specifically, it details the steps necessary to convert the raw data created by the DAM System into sleep duration and consolidation data, broken down into the LD, L, D, and DD phases of a behavior experiment.

MATERIALS

Equipment

Counting Macro (CM)
DAMFileScan Microsoft Excel

METHOD

CM: Importing and Preprocessing the Data

1. Import your data into the CM.
   i. Open the CM Excel file.
      Excel will provide a warning that the file contains macros. Choose to "Enable Macros."
   ii. Click on the tab farthest to the right entitled "Main."
   iii. Click on the main button entitled "Drosophila Activity Monitor Counting Macro."
   iv. In the Main Window that opens, click the "Import Data" button.
   v. Click the "Choose Directory" button to select the folder containing the fly-specific .txt files created by DAMFileScan.
   vi. In Section 4, highlight the board numbers you would like to import from the folder selected.
      If the txt files have the ".txt" extension in the file name, check the corresponding box.
   vii. Click the "Import" button.
      This will create board-specific tabs, containing fly-specific columns of data. The data are arranged as number of beam breaks per bin, in which each cell is a bin.

2. Provide run specifications for processing.
   i. After the data are imported, click the "Main Menu" button at the bottom of the window.
   ii. Choose a bin length at the top of the window; this should match the bin length specified in the DAMFileScan.
   iii. Choose a time for lights-on. We generally use 8:00 a.m.
   iv. To provide LD specifications, check the corresponding box.
   v. Choose a "Run date to start analysis." We generally use Day 2 (in which Day 1 is the day the boards were loaded). This allows time for the flies to adjust to their new environment.

vi. Choose the "# Days of Analysis." (For 5LD this is 4.)

> For a 5LD analysis, you can also choose "LD Sleep Analysis" from the "Standard Parameters" drop-down menu.

vii. To provide DD specifications, check the corresponding box.

viii. Choose a "Run date to start analysis." (For a standard 5LD 7DD run this would be 6.)

ix. Choose the "# Days of Analysis." (For 7DD this is 7.)

x. Enter the "Time of Last Data Collection." We generally collect our data after 2:00 p.m. and use 14:00:00 here.

> This value will be important for triaging dead or sick flies.

xi. Click the "OK" button.

3. Triage dead flies.

i. Click the "Fly Triage" button in the Main Window.

ii. To triage the flies that died during the LD phase click the "Label as "LD," don't delete" option button.

iii. In Section 2, choose the criteria by which the flies will be triaged. We generally use both "Total Range Selection Criteria" (i.e., activity levels during the experiment) and "End Range Selection Criteria" (i.e., activity levels after the experiment has ended).

> These criteria permit removal of all sick (i.e., lethargic) flies during and immediately after the experiment.
>
> This will fill in the bin ranges below based on the criteria specified in the Run Specs window (see Step 2).

iv. In Section 4, in the "Total Range" region, select "Activity/Bin" and enter the number of crossings per bin required to cross once per 5 min (e.g., if your bin length is 1 min, this value is 0.2).

v. In the "End Range" region, select "Activity/Bin" and enter 0 in the field.

vi. In Section 5, click the "…" button to specify a directory and file name for a fly triage summary file.

vii. Click the "GO" button. In the next window, verify the specs and click "OK."

viii. To triage flies for the DD phase, repeat the LD steps with the following exceptions.

ix. In Section 1, choose "Label as "DD," don't delete," and check the "Preserve previous triage labels" box.

> If unchecked, the flies triaged during LD will not be specified properly in the CM.

x. In Section 3, choose "Standard DD Sleep (Based on DD Run Specs)" from the drop-down menu and click the "Import Preset" button.

xi. In Section 5, enter a new file name and click the "GO" button.

> If the same file name is used the LD triage file will be replaced.

xii. Click the "Quit to Excel" button.

4. Enter Analysis Names for the Board Positions.

i. Click on the "MultiSleep" tab.

ii. In the left column, enter the genotype or other defining characteristic for each fly population.

iii. In the right column, enter the board number and positions associated with the name in the neighboring cell.

> Follow the example presented in cell C7. As a shortcut, highlight the Location cell of interest and click the "Add Board" button. Select the board of interest and click "OK" and the cell will be partially filled.

CM: Sleep Analysis

5. In the Main Window, click the "Sleep Analysis" button.

    Use the MultiSleep region; this will use the position information entered in the MutiSleep tab (see Step 4).

6. In the "Analysis Type" region, select "L/D/LD analysis" to analyze LD data, select "DD analysis" to analyze DD data. Both can be selected to analyze LD and DD simultaneously.

7. Click the "Import" button to import the specifications entered into the Run Specs window (see Step 2).

8. Click the "Choose Directory" button to select a directory in which a summary file and analysis name–specific files (see Step 4) will be saved.

9. Fill in a name in the "Enter name of summary book" field.

10. Click the "Analyze" button.

    For a large number of analysis names (i.e., >50) this can take an hour or more.

11. When the analysis is complete the word "Done" will appear in the bottom left-hand corner of the window, and the summary file will be open behind the window.

CM: Interpreting the Sleep Analysis Summary File

12. Open the summary file; it will be in a user-specified folder chosen when running the sleep analysis program.

13. Note the data are divided into eight tabs (if only LD or DD data were processed there will be fewer tabs).

    i. LD Analysis: data for several sleep/activity characteristics averaged for the LD portion of the experiment

    ii. L Analysis: data averaged for the L phase of the LD portion of the experiment

    iii. D Analysis: data averaged for the D phase of the LD portion of the experiment

    iv. DD Analysis: data averaged for the DD portion of the experiment

    v. LD Analysis, detailed: day-by-day data for tab i

    vi. L Analysis, detailed: day-by-day data for tab ii

    vii. D Analysis, detailed: day-by-day data for tab iii

    viii. DD Analysis, detailed: day-by-day data for tab iv

14. Note that the data in the tabs identified in Step 13i–iv are divided into 15 columns.

    i. Wake Bins (normalized)—the average number of bins/day in which the fly was identified as awake for the days of the experiment specified in the tab title

    ii. Bins of Sleep (normalized)—the average number of bins/day the fly was asleep

    iii. Min. of Sleep (normalized)—the average number of minutes of sleep/day

    iv. % Sleep—the percent of the time the fly was asleep

    v. Amount of Activity—the total number of beam crossings during the phase of interest

    vi. Wake Minutes (normalized)—the number of minutes/day the fly was awake

    vii. Waking Activity (Activity/Min)—the number of beam crossings/minute when the fly was awake

    viii. Total Bouts Awake—A wake bout is a period of activity flanked by periods of sleep (≥5 min of inactivity). The total number of wake bouts during this phase of the experiment.

    ix. Total Bouts Asleep—A sleep bout is a period of ≥5 min of inactivity flanked by periods of activity (wakefulness). The total number of sleep bouts during this phase of the experiment.

     x. ABL—Average Sleep Bout Length for this phase of the experiment

    xi. ABL Predicted day model—A detrending alogorithm diminishes any day-to-day trend in ABL.

    xii. ABL Detrended—the factor by which ABL was altered to get ABL Predicted day model

   xiii. CI—Consolidation Index is a weighted ABL, in which each sleep bout duration is squared and then averaged. This value has an advantage over ABL in that it will be much less influenced by potentially dubious short sleep bouts (5 min compared with 45 min).

   xiv. CI Predicted day model—A detrending alogorithm diminishes any day-to-day trend in CI.

    xv. CI Detrended—the factor by which CI was altered to get CI Predicted day model

15. Note the fly groups listed in the MultiSleep tab of the Counting Macro (see Step 4) are listed in column A. All values in each row are averages for this fly group.

16. Scroll down to observe the list of fly groups is repeated three times, showing the standard deviation, standard error, and *n* (number of flies that were not triaged for the phase of interest), respectively, for each value.

17. Scroll down to observe a black bar; below this bar the average values from above are repeated for each individual fly.

18. Click on a detailed analysis tab (e.g., "LD Analysis, detailed").

19. Note that the detailed analysis tabs show the same data as the first four tabs, but broken down into individual days.

    For each category, the individual days are listed first (left to right), with the last column providing a total average value under the heading of the category title.

CM: Sleep Eductions

This function provides hour-by-hour graphs for sleep duration.

20. Copy the information entered into the MultiSleep tab (see Step 4) and paste it into the MultiEducer tab.

21. In the Main Window, click the "Sleep Educer" button.

22. Click the "Input Specs" button.

23. Choose the bin length used, the day the eduction should start, and time of lights-on, and click the "OK" button.

    This will fill in the start bin information. To use only the LD information from the Run Specs window, click the "Import Analysis 1" button; to use only the DD information, click the "Import Analysis 2" button.

24. Choose the "Number of days to Educe" in the drop-down menu.

    This is the number of days the program will create a graph for. It will also create average LD and DD graphs.

25. In Section 2, choose the "Length of interval."

    This is the amount of time each data point on the graph will encompass. For example, if 1 h is selected, the graph will show how much this fly group slept on average per hour, for every hour of the experiment.

26. For the "Ignore flies that have been labeled in triage with:," check the boxes for the phase(s) being analyzed.

27. Select a directory to save the graphs (as Excel files) in by clicking the "..." button next to "3. Save directory:" under the "Multiple (from "MultiEducer" Sheet)" heading.

28. In Section 4, "Educe:," click the "Produce Eduction" button.

    The resulting files will contain four graphs and the necessary data points to recreate the graphs.

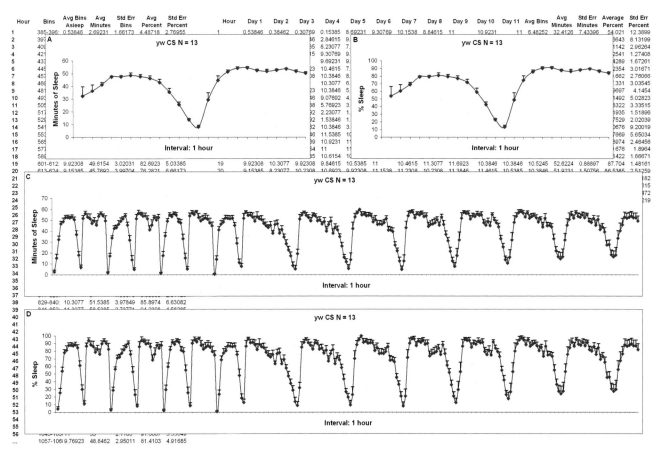

FIGURE 3. Screen shot of the sleep eduction output file examining yw x Canton-S males. (*A,B*) Average sleep duration per hour for the entire experiment (**4LD 7DD**) in (*A*) sleep minutes per hour and (*B*) percent sleep. (*C,D*) Average sleep duration per hour for the entire experiment with each hour displayed in (*C*) minutes of sleep and (*D*) percent sleep.

The bottom and middle graphs show sleep duration in minutes of sleep and percent sleep, respectively, per specified interval for the number of day specified (Fig. 3C,D). The top two graphs show the average sleep duration per specified interval for all the days of the experiment in minutes of sleep (Fig. 3A) or percent sleep (Fig. 3B).

CM: Latency to Sleep

This function provides the time lapse between lights-off and the first sleep bin.

29. Copy the information entered into the MultiSleep tab (see Step 4) and paste it into the Latency tab (or enter the specific groups of interest to be analyzed).

30. Check the appropriate Bin Length box.

31. Enter the time of Lights-Off (generally 20:00:00).

32. Enter the "Start analysis on run day:."

Day 1 is the day the boards were loaded. We generally begin analyses on Day 2.

33. Enter the number of days to gather latency-to-sleep data for.

34. Click the "Choose Directory" button and enter a location to save the latency to sleep files.

35. Click "Analyze."

36. Open the folder chosen to save the latency summary files in. Note there is a file for each group analyzed.

37. Open one of the latency files.

38. Note the data are arranged in an array, with all the individual flies in this group listed in column A and the days analyzed listed in row 2. The array shows the time lapse (min) between lights-off and the first sleep bout for each individual fly on each individual day.

39. The column after the last day gives individual fly averages for all the days of the analysis, followed by standard deviation and standard error.

40. The row after the last fly gives individual day averages for all the flies analyzed, followed by standard deviation and standard error.

41. The group average is listed in the cell at which the average column and average row intersect.

REFERENCES

Agosto J, Choi JC, Parisky KM, Stilwell G, Rosbash M, Griffith LC. 2008. Modulation of $GABA_A$ receptor desensitization uncouples sleep onset and maintenance in *Drosophila*. *Nat Neurosci* **11:** 354–359.

Bell-Pedersen D, Cassone VM, Earnest DJ, Golden SS, Hardin PE, Thomas TL, Zoran MJ. 2005. Circadian rhythms from multiple oscillators: Lessons from diverse organisms. *Nat Rev Genet* **6:** 544–556.

Cirelli C, Bushey D. 2008. Sleep and wakefulness in *Drosophila melanogaster*. *Ann NY Acad Sci* **1129:** 323–329.

Everson CA, Bergmann BM, Rechtschaffen A. 1989. Sleep deprivation in the rat: III. Total sleep deprivation. *Sleep* **12:** 13–21.

Hendricks JC, Finn SM, Panckeri KA, Chavkin J, Williams JA, Sehgal A, Pack AI. 2000. Rest in *Drosophila* is a sleep-like state. *Neuron* **25:** 129–138.

Huber R, Hill SL, Holladay C, Biesiadecki M, Tononi G, Cirelli C. 2004. Sleep homeostasis in *Drosophila melanogaster*. *Sleep* **27:** 628–639.

Konopka RJ, Benzer S. 1971. Clock mutants of *Drosophila melanogaster*. *Proc Natl Acad Sci* **68:** 2112–2116.

Nitabach MN, Taghert PH. 2008. Organization of the *Drosophila* circadian control circuit. *Curr Biol* **18:** R84–R93.

Raizen DM, Zimmerman JE, Maycock MH, Ta UD, You YJ, Sundaram MV, Pack AI. 2008. Lethargus is a *Caenorhabditis elegans* sleep-like state. *Nature* **451:** 569–572.

Shaw PJ, Cirelli C, Greenspan RJ, Tononi G. 2000. Correlates of sleep and waking in *Drosophila melanogaster*. *Science* **287:** 1834–1837.

van Swinderen B, Nitz DA, Greenspan RJ. 2004. Uncoupling of brain activity from movement defines arousal states in *Drosophila*. *Curr Biol* **14:** 81–87.

Zimmerman JE, Raizen DM, Maycock MH, Maislin G, Pack AI. 2008. A video method to study *Drosophila* sleep. *Sleep* **31:** 1587–1598.

32 Feeding Behavior of *Drosophila* Larvae

Ping Shen

Department of Cellular Biology, Biomedical and Health Sciences Institute, The University of Georgia, Athens, Georgia 30602

ABSTRACT

All animals have evolved highly optimized strategies to modulate their feeding behavior with respect to favorable and adverse circumstances. However, the regulatory mechanisms underlying the complexity of feeding behaviors, including food seeking, selection, and ingestion, remain poorly understood. The food responses of *Drosophila* larvae offer an excellent opportunity to study the genetic and neural regulation of feeding behavior. Compared with fed larvae, hungry larvae are more likely to display aggressive foraging, rapid food intake, compensatory feeding, and stress-resistant food procurement. It has been recently shown that the evolutionarily conserved insulin-like and neuropeptide Y–like signaling pathways coregulate diverse hunger-driven behaviors in fly larvae. Here I will summarize briefly the current understanding of the roles of these pathways and their associated neuronal networks in regulation of the motivation to feed. I will also describe a complementary set of behavioral paradigms, each designed to quantitatively assess a particular aspect of the hunger-driven food response. In combination, these assays help define the specific role of signaling molecules or neurons in the regulation of feeding behavior in foraging larvae.

INTRODUCTION

Food response by *Drosophila* larvae provides a useful platform to elucidate the molecular and cellular basis of complex food behaviors including food seeking, selection, and ingestion. Feeding larvae display diverse behavioral responses to food cues (Wu et al. 2003, 2005a,b; Kaun et al. 2008). Under favorable conditions, larvae prefer to feed on rich, palatable soft media that promote rapid growth. However, under prolonged food deprivation, these animals display a broad array of hunger-driven behaviors that favor short-term survival (Wu et al. 2005a,b; Lingo et al. 2007). In addition, the food-related behaviors of larvae are robust and readily quantifiable, providing excellent functional readouts of the neural processes that control complex food behaviors.

A conserved neural signaling network has been shown to regulate diverse hunger-driven behaviors in *Drosophila* larvae (Wu et al. 2005b; Gruninger et al. 2007; Melcher et al. 2007). Neurons that

505

produce *Drosophila* insulin-like peptides (dILPs), which are positively regulated by a cellular nutritional sensor (ribosomal S6 kinase), exert their influence by sensing the metabolic state and organizing the onset of hunger-driven behaviors (Wu et al. 2005a). Down- and up-regulation of dILP signaling activity mimic the effects of hunger and satiation on larval food intake, respectively. Another conserved signaling pathway in the network involves neuropeptide F (dNPF), an ortholog of human neuropeptide Y (Brown et al. 1999; Shen and Cai 2001; Garczynski et al. 2002). Neurons expressing the dNPF receptor NPFR1, which are negatively regulated by dILPs, selectively promote a self-preservative motivational state in fasting animals to seek and consume undesirable or nonpreferred foods (Wu et al. 2005a,b; Lingo et al. 2007). It has also been shown that hugin, another conserved neuropeptide related to Neuromedin U, regulates food choice behavior of *Drosophila* (Melcher and Pankratz 2005; Melcher et al. 2006). Together, these findings validate the use of the fly model to study the molecular and neural basis of feeding behavior.

LARVAL FEEDING RESPONSE TO SOFT AND HARD SUGAR MEDIA

Third-instar larvae fed *ad libitum* display active feeding activity on soft palatable food media (Wu et al. 2005a). Their feeding rate can be quantified by measuring the frequency of mouth-hook contraction. Under defined assay conditions, the baseline rate of feeding by young fed third instars (74 h after egg laying [AEL]) is quite consistent, and larval feeding rate increases gradually as food deprivation is prolonged. For example, food deprivation for 2 h has been shown to double the mouth-hook contraction rate on soft food. However, when exposed to hard sugar media (e.g., sugar solution embedded in agar), the feeding behavior of fed larvae is very different. To consume the same amount of food, larvae must work harder by pulverizing the agar block with their mouth hooks. Larvae do not display this feeding response to the solid media unless they are food deprived. The typical feeding profile of third-instar larvae (74 h AEL) on soft and hard media is shown in Figure 1. The comparison of parallel feeding responses to both soft and hard media provides a good measure of the hunger state of the larvae and their drive to consume and to work for food.

Several assays are available to analyze the response of larvae to food (Wu et al. 2003, 2005a,b; Lingo et al. 2007), each probing the larval food response from a different angle. The selection of such assays should be guided by one's research aims and approaches. It is often preferable to perform two or more different behavioral assays in parallel, because the combined use of different assays can lead to a more detailed understanding of the biological role of the gene or neurons of interest in the larval food response. The following series of protocols in this chapter describe, first, the preparation of *Drosophila* larvae, followed by their use in three larval feeding assays.

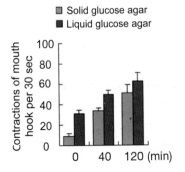

FIGURE 1. Larval feeding response to liquid and solid glucose media. Normal larvae (*w^1118^*, 74 h AEL) fed *ad libitum* display active feeding on the liquid but not solid food. Food deprivation induces rapid intake of the liquid food and promotes larval feeding response to the solid food. (Adapted from Wu et al. 2005a.)

Preparing *Drosophila* Larvae for Feeding Assays

Proper rearing of larvae is of paramount importance for all behavioral experiments. A time-tested procedure is described here to produce highly synchronized third-instar feeding larvae. To ensure synchronized growth of larvae, the rearing conditions must be optimized by controlling food quality, larval density, humidity, and temperature. Under such conditions, it takes about 72 h for a wild-type or *white* egg to develop into a third-instar larva. One reliable way to judge the age of third-instar larvae is to determine the time of the second molting (into the third instar). A feeding step is included to ensure the selection of young third instars that are actively foraging for behavioral assays.

MATERIALS

Reagents

Agar
> *Drosophila* agar, which is relatively inexpensive, may be used. However, its quality varies greatly, and each batch should be functionally tested against the Bacto Agar before use.

Apple juice (frozen concentrate)
Drosophila adult flies, male and female
Instant dry yeast (e.g., Fleischmann's brand)

Equipment

Dissecting microscope
Egg collection bottles
Incubator (25°C)
Microwave oven
Pasture pipettes
Petri dishes (35 mm and 100 mm)
Plastic fly bottles (180 mL)
Tweezers

METHOD

Preparation of Food Media

1. Prewarm the frozen apple juice to room temperature.

2. To make yeast paste, add 2 mL distilled water to 1 g of dry yeast and mix well.
   > Yeast paste should be moist and soft but not runny.

3. To make about 1 L of apple juice agar, add 700 mL distilled water to 24 g agar and mix well.
   > The exact amount of agar may need to be adjusted, depending on its quality.

4. Heat by microwaving until the agar is melted completely.

5. Add 300 mL concentrated apple juice immediately to the hot agar solution, and gently mix.

6. Dispense 4 mL of the agar to each dish with a 60-mL syringe.

> The agar dish can be the lid of a 40-mm Petri dish, which fits well with fly bottles (see below).

7. Store both yeast paste and apple juice agar dishes in sealed containers at 4°C.

Preparation of Synchronized Larvae

8. Introduce healthy adult flies (>30 females and >10 males) into a 180-mL plastic fly bottle.

> The bottle wall is punctured with a 21-gauge needle to generate small holes for ventilation.

9. To prepare a food dish, coat a dish containing apple juice agar with yeast paste; use this dish to cover the bottle top.

10. Change the food dish every day.

> If flies are transferred from laboratory stocks, it may take 2–3 d of prefeeding before flies begin to lay eggs at the peak rate.

11. Prewarm a new food dish (to 23°C–25°C), and use it to collect fresh eggs.

> It may take 2 h or longer to obtain the desired number of eggs.

12. Estimate the number of eggs (limit to <200) per dish.

13. Put the egg dish inside a 100-mm Petri dish together with a piece of moist tissue paper and incubate it at 25°C. Avoid stacking the plates during incubation.

14. After 2 d, when all larvae become second instars (L2), add fresh yeast paste to the dish. Estimate the number of eggs again to ensure there are <200 L2 larvae per dish.

15. Harvest (i.e., transfer into a Petri dish with a little water) young feeding third instars 74 h AEL or 2 h after the second molting. Handle larvae gently with water and a soft brush.

16. Feed larvae with yeast paste dyed with red food color for 30 min.

> This step allows one to eliminate inactive or wounded larvae.

17. Select those third instars whose gut is filled with red food.

> Second- and third-instar larvae differ in their mouth-hook anatomy and body size.

18. Rinse the fed larvae with water to remove any visible yeast paste on the body wall. They can either be used immediately (as fed larvae) or transferred to a thin layer of water for a defined time period (as food-deprived larvae). It is important that the body surface of the larvae is largely free of water before being transferred to assay plates. Excessive water may dilute food media or cause failures in the solid food assay.

Analysis of Feeding on Liquid Food

This test is designed for quantitative assessment of the food ingestion rate of individual larvae under different energy states. It provides a simple and reliable way to measure the graded modification of the baseline feeding rate of fed larvae as food deprivation is prolonged. The test is applicable to routine functional testing and larger-scale screening of genetic mutations and biologics that might affect food consumption.

MATERIALS

Reagents

Agar (batch tested; see Protocol 1)
Drosophila larvae (Protocol 1)
Petri dishes (60 mm)
Sugar (e.g., glucose)

Equipment

Dissecting microscope
Tweezers

METHOD

1. Prepare the assay plates by mixing 5 g glucose and 6 g agar with 45 mL distilled water. Mix well and soak for 4 h or overnight.

2. Add 6 g of the sugar agar paste to the Petri dish.

3. Tap the dish until the sugar agar paste is compact and evenly distributed (Fig. 2A).

4. Introduce 25 larvae onto the agar medium.

    Before transfer, larvae should be blotted dry using a dry brush or a small piece of soft tissue paper.

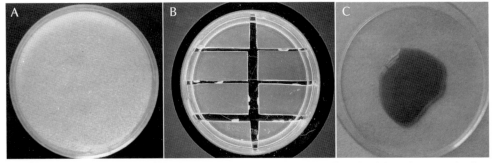

FIGURE 2. Assay plates used for quantification of larval feeding responses to liquid, solid, and quinine food media: 60-mm dish containing liquid food (*A*); a 40-mm dish containing solid food (*B*); a 60-mm dish containing quinine-yeast paste (*C*).

5. Wait for 1 min to allow the larvae to burrow into the agar paste.

6. Place the dish onto the microscope stage, turning the dish so that the bottom is facing the microscope lens.

> Do not use strong lighting.

7. Count the number of mouth-hook contractions for 10 or 30 sec.

> Avoid counting the larvae that feed at the edge of the agar paste.

See Troubleshooting.

TROUBLESHOOTING

Problem (Step 7): The mouth hook of larvae moves too fast to be accurately counted.

Solution: Make sure the agar paste is packed tightly to avoid air pockets. Another possible cause is that the agar paste is too runny. If so, slightly increase the amount of agar. It is strongly recommended to adjust the composition of the media so that the contraction frequencies of the mouth hook of fed and 2-h-deprived larvae are about 30 and 60 per 30 sec, respectively.

Problem (Step 7): The larvae cannot be seen very well.

Solution: This problem is likely a result of too much agar paste in the dish. Reduce the thickness of the agar paste.

Protocol 3

Analysis of Feeding on Solid Food

This test is designed for quantitative assessment of the willingness of individual larvae to procure solid food under different energy states. It provides a simple and reliable way to measure the graded modification of the baseline feeding rate of larvae as the period of food deprivation is increased. The test is applicable to routine functional testing and larger-scale screening of genetic mutations and biologics that might affect food consumption.

MATERIALS

Reagents

Agar (batch tested; see Protocol 1)
Drosophila larvae (Protocol 1)
Petri dishes (35 mm)
Sugar source (e.g., glucose)

Equipment

Dissecting microscope
Spatula (sharp-edged)
Tweezers

METHOD

1. Prepare assay plates by mixing 10 g glucose and 2.3 g agar with 90 mL distilled water.
2. Microwave the mix until agar is just melted.
   Avoid excessive water evaporation. Otherwise the agar will be too hard.
3. Add enough warm agar per dish to make the solid agar layer about 3.5 mm thick. Wait until agar has completely solidified.
4. Cut the solidified agar with a spatula (Fig. 2B) and allow it to air dry overnight. The appropriate drying time must be experimentally determined.
   Steps 1–4 should be performed 1 d before the behavior assay.
   See Troubleshooting for Steps 1–4.
5. Cover the dish until use.
6. Introduce 15 fed or fasting larvae (Protocol 1, Step 18) into the agar medium.
   Before transfer, larvae should be blotted dry using a dry brush or a small piece of soft tissue paper.
7. Wait for 5 min to allow larvae to settle.
   See Troubleshooting.
8. Place the dish onto the microscope stage, turning the dish so that the bottom is facing the microscope lens.
   Do not use strong lighting.

9. Randomly sample those larvae that are feeding by counting the number of mouth-hook contractions per 30 sec. Remove the larva immediately after counting and discard.

> A good way to minimize scoring movement-related mouth-hook contractions is to count only those contractions associated with persistent feeding (e.g., three or more consecutive scrapings by the mouth hooks). Do not count the larvae that feed inside freshly generated cracks, which occur frequently near the edge of the plate because of larval pushing against agar.

TROUBLESHOOTING

Problem (Steps 1–4): Neither fed nor fasted wild-type larvae feed on the solid medium.
Solution: This is probably because of the food being too hard. Reduce the drying time by testing several time points. If the problem persists, reduce the amount of agar in the medium.

Problem (Step 7): How do I know the hardness of the agar is appropriate?
Solution: The hardness of agar is right, if >50% fasting but virtually no fed larvae display persistent scraping of the agar at a given time. If overdrying is your frequent problem, it is advisable to make two batches of solid media, with one batch air-dried for a shorter time. In addition, air-drying of agar in a chamber of controlled humidity (e.g., 70%) can also be helpful to prevent overdrying.

Protocol 4

Analysis of Larval Feeding Response to Quinine-Adulterated Food

Omnivores, including humans, tend to avoid noxious or unfamiliar food. Fly larvae avoid bitter substances such as quinine (Wu et al. 2005b). As an increasing amount of quinine is added to an otherwise palatable liquid medium containing sugar and inactivated yeast, larvae become less likely to consume the tainted food. Larval feeding activity is assayed by scoring the percentage of larvae containing dyed food in a large portion of the midgut. Defensive foraging behavior can be modulated by physiological need and food deprivation significantly increases larval tolerance to quinine (Fig. 3). Therefore, the larval response to quinine food and its regulation by hunger provides a useful paradigm for elucidating the genetic and neural regulation of food choice that favors short-term survival. Other forms of stressors (e.g., cold temperature) can also be used to substitute quinine for such studies (Lingo et al. 2007).

The test described here is designed for quantitative assessment of the motivation of individual larvae to ingest quinine-adulterated food under different energy states. The test is applicable to routine functional testing and larger-scale screening of genetic mutations and biologics that might affect food consumption.

MATERIALS

CAUTION: See Appendix for proper handling of materials marked with <!>.

Reagents

Agar (batch tested; see Protocol 1)
Drosophila larvae (Protocol 1)
Drosophila yeast (Cat FLY8040, LabScientific, Inc., Livingston, NJ)
Ethanol (100%) <!>
Green food dye
Ice-chilled water
Quinine hydrochloride-2H$_2$O (Sigma Chemical Co., St. Louis, MO)

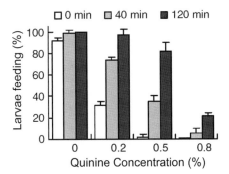

FIGURE 3. Larval feeding response to quinine-containing food. Wild-type larvae (74 h AEL) fed *ad libitum* displayed decreased feeding as the concentration of quinine is increased. Food deprivation promotes larval feeding of quinine food. (Adapted from Wu et al. 2005b.)

Equipment

Dissecting microscope
Petri dishes (60 mm)
Tweezers

METHOD

1. Prepare assay plates by adding 4 mL of distilled water to 2 g of dry yeast. Mix well and incubate in a boiling water bath for 1 h to inactivate yeast.

2. Dissolve 0.62 g of quinine hydrochloride-$2H_2O$ in 600 mL 100% ethanol to make the quinine stock.

3. To a small container, add, in succession, 5 g of the heat-inactivated yeast paste, 80 µL green food dye, and 200 µL ethanol containing the desired amount of quinine hydrochloride (e.g., 0, 0.06, or 0.12 g). Mix the three ingredients together until the dye is evenly distributed.

    This amount is sufficient for 8–10 assays.

    To avoid clumping of the dye, the food dye should not contact ethanol before mixing.

4. Add 0.5 g of the above yeast mix to the center of a 60-mm dish (Fig. 2C).

5. Introduce 20–40 larvae (Protocol 1, Step 18) to the edge of the yeast paste, and allow the larvae to feed for 20 min.

6. Terminate the feeding assay by adding ice-cold water to the dish and keeping it on ice.

7. Rinse larvae with cold water and count the larvae with green or white midgut.

    Larvae that fed persistently during the assay period should show readily visible green food in a large portion of the midgut.

    See Troubleshooting.

TROUBLESHOOTING

Problem (Step 7): Larvae seem to be insensitive to quinine.
Solution: This is most likely because of the sequestration or degradation of quinine by the active yeast. Make sure the yeast paste is completely heat-inactivated. Use Drosophila yeast instead of instant dry yeast (e.g., Fleischmann's brand), which appears to sequester or degrade quinine very efficiently.

DISCUSSION

The liquid food assay quantifies the initial feeding rate of individual fed or fasting larvae, providing a simple way to measure the motivational state of feeding in larvae. One potential drawback of this assay is that larvae tend to burrow into the soft medium, making it difficult to continuously video-record larval feeding activity over a relatively long period of time (e.g., >60 sec). One way to minimize larval burrowing is to reduce the amount of the soft medium, but exact conditions must be experimentally determined. On the other hand, quantification of persistent feeding activity can be achieved easily on solid food, which eliminates the larval burrowing problem. For example, one may be able to use the solid food assay to uncover fed mutants that display exuberant feeding activity or starved mutants that show attenuated feeding response over several minutes. Another useful feature of the solid food assay is that it appears to provide the functional readout of larval motivation to work for food (reward). The quinine food assay complements the liquid food assay in a different way. In this case, food remains rich and accessible but has an aversive taste. Like other animals, fruit flies also sample unfamiliar foods before feeding, which helps minimize risks associated with feeding (e.g., food poisons). Moreover, such a feeding decision is regulated by hunger (Melcher et al. 2007). So the quinine food assay may provide a useful avenue to assess the potential roles of genes and neurons in regulation of risk-prone behaviors.

The relatively simple *Drosophila* larva has a rich repertoire of food-related behaviors. If one is only interested in measuring food intake under different energy states, any assays that quantify food ingestion using colored dyes or radioactive isotopes or scoring feeding activity will probably work (King et al. 1955; Edgecomb et al. 1994; Brummel et al. 2004; Bross et al. 2005; Carvalho et al. 2005; Ja et al. 2007; Kaun et al. 2007). However, the food response of an animal is much more than eating or not eating. Careful selection of two or more complementary food behavior assays, like those described in this chapter, can provide much more information about the role of your favorite gene in fly feeding behavior. When considering the choice between adults and larvae for feeding studies, it is helpful to remember that obtaining a synchronized population of larvae that display largely uniform feeding activity is relatively easy, whereas the same cannot be said about the adults. The feeding behavior of male flies is more difficult to study than females because of their limited daily feeding activity. In addition, it still remains challenging to measure the food intake of adult flies under conditions more similar to the natural feeding environment (see Wong et al. 2009 for detailed discussion).

REFERENCES

Bross TG, Rogina B, Helfand SL. 2005. Behavioral, physical, and demographic changes in *Drosophila* populations through dietary restriction. *Aging Cell* **4:** 309–317.

Brown MR, Crim JW, Arata RC, Cai HN, Chun C, Shen P. 1999. Identification of a *Drosophila* brain-gut peptide related to the neuropeptide Y family. *Peptides* **20:** 1035–4102.

Brummel T, Ching A, Seroude L, Simon AF, Benzer S. 2004. *Drosophila* lifespan enhancement by exogenous bacteria. *Proc Natl Acad Sci* **101:** 12974–12979.

Carvalho GB, Kapahi P, Benzer S. 2005. Compensatory ingestion upon dietary restriction in *Drosophila melanogaster*. *Nat Methods* **2:** 813–815.

Edgecomb RS, Harth CE, Schneiderman AM. 1994. Regulation of feeding behavior in adult *Drosophila melanogaster* varies with feeding regime and nutritional state. *J Exp Biol* **197:** 215–235.

Garczynski SF, Brown MR, Shen P, Murray TF, Crim JW. 2002. Characterization of a functional neuropeptide F receptor from *Drosophila melanogaster*. *Peptides* **23:** 773–780.

Gruninger TR, LeBoeuf B, Liu Y, Rene L. 2007. Molecular signaling involved in regulating feeding and other motivated behaviors. *Mol Neurobiol* **35:** 1–20.

Ja WW, Carvalho GB, Mak EM, de la Rosa NN, Fang AY, Liong JC, Brummel T, Benzer S. 2007. Prandiology of *Drosophila* and the CAFE assay. *Proc Natl Acad Sci* **104:** 8253–8256.

Kaun KR, Chakaborty-Chatterjee M, Sokolowski MB. 2008. Natural variation in plasticity of glucose homeostasis and food intake. *J Exp Biol* **211:** 3160–3166.

King RC, Wilson LP. 1955. Studies with radiophosphorus in *Drosophila*. V. The phosphorous balance of adult females. *J Exp Zool* **130:** 71–82.

Lingo PR, Zhao Z, Shen P. 2007. Co-regulation of cold-resistant food acquisition by insulin- and neuropeptide Y-like systems in *Drosophila melanogaster*. *Neuroscience* **148:** 371–374.

Melcher C, Bader R, Walther S, Simakov O, Pankratz MJ. 2006. Neuromedin U and its putative *Drosophila* homolog hugin. *PLoS Bio.* **4:** e68. doi: 10.1371/journal.pbio.0040068.

Melcher C, Bader R, Pankratz MJ. 2007. Amino acids, taste circuits, and feeding behavior in *Drosophila*: Towards understanding the psychology of feeding in flies and man. *J Endocrinol* **192:** 467–472.

Melcher C, Pankratz MJ. 2005. Candidate gustatory interneurons modulating feeding behavior in the *Drosophila* brain. *PLoS Biol* **3:** e305. doi: 10.1371/journal.pone.0006063.

Shen P, Cai HN. 2001. *Drosophila* neuropeptide F mediates integration of chemosensory stimulation and conditioning of the nervous system by food. *J Neurobiol* **47:** 16–25.

Wong R, Piper MD, Wertheim B, Partridge L. 2009. Quantification of food intake in *Drosophila*. *PLoS ONE* **4(6):** e6063.

Wu Q, Wen T, Lee G, Park JH, Cai HN, Shen P. 2003. Developmental control of foraging and social behavior by the *Drosophila* neuropeptide Y-like system. *Neuron* **39:** 147–161.

Wu Q, Zhang Y, Xue J, Shen P. 2005a. Regulation of hunger-driven behaviors by neural ribosomal S6 kinase in *Drosophila*. *Proc Natl Acad Sci* **102:** 13289–13294.

Wu Q, Zhao Z, Shen P. 2005b. Regulation of aversion to noxious food by *Drosophila* neuropeptide Y- and insulin-like systems. *Nat Neurosci* **8:** 1350–1355.

Appendix: Cautions

Please note that the Cautions Appendix in this manual is not exhaustive. Readers should always consult individual manufacturers and other resources for current and specific product information. Chemicals and other materials discussed in text sections are not identified by the icon <!> used to indicate hazardous materials in the protocols. However, without special handling, they may be hazardous to the user. Please consult your local safety office or the manufacturer's safety guidelines for further information.

The following general cautions should always be observed.

- **Before beginning the procedure,** become completely familiar with the properties of substances to be used.

- **The absence of a warning** does not necessarily mean that the material is safe, because information may not always be complete or available.

- **If exposed** to toxic substances, contact your local safety office immediately for instructions.

- **Use proper disposal procedures** for all chemical, biological, and radioactive waste.

- **For specific guidelines on appropriate gloves to use,** consult your local safety office.

- **Handle concentrated acids and bases** with great care. Wear goggles and appropriate gloves. A face shield should be worn when handling large quantities.

  Do not mix strong acids with organic solvents because they may react. Sulfuric acid and nitric acid especially may react highly exothermically and cause fires and explosions.

  Do not mix strong bases with halogenated solvent because they may form reactive carbenes, which can lead to explosions.

- **Handle and store pressurized gas containers** with caution because they may contain flammable, toxic, or corrosive gases; asphyxiants; or oxidizers. For proper procedures, consult the Material Safety Data Sheet that must be provided by your vendor.

- **Never pipette** solutions using mouth suction. This method is not sterile and can be dangerous. Always use a pipette aid or bulb.

- **Keep halogenated and nonhalogenated** solvents separately (e.g., mixing chloroform and acetone can cause unexpected reactions in the presence of bases). Halogenated solvents are organic solvents such as chloroform, dichloromethane, trichlorotrifluoroethane, and dichloroethane. Nonhalogenated solvents include pentane, heptane, ethanol, methanol, benzene, toluene, N,N-dimethylformamide (DMF), dimethyl sulfoxide (DMSO), and acetonitrile.

- **Laser radiation**, visible or invisible, can cause severe damage to the eyes and skin. Take proper precautions to prevent exposure to direct and reflected beams. Always follow the manufacturer's safety guidelines and consult your local safety office. See caution below for more detailed information.

- **Flash lamps**, because of their light intensity, can be harmful to the eyes. They also may explode on occasion. Wear appropriate eye protection and follow the manufacturer's guidelines.

- **Photographic fixatives, developers, and photoresists** also contain chemicals that can be harmful. Handle them with care and follow the manufacturer's directions.

- **Power supplies and electrophoresis equipment** pose serious fire hazard and electrical shock hazards if not used properly.

- **Microwave ovens and autoclaves** in the laboratory require certain precautions. Accidents have occurred involving their use (e.g., when melting agar or Bacto Agar stored in bottles or when sterilizing). If the screw top is not completely removed and there is inadequate space for the steam to vent, the bottles can explode and cause severe injury when the containers are removed from the microwave or auto-clave. Always completely remove bottle caps before microwaving or autoclaving. An alternative method for routine agarose gels that do not require sterile agar is to weigh out the agar and place the solution in a flask.

- **Ultrasonicators** use high-frequency sound waves (16–100 kHz) for cell disruption and other purposes. This "ultrasound," conducted through air, does not pose a direct hazard to humans, but the associated high volumes of audible sound can cause a variety of effects, including headache, nausea, and tinnitus. Direct contact of the body with high-intensity ultrasound (not medical imaging equipment) should be avoided. Use appropriate ear protection and display signs on the door(s) of laboratories where the units are used.

- **Use extreme caution when handling cutting devices,** such as microtome blades, scalpels, razor blades, or needles. Microtome blades are extremely sharp! Use care when sectioning. If unfamiliar with their use, have an experienced user demonstrate proper procedures. For proper disposal, use the "sharps" disposal container in your lab. Discard used needles *unshielded*, with the syringe still attached. This prevents injuries and possible infections when manipulating used needles because many accidents occur while trying to replace the needle shield. Injuries may also be caused by broken pasteur pipettes, coverslips, or slides.

- **Procedures for the humane treatment of animals** must be observed at all times. Consult your local animal facility for guidelines. Animals, such as rats, are known to induce allergies that can increase in intensity with repeated exposure. Always wear a lab coat and gloves when handling these animals. If allergies to dander or saliva are known, wear a mask.

GENERAL PROPERTIES OF COMMON CHEMICALS

The hazardous materials list can be summarized in the following categories.

- Inorganic acids, such as hydrochloric, sulfuric, nitric, or phosphoric, are colorless liquids with stinging vapors. Avoid spills on skin or clothing. Spills should be diluted with large amounts of water. The concentrated forms of these acids can destroy paper, textiles, and skin and cause serious injury to the eyes.

- Inorganic bases, such as sodium hydroxide, are white solids that dissolve in water and under heat development. Concentrated solutions will slowly dissolve skin and even fingernails.

- Salts of heavy metals are usually colored, powdered solids that dissolve in water. Many of them are potent enzyme inhibitors and therefore toxic to humans and the environment (e.g., fish and algae).

- Most organic solvents are flammable volatile liquids. Avoid breathing the vapors, which can cause nausea or dizziness. Also avoid skin contact.

- Other organic compounds including organosulphur compounds, such as mercaptoethanol or organic amines, can have very unpleasant odors. Others are highly reactive and should be handled with appropriate care.

- If improperly handled, dyes and their solutions can stain not only your sample but also your skin and clothing. Some are also mutagenic (e.g., ethidium bromide), carcinogenic, and toxic.

- Nearly all names ending with "ase" (e.g., catalase, β-glucuronidase, or zymolyase) refer to enzymes. There are also other enzymes with nonsystematic names such as pepsin. Many of them are provided by manufacturers in preparations containing buffering substances, etc. Be aware of the individual properties of materials contained in these substances.

- Toxic compounds are often used to manipulate cells. They can be dangerous and should be handled appropriately.

- Be aware that several of the compounds listed have not been thoroughly studied with respect to their toxicological properties. Handle each chemical with appropriate respect. Although the toxic effects of a compound can be quantified (e.g., LD_{50} values), this is not possible for carcinogens or mutagens where one single exposure can have an effect. Also realize that dangers related to a given compound may also depend on its physical state (fine powder vs. large crystals/diethylether vs. glycerol/dry ice vs. carbon dioxide under pressure in a gas bomb). Anticipate under which circumstances during an experiment exposure is most likely to occur and how best to protect yourself and your environment.

HAZARDOUS MATERIALS

Note: In general, proprietary materials are not listed here. Kits and other commercial items as well as most anesthetics, sedatives, dyes, fixatives, embedding media, stains, herbicides, and fungicides are also not included. Anesthetics and antibiotics also require special care. Follow the manufacturer's safety guidelines that accompany these products.

Acetic acid (glacial) is highly corrosive and must be handled with great care. It may be a carcinogen. Liquid and mist cause severe burns to all body tissues. It may be harmful by inhalation, ingestion, or skin absorption. Wear appropriate gloves and goggles and use in a chemical fume hood. Keep away from heat, sparks, and open flame.

Acetone causes eye and skin irritation and is irritating to mucous membranes and upper respiratory tract. Do not breathe the vapors. It is also extremely flammable. Wear appropriate gloves and safety glasses. Keep away from heat, sparks, and open flame.

Alanine is irritating to the eyes, skin, and respiratory system. It may be harmful by inhalation, ingestion, or skin absorption. Wear appropriate gloves and safety glasses.

4-aminopyridine may be fatal if ingested. It is very toxic and harmful by inhalation, ingestion, and skin absorption. Wear appropriate gloves and safety goggles.

BaCl$_2$, *see* **Barium chloride**

Barium chloride (BACl$_2$) is toxic and harmful by inhalation, ingestion, or skin absorption. Wear appropriate gloves and safety goggles. Avoid breathing the dust.

Bleach (Sodium hypochlorite), NaOCl, is poisonous, can be explosive, and may react with organic solvents. It may be fatal by inhalation and is also harmful by ingestion and destructive to the skin. Wear appropriate gloves and safety glasses and use in a chemical fume hood to minimize exposure and odor.

Bromophenol blue may be harmful by inhalation, ingestion, or skin absorption. Wear appropriate gloves and safety glasses and use in a chemical fume hood.

CaCl$_2$, *see* **Calcium chloride**

Cacodylic acid is toxic and a possible carcinogen. It may cause heritable genetic damage and is harmful by inhalation, ingestion, or skin absorption. Wear appropriate gloves and safety glasses and use only in a chemical fume hood. Do not breathe the dust.

Cadmium chloride (CdCl$_2$) may be a carcinogen and is toxic. It is harmful by inhalation, ingestion, or skin absorption. Wear appropriate gloves and safety glasses and use only in a chemical fume hood. Do not breathe the dust. Avoid prolonged exposure.

Calcium chloride, CaCl$_2$, is hygroscopic and may cause cardiac disturbances. It may be harmful by inhalation, ingestion, or skin absorption. Do not breathe the dust. Wear appropriate gloves and safety goggles.

Carbon dioxide, CO$_2$, in all forms may be fatal by inhalation, ingestion, or skin absorption. In high concentrations, it can paralyze the respiratory center and cause suffocation. Use only in well-ventilated areas. In the form of dry ice, contact with carbon dioxide can also cause frostbite. Do not place large quantities of dry ice in enclosed areas such as cold rooms. Wear appropriate gloves and safety goggles.

CdCl$_2$, *see* **Cadmium chloride**

Cesium chloride, CsCl, may be harmful by inhalation, ingestion, or skin absorption. Wear appropriate gloves and safety glasses.

Cesium hydroxide, CsOH, may be harmful by inhalation, ingestion, or skin absorption. It is extremely destructive to the mucous membranes and upper respiratory tract; inhalation may be fatal. Do not breathe the dust. Wear appropriate gloves and safety glasses and always use in a chemical fume hood.

CHCl$_3$, *see* **Chloroform**

CH$_3$CH$_2$OH, *see* **Ethanol**

Chloroform, CHCl$_3$, is irritating to the skin, eyes, mucous membranes, and respiratory tract. It is a carcinogen and may damage the liver and kidneys. It is also volatile. Avoid breathing the vapors. Wear appropriate gloves and safety glasses and always use in a chemical fume hood.

Citric acid is an irritant and may be harmful by inhalation, ingestion, or skin absorption. It poses a risk of serious damage to the eyes. Wear appropriate gloves and safety goggles. Do not breathe the dust.

CO$_2$, *see* **Carbon dioxide**

CsCl, *see* **Cesium chloride**

CsOH, *see* **Cesium hydroxide**

DAB, *see* **3,3′-diaminobenzidine**

Desflurane may be harmful by inhalation, ingestion, or skin absorption. Acute exposure may lead to headaches, dizziness, drowsiness, unconsciousness, or death. Wear appropriate gloves and safety goggles.

3,3′-diaminobenzidine (DAB) is a carcinogen. Handle with extreme care. Avoid breathing vapors. Wear appropriate gloves and safety glasses and use in a chemical fume hood.

Diethyl ether, Et$_2$O or (C$_2$H$_5$)$_2$O, is extremely volatile and flammable. It is irritating to the eyes, mucous membranes, and skin. It is also a central nervous system depressant with anesthetic effects. It may be harmful by inhalation, ingestion, or skin absorption. Avoid breathing the vapors. Wear appropriate gloves and safety glasses and always use in a chemical fume hood. Explosive peroxides can form during storage or on exposure to air or direct sunlight. Keep away from heat, sparks, and open flame.

Dimethyl sulfoxide (DMSO) may be harmful by inhalation or skin absorption. It easily penetrates the skin and anything dissolved or mixed with it will be absorbed. Wear appropriate gloves and safety glasses and use in a chemical fume hood. DMSO is also combustible. Store in a tightly closed container. Keep away from heat, sparks, and open flame.

DMF, *see N,N-dimethylformamide*

DMSO, *see Dimethyl sulfoxide*

DPX is composed of Distyrene, a plasticizer, and xylene and is commercially available. Follow the manufacturer's guidelines for handling DPX.

EMS, *see Ethyl methanesulfonate*

Ethanol (EtOH), CH$_3$CH$_2$OH, is highly flammable and may be harmful by inhalation, ingestion, or skin absorption. Wear appropriate gloves and safety glasses. Keep away from heat, sparks, and open flame.

Ether, *see Diethyl ether*

Ethyl methanesulfonate (EMS) is a volatile organic solvent that is a mutagen and carcinogen. It is harmful if inhaled, ingested, or absorbed through the skin. Discard supernatants and washes containing EMS in a beaker containing 50% sodium thiosulfate. Decontaminate all material that has come in contact with EMS by treatment in a large volume of 10% (w/v) sodium thiosulfate. Use extreme caution when handling. When using undiluted EMS, wear protective appropriate gloves and use in a chemical fume hood. Store EMS in the cold. DO NOT mouth-pipette EMS. Pipettes used with undiluted EMS should not be too warm; chill them in the refrigerator before use to minimize the volatility of EMS. All glassware coming in contact with EMS should be immersed in a large beaker of 1 N NaOH or laboratory bleach before recycling or disposal.

Et$_2$O or (C$_2$H$_5$)$_2$O, *see Diethyl ether*

EtOH, *see Ethanol*

Formaldehyde, HCHO, is highly toxic and volatile. It is also a possible carcinogen. It is readily absorbed through the skin and is irritating or destructive to the skin, eyes, mucous membranes, and upper respiratory tract. Avoid breathing the vapors. Wear appropriate gloves and safety glasses and always use in a chemical fume hood. Keep away from heat, sparks, and open flame.

Formalin is a solution of formaldehyde in water. *See* **Formaldehyde**

Formamide is teratogenic. The vapor is irritating to the eyes, skin, mucous membranes, and upper respiratory tract. It may be harmful by inhalation, ingestion, or skin absorption. Wear appropriate gloves and safety glasses and always use a chemical fume hood when working with concentrated solutions of formamide. Keep working solutions covered as much as possible.

Glacial acetic acid, *see* **Acetic acid (glacial)**

Glutaraldehyde is toxic. It is readily absorbed through the skin and is irritating or destructive to the skin, eyes, mucous membranes, and upper respiratory tract. Wear appropriate gloves and safety glasses and always use in a chemical fume hood.

HCHO, *see* **Formaldehyde**

HCl, *see* **Hydrochloric acid**

HCON(CH$_3$)$_2$, *see N,N-dimethylformamide*

Heparin is an irritant and may act as anticoagulant subcutaneously or intravenously. It may be harmful by inhalation, ingestion, or skin absorption. Wear appropriate gloves and safety glasses.

Heptane may be harmful by inhalation, ingestion, or skin absorption. Wear appropriate gloves and safety glasses. It is extremely flammable. Keep away from heat, sparks, and open flame.

H$_2$O$_2$, *see* **Hydrogen peroxide**

H$_3$PO$_4$, *see* **Phosphoric acid**

Hydrochloric acid, HCl, is volatile and may be fatal if inhaled, ingested, or absorbed through the skin. It is extremely destructive to mucous membranes, upper respiratory tract, eyes, and skin. Wear appropriate gloves and safety glasses and use with great care in a chemical fume hood. Wear goggles when handling large quantities.

Hydrogen peroxide, H$_2$O$_2$, is corrosive, toxic, and extremely damaging to the skin. It may be harmful by inhalation, ingestion, and skin absorption. Wear appropriate gloves and safety glasses and use only in a chemical fume hood.

Imidazole is corrosive and may be harmful by inhalation, ingestion, or skin absorption. Wear appropriate gloves and safety glasses and use in a chemical fume hood.

Isoflurane is an irritant and may be harmful by inhalation, ingestion, or skin absorption. Chronic exposure may be harmful. Wear appropriate gloves and safety glasses.

Isopropanol is flammable and irritating. It may be harmful by inhalation, ingestion, or skin absorption. Wear appropriate gloves and safety glasses. Do not breathe the vapor. Keep away from heat, sparks, and open flame.

KCl, *see* **Potassium chloride**

KOH, *see* **Potassium hydroxide**

KOH/methanol, *see* **Potassium hydroxide**

Lead citrate is a potential cancer hazard and may cause possible death. There is a danger of cumulative effects. It is harmful by inhalation, ingestion, or skin absorption. Wear appropriate gloves and safety goggles and always use in a chemical fume hood. Do not breathe the dust.

LiCl, *see* **Lithium chloride**

Lithium chloride, LiCl, is an irritant to the eyes, skin, mucous membranes, and upper respiratory tract. It may be harmful by inhalation, ingestion, or skin absorption. Wear appropriate gloves, safety goggles, and use in a chemical fume hood. Do not breathe the dust.

Magnesium chloride, MgCl$_2$, may be harmful by inhalation, ingestion, or skin absorption. Wear appropriate gloves and safety glasses and use in a chemical fume hood.

Magnesium sulfate, MgSO$_4$, presents chronic health hazards and affects the central nervous system and the gastrointestinal tract. It may be harmful by inhalation, ingestion, or skin absorption. Wear appropriate gloves and safety glasses and use in a chemical fume hood.

Maleic acid is toxic and harmful by inhalation, ingestion, or skin absorption. Reaction with water or moist air can release toxic, corrosive, or flammable gases. Do not breathe the vapors or dust. Wear appropriate gloves and safety glasses.

MCH, *see* **4-methylcyclohexanol**

MeOH or H$_3$COH, *see* **Methanol**

Methanesulfonic acid causes burns and is extremely destructive to the mucous membranes and upper respiratory tract. It is harmful by inhalation, ingestion, or skin absorption. Wear appropriate gloves and safety goggles. Do not breathe the vapor or mist.

Methanol, MeOH or H$_3$COH, is toxic, can cause blindness, and is highly flammable. It may be harmful by inhalation, ingestion, or skin absorption. Adequate ventilation is necessary to limit exposure to vapors. Avoid inhaling these vapors. Wear appropriate gloves and safety goggles and use only in a chemical fume hood.

4-methylcyclohexanol (MCH) is combustible and may cause nervous system depression. It may be harmful by inhalation, ingestion, or skin absorption. Wear appropriate gloves and safety goggles. Keep away from heat, sparks, and open flame. Do not breathe the vapor or dust.

Methyl 4-hydroxybenzoate (Nipagin) is an irritant and may be harmful by inhalation, ingestion, or skin absorption. Wear appropriate gloves and safety glasses.

Methylparaben (*p*-hydroxymethylbenzoate), *see* **Methyl 4-hydroxybenzoate (Nipagin)**

MgCl$_2$, *see* **Magnesium chloride**

MgSO$_4$, *see* **Magnesium sulfate**

Myristic acid may be harmful by inhalation, ingestion, or skin absorption. Wear appropriate gloves and safety glasses.

Na$_2$CO$_3$, *see* **Sodium carbonate**

NaN$_3$, *see* **Sodium azide**

NaOCl, *see* **Bleach (Sodium hypochlorite)**

NaOH, *see* **Sodium hydroxide**

Nipagin, *see* **Methyl 4-hydroxybenzoate**

***N,N*-dimethylformamide (DMF), HCON(CH$_3$)$_2$,** is a possible carcinogen and is irritating to the eyes, skin, and mucous membranes. It can exert its toxic effects through inhalation, ingestion, or skin absorption. Chronic inhalation can cause liver and kidney damage. Wear appropriate gloves and safety glasses and use in a chemical fume hood.

NPG, *see* ***n*-propyl gallate**

***n*-propyl gallate (NPG)** is related to benzoic acid. Benzoic acid is an irritant and may be harmful by inhalation, ingestion, or skin absorption. Wear appropriate gloves and safety glasses. Do not breathe the dust.

***N*-tris(hydroxymethyl)methyl-2-aminoethane sulfonic acid (TES)** may be harmful by inhalation, ingestion, or skin absorption. Wear appropriate gloves and safety goggles.

OCT is composed of polyvinyl alcohol, polyethylene glycol, and dimethyl benzyl ammonium chloride. Follow the manufacturer's guidelines for handling OCT.

Octanol is highly flammable. Keep away from heat, sparks, and open flame. It may be harmful by inhalation, ingestion, or skin absorption. Wear appropriate gloves and safety goggles.

OsO₄, *see* **Osmium tetroxide**

Osmium tetroxide (osmic acid), OsO₄, is highly toxic if inhaled, ingested, or absorbed through the skin. Vapors can react with corneal tissues and cause blindness. There is a possible risk of irreversible effects. Wear appropriate gloves and safety goggles and always use in a chemical fume hood. Do not breathe the vapors.

Paraformaldehyde is highly toxic and may be fatal. It may be a carcinogen. It is readily absorbed through the skin and is extremely destructive to the skin, eyes, mucous membranes, and upper respiratory tract. Avoid breathing the dust or vapor. Wear appropriate gloves and safety glasses and use in a chemical fume hood. Keep away from heat, sparks, and open flame.

Phenol is extremely toxic, highly corrosive, and can cause severe burns. It may be harmful by inhalation, ingestion, or skin absorption. Wear appropriate gloves, goggles, protective clothing, and always use in a chemical fume hood. Rinse any areas of skin that come in contact with phenol with a large volume of water and wash with soap and water; do not use ethanol!

Phosphoric acid, H₃PO₄, is highly corrosive and is extremely destructive to the tissue of the mucous membranes and upper respiratory tract, eyes, and skin. It is harmful by inhalation, ingestion, or skin absorption. Wear appropriate gloves and safety glasses. Do not breathe the vapors.

Picric acid powder (Trinitrophenol) is caustic and potentially explosive if it is dissolved and then allowed to dry out. Care must be taken to ensure that stored solutions do not dry out. Handle all concentrated acids with great care. It is also highly toxic and may be harmful by inhalation, ingestion, or skin absorption. Wear appropriate gloves and safety goggles and use in a chemical fume hood.

Potassium chloride, KCl, may be harmful by inhalation, ingestion, or skin absorption. Wear appropriate gloves and safety glasses.

Potassium gluconate may be harmful by inhalation, ingestion, or skin absorption. Wear appropriate gloves and safety glasses.

Potassium hydroxide, KOH, and **KOH/methanol,** are highly toxic and may be fatal if swallowed. They may be harmful by inhalation, ingestion, or skin absorption. Solutions are corrosive and can cause severe burns. They should be handled with great care. Wear appropriate gloves and safety goggles.

Propionic acid is highly corrosive and causes burns to any area of contact. It is flammable in both liquid and vapor forms and may be harmful by inhalation, ingestion, or skin absorption. Wear appropriate gloves and safety goggles and use only with adequate ventilation. Keep away from heat, sparks, and open flame.

Propylene oxide is highly flammable, toxic, and may be carcinogenic. High concentrations are extremely destructive to the mucous membranes and upper respiratory tract. It may be harmful by inhalation, ingestion, or skin absorption. Wear appropriate gloves and safety glasses and use only in a chemical fume hood. Keep away from heat, sparks, and open flame.

Sodium azide, NaN₃, is highly poisonous. It blocks the cytochrome electron transport system. Solutions containing sodium azide should be clearly marked. It may be harmful by inhalation, ingestion, or skin absorption. Wear appropriate gloves and safety goggles and handle it with great care. Sodium azide is an oxidizing agent and should not be stored near flammable chemicals.

Sodium carbonate, Na₂CO₃, may be harmful by inhalation, ingestion, or skin absorption. Wear appropriate gloves and safety glasses and use in a chemical fume hood.

Sodium citrate, *see* **Citric acid**

Sodium hydroxide, NaOH, and **solutions containing NaOH,** are highly toxic and caustic and should be handled with great care. Wear appropriate gloves and a face mask. All other concentrated bases should be handled in a similar manner.

TEA, *see* **Triethylamine**

TEAC, *see* **Tetraethylammonium chloride**

TES, *see* **N-tris(hydroxymethyl)methyl-2-aminoethane sulfonic acid**

Tetraethylammonium chloride (TEAC) may cause allergic skin reaction and may be harmful by inhalation, ingestion, or skin absorption. It is irritating to the mucous membranes and upper respiratory tract. Do not breathe the dust. Wear appropriate gloves and safety glasses.

Tetrodotoxin (TTX) is one of the most toxic substances known to man. Death can occur within 30 min. It is extremely harmful by inhalation, ingestion, or skin absorption. Wear appropriate gloves and safety goggles and use in a chemical fume hood. Do not breathe the dust.

Triethylamine (TEA) is highly toxic and flammable. It is extremely corrosive to the mucous membranes, upper respiratory tract, eyes, and skin. It may be harmful by inhalation, ingestion, or skin absorption. Wear appropriate gloves and safety glasses and use in a chemical fume hood. Keep away from heat, sparks, and open flame.

Trinitrophenol, *see* **Picric acid powder**

Triton X-100 causes severe eye irritation and burns. It may be harmful by inhalation, ingestion, or skin absorption. Wear appropriate gloves and safety goggles. Do not breathe the vapor.

TTX, *see* **Tetrodotoxin**

Uranyl acetate is toxic if inhaled, ingested, or absorbed through the skin. Wear appropriate gloves and safety glasses and use in a chemical fume hood.

Xylene is flammable and may be narcotic at high concentrations. It may be harmful by inhalation, ingestion, or skin absorption. Wear appropriate gloves and safety glasses and use only in a chemical fume hood. Keep away from heat, sparks, and open flame.

Index

Page references followed by f denote figures; those followed by t denote tables.